Computational Techniques for Process Simulation and Analysis Using MATLAB®

Computational Techniques for Process Simulation and Analysis Using MATLAB®

Niket S. Kaisare

CRC Press
Taylor & Francis Group
Boca Raton London New York

CRC Press is an imprint of the
Taylor & Francis Group, an **informa** business

CRC Press
Taylor & Francis Group,
6000 Broken Sound Parkway NW, Suite 300,
Boca Raton, FL 33487-2742

© 2018 by Taylor & Francis Group, LLC
CRC Press is an imprint of Taylor & Francis Group, an Informa business

No claim to original U.S. Government works

Printed on acid-free paper

International Standard Book Number-13: 978-1-1387-4608-4 (Paperback)
International Standard Book Number-13: 978-1-4987-6211-3 (Hardback)

Visit the Taylor & Francis Web site at
http://www.taylorandfrancis.com

and the CRC Press Web site at
http://www.crcpress.com

To Pradnya for her unending love, patience, and support
and
To my parents, Komal and Satish Kaisare

Contents

Preface, xix

Author, xxiii

CHAPTER 1 ■	Introduction	1
1.1	OVERVIEW	1
	1.1.1 A General Model	1
	1.1.2 A Process Example	2
	1.1.3 Analysis of Dynamical Systems	3
1.2	STRUCTURE OF A MATLAB® CODE	3
	1.2.1 Writing Our First MATLAB® Script	5
	1.2.2 MATLAB® Functions	7
	1.2.3 Using Array Operations in MATLAB®	9
	1.2.4 Loops and Execution Control	10
	1.2.5 Section Recap	11
1.3	APPROXIMATIONS AND ERRORS IN NUMERICAL METHODS	12
	1.3.1 Machine Precision	12
	1.3.2 Round-Off Error	14
	1.3.3 Taylor's Series and Truncation Error	15
	1.3.4 Trade-Off between Truncation and Round-Off Errors	18
1.4	ERROR ANALYSIS	20
	1.4.1 Convergence and Stability	20
	1.4.2 Global Truncation Error	21
1.5	OUTLOOK	23

Section I **Dynamic Simulations and Linear Analysis**

Chapter 2 ■ Linear Algebra   27

2.1 INTRODUCTION   27

  2.1.1 Solving a System of Linear Equations   27

  2.1.2 Overview   28

2.2 VECTOR SPACES   30

  2.2.1 Definition and Properties   30

  2.2.2 Span, Linear Independence, and Subspaces   32

  2.2.3 Basis and Coordinate Transformation   34

    2.2.3.1 *Change of Basis*   34

  2.2.4 Null (Kernel) and Image Spaces of a Matrix   35

    2.2.4.1 *Matrix as Linear Operator*   35

    2.2.4.2 *Null and Image Spaces in MATLAB®*   39

2.3 SINGULAR VALUE DECOMPOSITION   41

  2.3.1 Orthonormal Vectors   41

  2.3.2 Singular Value Decomposition   42

  2.3.3 Condition Number   47

    2.3.3.1 *Singular Values, Rank, and Condition Number*   47

    2.3.3.2 *Sensitivity of Solutions to Linear Equations*   47

  2.3.4 Directionality   51

2.4 EIGENVALUES AND EIGENVECTORS   54

  2.4.1 Orientation for This Section   54

  2.4.2 Brief Recap of Definitions   54

  2.4.3 Eigenvalue Decomposition   56

  2.4.4 Applications   58

    2.4.4.1 *Similarity Transform*   62

    2.4.4.2 *Linear Differential Equations*   63

    2.4.4.3 *Linear Difference Equations*   64

2.5 EPILOGUE   65

EXERCISES   67

Chapter 3 ■ Ordinary Differential Equations: Explicit Methods   69

3.1 GENERAL SETUP   69

  3.1.1 Some Examples   69

  3.1.2 Geometric Interpretation   72

3.1.3	Euler's Explicit Method	74
3.1.4	Euler's Implicit Method	76
3.1.5	Stability and Step-Size	78
	3.1.5.1 *Stability of Euler's Explicit Method*	78
	3.1.5.2 *Error and Stability of Euler's Implicit Method*	79
3.1.6	Multivariable ODE	80
	3.1.6.1 *Nonlinear Case*	81
3.2	SECOND-ORDER METHODS: A JOURNEY THROUGH THE WOODS	82
3.2.1	Some History	82
3.2.2	Runge-Kutta (RK-2) Methods	83
	3.2.2.1 *Derivation for RK-2 Methods*	83
	3.2.2.2 *Heun's Method*	84
	3.2.2.3 *Other RK-2 Methods*	86
3.2.3	Step-Size Halving: Error Estimate for RK-2	87
3.2.4	Richardson's Extrapolation	89
3.2.5	Other Second-Order Methods (*)	91
	3.2.5.1 *Trapezoidal Rule: An Implicit Second-Order Method*	91
	3.2.5.2 *Second-Order Adams-Bashforth Methods*	92
	3.2.5.3 *Predictor-Corrector Methods*	92
	3.2.5.4 *Backward Differentiation Formulae*	93
3.3	HIGHER-ORDER RUNGE-KUTTA METHODS	93
3.3.1	Explicit Runge-Kutta Methods: Generalization	93
3.3.2	Error Estimation and Embedded RK Methods	97
	3.3.2.1 *MATLAB® Solver* `ode23`	100
3.3.3	The Workhorse: Fourth-Order Runge-Kutta	101
	3.3.3.1 *Classical RK-4 Method(s)*	102
	3.3.3.2 *Kutta's 3/8th Rule RK-4 Method*	103
3.4	MATLAB® ODE45 SOLVER: OPTIONS AND PARAMETERIZATION	103
3.5	CASE STUDIES AND EXAMPLES	105
3.5.1	An Ideal PFR	106
	3.5.1.1 *Simulation of PFR as ODE-IVP*	106
	3.5.1.2 *Numerical Integration for PFR Design*	108
	3.5.1.3 *Comparison of ODE-IVP with Integration*	110
3.5.2	Multiple Steady States: Nonisothermal CSTR	111
	3.5.2.1 *Model and Problem Setup*	111

		3.5.2.2	*Simulation of Transient CSTR*	113
		3.5.2.3	*Step Change in Inlet Temperature*	115
	3.5.3	Hybrid System: Two-Tank with Heater		116
	3.5.4	Chemostat: Preview into "Stiff" System		120
3.6	EPILOGUE			125
EXERCISES				125

CHAPTER 4 ▪ Partial Differential Equations in Time   127

4.1	GENERAL SETUP			127
	4.1.1	Classification of PDEs		128
	4.1.2	Brief History of Second-Order PDEs		128
	4.1.3	Classification of Second-Order PDEs and Practical Implications		129
		4.1.3.1	*Elliptic PDE*	129
		4.1.3.2	*Hyperbolic PDE*	130
		4.1.3.3	*First-Order Hyperbolic PDEs*	131
		4.1.3.4	*Parabolic PDE*	132
	4.1.4	Initial and Boundary Conditions		132
4.2	A BRIEF OVERVIEW OF NUMERICAL METHODS			133
	4.2.1	Finite Difference		133
	4.2.2	Method of Lines		134
	4.2.3	Finite Volume Methods		134
	4.2.4	Finite Element Methods		135
4.3	HYPERBOLIC PDE: CONVECTIVE SYSTEMS			135
	4.3.1	Finite Differences in Space and Time		136
		4.3.1.1	*Upwind Difference in Space*	136
		4.3.1.2	*Forward in Time Central in Space (FTCS) Differencing*	138
		4.3.1.3	*Lax-Friedrichs Scheme*	139
		4.3.1.4	*Higher-Order Methods*	139
	4.3.2	Crank-Nicolson: Second-Order Implicit Method		140
		4.3.2.1	*Preview of Numerical Solution*	141
	4.3.3	Solution Using Method of Lines		141
		4.3.3.1	*MoL with Central Difference in Space*	142
		4.3.3.2	*MoL with Upwind Difference in Space*	145
	4.3.4	Numerical Diffusion		149

4.4	PARABOLIC PDE: DIFFUSIVE SYSTEMS	150
	4.4.1 Finite Difference in Space and Time	152
	4.4.2 Crank-Nicolson Method	153
	4.4.3 Method of Lines Using MATLAB® ODE Solvers	154
	4.4.3.1 *MoL with Central Difference in Space*	154
	4.4.4 Methods to Improve Stability	157
4.5	CASE STUDIES AND EXAMPLES	157
	4.5.1 Nonisothermal Plug Flow Reactor	157
	4.5.2 Packed Bed Reactor with Multiple Reactions	164
	4.5.3 Steady Graetz Problem: Parabolic PDE in Two Spatial Dimensions	170
	4.5.3.1 *Heat Transfer in Fluid Flowing through a Tube*	170
	4.5.3.2 *Effect of Velocity Profile*	174
	4.5.3.3 *Calculation of Nusselt Number*	174
4.6	EPILOGUE	176
	EXERCISES	177

Chapter 5 ■	Section Wrap-Up: Simulation and Analysis	179
5.1	BINARY DISTILLATION COLUMN: STAGED ODE MODEL	181
	5.1.1 Model Description	181
	5.1.2 Model Equations and Simulation	183
	5.1.3 Effect of Parameters: Reflux Ratio and Relative Volatility	185
5.2	STABILITY ANALYSIS FOR LINEAR SYSTEMS	186
	5.2.1 Motivation: Linear Stability Analysis of a Chemostat	187
	5.2.1.1 *Phase Portrait at the Steady State*	190
	5.2.1.2 *Trivial Steady State and Analysis*	190
	5.2.2 Eigenvalues, Stability, and Dynamics	191
	5.2.2.1 *Dynamics When Eigenvalues Are Real and Distinct*	192
	5.2.2.2 *An Example*	197
	5.2.2.3 *Summary*	197
	5.2.3 Transient Growth in Stable Linear Systems	198
	5.2.3.1 *Defining Normal and Nonnormal Matrices*	198
	5.2.3.2 *Analysis of Nonnormal Systems*	199
5.3	COMBINED PARABOLIC PDE WITH ODE-IVP: POLYMER CURING	201

5.4   TIME-VARYING INLET CONDITIONS AND PROCESS DISTURBANCES   208

    5.4.1   Chemostat with Time-Varying Inlet Flowrate   208

    5.4.2   Zero-Order Hold Reconstruction in Digital Control   212

5.5   SIMULATING SYSTEM WITH BOUNDARY CONSTRAINTS   215

    5.5.1   PFR with Temperature Profile Specified   216

5.6   WRAP-UP   219

EXERCISES   219

SECTION II   **Linear and Nonlinear Equations and Bifurcation**

CHAPTER 6 ■ Nonlinear Algebraic Equations   225

6.1   GENERAL SETUP   225

    6.1.1   A Motivating Example: Equation of State   226

6.2   EQUATIONS IN SINGLE VARIABLE   227

    6.2.1   Bisection Method   228

    6.2.2   Secant and Related Methods   233

        6.2.2.1   *Regula-Falsi: Method of False Position*   235

        6.2.2.2   *Brent's Method*   235

    6.2.3   Fixed Point Iteration   236

    6.2.4   Newton-Raphson in Single Variable   238

    6.2.5   Comparison of Numerical Methods   240

6.3   NEWTON-RAPHSON: EXTENSIONS AND MULTIVARIATE   241

    6.3.1   Multivariate Newton-Raphson   241

    6.3.2   Modified Secant Method   245

    6.3.3   Line Search and Other Methods   247

6.4   MATLAB® SOLVERS   249

    6.4.1   Single Variable Solver: `fzero`   249

    6.4.2   Multiple Variable Solver: `fsolve`   250

6.5   CASE STUDIES AND EXAMPLES   253

    6.5.1   Recap: Equation of State   253

    6.5.2   Two-Phase Vapor-Liquid Equilibrium   253

        6.5.2.1   *Bubble Temperature Calculation*   254

        6.5.2.2   *Dew Temperature Calculation*   254

        6.5.2.3   *Generating the T–x–y Diagram*   255

    6.5.3   Steady State Multiplicity in CSTR   257

6.5.4   Recap: Chemostat   261

6.5.5   Integral Equations: Conversion from a PFR   262

     6.5.5.1   *First-Order Kinetics*   263

     6.5.5.2   *Complex Kinetics*   266

6.6   EPILOGUE   268

EXERCISES   271

CHAPTER 7   ■   Special Methods for Linear and Nonlinear Equations   273

7.1   GENERAL SETUP   273

7.1.1   Ordinary Differential Equation–Boundary Value Problems   274

7.1.2   Elliptic PDEs   274

7.1.3   Outlook of This Chapter   275

7.2   TRIDIAGONAL AND BANDED SYSTEMS   275

7.2.1   What Is a Banded System?   275

     7.2.1.1   *Tridiagonal Matrix*   276

7.2.2   Thomas Algorithm a.k.a TDMA   276

     7.2.2.1   *Heat Conduction Problem*   277

     7.2.2.2   *Thomas Algorithm*   281

7.2.3   ODE-BVP with Flux Specified at Boundary   285

7.2.4   Extension to Banded Systems   288

7.2.5   Elliptic PDEs in Two Dimensions   289

7.3   ITERATIVE METHODS   290

7.3.1   Gauss-Siedel Method   291

7.3.2   Iterative Method with Under-Relaxation   295

7.4   NONLINEAR BANDED SYSTEMS   296

7.4.1   Nonlinear ODE-BVP Example   296

     7.4.1.1   *Heat Conduction with Radiative Heat Loss*   297

7.4.2   Modified Successive Linearization–Based Approach   298

7.4.3   Gauss-Siedel with Linearization of Source Term   302

7.4.4   Using `fsolve` with Sparse Systems   304

7.5   EXAMPLES   304

7.5.1   Heat Conduction with Convective or Radiative Losses   304

7.5.2   Diffusion and Reaction in a Catalyst Pellet   305

7.5.2.1   *Linear System and Thiele Modulus*   *305*

7.5.2.2   *Langmuir-Hinshelwood Kinetics in a Pellet*   *308*

7.6   EPILOGUE   311

EXERCISES   311

CHAPTER 8   ■   Implicit Methods: Differential and Differential Algebraic Systems   313

8.1   GENERAL SETUP   313

8.1.1   Stiff System of Equation   313

8.1.1.1   *Stiff ODE in Single Variable*   *315*

8.1.2   Implicit Methods for Distributed Parameter Systems   316

8.1.3   Differential Algebraic Equations   316

8.2   MULTISTEP METHODS FOR DIFFERENTIAL EQUATIONS   317

8.2.1   Implicit Adams-Moulton Methods   318

8.2.2   Higher-Order Adams-Moulton Method   319

8.2.3   Explicit Adams-Bashforth Method   320

8.2.4   Backward Difference Formula   322

8.2.5   Stability and MATLAB® Solvers   325

8.2.5.1   *Explicit Adams-Bashforth Methods*   *325*

8.2.5.2   *Implicit Euler and Trapezoidal Methods*   *325*

8.2.5.3   *Implicit Adams-Moulton Methods of Higher Order*   *325*

8.2.5.4   *BDF/NDF Methods*   *325*

8.2.5.5   *MATLAB® Nonstiff Solvers*   *326*

8.2.5.6   *MATLAB® Stiff Solvers*   *326*

8.3   IMPLICIT SOLUTIONS FOR DIFFERENTIAL EQUATIONS   327

8.3.1   Trapezoidal Method for Stiff ODE   327

8.3.1.1   *Adaptive Step-Sizing*   *329*

8.3.1.2   *Multivariable Example*   *330*

8.3.2   Crank-Nicolson Method for Hyperbolic PDEs   331

8.3.2.1   *Exploiting Sparse Structure for Efficient Simulation*   *337*

8.4   DIFFERENTIAL ALGEBRAIC EQUATIONS   337

8.4.1   An Introductory Example   338

8.4.1.1   *Direct Substitution*   *338*

8.4.1.2   *Formulating and Solving a DAE*   *339*

8.4.2   Index of a DAE and More Examples   340

    8.4.2.1   *Example 2: Pendulum in Cartesian Coordinate System*   341

    8.4.2.2   *Example 3: Heterogeneous Catalytic Reactor*   341

8.4.3   Solution Methodology: Overview   342

    8.4.3.1   *Solving Algebraic Equation within ODE*   343

    8.4.3.2   *Combined Approach*   345

8.4.4   Solving Semiexplicit DAEs Using `ode15s` in MATLAB®   348

8.5   CASE STUDIES AND EXAMPLES   351

8.5.1   Heterogeneous Catalytic Reactor: Single Complex Reaction   351

8.5.2   Flash Separation/Batch Distillation   353

8.6   EPILOGUE   359

EXERCISES   360

CHAPTER 9 ■ Section Wrap-Up: Nonlinear Analysis   363

9.1   NONLINEAR ANALYSIS OF CHEMOSTAT: "TRANSCRITICAL"
    BIFURCATION   364

9.1.1   Steady State Multiplicity and Stability   364

9.1.2   Phase-Plane Analysis   365

9.1.3   Bifurcation with Variation in Dilution Rate   366

9.1.4   Transcritical Bifurcation   368

9.2   NONISOTHERMAL CSTR: "TURNING-POINT" BIFURCATION   372

9.2.1   Steady States: Graphical Approach   372

9.2.2   Stability Analysis at Steady States   374

9.2.3   Phase-Plane Analysis   376

9.2.4   Turning-Point Bifurcation   377

9.3   LIMIT CYCLE OSCILLATIONS   379

9.3.1   Oscillations in Linear Systems   379

9.3.2   Limit Cycles: van der Pol Oscillator   381

    9.3.2.1   *Relaxation vs. Harmonic Oscillations*   382

9.3.3   Oscillating Chemical Reactions   383

9.4   SIMULATION OF METHANOL SYNTHESIS IN TUBULAR REACTOR   387

9.4.1   Steady State PFR with Pressure Drop   388

    9.4.1.1   *Reaction Kinetics*   388

    9.4.1.2   *Input Parameters and Initial Processing*   389

    9.4.1.3   *Steady State PFR Model*   391

9.4.2   Transient Model   394

9.5	TRAJECTORY OF A CRICKET BALL	398
	9.5.1 Solving the ODE for Trajectory	399
	9.5.2 Location Where the Ball Hits the Ground	400
	9.5.3 Animation	403
9.6	WRAP-UP	405
EXERCISES		405

SECTION III **Modeling of Data**

CHAPTER 10 ■ Regression and Parameter Estimation   409

10.1	GENERAL SETUP	409
	10.1.1 Orientation	410
	10.1.2 Some Statistics	411
	10.1.3 Some Other Considerations in Regression	413
10.2	LINEAR LEAST SQUARES REGRESSION	413
	10.2.1 Fitting a Straight Line	413
	10.2.2 General Matrix Approach	415
	10.2.3 Goodness of Fit	418
	10.2.3.1 Maximum Likelihood Solution	418
	10.2.3.2 Error and Coefficient of Determination	419
10.3	REGRESSION IN MULTIPLE VARIABLES	423
	10.3.1 General Multilinear Regression	423
	10.3.2 Polynomial Regression	424
	10.3.3 Singularity and SVD	428
10.4	NONLINEAR ESTIMATION	429
	10.4.1 Functional Regression by Linearization	429
	10.4.2 MATLAB® Solver: Linear Regression	432
	10.4.3 Nonlinear Regression Using Optimization Toolbox	434
10.5	CASE STUDIES AND EXAMPLES	437
	10.5.1 Specific Heat: Revisited	437
	10.5.2 Antoine's Equation for Vapor Pressure	438
	10.5.2.1 Linear Regression for Benzene	439
	10.5.2.2 Nonlinear Regression for Ethylbenzene	440

10.5.3   Complex Langmuir-Hinshelwood Kinetic Model   441

10.5.3.1   *Case 1: Experiments Performed at Single Concentration of B*   442

10.5.3.2   *Case 2: Experiments Performed at Different Initial Concentrations of B*   444

10.5.4   Reaction Rate: Differential Approach   445

10.6   EPILOGUE   448

10.6.1   Summary   448

10.6.2   Data Tables   449

EXERCISES   450

APPENDIX A: MATLAB® PRIMER, 451

APPENDIX B: NUMERICAL DIFFERENTIATION, 475

APPENDIX C: GAUSS ELIMINATION FOR LINEAR EQUATIONS, 485

APPENDIX D: INTERPOLATION, 499

APPENDIX E: NUMERICAL INTEGRATION, 511

BIBLIOGRAPHY, 527

INDEX, 529

Preface

S TUDENTS TODAY ARE EXPECTED to know one or more of the several computing or simulation tools as part of their curriculum, due to their widespread use in the industry.
MATLAB® has become one of the prominent languages used in research and industry.
MATLAB is a numerical computing environment that is based on a MATLAB scripting
language. MathWorks, the makers of MATLAB, describe it as "the language of technical
computing." The focus of this book will be to highlight the *use* of MATLAB in technical
computing or, more specifically, in solving problems in the *analysis and simulation of processes* of interest to engineers.

This is intended to be an intermediate-level book, geared toward postgraduate students,
practicing engineers, and researchers who use MATLAB. It provides advanced treatment
of topics relevant to modeling, simulation, and analysis of dynamical systems. Although
this is not an introductory MATLAB or numerical techniques textbook, it may however
be used as a companion book for introductory courses. For the sake of completeness, a
primer on MATLAB as well as introduction to some numerical techniques is provided in
the Appendices. Since mid-2000s,we have always used MATLAB in electives in IIT Madras.
The popularity of MATLAB among students led us to start a *core* undergraduate (sophomore) and a postgraduate (first-year masters) laboratory. Since 2016, I have started teaching a massive open online course (MOOC) on MATLAB programming on the NPTEL
platform.* The first two years of this course had over 10,000 enrolled students. Needless to
say, MATLAB has become an important tool in teaching and research. The focus of all the
above courses is to introduce students to MATLAB as a numerical methods tool. Some of
the students who complete these courses inquire about the next-level courses that would
help them *apply* MATLAB skills to solve engineering problems. This book may also be used
for this purpose. In introductory courses, a significant amount of time is spent in developing the background for numerical methods itself. In our effort to make the treatment general and at a beginner's level, we eschew real-world examples in favor of abstracted ones.
For example, we would often introduce a second-order ODE using a generic formulation,
such as $y'' + ay' + b(y - c) = 0$. A sophomore who hasn't taken a heat transfer course may
not yet appreciate a "heating in a rod" problem. An intermediate-level text means that it
is more valuable to use a real example, such as $T'' + r^{-1}T' + \beta(T - T_a) = 0$. The utility of such

---

* NPTEL stands for National Programme for Technology Enhanced Learning and is a Government of India–funded initiative
  to bring high-quality engineering and science courses on an online (MOOC) platform to enhance students' learning.

an approach cannot be understated, since it allows the freedom to introduce some of the complexity that engineers, scientists, and researchers face in their work.

The value of using real-world examples was highlighted during my experience in industrial R&D, where we used MATLAB extensively. We needed to interface with cross-functional teams: engineering, implementation, and software development teams. Individuals came from a wide range of backgrounds. These interactions exposed me to a new experience: Your work must be understood by people with very different backgrounds, who may not speak the same technical language. The codes had to bridge the "language barrier" spoken in different teams, and the codes were to be combined with a reasonably intuitive interface. I have tried to adopt some of these principles in this book, without moving too far from the more common pedagogy in creating such a book.

Thus, a practically *oriented text* that caters to an intermediate-level audience is my objective in writing this book.

ORIGIN OF THIS BOOK

There are several excellent books on numerical techniques for engineers. Laurene Forsell's book, *Numerical Methods Using MATLAB*, provides a MATLAB-based approach to learning numerical techniques. The books on numerical techniques by Chapra and Canale and by S.K. Gupta are excellent undergraduate textbooks, which introduce undergraduates to this subject for the first time. Thus, their focus is conceptual understanding of numerical techniques themselves. While undergraduate teaching is in good stead, a textbook that covers *core* requirements for a balanced postgraduate curriculum is missing. Such a book will also be useful to practicing engineers, scientists, and researchers who use MATLAB.

This book is borne out of my experience in teaching a postgraduate course called *Process Analysis and Simulation*, postgraduate lab in *Process Simulation*, and theory of computational techniques. They provide the first-year postgraduates the basis to tackle research problems in their theses. The former course takes a balanced focus on modeling, simulation, and analysis of chemical process systems, while the process simulation lab gives them a numerical methods perspective. Postgraduate-level books, such as the evergreen *Numerical Recipes* by Press et al., are rather advanced and focused on numerical methods. On the other hand, the book by Strogatz on nonlinear dynamics or other similar books are not general enough for the needs of an audience interested in simulations. A "bridge" book, which assumes some familiarity with undergraduate material, but still covers the basics, is missing.

Having said this, I do not intend this to be a postgraduate numerical methods text. This book aims to introduce students and practitioners to simulation and analysis of process systems in MATLAB. We often find it difficult to connect the numerical tool to the physical analysis of a system. This book intends to bring in a strong process simulation treatment to linear stability and nonlinear analysis.

Thus, this book intends to bring a practical approach to expounding theories: both numerical aspects of stability and convergence, as well as linear and nonlinear analysis of systems. The "process" is the focus. Numerical methods are introduced insofar as is essential to make a judicious choice of algorithms for simulation and analysis.

PREREQUISITES

Since this is a postgraduate-level text, some familiarity with an undergraduate-level numerical techniques or an equivalent course is assumed, though we will review all the relevant concepts at the appropriate stage. So, the students are not expected to remember the details or nuances of "Newton-Raphson" or "Runge-Kutta" methods, but this book is not the first time they hear these terms.

Some familiarity with coding (MATLAB, Fortran, C++, Python, or any language) will be useful, but not a prerequisite. MATLAB primer is provided in the Appendix for first-time users of MATLAB. Finally, with respect to writing MATLAB code, I focus on "doing it right the first time" approach—by bringing in good programming practices that I have learnt over the years. Things like commenting and structuring your code, scoping of variables, etc., are also covered, not as an afterthought but as an integral part of the discussion. However, these are dealt with more informally than a "programming language" course.

HOW THIS BOOK IS LAID OUT

This book derives examples from three different courses I have taught: (i) Numerical Methods, (ii) Process Analysis and Simulation, and (iii) Computational Programming Lab. It is structured so that it may be used for any of the three courses. Each chapter deals with one approach to solving computational problems (e.g., ODEs, PDEs, nonlinear equations, etc.), culminating in case studies that utilize the concepts discussed in the chapter.

I have split the book into three sections, which are laid out with a "Process Analysis" viewpoint: Section I covers system dynamics and linear system analysis; Section II covers solving nonlinear equations, including differential algebraic equations (DAEs); and Section III covers function approximation and optimization for modeling of data. The following table summarizes the various chapters in the book:

Basics	Chapter 1 Introduction	Appendix A MATLAB® Primer		
Appendices		Appendix B Differentiation	Appendix C Linear Equations	Appendix E Integration
Section I. Dynamics	Chapter 2 Linear Algebra	Chapter 3 ODE-IVP	Chapter 4 Transient PDEs	Chapter 5 Simulation
Section II. Equations	Chapter 6 Nonlinear Equations	Chapter 7 Special Methods (ODE-BVP/PDE)	Chapter 8 Implicit Methods (DAEs)	Chapter 9 Nonlinear Analysis
Section III. Data Fitting	Appendix D Interpolation	Chapter 10 Regression		

LAYOUT FOR PROCESS ANALYSIS

The layout of this book is largely based on the postgraduate-level process simulation and analysis course. The material I cover in this course is chronologically as laid out in the book. The course starts with an introduction to the role of simulation and analysis in engineering, and a primer on MATLAB. Thereafter, I introduce concepts in linear algebra (Chapter 2), ODEs (Chapter 3), and solving hyperbolic and parabolic PDEs (Chapter 4). Problems in either linear analysis or dynamical simulations (Chapter 5) typically form mid-term projects for students. The second part of the course also follows a similar structure, with nonlinear equations (Chapter 6), ODE-BVPs and elliptic PDEs (Chapter 7), and DAEs (Chapter 8) providing the adequate background for end-semester projects involving nonlinear analysis and bifurcation (Chapter 9). I have added Chapter 10 (Parameter Estimation) for the sake of completion.

LAYOUT FOR NUMERICAL METHODS

This book may also be used for an advanced numerical methods course. In such a case, I suggest treating the material column-wise. This course may start with the first row to cover the basics (Introduction, MATLAB Primer, Differentiation, Integration, and Linear Equations). Thereafter, Chapter 2 and Chapter 6 may be covered, to equip students to solve linear and nonlinear equations. Chapters 3 and 7 cover ODE-IVP and ODE-BVP, respectively, followed by Parameter Estimation (Chapter 10). A four-credit course may also cover PDEs (Chapter 4). Typically, Chapters 5 and 9 will be beyond the scope of such a course.

LAYOUT FOR NUMERICAL DIFFERENTIAL EQUATIONS

The shaded chapters (Chapters 3, 4, 5, 7, 8, and 9), along with appendices on numerical differentiation and integration, can form a numerical differential equations course.

FOR PRACTICING ENGINEER OR NEW RESEARCHER

A practicing engineer or researcher can embark on a self-guided journey through case studies and examples covered in this book. This includes not only the case studies analyzed in Chapters 5 and 9 but also the ones discussed in other chapters (penultimate section in the other chapters).

MATLAB® is a registered trademark of The MathWorks, Inc. For product information, please contact:

The MathWorks, Inc.
3 Apple Hill Drive
Natick, MA 01760-2098 USA
Tel: 508-647-7000
Fax: 508-647-7001
E-mail: info@mathworks.com
Web: www.mathworks.com

Author

Dr. Niket S. Kaisare is currently an associate professor in the Department of Chemical Engineering at IIT Madras. He received his PhD in chemical engineering from Georgia Institute of Technology, working in the area of model-based advanced process control. He then joined the Department of Chemical Engineering at the University of Delaware, where he worked on multiscale modeling of reacting flows in microreactors, as a postdoctoral researcher. After this, he joined IIT Madras as assistant professor in 2007. While in IIT Madras, he taught several courses in process analysis and simulation, computational techniques, process simulation laboratory, and advanced control theory. MATLAB was used extensively in most of these courses. He has also taught an online course called "MATLAB Programming for Numerical Computations" as a part of NPTEL (National Programme for Technology Enhanced Learning). This course was popular, with more than ten thousand students enrolling in it.

He spent three years, from mid-2011 to 2014, in industrial R&D. During this stint, he worked on numerous simulation problems related to modeling of vehicle catalytic convertors, cryogenic hydrogen storage, monitoring and control of oil and gas wells, and automation engineering. As a part of the R&D team, he used MATLAB extensively and spent a significant part of his time interfacing with engineering and development teams.

He has extensive experience working in MATLAB and FORTRAN as well as simulation softwares Fluent and Comsol. He also has good working experience with various other simulation tools, such as Aspen-Plus/Unisim, Gaussian, and Abacus. His current research program is focused on "multiscale modeling, analysis, and control of catalytic microreactors for energy- and fuel-processing applications."

CHAPTER 1

Introduction

1.1 OVERVIEW

1.1.1 A General Model

This book is targeted toward postgraduate students, senior undergraduates, researchers, and practicing engineers to provide them with a practical guide for using MATLAB® for process simulation and numerical analysis. MATLAB was listed among the top ten programming languages by the *IEEE Spectrum* magazine in 2015 (a list that was topped by Java, followed by C and C++). While the basics of MATLAB can be learnt through various sources, the focus of this book is on the *analysis and simulation of processes* of interest to engineers.

The terms "analysis" and "simulation" are generic terms that define a rather broad spectrum of problems and solution techniques. Engineering is a discipline that deals with the transformation of raw material, momentum, or energy. Thus, this book will focus on those process examples where the variables of interest vary with time and/or space, including the relationship of these state variables with their properties. I will use an example of a reactor-separator process in Section 1.1.2 to illustrate this. While this is a chemical engineering example, the treatment in this book is general enough for other engineering and science disciplines to also find it useful.

The problems mentioned above that are considered in this book include ordinary and partial differential equations (ODEs and PDEs), algebraic equations (either linear or nonlinear), or combinations thereof. The three sections of this book are organized based on the computational methodology and analysis tools that will be used for the respective problems.

Section I of this book includes Chapters 2 through 5 and deals with ODE-IVPs (initial value problems) as well as the problems that can be converted into a standard form that can be solved with ODE-IVP tools. A generic ODE-IVP is of the type

$$\frac{d\mathbf{y}}{dt} = \mathbf{f}\left(t, \mathbf{y}; \boldsymbol{\phi}\right) \tag{1.1}$$

where
t is an independent variable
$\mathbf{y} \in \mathcal{R}^n$ is a vector of dependent variables
$\boldsymbol{\phi}$ represents parameters

Examples include simulations of level and temperature in a stirred tank, simulations of a reactor, mass-spring-damper, pendulum (in cylindrical coordinates), and others.

Section II of this book includes Chapters 6 through 9 and deals with the problems of the type

$$0 = \mathbf{g}(\mathbf{x};\boldsymbol{\phi}) \tag{1.2}$$

where

$\mathbf{x} \in \mathcal{R}^m$ is a vector of dependent variables

$\boldsymbol{\phi}$ represents parameters

Nonlinear algebraic equations, such as Equation 1.2, fall under this category. Moreover, ODE-BVPs (boundary value problems) and several PDEs are also converted into the form of Equation 1.2. Section II will not only cover techniques to solve algebraic equations but also expound methods to convert ODEs/PDEs to this form. A combination of Equations 1.1 and 1.2, called differential algebraic equations (DAEs), is covered in Chapter 8. Chapters 5 and 9 are the concluding chapters of the first two sections. They build on the concepts from the preceding chapters in the respective sections for the analysis of dynamical systems and provide an introduction to advanced topics in simulations.

Finally, Chapter 10, included in Section III, deals with the parameter estimation problem, that is, to compute the parameter vector, $\boldsymbol{\phi}$, that best fits the experimental data.

1.1.2 A Process Example

I will use an example from a typical chemical process to motivate the discussion. Figure 1.1 shows a flow sheet of a typical process with a recycle. A reaction of the type A → B takes place in the plug flow reactor (PFR); the two species are separated in the distillation column; product B is obtained as the bottoms product, whereas the lighter species A is recycled back.

The PFR is modeled using the following ODE:

$$F\frac{dx_A}{dV} = -r(x_A), \quad x_A\big|_{V=0} = x_{\text{in}} \tag{1.3}$$

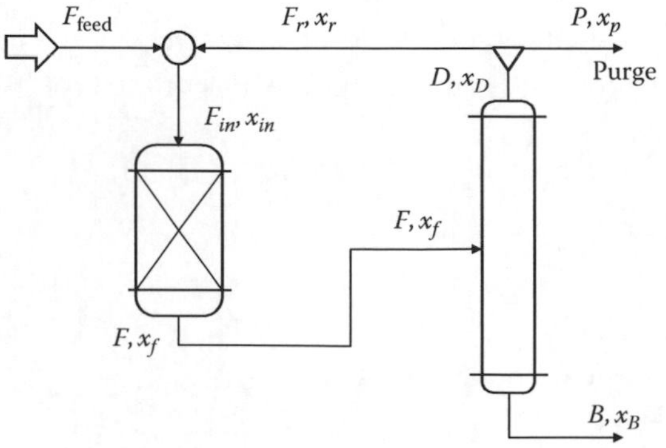

FIGURE 1.1   A typical process consisting of a reactor and a separator, with a recycle.

The reactor outlet conditions are obtained by solving the ODE-IVP above. ODE-IVP problems are covered in Chapter 3 of this book. If a dynamic response of the PFR is required, the resulting model is a PDE, where the state variable of interest varies in both space and time. Solutions to transient PDEs are covered in Chapter 4. Advanced topics in simulation are presented in Chapter 5, for example, when the inlet conditions or model parameters vary with time and/or space.

The distillation column consists of N nonlinear algebraic equations in N unknowns (mole fractions on each tray). For example, one of the model equations for the kth tray is given by

$$0 = \left(L_{i-1}x_{i-1} - L_i x_i\right) + \left(V_{i+1}y_{i+1} - V_i y_i\right) \quad \text{where } y_i = \frac{\alpha x_i}{1+(\alpha-1)x_i} \tag{1.4}$$

Such balance equations are written for each ideal stage of the distillation column, resulting in N nonlinear algebraic equations that need to be solved simultaneously to obtain N variables. These are further discussed in Chapter 6.

Axial dispersion is neglected while deriving the model (Equation 1.3). Inclusion of the axial dispersion term converts this IVP to a BVP, which is covered in Chapter 7. Discretizing the ODE-BVP results in a set of equations with a special matrix structure. Mass transfer limitations result in DAEs, which are covered in Chapter 8.

1.1.3 Analysis of Dynamical Systems

There is an equal amount of focus on the analysis of dynamical systems. To this end, Chapter 2 revisits concepts in linear algebra. I present a more contemporary treatment of linear algebra. Chapter 5 presents transient analysis of dynamical systems and their linear stability behavior. In addition to stability and dynamics based on eigenvalues of the linear dynamical system, the topic of transient growth in nonnormal systems is discussed. Related concepts of directionality and analysis using eigenvalue and singular value decompositions are discussed.

Chapter 9, which wraps up Section II of this book, is dedicated to nonlinear analysis and bifurcation. Well-known examples of stirred reactor, chemostat, mass-spring-damper system, and van der Pol oscillator will be used in this chapter. Chapter 10 is devoted to an important topic of parameter estimation.

Furthermore, advanced topics in efficient simulation and analysis are also presented. These include handling time-varying inputs and boundary constraints (Chapter 3), combination of ODE and PDE (Chapter 5), and a fun example of tracking the trajectory of a cricket ball (or baseball) with simulation and animation.

Before we get to these examples, I will review good practices and structuring of MATLAB codes as well as errors in numerical approaches.

1.2 STRUCTURE OF A MATLAB® CODE

A brief introduction to some programming practices specific to MATLAB is covered in this section. The intention of this section is to introduce the reader to *good MATLAB programming practices*, rather than "Introduction to MATLAB." A basic primer on using

MATLAB is instead provided in Appendix A. This book follows the principle of "learn it right the first time." Good programming hygiene, in writing MATLAB codes, is evangelized and implemented right from the first example. The book follows another principle that the best way to learn programming is through extensive practice. MathWorks, the parent company that develops MATLAB, has good introductory video tutorials, available at: http://in.mathworks.com/products/matlab/videos.html.* A beginner may want to start with their "Getting Started" videos.† I also have an introductory MOOC course on using MATLAB for numerical computations on National Programme for Technology Enhanced Learning.‡

Figure 1.2 shows a screenshot of MATLAB window. The main section contains two windows: MATLAB editor at the top and MATLAB command window at the bottom. The MATLAB editor currently shows the MATLAB file `firstFlowSheet.m`, which is a "driver *script*" to simulate the reactor-separator flow sheet described above. Line number 13 shows the following statement:

```
[F,x,err] = solveFlowSheet(Ffeed,Vpfr,purge,initVal);
```

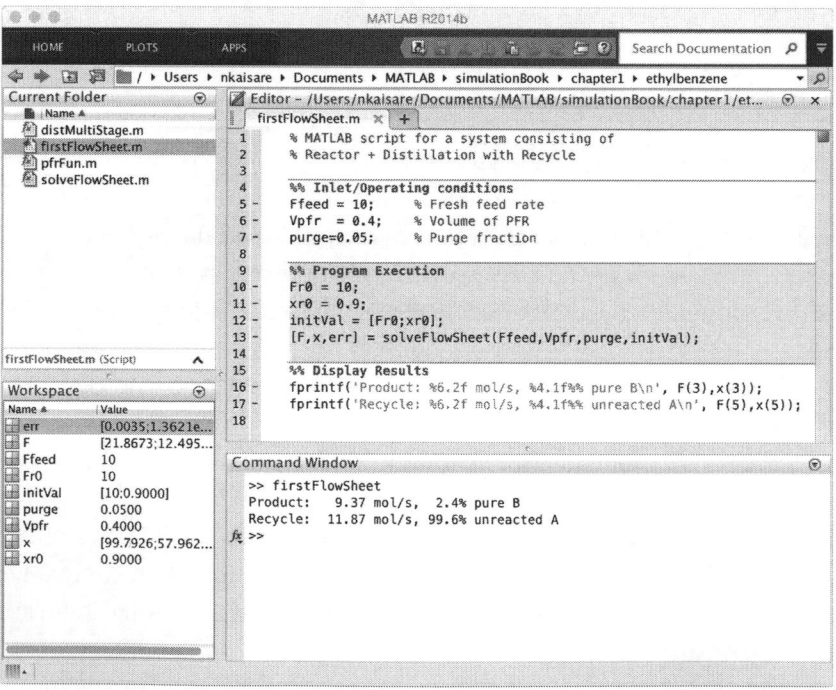

FIGURE 1.2   Screenshot of MATLAB® window.

A careful look at the directory listing (left-top window) shows a file `solveFlowSheet.m`. This is a MATLAB *function* file. A brief description of MATLAB script and function files is presented in Appendix A (Section A.6). In this example, this function takes in the input and initial conditions for the flow sheet and computes all the flowrates and mole fractions. These are returned to the calling program and are captured in variables `F` and `x` (which are standard notations for flowrate and mole fraction, respectively). Some of the basic principles of programming in MATLAB, illustrated in Figure 1.2, include the following:

- Codes are *sectioned* using this new feature* in MATLAB editor. Each program has input, executing, and output blocks. Sanctity of these blocks is maintained, as far as possible.

- Codes are well-commented.

- Codes are modular, where each MATLAB function is intended to perform a specific task.

- Appropriate use of MATLAB functions and scripts.

- Using standard or descriptive names for variables.

- Careful attention to variable definitions and scoping.

These aspects of MATLAB coding are described in this section. Another aspect discussed presently is to use the powerful matrix and linear algebra capabilities to write efficient and highly readable codes. Some of these features will be described using factorial and Maclaurin series expansion of exponent:

$$e^x = 1 + x + \frac{x^2}{2!} + \cdots + \frac{x^n}{n!} \tag{1.5}$$

1.2.1 Writing Our First MATLAB® Script

The first MATLAB script we will write is to compute a factorial. The script is also written to interact with the user for input/output. The reader is encouraged to view the corresponding tutorial video at http://in.mathworks.com/videos/writing-a-matlab-program-69023.html.

Example 1.1 Computing a Factorial

Problem: Write a MATLAB script to compute a factorial of a number input by the user.

 Solution: The script file for computing factorial is given below:

```
%% User Inputs
n = input('Please enter a number: ');
```

---

* Since MATLAB 2009. So, this feature is not really new!

```
% Checking for non-negative number input by user
if (n<0)
    error('Negative numbers are not allowed');
end


%% Computing factorial
fact = 1;
for i = 1:n
    fact = fact*i;
end


%% Display Results
disp([num2str(n), '! is ', num2str(fact)]);
```

The code consists of three sections: (i) user input section, where the input parameters for the code are specified; (ii) computing section, which forms the main core of the code; and (iii) results/output section, that is, the last line where the result is displayed. The lines starting with "%%" (two percentage signs) as the first two characters in a line indicate the start of a new section. Sectioning helps keep the code clean and easy to understand. Additionally, comments are provided in the code to help readability.

PERSONAL EXPERIENCE

While this may seem unnecessary for short codes, good coding habits need to be developed right from the start. It is generally accepted that commenting makes codes readable and shareable with others. However, advantages of structuring and commenting your code go beyond that. Comments help us understand *our own codes* better and faster. It helps in coding by making one think about the skeletal structure of the code before they start writing the code, thus reducing the number of errors. It also reduces the time taken for debugging a code.

The above code is how most students write their first code in MATLAB. However, this does not take advantage of the powerful array operations provided by MATLAB. For example, the command prod(1:n) will calculate the product of vector [1:n], which is nothing but the factorial itself. Indeed, the entire computing section can be replaced by a single line:

```
fact = prod(1:n);
```

If factorials for all integers from 1 through n are desired, then it is more efficient to use cumprod(1:n). This will return a $1 \times n$ vector, whose ith element is $i!$. At the command prompt, type 'help cumprod' to understand its use.

1.2.2 MATLAB® Functions

A MATLAB script, shown in Example 1.1, is a MATLAB file that contains statements that are executed sequentially by MATLAB (as if they were typed on the command prompt). The script is executed by calling it using the file name. The script shares workspace with the entity that calls it; in most cases, that is the command window workspace. MATLAB functions, on the other hand, are intended to execute commands (which may call other functions or subfunctions) for a specific purpose, with a set of input and output parameters. The variables defined in a function have a local scope. In other words, these variables are not available in the MATLAB workspace. The variables defined in the script, on the other hand, are available in the main MATLAB workspace.

The first line in a MATLAB function is the function definition:

```
function [out1, out2, ...] = funName(in1, in2, ...)
```

Here, `function` is a keyword that defines function and `funName` is the name of the function. The function command also defines a comma-separated list of input and output variables. The name of the m-file containing the function should be the same as the function name.* If the file name is different than the function name, the function is known to MATLAB through its file name. It is a good programming practice to use the same name for the function and the m-file.

Let us redo the factorial problem of Example 1.1 using the function.

Example 1.2 Example 1.1 Revisited

Problem: Write a MATLAB function to compute the factorial of *n*.

Solution: The MATLAB function is given below. It may be compared with the script file from Example 1.1.

```
function factVal = myFact(n)
% Function to calculate factorial of n
% Function usage:
%     factVal = myFact(n)
if (n<0)
    error('Negative numbers are not allowed');
else
    factVal = prod(1:n);
end
```

The above function is saved in a file, `myFact.m`, which has the same name (without the extension ".m") as the function name. `myFact` calculates the factorial of *n* when invoked at command prompt:

```
>> fact = myFact(5)
fact =
    120
```

---

* Exception is a nested or daughter function, which will not be considered in this book.

When the first example is executed, the commands in the script file are executed in the same *workspace* as the MATLAB command workspace. Thus, all the variables, n, fact, and i, are available in the command workspace. If, on the other hand, function myFact is used, only the variable fact appears on the MATLAB workspace. If user interaction is required similar to that in Example 1.1, a *separate MATLAB script* may be written as

```
% File: factDriver.m
% User Input
n = input('Please enter a number: ');
% Calculate factorial and display
disp([num2str(n), '! is ', num2str(myFact(n))]);
```

Following are a few tips regarding functions and scripts. Let us take an analogy with C++. When it comes to large projects, I treat a "driver" script similar to the void main{...} of C++. While this distinction is actually incorrect, it is a good programming practice in MATLAB to have a single driver script, whereas all other tasks are executed in individual functions.

Alternatively, the way to think about scripts vs. functions (usually for stand-alone problems) is to look at the *purpose* of the file. If the file executes a sequence of commands toward a particular end, a script is perhaps more suitable, for example, the computation of factorial when the purpose is to compute the factorial. Function, on the other hand, may be used

- To define a function $f(x, y, ...)$ of one or more variables

- To calculate output variable(s) or properties as a function of input conditions, such as reaction rate $r(T, C_i)$ and specific heat $c_p(T)$

- To define a function $f(x, y, ...)$ to be passed as an argument to a MATLAB solver

- To perform computations for individual task, unit operation, equipment, or system (such as the PFR or distillation column in the flow sheet Figure 1.1)

The above are good practices and not really a syntactic requirement from MATLAB. Thus, if one's purpose is to compute a factorial, Example 1.1 should be used; instead, if the purpose is to *use* the factorial in Maclaurin series expansion, Example 1.2 is the chosen alternative.

With this background, the following example will show the computation of Maclaurin series expansion of e^x as per Equation 1.5.

Example 1.3 Maclaurin Expansion of Exponent

Problem/Solution: The following code computes e^2 using Maclaurin series up to x^{10} terms:

```
%% Input values
x = 2;
n = 10;
```

```
%% Calculation of e^x
expVal = 1.0;
for i = 1:n
    currTerm = x^i/myFact(i);    % Calculate i-th term
    expVal = expVal + currTerm;
end

%% Displaying results
disp(['exp(2) = ', num2str(expVal)]);
```

The result of executing of the above script is:

```
exp(2) = 7.3889
```

The above example is shown for pedagogical purposes. This is a very inefficient way of coding; the example is primarily used to introduce the reader to the difference between functions and scripts and to show how to use functions (myFact.m in this example) in MATLAB. The inefficiency of the code is that a large number of computations were unnecessarily used in calculating the value of currTerm in each for loop iteration. A better way to do this is a *recursive* computation:

```
currTerm = currTerm*(x/i);
```

An astute reader will also recognize the use of brackets here, to reduce the chance of round-off error.

1.2.3 Using Array Operations in MATLAB®

If a user has experience with coding in C++, Java, or FORTRAN, this is the method employed to compute the functions using Maclaurin series. However, the most intermediate and advanced users of MATLAB will exploit highly powerful matrix operations that MATLAB natively provides. Consider the vector:

$$v = \left[\frac{x}{1} \; \frac{x}{2} \; \frac{x}{3} \; \cdots \; \frac{x}{n} \right]$$

MATLAB function cumprod (see Appendix A for a MATLAB Primer) computes cumulative product. Thus, first element is itself, second is a product of the first two elements, third is a product of first three elements, and so on:

$$\mathrm{cumprod}(v) = \left[\frac{x}{1} \quad \frac{x^2}{1 \cdot 2} \quad \frac{x^3}{1 \cdot 2 \cdot 3} \quad \cdots \quad \frac{x^n}{n!} \right]$$

Example 1.3 (Continued)

The entire code for computing the Maclaurin series of e^x can be written as

```
expVal = 1.0 + sum(cumprod(x./[1:n]));
```

The term in the inner brackets, `x./[1:n]`, generates the vector v. Note the use of element-wise division operator " `./` "—a unique feature of MATLAB's matrix capabilities. The result of `cumprod` is all terms in the series expansion but leading 1.

As an exercise, compute the Maclaurin series expansions of

$$\sin(x) = x - \frac{x^3}{3!} + \frac{x^5}{5!} + \cdots$$

$$\cos(x) = 1 - \frac{x^2}{2!} + \frac{x^4}{4!} - \cdots$$

using array operations, without using loops. **Hint:**

$$\text{cumprod}(-v) = \left[\begin{array}{ccccc} \dfrac{-x}{1} & \dfrac{x^2}{1 \cdot 2} & \dfrac{-x^3}{1 \cdot 2 \cdot 3} & \cdots & \dfrac{(-1)^n x^n}{n!} \end{array} \right]$$

1.2.4 Loops and Execution Control

An introduction to `for/while` loops and `if-then-else` conditions is provided in Appendix A (Section A.6). In this section, we will use these statements to write our first code for an *iterative* numerical technique, which is one of the most important classes of numerical techniques. Héron's method, described in the first century AD, is arguably one of the oldest iterative numerical techniques. It is used for computing the square root of 2 (or any number n). The working principle of the method is that for any positive real number, x, the solution $\sqrt{2}$ always lies between x and $2/x$. Starting with an arbitrary guess, $x^{(i)}$, the next guess is taken as arithmetic mean of $x^{(i)}$ and $2/x^{(i)}$.

Example 1.4 Héron's Method

Problem/Solution: Starting with an initial guess of $x^{(0)} = 2$, use 10 iterations of Héron's method:

$$x^{(i+1)} = \frac{1}{2}\left(x^{(i)} + \frac{2}{x^{(i)}} \right)$$

```
% Square root of 2 using Heron's algorithm
x = 2;      % Initial guess
```

```
%% Iterations
for i = 1:10
    x(i+1) = (x(i)+2/x(i))/2;
end
%% Display results
disp(['Herons Square Root: ', num2str(x(end))]);
```

All the intermediate values $x^{(i)}$ can be displayed as below:

```
>> disp(x)
    2.0000    1.5000    1.4167    1.4142    1.4142    1.4142
```

Starting with an initial guess of 2.0, the value of x rapidly converges to 1.4142 in the fifth iteration. Taking the difference of the consecutive terms of x gives us

```
>> disp(diff(x));
-0.50    -0.0833    -0.0025    -2.124e-6    -1.595e-12    0
```

The fourth iteration result of 1.4142 differs from the previous by 2.124×10^{-6}. Often, this is considered as sufficiently accurate for a numerical technique.

1.2.5 Section Recap

We started this section with a quick view of the structure of a typical MATLAB code in Figure 1.2. The importance of creating a good skeletal structure for MATLAB codes was discussed. Sectioning and comments are very useful in this aspect to keep the code clean, readable, and easy to debug. The importance of having clear demarcation between (1) input/parameter definition, (2) computation, and (3) output/wrap-up sections in a code cannot be emphasized enough. It is far too common to see programmers define their model parameters at multiple points in a code. This was emphasized by using input-computation-output sectioning in smallest of projects (see Example 1.1 or Example 1.3). Example 1.2 was used to highlight the use of a function. Unlike a project driver script file, a function gets its inputs as input arguments, and returns output arguments. Thus, most functions primarily have a computation section.

The choice of variable names used is also important. When the context was clear, variable names x or i were used. However, it is best to have more descriptive names, such as `currTerm` to represent "current term" and `expVal` to represent "value of exponential." Also note capitalization of the first letter of an English word in the variable name to improve readability. A variable `currTerm` is more easily readable than `currterm`.

Finally, as users will see later in this book, the use of `global` for sharing variables between functions or workspaces is avoided. We pay attention to variable scoping and treat workspaces as sacrosanct.

1.3 APPROXIMATIONS AND ERRORS IN NUMERICAL METHODS

We have considered two examples of numerical methods in the previous section: e^x from truncated infinite series and calculation of $x = \sqrt{2}$ using an *iterative* algorithm. The true value of $e^2 = 7.3890561$. The value computed using the truncated Maclaurin series was 7.3889. The difference between the two values is the error in numerical computation using the numerical approximation. Before proceeding further, we will now formally define *error*.

DEFINITION: ERRORS

The true value of a variable x is represented as x^* while the current approximation as $x^{(i)}$. The absolute (true) error is defined as

$$E = \left| x^* - x^{(i)} \right|$$

and the relative (true) error is defined as

$$\bar{E} = \left| \frac{\left(x^* - x^{(i)} \right)}{x^*} \right|$$

Typically, the true solution is not known *a priori*. Hence, the following two definitions of approximation errors are defined. Herein, the error is defined based on the difference between subsequent approximations of the solution. Thus

$$e = \left| x^{(i+1)} - x^{(i)} \right| \quad \text{and} \quad \bar{e} = \left| \frac{x^{(i+1)} - x^{(i)}}{x^{(i)}} \right|$$

EXAMPLE

With these definitions, the absolute error in the calculation of e^2 using Maclaurin series through x^{10} terms is

$$E = 6.139 \times 10^{-5}$$

Consider the fourth iteration value in Héron's algorithm of 1.4142 (x(5) in Example 1.4, since x(1) is the initial guess). The true and approximation errors are

$$E^{(4)} = \left| \sqrt{2} - x^{(4)} \right| = 1.595 \times 10^{-12} \quad \text{and} \quad e^{(4)} = \left| x^{(4)} - x^{(3)} \right| = 2.124 \times 10^{-6}$$

The error occurs due to two reasons: (1) truncation of the infinite series to finite number of terms and (2) finite precision representation leading to rounding off.

1.3.1 Machine Precision

A computer is a finite precision machine. While integers are exactly represented in a computer representation, real number representation is inexact. Real numbers are represented using a finite number of bits (or binary digits) in *floating-point representation* in a computer.

Consider two real numbers $r_1, r_2 \in \mathcal{R}$; there are infinite number of real numbers between r_1 and r_2. However, in computer representation of real numbers, there are only a finite number of real numbers between r_1 and r_2. To understand this concept better, let us consider an example of decimal floating-point representation of real numbers. Consider that our "mind-computer" represents a positive real number using six digits: five decimal digits and a single signed decimal exponent. The five decimal digits form *mantissa m*, whereas the sixth digit is the signed exponent. The floating-point representation of this number may be written as

$$\underbrace{0.\text{xxxxx}}_{m} \times 10^{e-5}$$

The mantissa lies between 1 and 1/base (base=10 in decimal system), that is, $m \in [0, 1)$. The digit in the exponent, e, also takes the value between 0–9; thus with a bias of –5 shown above, the exponent takes values from $10^{-5} - 10^4$. The number 314.27 is represented as

$$0.31427 \times 10^{8-5}$$

The next number that can be represented in this system is therefore

$$0.31428 \times 10^{8-5}$$

The decimal computer cannot represent any other number between the two numbers, 314.27 and 314.28. With the five-digit mantissa, the number 314.27<u>12</u> cannot be distinguished from 314.27. Thus, the last two digits get chopped off as a consequence of finite precision representation of a number. In summary, the *floating-point representation* is of the form

$$m \times b^{e-F} \tag{1.6}$$

where
    m is the mantissa, with $1/b \leq m < 1$
    e is the exponent
    b is the base (2 in binary; 10 in decimal)
    F is the exponent bias (fixed for a computer)

Thus, real number representation in a computer has a "least count." This least count, when normalized, depends on the mantissa and is known as *machine precision*.

DEFINITION: MACHINE PRECISION

Machine precision ϵ is defined as the maximum relative error in the floating-point representation of real numbers. In practice, it is the smallest positive real number such that the number $(1 + \epsilon)$ can be distinguished from 1.

MATLAB uses double precision numbers with a word length of 64 bits: 1 sign bit, 52 mantissa bits, and 11 exponent bits. The machine precision in MATLAB is obtained using the keyword eps. It is typically 2^{-52}. Consider the following statements typed on the command prompt:

```
>> x = 1+2^-52;
>> disp(x-1.0)
    2.2204e-16
>> x = 1+2^-53;
>> disp(x-1.0)
    0
```

The second result is a consequence of the "least count" in MATLAB. Example 1.5 below contains a code to recursively compute machine precision in MATLAB. Starting with an initial value of $\epsilon = 1$, the following procedure is adopted. The value of ϵ is halved; $(1+\epsilon)$ is computed and compared with 1. If the two cannot be distinguished, the value of ϵ has fallen below the machine precision; else, the above procedure is repeated until 1 and $(1+\epsilon)$ are undistinguishable.

Example 1.5 MATLAB® Code to Compute Machine Precision

```
% Code for finding machine precision
epsilon = 1.0;
iFlag = true;
while(iFlag)
    epsilon = epsilon/2.0;
    testNum = 1+epsilon;
    if (testNum-1.0 == 0)
        iFlag = false;
    end
end
disp(['Machine Precision = ', num2str(epsilon*2)]);
```

1.3.2 Round-Off Error

The finite precision of a computer leads to *round-off errors*, errors that result due to inexact representation of real numbers in a computer. Note that a computer chops off the additional digits (thus, 314.27<u>12</u> and 314.27<u>72</u> are both represented as 314.27). Continuing with the example of six-digit decimal floating-point representation, the following example highlights the origins of round-off errors. The values of x satisfying quadratic equation $x^2 - bx + c = 0$ are

$$x = \frac{b \pm \sqrt{b^2 - 4c}}{2}$$

Example 1.6 Quadratic Equations

Problem: Solve the quadratic equation $x^2 - 97x + 1 = 0$ using floating-point representation. The true solutions of the equation are 96.98968 and 0.01031037.

Solution: Consider the term $p = \sqrt{b^2 - 4c}$:

$$p = \sqrt{(0.97e2)^2 - (0.4e1)} = \sqrt{0.94090e4 - 0.00040e4}$$

The digits of the mantissa are moved during subtraction so that the two numbers have the same exponent. If one of the numbers is much smaller than the other, *precision is lost*.

In this example, $p = \sqrt{0.94050e4} = 0.96979e2$.

The value of p is obtained accurately for five significant digits. The two solutions therefore are

$$x = 0.5\left(0.97000e2 \pm 0.96979e2\right) = \begin{cases} 0.96989 \times 10^2 \\ 0.10500 \times 10^{-1} \end{cases}$$

Note the second solution, true value is 0.01031 and computed value is 0.01050. This is due to the fact that the values of b and p are fairly close to each other; thus chopping off the trailing digits in p before subtraction results in a rather large *round-off error*.

An alternative formula for the second solution is

$$x = \frac{2c}{b + \sqrt{b^2 - 4c}}$$

$$x = 0.20000e1 \div \left(0.97000e2 + 0.96979e2\right) = 0.10310 \times 10^{-1}$$

Thus, avoiding subtraction between two close numbers results in better error behavior.

The above example demonstrates how the machine precision gives rise to round-off errors. The computer-stored value x is related to the true value $\bar{x}$ as

$$x = \bar{x}\left(1 + \varepsilon_{\text{mach}}\right)$$

As we shall see in Section 1.3.4, subtraction of two very close numbers results in errors in numerical differentiation. Before that, we will introduce the other type of error, *truncation error*.

1.3.3 Taylor's Series and Truncation Error

Consider Taylor's series expansion:

$$f(a+h) = f(a) + hf'(a) + \frac{h^2}{2!}f''(a) + \cdots + \frac{h^n}{n!}f^{[n]}(a) + E_n \tag{1.7}$$

Taylor's series expansion is used, either directly or indirectly, for the derivation of a large number of numerical algorithms described in this text. Finite number of terms in Taylor's series are retained (e.g., 10 terms were used in computing e^2). The error resulting from truncating an infinite series in deriving a numerical approximation is known as *truncation error*. The nth order error term is

$$E_n = \frac{h^{n+1}}{(n+1)!} f^{[n+1]}(\xi), \quad \text{where } \xi \in \left[a, (a+h)\right]$$

This is often represented as

$$\mathcal{O}\left(h^{n+1}\right)$$

and the error is said to be *of the order of h^{n+1}*.

The greater the number of terms retained in Taylor's series, the smaller is the error; the smaller the value of h chosen, the smaller is the error. Recall that the error in computing e^2 using series expansion until the 10th term was $E = 6.139 \times 10^{-5}$.

We demonstrate the concept of order of accuracy and error using the example of numerical integration. Classical Newton-Cotes integration formulae (see Appendix E for more information on numerical integration), namely, the trapezoidal rule and Simpson's 1/3rd rule, will be used to demonstrate this. An integral can be approximated using these rules as

$$\text{Trapezoidal rule: } h = (b - a) \quad \int_a^b f(x)\,dx = \frac{h}{2}\left[f_a + f_b\right] + \mathcal{O}\left(h^3\right) \tag{1.8}$$

$$\text{Simpson's 1/3rd rule: } \hat{h} = (b - a)/2 \quad \int_a^b f(x)\,dx = \frac{h}{3}\left[f_a + 4f_{a+\hat{h}} + f_b\right] + \mathcal{O}\left(\hat{h}^5\right) \tag{1.9}$$

The last term in both the equations indicate how the *truncation errors* in using the above formula depend on the step-size, h. Consider the numerical example below that illustrates truncation error and order of accuracy.

Example 1.7 Truncation Error in Numerical Integration

Problem: Use the trapezoidal and Simpson's 1/3rd rules to compute (for different values of h):

$$\int_0^h \frac{1}{1+x^2}\,dx$$

The true solution is $\tan^{-1}(h)$. The aim is to compute errors in numerical integration.

Solution: The code for calculating the numerical integral is given below. The integral is computed for four different values of step-size *h*. Although the first instinct is to use a `for` loop to loop over different values of *h*, we will exploit the powerful array operations in MATLAB to write an efficient code:

```
% Integral using Newton-Cotes (single application)
h = [1, 0.1, 0.01, 0.001];    % Various step sizes
f1 = 1;                        % f(a) = f(0)


%% Trapezoidal Rule
trueInt = atan(h);        % True values of integral
f2 = 1./(1+h.^2);         % f(a+h)
trapInt = h/2.*(f1+f2);
trapErr = abs(trueInt-trapInt);


%% Simpson's 1/3rd Rule
trueInt = atan(2*h);      % True value of integral
f3 = 1./(1+(2*h).^2);     % f(a+2h)
simpInt = h/3.*(f1+4*f2+f3);
simpErr = abs(trueInt-simpInt);


%% Plotting results
loglog(h,trapErr,'-b', h,simpErr,'--r');
xlabel('step size, h'); ylabel('error');
```

The `atan` function used in line 5 returns a vector of the same size as vector h, wherein each element is $\tan^{-1}(\cdot)$ of the corresponding element of h. The next line contains computation of $f(a+h)=f(h)$ for all values of h. Notice the use of element wise division ($./$) and power ($.\^$) operators.

Figure 1.3 shows log-log plot of error in the numerical computation of integral vs. step-size *h*. Both of these are approximately straight lines. The error is lower with Simpson's 1/3rd rule and falls faster with step-size compared to the trapezoidal rule. The step-size falls by three orders of magnitude. Corresponding error using the trapezoidal rule falls by nine orders of magnitude (from $\sim 10^{-1}$ to $\sim 10^{-10}$). Similarly, error in Simpson's 1/3rd rule falls by 15 orders of magnitude (from $\sim 10^{-1}$ to $\sim 10^{-16}$).

Alternatively, one can plot $\log(E)$ vs. $\log(h)$ on a linear plot. The straight line for the trapezoidal rule has a slope of approximately 3, whereas that for Simpson's 1/3rd rule is 5. In other words, $\log(e_{trap}) \sim 3\log(h) = \log(h^3)$ and $\log(e_{simp}) \sim 5\log(\hat{h}) = \log(\hat{h}^5)$. Note that this corresponds to the order of the error term in Equations 1.6 and 1.7.

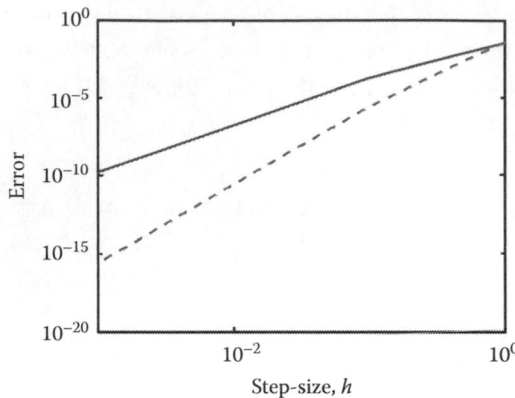

FIGURE 1.3 Truncation error vs. step-size for single application of the trapezoidal (solid) and Simpson's 1/3rd rules (dashed).

1.3.4 Trade-Off between Truncation and Round-Off Errors

Consider the example of a numerical approximation of a derivative, given by

$$f'(x)\big|_{x_i} = \frac{f(x_i + h) - f(x_i)}{h} + \mathcal{O}(h) \tag{1.10}$$

$$f'(x)\big|_{x_i} = \frac{f(x_i + h) - f(x_i - h)}{2h} + \mathcal{O}(h^2) \tag{1.11}$$

Equation 1.10 is known as forward difference approximation, whereas Equation 1.11 is the central difference. Refer to Appendix B for more details on numerical derivatives.

Example 1.8 Numerical Derivatives and Truncation Error

Problem: Find the numerical derivative of $f(x) = \tan^{-1}(x)$ at $x = 1$ using forward and central difference formulae for different values of h. True solution is $f'(x) = \left(1 + x^2\right)^{-1}$.

    Solution: This problem will be solved in two parts. First, numerical derivatives will be obtained for $h = 0.01$ and $h = 0.0001$ using both the methods. Using forward difference

$$f'(x) \approx 0.497508 \quad \text{and} \quad f'(x) \approx 0.499975$$

$$e_{0.01} = 2.492 \times 10^{-3} \quad \text{and} \quad e_{0.0001} = 2.4999 \times 10^{-5}$$

Thus, decreasing the step-size h by two orders of magnitude resulted in a reduction in the error by two orders of magnitude. That is because the error is proportional to h^1.

    Consider the results for central difference:

$$e_{0.01} = 8.333 \times 10^{-6} \quad \text{and} \quad e_{0.0001} = 8.332 \times 10^{-10}$$

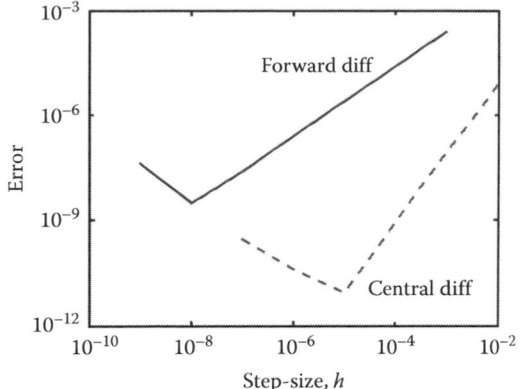

FIGURE 1.4   Effect of step-size on error in computing numerical derivative using forward and central difference formulae.

Not only is the error in the central difference formula lower than forward difference but reducing the step-size by two orders of magnitude improves the error by *four* orders of magnitude: a consequence of error being proportional to h^2 for central differences.

Figure 1.4 shows the error vs. step-size plot (on a log-log scale) using forward and central difference formulae to compute the numerical derivative of $\tan^{-1}(x)$. The following code was used for this purpose:

```
% Numerical derivatives with forward/central differences
x = 1;
trueVal = 1/(1+x^2);

%% Forward difference
hFwd = 10.^[-9:-3];
df_fwd = (atan(x+hFwd)-atan(x)) ./ (hFwd);
errFwd = abs(df_fwd-trueVal);

%% Central difference
hCtr = 10.^[-7:-2];
df_ctr = (atan(x+hCtr)-atan(x-hCtr)) ./ (2*hCtr);
errCtr = abs(df_ctr-trueVal);

%% Plotting
loglog(hFwd,errFwd,'-b',hCtr,errCtr,'--r');
```

The above example highlights the trade-off between truncation and round-off errors. The truncation error decreases as step-size is reduced, whereas the round-off error increases at very low step-sizes. Thus, there is an optimal step-size at which the net error is minimum. In case of numerical differentiation, the optimal choice of step-size for forward and central difference formulae is

$$h_{\text{opt}}^{\text{fwd}} = \varepsilon_{\text{mach}}^{1/2} \quad \text{and} \quad h_{\text{opt}}^{\text{fwd}} = \varepsilon_{\text{mach}}^{1/3} \tag{1.12}$$

where $\varepsilon_{\text{mach}}$ is the machine precision. Therefore, the optimal values of step-sizes for forward and central difference formulae are $\sim 10^{-8}$ and $\sim 10^{-5}$, respectively. When the step-size is greater than these respective values, slopes of e_h vs. h curves are 1 and 2, respectively, for the two formulae. Note that this slope corresponds to the order of accuracy, $\mathcal{O}(h^1)$ and $\mathcal{O}(h^2)$, respectively.

1.4 ERROR ANALYSIS

A basic understanding of some key numerical issues is important to make an appropriate choice of numerical method used to solve *process simulation* problems. Some of these concepts are introduced presently. Numerical methods broadly work in three ways. First is the direct way of computation, such as approximation of e^2 from truncated Taylor's series or numerical derivatives shown above. Second is an iterative way, such as the Héron's method in computing $\sqrt{2}$. Third is step-wise recursive method, such as multiple applications of Newton-Cotes integration formulae (see Appendix E for details). Solutions of differential equations (see Chapter 3, for example), optimization, etc., also fall under this category.

Error analysis introduced in this section is applicable to iterative and recursive methods. A general introduction to some key concepts will be presented. As defined before, $x^{(i)}$ is the current numerical solution, and x^* is the true solution. We can write $x^{(i)} = x^* + E^{(i)}$, where $E^{(i)}$ is the error. As we have seen in the examples before, the numerical solution changes with the step-size and/or number of recursions. The next value $x^{(i+1)}$ is computed as some function of $x^{(i)}$:

$$x^{(i+1)} = f\left(x^{(i)}\right) = f\left(x^* + E^{(i)}\right)$$

Thus, it is possible to write

$$E^{(i+1)} = \mathcal{F}\left(E^{(i)}; x^*, x^{(i)}\right)$$

indicating that the error depends on the previous error (as also on the true solution and numerical solution). Issues regarding behavior of the numerical solution $x^{(i)}$ and error $E^{(i)}$ are discussed.

1.4.1 Convergence and Stability

Convergence and stability are properties of interest to iterative and recursive methods. A numerical method is said to be convergent if the numerical errors $E^{(i)}$ go to zero as i increases. In other words, as the number of iterations increase or step-size decreases, the solution $x^{(i)}$ approaches the true solution x^*. Stability is a related property. The error $E^{(i)}$ is a result of incorrect initial conditions and round-off errors. A method is said to be stable if the effect of these errors do not grow with the application of the numerical method. In other words, $E^{(i)}$ should decrease (or at least not increase in unbounded manner) as i increases.

An important consideration in choosing a numerical method is its convergence behavior. All the examples considered so far were stable and convergent. The next consideration is "how fast does the numerical method converge." The order of convergence of an iterative numerical method is said to be n if

$$E^{(i+1)} = \kappa \left[E^{(i)} \right]^n$$

Consider the errors in Example 1.4. Subsequent iterations of Héron's algorithm result in the following errors:

$$e^{(i)} = \left\{ 5 \times 10^{-1} \quad 8.33 \times 10^{-2} \quad 2.5 \times 10^{-3} \quad 2.12 \times 10^{-6} \quad 1.6 \times 10^{-12} \quad 0 \right\}$$

It can be inferred that the algorithm has a *quadratic order of convergence*, that is, $n = 2$. Besides, the constant κ is also important. For example, when $n = 1$ (linearly convergent method), the numerical method is convergent if $|\kappa| < 1$.

The terms stability and convergence have different meanings. Stability refers to whether or not errors grow with iterations or recursions, whereas convergence means whether the numerical solution reaches the true solution.

1.4.2 Global Truncation Error

Let us revisit the numerical integration to motivate this section. The accuracy of numerical integration can be increased by dividing the region into multiple intervals and using multiple applications of the trapezoidal and Simpson's 1/3rd rules, which is given by

$$\text{Trapezoidal rule}: \int_a^b f(x)\,dx = \frac{h}{2}\left[f_1 + 2f_2 + 2f_3 + \cdots + 2f_n + f_{n+1} \right] \tag{1.13}$$

$$\text{Simpson's 1/3rd rule}: \int_a^b f(x)\,dx = \frac{h}{3}\left[f_1 + 4f_2 + 2f_3 + 4f_4 + \cdots + 4f_n + f_{n+1} \right] \tag{1.14}$$

In the above, $h = (b-a)/n$. We choose n as an even number. While number of intervals is n, there are $(n+1)$ nodes, with $f_1 = f(a), f_{i+1} = f(a+ih)$ and $f_{n+1} = f(b)$.

Let us take the last example as a segue into using numerical solutions for the design, simulation, and analysis of process systems. The design equation of a PFR is given by

$$V = \frac{F_0}{kC_0^m} \int_0^a \frac{dX}{r(X)} \tag{1.15}$$

where X is the conversion, F_0 is the inlet flowrate and the reaction rate is given by $kC_0^m r(X)$. The design problem is to find the volume V of the PFR for the desired conversion, a.

Example 1.9 Newton-Cotes Formulae for Design of a PFR

Problem: Use multiple applications of the trapezoidal and Simpson's 1/3rd rules to compute the volume of PFR required for 80% conversion, when Equation 1.15 simplifies to

$$V = 0.5 \int_0^{0.8} \frac{dX}{(1-X)^{1.25}}$$

Solution: The true solution is given by $V_{true} = 2[(1-a)^{-0.25} - 1] = 0.9907 \text{ m}^3$.
The following code is written to calculate the volume using n intervals:

```
% Newton Cotes for PFR Design
b = 0.8;                              % Conversion
trueVal = 2 * ((1-b)^-0.25 - 1);  % PFR volume
N = [8, 32, 128, 512, 800];       % # of intervals

%% Solving design equation
for i = 1:length(N)
    n = N(i);      h = b/n;
    x = [0:h:b];
    fx = 0.5 ./ (1-x).^1.25;
    % Trapezoidal Rule
    V_trap(i) = h/2*(fx(1)+fx(n+1)) + h*sum(fx(2:n));
    % Simpson's 1/3 Rule
    V_simp(i)  = h/3*(fx(1)+fx(n+1)) + ...
        h/3*4*sum(fx(2:2:n)) + h/3*2*sum(fx(3:2:n));
end

%% Display results
trapErr = abs(trueVal-V_trap);
simpErr = abs(trueVal-V_simp);
loglog(b./N, [trapErr;simpErr]);
```

Figure 1.5 shows the effect of step-size on error for the two methods. The step-size h is decreased by two orders of magnitude. The error using trapezoidal rule reduces from 10^{-2} to 10^{-6}, that is, by four orders of magnitude, whereas the error using Simpson's rule decreases by eight orders of magnitude.

The PFR volumes obtained using the two numerical methods are

```
V_trap = 1.0091    0.9919    0.9908    0.9907    0.9907
V_simp = 0.9922    0.9907    0.9907    0.9907    0.9907
```

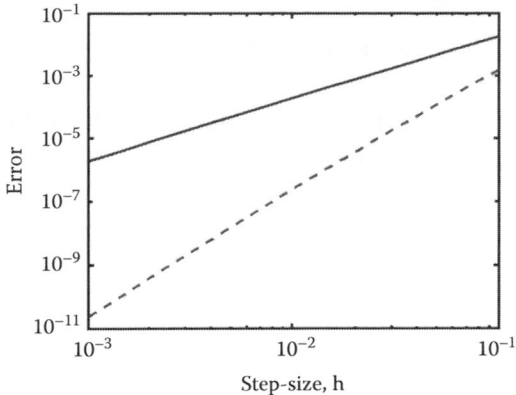

FIGURE 1.5   Error analysis for the trapezoidal and Simpson's 1/3rd rules.

Example 1.9 shows the application of the numerical integration technique to solve a PFR design problem. The solution obtained is accurate even with a reasonably small number of intervals used. The main intention of this example was, however, to demonstrate error analysis for the integration formulae. According to Equation 1.8, the truncation errors in the trapezoidal rule are $\mathcal{O}\left(h^3\right)$. However, Figure 1.5 shows that the error vs. step-size curve has a slope of 2. This is because the *global truncation error* incurred in multiple applications of the trapezoidal rule is of the order h^2.

Equations 1.8 and 1.9 showed the local truncation error, error resulting from a single application of the numerical technique. When multiple applications of the numerical technique are used, we encounter the local error, as well as an error resulting from the accumulation of errors until the current point. The net effect of this is to typically reduce the order of accuracy of the numerical technique. In case of the two Newton-Cotes formulae, global truncation errors are given by

$$E_{\text{trap}}^{(\text{gte})} \propto h^2 f''(\xi), \quad E_{\text{simp}}^{(\text{gte})} \propto h^4 f''''(\xi)$$

1.5 OUTLOOK

Example 1.9 was our first glimpse into Process Simulation using MATLAB, where we used numerical integration to compute the volume of the PFR. Section 1.4.2 started with a discussion on an integration problem (Equations 1.13 and 1.14), followed by a practical example (design of PFR), and ending with Example 1.9 showing the solution and results in MATLAB. We will follow a similar structure in most of the chapters: First, the numerical techniques of importance will be discussed, followed by a concise description of the problem to be simulated or analyzed, and ending with the application of those techniques to solve the problem.

The chapters are organized based on the *approach* to solving the problem. In this context, the treatment in this book differs from introductory textbooks. For example, Chapter 3 covers ODE-IVPs of nonstiff systems. On the other hand, ODE-BVPs are covered in Chapter 7 along with elliptic PDEs since they both involve systems that are diffusive in one or more dimensions, respectively. Solutions of stiff ODEs and DAEs, which require a similar set of tools for simulation and analysis, are covered in Chapter 8. Numerical techniques that are more generic in nature, such as numerical differentiation, integration, and solving linear equations, are covered in the Appendices.

I

Dynamic Simulations and Linear Analysis

CHAPTER 2

Linear Algebra

2.1 INTRODUCTION

Linear algebra is typically introduced in an introductory engineering mathematics course in the context of solving a set of n linear equations in n unknowns. Consider the following example:

$$x + 2y = 1$$
$$x - y = 4 \tag{2.1}$$

The above set of linear equations is written in the standard form as

$$\underbrace{\begin{bmatrix} 1 & 2 \\ 1 & -1 \end{bmatrix}}_{A} \underbrace{\begin{bmatrix} x \\ y \end{bmatrix}}_{\mathbf{x}} = \underbrace{\begin{bmatrix} 1 \\ 4 \end{bmatrix}}_{\mathbf{b}} \tag{2.2}$$

The geometric interpretation that is commonly discussed in introductory courses is that the solution $(3, -1)$ is the point of intersection of the two lines. This is a useful way of introducing the problem of solving linear equations. However, as explained by Strang in his *Linear Algebra* text, a more powerful and versatile visualization is through vector spaces.

Before moving to introduce vectors and vector spaces, I will very briefly review some results in the solution of a system of linear equations.

2.1.1 Solving a System of Linear Equations

Let us extend Equations 2.1 and 2.2 to a system of equations with n unknowns. A general linear equation in n unknowns is written as

$$a_{i1}x_1 + a_{i2}x_2 + \cdots + a_{in}x_n = b_i \tag{2.3}$$

We will define the solution vector

$$\mathbf{x} = \begin{bmatrix} x_1 \\ x_2 \\ \vdots \\ x_n \end{bmatrix} \tag{2.4}$$

and the linear equation (2.3) may then be written as

$$\begin{bmatrix} a_{i1} & a_{i2} & \cdots & a_{in} \end{bmatrix} \mathbf{x} = b_i \tag{2.5}$$

There may be n such equations in n unknowns. All such equations are then written together and converted into the following standard form:

$$\underbrace{\begin{bmatrix} a_{11} & a_{12} & \cdots & a_{1n} \\ a_{21} & a_{22} & \cdots & a_{2n} \\ \vdots & \ddots & \ddots & \vdots \\ a_{n1} & a_{n2} & \cdots & a_{nn} \end{bmatrix}}_{A} \underbrace{\begin{bmatrix} x_1 \\ x_2 \\ \vdots \\ x_n \end{bmatrix}}_{\mathbf{x}} = \underbrace{\begin{bmatrix} b_1 \\ b_2 \\ \vdots \\ b_n \end{bmatrix}}_{\mathbf{b}} \tag{2.6}$$

Here, the boldface small letters are used to represent vectors. Vectors can be both row and column vectors. However, for consistency of notation, we will write vectors as $n \times 1$ column vectors (unless specified otherwise). The vectors $\mathbf{x}$ and $\mathbf{b}$ in the above equation are composed of n elements. Each element, represented as nonbold characters with a subscript, is a real scalar. We will define vectors and vector spaces in the next section.

The view in Equation 2.5 is a row-wise view of linear equations. The necessary and sufficient condition for the set of linear equations (2.6) to have a unique solution is that determinant of A should be nonzero. An equivalent statement is that a unique solution to Equation 2.6 exists if and only if $\text{rank}(A) = n$. If this condition is met, inverse of A exists and the solution is given by

$$\mathbf{x} = A^{-1}\mathbf{b} \tag{2.7}$$

Numerical methods to solve linear equations are not discussed in this chapter. Instead, the Gauss Elimination and LU decomposition methods are provided in Appendix C for ready reference. Relevant MATLAB® commands are summarized in Section 2.5 at the end of this chapter.

2.1.2 Overview

While solving linear equations is one application, in general linear algebra finds widespread use in engineering (as also in science, economics, and other fields). Theories are often well developed for linear systems, and detailed analysis can provide insights into system behavior. It is common to linearize a nonlinear system using Taylor's series expansion;

the properties of the linearized system give a good qualitative picture of the behavior of the original nonlinear system. For example, insights into stability, controllability, and dynamics can be obtained by linear system analysis. Nonlinear equations and optimization problems may be solved by successively linearizing the system. Linear algebra is also useful for parameter estimation using least squares approach.

The discussion in this chapter will be cast in one of the following three ways. All these are equivalent but are typically seen in different types of problems. First, the problem is cast in terms of m linear correlations between n elements of a vector:

$$\underbrace{\begin{bmatrix} a_{11} & a_{12} & \cdots & a_{1n} \\ a_{21} & a_{22} & \cdots & a_{2n} \\ \vdots & \ddots & \ddots & \vdots \\ \vdots & \ddots & \ddots & \vdots \\ a_{m1} & a_{m2} & \cdots & a_{mn} \end{bmatrix}}_{A} \underbrace{\begin{bmatrix} x_1 \\ x_2 \\ \vdots \\ x_n \end{bmatrix}}_{\mathbf{x}} = \underbrace{\begin{bmatrix} b_1 \\ b_2 \\ \vdots \\ \vdots \\ b_m \end{bmatrix}}_{\mathbf{b}} \tag{2.8}$$

When $m = n$, this is same as Equation 2.6 encountered in solving linear equations. In least squares parameter estimation problems, the number of rows is greater, that is, $m > n$. In either cases, one intends to *solve* the problem to obtain $\mathbf{x}$ given the right-hand side vector $\mathbf{b}$.

The second case is where a linear correlation exists between an input vector, $\mathbf{x}$, and an output vector, $\mathbf{y}$. Such a linear mapping is mathematically represented as matrix multiplication:

$$\underbrace{\begin{bmatrix} b_1 \\ b_2 \\ \vdots \\ \vdots \\ b_m \end{bmatrix}}_{\mathbf{y}} = \underbrace{\begin{bmatrix} a_{11} & a_{12} & \cdots & a_{1n} \\ a_{21} & a_{22} & \cdots & a_{2n} \\ \vdots & \ddots & \ddots & \vdots \\ \vdots & \ddots & \ddots & \vdots \\ a_{m1} & a_{m2} & \cdots & a_{mn} \end{bmatrix}}_{A} \underbrace{\begin{bmatrix} x_1 \\ x_2 \\ \vdots \\ x_n \end{bmatrix}}_{\mathbf{x}} \tag{2.9}$$

The structure of matrix multiplication, $A\mathbf{x}$, is the same in both Equations 2.8 and 2.9. However, in this problem, we are interested in *analyzing* (and perhaps predicting) how the output responds to changes in the input vector $\mathbf{x}$.

The third case is seen as solving a linear system of differential or difference equations:

$$\frac{d}{dt} \underbrace{\begin{bmatrix} b_1 \\ b_2 \\ \vdots \\ b_n \end{bmatrix}}_{\mathbf{y}} = \underbrace{\begin{bmatrix} a_{11} & a_{12} & \cdots & a_{1n} \\ a_{21} & a_{22} & \cdots & a_{2n} \\ \vdots & \ddots & \ddots & \vdots \\ a_{n1} & a_{n2} & \cdots & a_{nn} \end{bmatrix}}_{A} \underbrace{\begin{bmatrix} x_1 \\ x_2 \\ \vdots \\ x_n \end{bmatrix}}_{\mathbf{y}} \tag{2.10}$$

Unlike the second case, here the n-dimensional $\mathbf{y}$-space maps onto itself.

The three problems are somewhat equivalent numerically and require a similar repertoire of tools for analysis. However, I will bring out different practical significance for the three problems so that these tools may be interpreted in relation to engineering problems of interest.

We need to take a column-wise view of the linear equations (2.8), (2.9), or (2.10). This calls for definition of vector spaces, which will be covered in Section 2.2. Solution of linear equations will be interpreted in that context. Thereafter, Sections 2.3 and 2.4 introduce singular value and eigenvalue decompositions. We will thus build some background in this chapter, which would be then used in some of the subsequent discussions in this book.

In summary, linear algebra is a versatile and practically relevant field. It is therefore apt to start this book with a brief discussion on linear algebra. My treatment in this chapter is inspired heavily by the online course and textbook by Prof. Gilbert Strang.

2.2 VECTOR SPACES

2.2.1 Definition and Properties

Let us first define a vector. A *vector* may be defined as an ordered list of scalars. A vector $\mathbf{x} \in \mathcal{R}^n$ is a column vector with n elements (real numbers). Geometrically, the vector $\mathbf{x} \in \mathcal{R}^n$ is a point in n-dimensional Euclidean space.

We visualize a vector in the form of a line starting from the origin toward this point in space, which we often associate with a certain magnitude and direction. The term "magnitude" may be generalized in the form of *vector norm*: *Norm* is a positive scalar that is indicative of the size of the vector. For example, the two-norm of a vector is what we associate as the Euclidean distance:

$$\|\mathbf{x}\|_2 = \sqrt{x_1^2 + x_2^2 + \cdots + x_n^2} \tag{2.11}$$

A vector space consists of a set of vectors, along with rules of addition, and scalar multiplication:

$$\mathbf{x} + \mathbf{y} = \begin{bmatrix} x_1 + y_1 \\ x_2 + y_2 \\ \vdots \\ x_n + y_n \end{bmatrix}, \quad c\mathbf{x} = \begin{bmatrix} cx_1 \\ cx_2 \\ \vdots \\ cx_n \end{bmatrix}$$

Based on the above definition of vector space, several rules apply since a vector space is a linear space. These rules are well known to the reader as commutative and associative laws for addition

$$\mathbf{x} + \mathbf{y} = \mathbf{y} + \mathbf{x}, \quad (\mathbf{x} + \mathbf{y}) + \mathbf{z} = \mathbf{x} + (\mathbf{y} + \mathbf{z}) \tag{2.12}$$

and laws for scalar multiplication

$$c(\mathbf{x} + \mathbf{y}) = c\mathbf{x} + c\mathbf{y}, \quad (c_1 + c_2)\mathbf{x} = c_1\mathbf{x} + c_2\mathbf{x} \tag{2.13}$$

and that there exists a unique zero vector in the vector space, such that

$$\mathbf{x} + \mathbf{0} = \mathbf{x}, \quad \mathbf{x} + (-\mathbf{x}) = \mathbf{0} \tag{2.14}$$

With these definitions of vectors and vector spaces in place, let us revisit Equation 2.2. An alternative interpretation uses the concept of vector spaces just defined:

$$\mathbf{v}_1 x + \mathbf{v}_2 y = \mathbf{b} \tag{2.15}$$

where $\mathbf{v}_1 = \begin{bmatrix} 1 \\ 1 \end{bmatrix}$ and $\mathbf{v}_2 = \begin{bmatrix} 2 \\ -1 \end{bmatrix}$ are the column vectors that constitute matrix A. In other words, $\mathbf{v}_i$ is the ith column of the matrix. Thus, the aim of solving linear equations can now be expressed as finding the values of "coefficients" x and y such that the vector $\mathbf{b}$ on the right-hand side is a *linear combination* of the vectors $\mathbf{v}_1$ and $\mathbf{v}_2$. As shown in Figure 2.1a, this is like "completing a parallelogram" that we studied in high school.

Consider, instead, if the second equation changed so that the two linear equations are

$$x + 2y = 1$$

$$2x + 4y = 4$$

then the system of equations does not have a solution. The two column vectors in this case are

$$\mathbf{w}_1 = \begin{bmatrix} 1 \\ 2 \end{bmatrix} \quad \text{and} \quad \mathbf{w}_2 = \begin{bmatrix} 2 \\ 4 \end{bmatrix}$$

These two vectors lie on the same line from the origin. In fact, $\mathbf{w}_2 = 2\mathbf{w}_1$. As shown in Figure 2.1b, since the vector $\mathbf{b}$ does not lie along the same line, there is no solution. If instead, the second equation was changed so that

$$x + 2y = 1$$

$$2x + 4y = 2$$

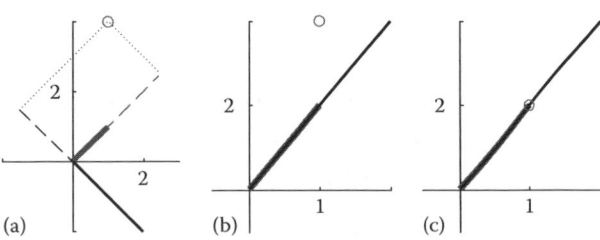

FIGURE 2.1 Demonstration of case with (a) unique solution, (b) no solution, and (c) infinite solutions. The two thick lines represent the two column vectors of matrix A and the symbol represents the right-hand side vector, $\mathbf{b}$ or $\mathbf{d}$.

there will be infinitely many solutions. This is because for this case, the three vectors

$$\mathbf{w}_1 = \begin{bmatrix} 1 \\ 2 \end{bmatrix}, \quad \mathbf{w}_2 = \begin{bmatrix} 2 \\ 4 \end{bmatrix}, \quad \mathbf{d} = \begin{bmatrix} 1 \\ 2 \end{bmatrix}$$

lie along the same line from the origin. This is shown in Figure 2.1c. It is easy to see that two solutions are $(1,0)$ and $(0,0.5)$. Likewise, infinitely many solutions for the two equations can be obtained.

As will be discussed presently, the three geometric views in Figure 2.1 are generalizable and powerful ways to introduce key linear algebra concepts. The utility of linear equation solving in this chapter was limited to introducing *geometrically* the concept of vector spaces. I will not discuss linear equation solving further, but use this visualization to understand some key concepts.

2.2.2 Span, Linear Independence, and Subspaces

The left-hand side of Equation 2.15 represents a linear combination of two vectors, $\mathbf{v}_1$ and $\mathbf{v}_2$. The first definition is that of span. *Span of vectors* is defined as a vector space formed by all linear combinations of those vectors. Thus, span$\{\mathbf{v}_1\}$ is a line and span$\{\mathbf{v}_1, \mathbf{v}_2\}$ is a plane. As seen in Figure 2.1a, vectors $\mathbf{v}_1$ and $\mathbf{v}_2$ span the entire $\mathcal{R}^2$ space.

Generalizing this definition, let $\mathbf{x}_1, \mathbf{x}_2, \ldots \in \mathcal{R}^n$ be vectors in n-dimensional space. Then

$$\text{span}\{\mathbf{x}_1, \ldots, \mathbf{x}_k\} = c_1\mathbf{x}_1 + c_2\mathbf{x}_2 + \cdots + c_k\mathbf{x}_k \tag{2.16}$$

where $c_i \in R$ are real-valued scalars.

This brings us to the concept of *linear independence*. A set of vectors is said to be linearly independent if no vector in that set can be expressed as a linear combination of the remaining vectors. Mathematically, $\mathbf{x}_1, \mathbf{x}_2, \ldots \in \mathcal{R}^n$ are linearly independent if

$$c_1\mathbf{x}_1 + c_2\mathbf{x}_2 + \cdots + c_k\mathbf{x}_k = 0 \quad \text{implies } c_i = 0 \forall i \tag{2.17}$$

In other words, a set of vectors is linearly *dependent* if one (or more) of the vectors can be represented as a linear combination of the remaining vectors. Thus, the original vectors

$$\mathbf{v}_1 = \begin{bmatrix} 1 \\ 1 \end{bmatrix} \quad \text{and} \quad \mathbf{v}_2 = \begin{bmatrix} 2 \\ -1 \end{bmatrix}$$

from Figure 2.1a are linearly independent because neither of them can be expressed as a linear combination of the other. Considering Equation 2.17, the only way they can sum to **0** is if both c_1 and c_2 are zero. On the other hand, the vectors

$$\mathbf{w}_1 = \begin{bmatrix} 1 \\ 2 \end{bmatrix} \quad \text{and} \quad \mathbf{w}_2 = \begin{bmatrix} 2 \\ 4 \end{bmatrix}$$

from Figure 2.1b are linearly dependent.

Now we are equipped to understand span a bit better. In the second example, span$\{\mathbf{w}_1, \mathbf{w}_2\}$ is a line, whereas span$\{\mathbf{v}_1, \mathbf{v}_2\}$ is the entire 2D space. The difference between Figure 2.1b and c is that the vector $\mathbf{b}$ cannot be expressed as a linear combination of $\mathbf{w}_1$ and $\mathbf{w}_2$, but there exists at least one set of scalars (in fact, there are infinitely many such scalars) such that

$$\mathbf{d} = c_1 \mathbf{w}_1 + c_2 \mathbf{w}_2$$

Another way of stating this is that vector $\mathbf{d}$ lies in the span$\{\mathbf{w}_1, \mathbf{w}_2\}$, that is

$$\mathbf{d} \in \text{span}\{\mathbf{w}_1, \mathbf{w}_2\}$$

which is said to be a *subspace* of the $\mathcal{R}^2$ space. A subspace is not merely a subset of a vector space; it is also a vector space itself. Thus, a line that does not pass through the origin is not a subspace. This is because it violates the rules in (2.14)* since the zero vector, $\mathbf{0}$, does not lie on this line.

Although the discussion above used two vectors in $\mathcal{R}^2$ space, the same arguments are applicable to higher dimensional vector spaces as well. For example, vectors

$$\mathbf{v}_1 = \begin{bmatrix} 1 \\ 1 \\ 1 \\ 1 \end{bmatrix} \quad \text{and} \quad \mathbf{v}_2 = \begin{bmatrix} 2 \\ -1 \\ 0 \\ 1 \end{bmatrix}$$

are two linearly independent vectors in $\mathcal{R}^4$ space, whereas span$\{\mathbf{v}_1, \mathbf{v}_2\}$ will be a 2D subspace of $\mathcal{R}^4$ space.

Consider another case where

$$\mathbf{v}_1 = \begin{bmatrix} 1 \\ 2 \end{bmatrix}, \quad \mathbf{v}_2 = \begin{bmatrix} 2 \\ -1 \end{bmatrix}, \quad \mathbf{v}_3 = \begin{bmatrix} 0 \\ 1 \end{bmatrix}$$

We had seen earlier that vectors $\mathbf{v}_1$ and $\mathbf{v}_2$ span the entire $\mathcal{R}^2$ space. Hence, $\mathbf{v}_3$ can be expressed as a linear combination of $\mathbf{v}_1$ and $\mathbf{v}_2$ (one can verify that $\mathbf{v}_3 = 0.4\mathbf{v}_1 - 0.2\mathbf{v}_2$). Thus, span$\{\mathbf{v}_1, \mathbf{v}_2, \mathbf{v}_3\}$ is still the 2D $\mathcal{R}^2$ space.

Finally, I end this subsection with the definition of dimension:

Dimension of a subspace is defined as the number of linearly independent vectors that span the subspace.

---

* It violates some other rules also, but it is obvious that (2.6) is violated since $\mathbf{0}$ is not on the line.

2.2.3 Basis and Coordinate Transformation

Let $\mathbf{v}_1, \mathbf{v}_2, \ldots, \mathbf{v}_k \in \mathcal{R}^n$ be k vectors in an n-dimensional space. Based on the above discussions, it would be clear that if $k > n$, then the vectors are linearly dependent. At most, n vectors in $\mathcal{R}^n$ space can be linearly independent.

If vectors $\mathbf{v}_1, \mathbf{v}_2, \ldots, \mathbf{v}_n \in \mathcal{R}^n$ are linearly independent, then span$\{\mathbf{v}_1, \mathbf{v}_2, \ldots, \mathbf{v}_n\}$ is the $\mathcal{R}^n$ space itself. Thus, any vector $\mathbf{x} \in \mathcal{R}^n$ can be represented *uniquely* as a linear combination:

$$\mathbf{x} = \alpha_1 \mathbf{v}_1 + \alpha_2 \mathbf{v}_2 + \cdots + \alpha_n \mathbf{v}_n \tag{2.18}$$

The proof for this statement is left as an exercise. Therefore, the vectors $\{\mathbf{v}_1, \mathbf{v}_2, \ldots, \mathbf{v}_n\}$ together can form the *basis* of $\mathcal{R}^n$. The matrix formulation

$$\mathbf{x} = \begin{bmatrix} \mathbf{v}_1 & \mathbf{v}_2 & \cdots & \mathbf{v}_n \end{bmatrix} \begin{bmatrix} \alpha_1 \\ \alpha_2 \\ \vdots \\ \alpha_n \end{bmatrix} \tag{2.19}$$

is equivalent to Equation 2.18. This works exactly in the same manner in which Equation 2.15 was obtained from Equation 2.1. The proof for the above is also left as an exercise.

Basis vectors are therefore a set of linearly independent vectors that span the vector space. The set of coordinate vectors

$$\mathbf{e}_1 = \begin{bmatrix} 1 \\ 0 \\ \vdots \\ 0 \end{bmatrix}, \quad \mathbf{e}_2 = \begin{bmatrix} 0 \\ 1 \\ \vdots \\ 0 \end{bmatrix}, \quad \cdots, \quad \mathbf{e}_n = \begin{bmatrix} 0 \\ \vdots \\ 0 \\ 1 \end{bmatrix} \tag{2.20}$$

is called *natural basis*.

2.2.3.1 Change of Basis

Let vector $\mathbf{x} \in \mathcal{R}^n$ be represented as

$$\mathbf{x} = \begin{bmatrix} x_1 \\ x_2 \\ \vdots \\ x_n \end{bmatrix}$$

in the natural basis. Let us say we want to change the basis to the vectors $\{\mathbf{v}_1, \mathbf{v}_2, \ldots, \mathbf{v}_n\}$. This is possible, since these are linearly independent vectors that span the $\mathcal{R}^n$ space. The vector $\mathbf{x}$

is represented by Equation 2.19. Since the vector in the natural basis and $\{\mathbf{v}_1, \mathbf{v}_2, \ldots, \mathbf{v}_n\}$ basis is the same point, the following equality holds:

$$
\begin{bmatrix} \mathbf{e}_1 & \mathbf{e}_2 & \cdots & \mathbf{e}_n \end{bmatrix}
\begin{bmatrix} x_1 \\ x_2 \\ \vdots \\ x_n \end{bmatrix}
= \underbrace{\begin{bmatrix} \mathbf{v}_1 & \mathbf{v}_2 & \cdots & \mathbf{v}_n \end{bmatrix}}_{T}
\begin{bmatrix} \alpha_1 \\ \alpha_2 \\ \vdots \\ \alpha_n \end{bmatrix}
\tag{2.21}
$$

The first matrix on the left-hand side is the identity matrix. Thus, coordinate transformation to the new basis is given by

$$
\begin{bmatrix} \alpha_1 \\ \alpha_2 \\ \vdots \\ \alpha_n \end{bmatrix}
= T^{-1}
\begin{bmatrix} x_1 \\ x_2 \\ \vdots \\ x_n \end{bmatrix}
\tag{2.22}
$$

where the coordinate transformation matrix T is an $n \times n$ matrix whose n columns are the n basis vectors. Change of basis is an important operation that we will use in the analysis of linear systems and stiff ordinary differential equations (ODEs) (in Chapters 5 and 8, respectively).

2.2.4 Null (Kernel) and Image Spaces of a Matrix

2.2.4.1 Matrix as Linear Operator

The discussion so far focused on vectors and vector spaces. Let us now turn to looking at the matrix A. Let $\mathbf{x} \in \mathcal{X}$ be n-dimensional *input* vector and $\mathbf{y} \in \mathcal{Y}$ be an m-dimensional *output* vector. In general, a model that correlates the inputs and outputs (at steady state) is given by

$$
\mathbf{y} = \mathbf{f}(\mathbf{x})
\tag{2.23}
$$

The above is a *nonlinear* mapping between the inputs and outputs of the system. A *linear* mapping between the inputs and outputs will be given by[*]

$$
\mathbf{y} = A\mathbf{x}
\tag{2.24}
$$

---

[*] The mapping, $\mathbf{y} = A\mathbf{x} + \mathbf{b}$ is known as affine mapping, and not a linear mapping. If $\mathbf{y}_1 = A\mathbf{x}_1 + \mathbf{b}$ and $\mathbf{y}_2 = A\mathbf{x}_2 + \mathbf{b}$, then

$$
A(c_1\mathbf{x}_1 + c_2\mathbf{x}_2) + \mathbf{b} \neq c_1\mathbf{y}_1 + c_2\mathbf{y}_2
$$

Although affine functions are not linear, it is easy to convert them to linear:

$$
\underbrace{\mathbf{y} - \mathbf{b}_2}_{\bar{\mathbf{y}}} = A(\underbrace{\mathbf{x} + \mathbf{b}_1}_{\bar{\mathbf{x}}})
$$

where $\mathbf{b} = A\mathbf{b}_1 + \mathbf{b}_2$.

This brings us to an important result. Let $\mathcal{X}$ be an n-dimensional vector space and $\mathcal{Y}$ be an m-dimensional vector space. A linear operation between $\mathcal{X}$ to $\mathcal{Y}$ can be represented as a matrix multiplication, as shown in Equation 2.24.

The blending operation example is provided in the box below to show an example of linear operator and its matrix representation.

Example 2.1 Blending Operation Example

Blending is an important operation in refineries. Raw materials from various sources differ in their quality. They are blended together, along with a makeup stream, in order to meet both quantity and product specification requirements. Let us consider a simplified case where two streams are being mixed with a third makeup stream, and the final requirement from the blending operation is to obtain the output stream at a certain production rate (say, F kg/minute) with a specific amount of the primary component (given by mass fraction ω in the outlet stream). At steady state, the overall mass balance

$$F = F_1 + F_2 + F_3 \tag{2.25}$$

and balance on the primary component

$$F\omega = F_1\omega_1 + F_2\omega_2 + F_3\omega_3 \tag{2.26}$$

give a mathematical model for the operation.

The input vector

$$\mathbf{x} = \begin{bmatrix} F_1 \\ F_2 \\ F_3 \end{bmatrix}$$

consists of the three mass flowrates. If the desired output is in terms of total mass flow and composition of the primary component, the output vector consists of

$$\tilde{\mathbf{y}} = \begin{bmatrix} F \\ \omega \end{bmatrix}$$

The first model equation (2.25) is linear, whereas Equation 2.26 is nonlinear:

$$\omega = \frac{1}{F}\left(F_1\omega_1 + F_2\omega_2 + F_3\omega_3\right)$$

Thus, the mapping between $\mathbf{x}$ and $\tilde{\mathbf{y}}$ is nonlinear.

It is possible, however, to define the output vector as total mass flow and mass flow of the primary component F_{prim} at the outlet. Thus

$$\mathbf{y} = \begin{bmatrix} F \\ F_{\text{prim}} \end{bmatrix}$$

and the *linear operator* that models the steady state blending operation is

$$\mathbf{y} = \underbrace{\begin{bmatrix} 1 & 1 & 1 \\ \omega_1 & \omega_2 & \omega_3 \end{bmatrix}}_{C} \mathbf{x} \qquad (2.27)$$

The matrix C is a linear mapping between the three inlet flowrates and the net production rates from the blending system. It is represented as

$$C : \mathcal{X} \to \mathcal{Y}$$

where $\mathcal{X}$ and $\mathcal{Y}$ are 3D and 2D spaces, respectively.

Although I have stated this before, it bears repeating. The linear equation, $A\mathbf{x} = \mathbf{b}$, was introduced to motivate the concept of vectors and vector spaces. Thereafter, Section 2.2.2 introduced the concept of span and linear independence; Section 2.2.3 introduced the concept of basis; and this subsection introduced the concept of matrix as a linear operator. It would help to "upgrade our thinking" in terms of input vector $\mathbf{x}$, output vector $\mathbf{y}$, and linear transformation represented by the matrix A.

Null space or *Kernel* of a matrix is defined as the complete set of vectors $\mathbf{x} \in \mathcal{X}$ such that their linear transformation under matrix A maps to the zero vector. Mathematically

$$ker(A) = \{\mathbf{x} : A\mathbf{x} = \mathbf{0}\} \qquad (2.28)$$

In previous subsections, we have defined the column vectors of the matrix as the m vectors that form individual columns of the matrix. Consider the case when the two column vectors, $\mathbf{v}_1$ and $\mathbf{v}_2$, were linearly independent. In such a case

$$\underbrace{\begin{bmatrix} 1 & 2 \\ 1 & -1 \end{bmatrix}}_{A} \mathbf{x} = \mathbf{0}$$

has a unique solution, $\mathbf{x} = \mathbf{0}$. Thus, when the column vectors of A are linearly independent, the null space of the matrix is a single point, the origin.

The second case is when the two column vectors, $\mathbf{w}_1$ and $\mathbf{w}_2$, are linearly dependent. The set of $\mathbf{x}$ that satisfies the equation

$$\underbrace{\begin{bmatrix} 1 & 2 \\ 2 & 4 \end{bmatrix}}_{A} \mathbf{x} = \mathbf{0}$$

includes all the vectors

$$\begin{bmatrix} -2a \\ a \end{bmatrix}$$

The physical consequence (or interpretation) of null space is as follows: Any input set in the null space will lose its information when acted upon by the linear transformation A.

Note that the null space or kernel is a *subspace* of the input space.

Image space of a matrix is the set of output vectors $\mathbf{y} \in \mathcal{Y}$ that are obtained by the transformation of all $\mathbf{x} \in \mathcal{X}$ through the matrix A. Stated in a different way, the image space of matrix A is all $\mathbf{y}$ that satisfy

$$\text{im}\{A\} = \{\mathbf{y}| \ \mathbf{y} = A\mathbf{x} : \mathbf{x} \in \mathcal{X}\} \tag{2.29}$$

The image space for

$$\mathbf{y} = \underbrace{\begin{bmatrix} 1 & 2 \\ 1 & -1 \end{bmatrix}}_{A} \mathbf{x}$$

is the entire 2D space. Consider the other system

$$\mathbf{y} = \underbrace{\begin{bmatrix} 1 & 2 \\ 2 & 4 \end{bmatrix}}_{A} \mathbf{x}$$

which can then be written as

$$\mathbf{y} = x_1 \begin{bmatrix} 1 \\ 2 \end{bmatrix} + x_2 \begin{bmatrix} 2 \\ 4 \end{bmatrix}$$

Thus, the image space of this matrix is the 1D subspace along the vector $\mathbf{w}_1 = [1 \quad 2]^T$.

IMAGE SPACE AND COLUMN SPACE

In Section 2.2.1, we introduced $\mathbf{v}_1$ and $\mathbf{v}_2$, and $\mathbf{w}_1$ and $\mathbf{w}_2$ as the *column* vectors of matrix A, and defined *span* of vectors in Section 2.2.2. I have deferred defining the *column space* of a matrix until this point.

Column space of a matrix is defined as the span of its column vectors. Thus, column space for the two examples from Section 2.2.1 are

$$c_1\mathbf{v}_1 + c_2\mathbf{v}_2 \quad \text{or} \quad c_1\mathbf{w}_1 + c_2\mathbf{w}_2$$

respectively. Thus, the condition that the equation $A\mathbf{x} = \mathbf{b}$ has a solution is as follows: The vector $\mathbf{b}$ is in the column space of A.

Now, look at the definition of *image space*: It is the set of all *output* vectors $\mathbf{y}$ such that

$$\mathbf{y} = x_1\mathbf{v}_1 + x_2\mathbf{v}_2 \quad \left(\text{for the first example}\right)$$

$$\mathbf{y} = x_1\mathbf{w}_1 + x_2\mathbf{w}_2 \quad \left(\text{for the second example}\right)$$

Notice that the two definitions are similar since (x_1, x_2) or (c_1, c_2) are real numbers. So, what is the difference between image space and column space?

One way to look at them is to realize that image and column space are *numerically* the same: They are both the linear combinations of the column vectors of a matrix.

Another way to look at this is the *physical interpretation*. The term "column space" was invoked when we discussed solving a linear equation: One or more solutions exist when the vector on the right-hand side of Equation 2.2 lies in the column space of A. Figure 2.1 illustrates the statement "the equation $A\mathbf{x} = \mathbf{b}$ has solution(s) if the vector $\mathbf{b}$ lies in the column space of matrix A."

On the other hand, "image space" does not talk about solving linear equations; instead it is introduced in the context of *linear transformation*. When the input vector $\mathbf{x}$ is acted upon by a linear operation, the corresponding output vector lies in the image space. In the top panels of Figure 2.2, the image of any arbitrary vector in $\mathcal{X}$ may lie anywhere on the 2D space. So, the image space is the entire $\mathcal{R}^2$ space.

In the bottom panels, the image space of the matrix A was $\text{im}(A) = [1 \quad 2]^T$. Note that all the dots lie on this line. Furthermore, $\text{ker}(A) = [-2 \quad 1]^T$, implying all the open circles on the left-bottom panel map to the origin in the $\mathcal{Y}$-space on the right-bottom panel. The example also illustrates the fact that the image space is a subspace of the output space, $\mathcal{Y}$, whereas the null space is a subspace of the input space, $\mathcal{X}$.

2.2.4.2 Null and Image Spaces in MATLAB®

With the matrix A defined as above, the null space of a matrix can be obtained in MATLAB as

```
>> disp(null(A))
   -0.8944
    0.4472
```

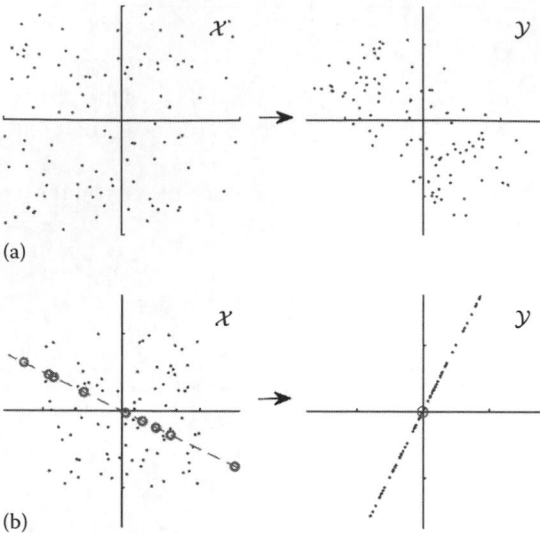

(a)

(b)

FIGURE 2.2 Linear transformation from $\mathcal{X} \to \mathcal{Y}$ for the two different A matrices. (a) The column vectors of A matrix are linearly independent; (b) the column vectors are linearly dependent. Dots/circles on the $\mathcal{X}$-space on the left map to dots/circles on the $\mathcal{Y}$-space on the right.

Note that the above is same as the vector

$$\begin{bmatrix} -2a \\ a \end{bmatrix}$$

(where $a = 1/\sqrt{5}$). The image space can be obtained using the command orth. The image space is also called range of the matrix.

```
>> disp(orth(A))
    -0.4472
    -0.8944
```

Example 2.1 Revisited: Blending Operation Example

Consider the example of blending operations. First point to note is that if we implement the constraint that the flowrates must be positive, the sets $\mathcal{X}$ and $\mathcal{Y}$ are no longer vector spaces (because, for example, $-\mathbf{x} \notin \mathcal{X}$ and $-\mathbf{y} \notin \mathcal{Y}$).

Case 1: For this exercise, let us abandon the requirement that $F_i \geq 0$. To still give it a "physical meaning," let us undertake a pedagogical exercise that if any of the flowrates, F_i, are negative, it implies that the flow is taken out from the blender. Consider the case where the first stream does not contain any primary component, the second stream has $\omega_2 = 0.5$, and the third stream is the pure primary component. Thus, in Equation 2.27

$$C = \begin{bmatrix} 1 & 1 & 1 \\ 0 & 0.5 & 1 \end{bmatrix}$$

The image space, im(C), is the entire $\mathcal{R}^2$ space, since it is possible to get any net and primary component flowrates through linear combinations of F_1, F_2, and F_3. What is the null space, ker(C)? It is the combination of the three flowrates such that both the net flowrate, F, and primary component flowrate, F_{prim}, are zero. It is easy to see that for the primary component flow to be zero, $F_3 = a$, $F_2 = -2a$. If the flowrate $F_1 = a$, we can see that the net flowrate $F = 0$. Thus, the null space is given by $\mathbf{x} = \begin{bmatrix} a & -2a & a \end{bmatrix}^T$. The unit vector in that direction is therefore

```
>> disp(null(C))
    0.4082
   -0.8165
    0.4082
```

Case 2: For this case, let us consider that all the three streams have the same composition of the primary component, that is, $\omega_1 = \omega_2 = \omega_3 = q$. It should be clear that mixing of these three streams will lead to the outlet stream with the same composition, q. Thus, the net flowrate will be, $F = F_1 + F_2 + F_3$, and the net flowrate of the primary component, $F_{\text{prim}} = qF$. Thus, the image space will be 1D subspace of $\mathcal{Y}$, such that

$$\text{im}(C) = \begin{bmatrix} 1 \\ q \end{bmatrix} F$$

While it is possible to obtain any value of primary component flowrate, F_{prim}, the ratio of this to the net flowrate (i.e., mass fraction) will always be q.

It can be verified using `orth` command in MATLAB that the image space is given by

$$\text{im}(C) = \text{span}\left\{ \begin{bmatrix} 1/\sqrt{1+q^2} \\ q/\sqrt{1+q^2} \end{bmatrix} \right\}$$

2.3 SINGULAR VALUE DECOMPOSITION

Singular value decomposition (SVD) and eigenvalue decomposition are very important concepts that will be useful. Before discussing these, the following subsection presents a few important definitions.

2.3.1 Orthonormal Vectors

Let $\mathbf{x}_1 \in \mathcal{R}^n$ and $\mathbf{x}_2 \in \mathcal{R}^n$ be two vectors in n-dimensional space. The *dot product* or the *inner product* of the two vectors is

$$\langle \mathbf{x}_1, \mathbf{x}_2 \rangle = \mathbf{x}_1^T \mathbf{x}_2 = \mathbf{x}_2^T \mathbf{x}_1 = \langle \mathbf{x}_2, \mathbf{x}_1 \rangle \tag{2.30}$$

A vector is a *unit vector* if its inner product with itself equals one. In the last example, the basis for the image space was defined as

$$
\begin{bmatrix} 1/\sqrt{1+q^2} \\ q/\sqrt{1+q^2} \end{bmatrix}
$$

and one can easily verify that this is a unit vector.

Two vectors are said to be *orthogonal vectors* if their inner product is zero, that is, $\mathbf{x}_1^T \mathbf{x}_2 = 0$. Furthermore, if both the vectors are unit vectors as well, they are said to be *orthonormal*.

Orthonormality is a useful property. Recall that there were multiple possibilities for defining the basis for image and null spaces. However, the commands `orth` and `null` resulted in a specific type of basis vectors: They were all mutually orthonormal. Vectors $\mathbf{x}_1, \ldots, \mathbf{x}_k \in \mathcal{R}^n$ are said to be mutually orthonormal iff they satisfy the condition

$$
\mathbf{x}_i^T \mathbf{x}_j = \begin{cases} 1 & \text{if } i = j \\ 0 & \text{otherwise} \end{cases} \tag{2.31}
$$

An orthogonal matrix is a square matrix whose column vectors are orthonormal to each other. Thus, since

$$
A = \begin{bmatrix} \mathbf{v}_1 & \cdots & \mathbf{v}_n \end{bmatrix}, \quad \text{where } \mathbf{v}_1, \ldots, \mathbf{v}_n \text{ are mutually orthonormal vectors,} \tag{2.32}
$$

from the definition (2.31) of orthonormal vectors, an orthogonal matrix satisfies

$$
AA^T = A^T A = I \tag{2.33}
$$

Thus, for an orthogonal matrix

$$
A^T = A^{-1} \tag{2.34}
$$

The three properties, (2.32), (2.33), and (2.34), are key properties of orthogonal matrices.

Orthogonal Subspaces: Two subspaces are said to be orthogonal if any vector on one subspace is orthogonal to any other vector on the other subspace.

Remark: Since any vector in a subspace can be represented as a linear combination of its basis vectors, it is sufficient to ensure that basis vectors of the two subspaces are orthogonal to each other.

2.3.2 Singular Value Decomposition

Let us start this subsection with an example of linear transformation between input and output vector spaces, which is represented as matrix multiplication. The change of basis through transformation matrix was given in Equation 2.22. For the purpose of this discussion, let us restrict transformation to orthonormal basis set. The orthonormal basis set that spans the input space forms the n columns of input transformation matrix:

$$T_x = \begin{bmatrix} \mathbf{v}_1 & \mathbf{v}_2 & \cdots & \mathbf{v}_n \end{bmatrix} \tag{2.35}$$

and the basis set that spans the output space forms the m columns of the output transformation matrix:

$$T_y = \begin{bmatrix} \mathbf{u}_1 & \mathbf{u}_2 & \cdots & \mathbf{u}_m \end{bmatrix} \tag{2.36}$$

Since the transformation matrices are chosen to be orthogonal

$$T^{-1} = T^T$$

Recall that the relationship between the original vector $\mathbf{x}$ and the transformed vector (in the new basis set in (2.35)) is

$$\hat{\mathbf{x}} = T_x^T \mathbf{x} \tag{2.37}$$

and

$$\hat{\mathbf{y}} = T_y^T \mathbf{y} \tag{2.38}$$

These can be substituted in the original linear operation $\mathbf{y} = A\mathbf{x}$ to get

$$\hat{\mathbf{y}} = \underbrace{T_y^T A T_x}_{\hat{A}} \hat{\mathbf{x}} \tag{2.39}$$

The above expression is valid for any appropriate set of orthonormal basis vectors. The coordinate transformation is most effective for a "special" set of basis vectors.

Consider

$$A = \begin{bmatrix} 1 & 2 \\ 2 & 4 \end{bmatrix}$$

for which the geometric interpretation of null and image spaces was shown in Figure 2.2b. The image space for this matrix was $\mathrm{im}(A) = \mathrm{span}\left\{ \begin{bmatrix} 1 & 2 \end{bmatrix}^T \right\}$. Let us choose this vector as one of the basis, $\mathbf{u}_1$, for the output space. Since we want to choose orthonormal basis vector

$$\mathbf{u}_1 = \begin{bmatrix} 1/\sqrt{5} \\ 2/\sqrt{5} \end{bmatrix} \tag{2.40}$$

the other basis vector is the unit vector perpendicular to $\mathbf{u}_1$

$$\mathbf{u}_2 = \begin{bmatrix} -2/\sqrt{5} \\ 1/\sqrt{5} \end{bmatrix} \tag{2.41}$$

and the output transformation matrix is as given in Equation 2.36.

In the input space, let one of the basis vectors ($\mathbf{v}_2$) be the vector that spans the null space and the other vector ($\mathbf{v}_1$) be orthogonal to the null space. Thus

$$\mathbf{v}_2 = \begin{bmatrix} -2a/\sqrt{5a^2} \\ a/\sqrt{5a^2} \end{bmatrix} = \begin{bmatrix} -2/\sqrt{5} \\ 1/\sqrt{5} \end{bmatrix} \tag{2.42}$$

$$\mathbf{v}_1 = \begin{bmatrix} 1/\sqrt{5} \\ 2/\sqrt{5} \end{bmatrix} \tag{2.43}$$

and T_x is given by Equation 2.35:

$$T_x = \begin{bmatrix} 1/\sqrt{5} & -2/\sqrt{5} \\ 2/\sqrt{5} & 1/\sqrt{5} \end{bmatrix} \tag{2.44}$$

With these defined, we may now compute

$$\hat{A} = T_y^T A T_x = \begin{bmatrix} 5 & 0 \\ 0 & 0 \end{bmatrix} \tag{2.45}$$

The above matrix in the transformed input/output spaces implies

$$\begin{bmatrix} \hat{y}_1 \\ \hat{y}_2 \end{bmatrix} = \begin{bmatrix} 5\hat{x}_1 \\ 0 \end{bmatrix} \tag{2.46}$$

The above equation makes the following rules quite clear: (i) Any point in the $\mathcal{X}$ space gets mapped along $\hat{y}_1$ in the output space; (ii) $\hat{y}_2 = 0$ implies that no point in the $\mathcal{X}$ spaces is mapped to this subspace; (iii) only the information in $\hat{x}_1$ subspace is retained; whereas (iv) the information along $\hat{x}_2$ is lost. Thus, $\mathbf{u}_1$ is the image space and $\mathbf{u}_2$ is the subspace orthogonal to it; and $\mathbf{v}_2$ is the null space and $\mathbf{v}_1$ is the subspace orthogonal to the null space.

We are now ready to generalize the above representation to any $m \times n$ matrix A through the concept of *SVD*, which states that any matrix A can be represented as

$$A = U\Sigma V^T \tag{2.47}$$

where

$$U = \begin{bmatrix} \mathbf{u}_1 & \mathbf{u}_2 & \cdots & \mathbf{u}_m \end{bmatrix} \tag{2.48}$$

and

$$V = \begin{bmatrix} \mathbf{v}_1 & \mathbf{v}_2 & \cdots & \mathbf{v}_n \end{bmatrix} \tag{2.49}$$

are orthogonal matrices. The matrix Σ is an $m \times n$ matrix such that its diagonal elements are nonnegative (i.e., positive or zero) and all the other elements are zero. The diagonal elements of matrix Σ are called *singular values*. The singular values are ordered, that is, $\sigma_1 \geq \sigma_2 \geq \cdots$. The number of singular values is exactly equal to m or n, whichever is lower. Some of these singular values can be zero.

The MATLAB command for singular value decomposition is svd.

When A is a square matrix, the SVD may be written as

$$A = \underbrace{\begin{bmatrix} \mathbf{u}_1 & \cdots & \mathbf{u}_n \end{bmatrix}}_{U} \underbrace{\begin{bmatrix} \sigma_1 & & \\ & \ddots & \\ & & \sigma_n \end{bmatrix}}_{\Sigma} \underbrace{\begin{bmatrix} \mathbf{v}_1 & \cdots & \mathbf{v}_n \end{bmatrix}}_{V}^{T} \tag{2.50}$$

where all the nondiagonal elements of Σ are zero.

When A is a tall matrix, that is, $m > n$,

$$A = \underbrace{\begin{bmatrix} \mathbf{u}_1 & \cdots & \mathbf{u}_m \end{bmatrix}}_{U} \underbrace{\begin{bmatrix} \sigma_1 & & \\ & \ddots & \\ & & \sigma_n \\ 0 & \cdots & 0 \\ 0 & \cdots & 0 \end{bmatrix}}_{\Sigma} \underbrace{\begin{bmatrix} \mathbf{v}_1 & \cdots & \mathbf{v}_n \end{bmatrix}}_{V}^{T} \tag{2.51}$$

Likewise, when $m < n$, the matrix of singular values is given by

$$\Sigma = \begin{bmatrix} \sigma_1 & & & 0 & 0 \\ & \ddots & & \vdots & \vdots \\ & & \sigma_n & 0 & 0 \end{bmatrix} \tag{2.52}$$

Example 2.1 Revisited: Blending Operation Example

Previously, we had considered two different linear operator matrices for the blending system example:

$$C = \begin{bmatrix} 1 & 1 & 1 \\ 0 & 0.5 & 1 \end{bmatrix}, \quad C' = \begin{bmatrix} 1 & 1 & 1 \\ 0.5 & 0.5 & 0.5 \end{bmatrix}$$

(In the latter case, I have chosen $q = 0.5$.)

Recall that the image space of C was the entire $\mathcal{R}^2$ space, whereas the null space was

$$\mathbf{x} = \begin{bmatrix} a \\ -2a \\ a \end{bmatrix}$$

Executing the SVD in MATLAB yields

$$A = \begin{bmatrix} -0.8671 & -0.4981 \\ -0.4981 & 0.8671 \end{bmatrix} \begin{bmatrix} 1.9651 & 0 & 0 \\ 0 & 0.6233 & 0 \end{bmatrix} \begin{bmatrix} -0.4413 & -0.7991 & 0.4082 \\ -0.5680 & -0.1035 & -0.8165 \\ -0.6947 & 0.5922 & 0.4082 \end{bmatrix}^T$$

Since both the singular values are nonzero, the image space is the complete $\mathcal{R}^2$ space. Since there can be at most two positive singular values, the null space will be non-empty. As can be seen from the matrix V, the last column is the null space that we obtained earlier.

A similar exercise can be done on the other matrix

$$C' = \begin{bmatrix} 1 & 1 & 1 \\ 0.5 & 0.5 & 0.5 \end{bmatrix}$$

as well. The reader can verify that the following results agree with the discussion regarding the image and null space of this matrix as well.

```
>> [U,S,V]=svd(A)
U =
    -0.8944    -0.4472
    -0.4472     0.8944
S =
     1.9365     0     0
          0     0     0
V =
    -0.5774    -0.8165    -0.0000
    -0.5774     0.4082    -0.7071
    -0.5774     0.4082     0.7071
```

2.3.3 Condition Number

2.3.3.1 Singular Values, Rank, and Condition Number

Let us summarize the results of the discussion on SVD. Let $A := \mathcal{R}^n \rightarrow \mathcal{R}^m$ be an $m \times n$ matrix with ℓ singular values, $\{\sigma_1, \ldots, \sigma_\ell\}$. Here, $\ell \leq \min(m, n)$. The two orthogonal matrices may be written as

$$U = \begin{bmatrix} \mathbf{u}_1 & \cdots & \mathbf{u}_\ell | \cdots & \mathbf{u}_m \end{bmatrix} \quad V = \begin{bmatrix} \mathbf{v}_1 & \cdots & \mathbf{v}_\ell | \cdots & \mathbf{v}_n \end{bmatrix} \tag{2.53}$$

The first ℓ column vectors of U span the image space and the remaining $(m - \ell)$ column vectors form the subspace orthogonal to the image space. The latter subspace is "unreachable" on linear operation through A. Similarly, the last $(n - \ell)$ column vectors of V span the null space. The information along the null space is lost when data is operated upon by linear operator A. The first ℓ column vectors of V span a subspace that is orthogonal to the null space of A.

Rank of a matrix is defined as the number of linearly independent rows of a matrix. The concept of rank is introduced in introductory linear algebra courses while discussing whether linear equation, $A\mathbf{x} = \mathbf{b}$, has a unique solution. The concept of rank is useful for a broader range of problems. In fact, *row rank* and *column rank* of a matrix are the number of linearly independent rows and columns of a matrix, respectively. Thus, the column rank of a matrix equals the number of linearly independent column vectors, which in turn, equals the dimension of the image space of the matrix. It can be proved that row rank and column rank of a matrix are equal (the proof is left as student exercise).

Since $\text{im}(A) = \text{span}\{\mathbf{u}_1, \ldots, \mathbf{u}_\ell\}$, rank is also equal to the number of nonzero singular values of A. Furthermore, if $\ell = m$, the image space is the entire m-dimensional vector space of outputs; else, it is an ℓ-dimensional subspace of the $\mathcal{R}^m$ output space. Equivalently, if $\ell = n$, the null space of A is just the origin.

While rank is a useful criterion, it does not tell us much about the relative values of $\sigma_1, \ldots, \sigma_\ell$. *Condition number* is the ratio of the largest and smallest singular values, which is an important parameter to analyze a linear system. Formally, condition number is defined as the ratio of the two-norm of A to the two-norm of A^{-1}, which simply means that condition number is the ratio of maximum singular values of A and A^{-1}:

$$c(A) = \frac{\sigma_{\max}(A)}{\sigma_{\max}(A^{-1})} = \frac{\sigma_{\max}(A)}{\sigma_{\min}(A)} \tag{2.54}$$

Condition number is often invoked in the context of solving a set of linear equations. It indicates how sensitive the solution will be to small errors in data. It is useful in linear least squares problems (this will be dealt with in Chapter 10). It also indicates directionality of the system. The *implications* of condition number in these scenarios will be discussed in the remainder of this subsection.

2.3.3.2 Sensitivity of Solutions to Linear Equations

We will start with a "classical" example to motivate discussion in this section.

Example 2.2 Condition Number and Solution of Linear Equations

Consider the linear equation, $Ax = \mathbf{b}$. A "classical" example of ill-conditioned matrix is

$$A = \begin{bmatrix} 1 & 2 \\ 2 & 4.001 \end{bmatrix}$$

This is a classic example that is used to demonstrate the importance of condition number. The condition number can be computed in MATLAB using

```
>> cond(A)
ans =
   2.5008e+04
```

which indicates a highly "ill-conditioned" matrix. The reason for this would be clear to the reader. If the element $A(2,2) = 4$, we notice that the *rank* of the matrix will be 1, the matrix would not be invertible and a unique solution to linear equation would not exist. However, with the A matrix given in this problem, the linear equation has a unique solution for any value of $\mathbf{b}$.

When $\mathbf{b} = [1 \quad 2]^T$, the solution of the linear equation is

```
>> x=inv(A)*b;
x =
     1
     0
```

The command `inv(A)` obtains A^{-1}.

Let us say there was a 1% error in $\mathbf{b}$ vector. For

$$\mathbf{b} = \begin{bmatrix} 1.01 \\ 2 \end{bmatrix}, \quad \mathbf{x} = \begin{bmatrix} 41.01 \\ -20 \end{bmatrix}$$

The solution has changed significantly with a rather small change in $\mathbf{b}$. Thus, condition number is an indicator of the *sensitivity of the solution* of linear equations to errors in data. When faced with highly ill-conditioned matrix, one should reformulate the problem to avoid this issue, instead of finding "improved" solution algorithms. Condition number is a property of the system and not of the solution technique.

There is another part to the discussion of ill-conditioned matrices. The SVD of matrix A from Example 2.2 gives

$$A = \begin{bmatrix} -0.4471 & -0.8945 \\ -0.8945 & 0.4471 \end{bmatrix} \begin{bmatrix} 5.0008 & 0 \\ 0 & 0.0002 \end{bmatrix} \begin{bmatrix} -0.4471 & -0.8945 \\ -0.8945 & -0.4471 \end{bmatrix}^T \tag{2.55}$$

The value of solution was highly sensitive to initial conditions, when $\mathbf{b} = [1 \quad 2]^T$, that is, when $\mathbf{b}$ was along or "close to" the vector $\mathbf{u}_1$. If instead

$$\mathbf{b} = \begin{bmatrix} 1 \\ 0 \end{bmatrix}, \quad \mathbf{x} = \begin{bmatrix} 4001.0 \\ -2000.0 \end{bmatrix}$$

the solution $\mathbf{x}$ has a significantly larger magnitude than the right-hand side $\mathbf{b}$. As one moves the value of b_2 from 2.0 to 0.0, the solution has moved from $x_1 = 1.0$ to $x_1 = 4001$. The smallest singular value of A becomes the largest singular value of A^{-1}. Hence, the resulting solution becomes very large.

A large value of solution vector is not the hallmark of an ill-conditioned matrix. The hallmark of an ill-conditioned matrix is high sensitivity to errors. Consider the following matrix:

$$A = \begin{bmatrix} 1000 & 2000 \\ 2000 & 4001 \end{bmatrix}$$

Note that if σ_i are singular values of A, $(\alpha\sigma_i)$ are singular values of $[\alpha A]$. Thus, the condition number of the above matrix remains the same as that in Example 2.2. The solution with this matrix for different values of $\mathbf{b}$ (when $\mathbf{b}$ is in the column space of $\mathbf{u}_1$) is as follows:

$$\mathbf{b} = \begin{bmatrix} 1 \\ 2 \end{bmatrix}, \quad \mathbf{x} = \begin{bmatrix} 0.001 \\ 0 \end{bmatrix}$$

$$\mathbf{b} = \begin{bmatrix} 1.01 \\ 2 \end{bmatrix}, \quad \mathbf{x} = \begin{bmatrix} 0.041 \\ -0.020 \end{bmatrix}$$

The sensitivity of the solution to errors is the key factor. Note that again the value of x_1 has changed by a factor of 40, even though the values of $\mathbf{x}$ are very small.

We can use SVD to analyze this behavior. Transforming the coordinate systems to the basis set represented by column vectors of V for the vector $\mathbf{x}$ and U for the vector $\mathbf{b}$, we get the following:

$$\left(U\Sigma V^T\right)\mathbf{x} = \mathbf{b} \quad \underbrace{\left(V^T\mathbf{x}\right)}_{\hat{\mathbf{x}}} = \Sigma^{-1}\underbrace{\left(U^T\mathbf{b}\right)}_{\hat{\mathbf{b}}} \tag{2.56}$$

In the new basis set, this equation becomes

$$\hat{x}_1 = \left(\sigma_1^{-1}\right)\hat{b}_1 \quad \hat{x}_2 = \left(\sigma_2^{-1}\right)\hat{b}_2 \tag{2.57}$$

In Example 2.2, the two values (see Equation 2.55) are

$$\left(\sigma_1^{-1}\right) = 0.2, \quad \left(\sigma_2^{-1}\right) = 5000.8$$

Let us consider the case when $\mathbf{b} \in \text{span}\{\mathbf{u}_1\}$

$$\mathbf{b} = \begin{bmatrix} 1 \\ 2.0004 \end{bmatrix} \Rightarrow \hat{\mathbf{b}} = \begin{bmatrix} -2.236 \\ 0 \end{bmatrix}$$

and the value of $\hat{x}_1 = -0.4471$. In the original coordinate system, the solution is therefore

$$\mathbf{x} = V\hat{\mathbf{x}} = \begin{bmatrix} 0.2 \\ 0.4 \end{bmatrix} \tag{2.58}$$

When the value of $\mathbf{b}$ is moved slightly away,

$$\hat{\mathbf{b}} = \begin{bmatrix} -2.236 \\ \varepsilon \end{bmatrix}$$

yielding $\hat{x}_1 = -0.4471$ and $\hat{x}_2 = 5000.8\varepsilon$. For example, when $\mathbf{b} = \begin{bmatrix} 1 & 2 \end{bmatrix}^T$ as in Example 2.2, $\varepsilon = -1.789 \times 10^{-4}$, which yields $\hat{x}_2 = -0.8945$. Thus

$$\hat{x} = \begin{bmatrix} -0.4771 \\ -0.8495 \end{bmatrix} \Rightarrow \mathbf{x} = \begin{bmatrix} 1 \\ 0 \end{bmatrix}$$

Example 2.2 (Continued) Summary of Linear Equation Solving Results

Let us summarize the above results in the context of solving linear equations:

$$\mathbf{b} = \begin{bmatrix} 1 \\ 2.0004 \end{bmatrix} \Rightarrow \mathbf{x} = A^{-1}\mathbf{b} = \begin{bmatrix} 0.2 \\ 0.4 \end{bmatrix}$$

$$\mathbf{b} = \begin{bmatrix} 1 \\ 2 \end{bmatrix} \Rightarrow \mathbf{x} = A^{-1}\mathbf{b} = \begin{bmatrix} 1 \\ 0 \end{bmatrix}$$

The summary of the above results in modified coordinate spaces:

$$\hat{\mathbf{b}} = U^T\mathbf{b} = \begin{bmatrix} -2.236 \\ 0 \end{bmatrix} \Rightarrow \hat{\mathbf{x}} = S^{-1}\hat{\mathbf{b}} = \begin{bmatrix} -0.4471 \\ 0 \end{bmatrix}$$

$$\hat{\mathbf{b}} = U^T \mathbf{b} = \begin{bmatrix} -2.236 \\ 0.0002 \end{bmatrix} \Rightarrow \hat{\mathbf{x}} = S^{-1}\hat{\mathbf{b}} = \begin{bmatrix} -0.4471 \\ -0.8495 \end{bmatrix}$$

The last step above is the reason for the sensitivity of the solution to errors. A very small change in $\mathbf{b}$ resulted in a significant change in $\hat{x}_2$ from 0 to -0.8495 (which is double the value of $\hat{x}_1$). Therefore, the final solution changes significantly:

$$\mathbf{x} = V\hat{\mathbf{x}} = \begin{bmatrix} 0.2 \\ 0.4 \end{bmatrix}$$

$$\mathbf{x} = V\hat{\mathbf{x}} = \begin{bmatrix} 1 \\ 0 \end{bmatrix}$$

The above discussion would also make it clear that varying $\mathbf{b}$ along the $\mathbf{u}_1$ direction will result in only a proportional change in the solution. For example

$$\mathbf{b} = \begin{bmatrix} 1.1 \\ 2.2004 \end{bmatrix} \Rightarrow \mathbf{x} = A^{-1}\mathbf{b} = \begin{bmatrix} 0.22 \\ 0.44 \end{bmatrix}$$

A standard interpretation of the solution of linear equations with ill-conditioned matrix is that of "intersection of two nearly parallel lines." However, this geometric picture is limiting, especially when one expands to higher dimensions. In contrast, the approach of expressing the linear equation in context of SVD provides new and interesting insights into its solution, which are easily generalizable.

Consider a $n \times n$ matrix, whose singular values, $(\sigma_1, ..., \sigma_\ell)$ are of similar order of magnitude, whereas $\sigma_\ell / \sigma_{\ell+1} \gg 1$. When the matrix is inverted, σ_i^{-1} will be comparatively low for the first ℓ elements, whereas it would be high for the remaining $(\ell+1)$ to n elements. Thus, small values of $\{\hat{b}_{\ell+1}, ..., \hat{b}_n\}$ will map as large values in the solution $\mathbf{x}$. If errors in $\mathbf{b}$ are in the subspace spanned by the first ℓ column vectors of U, the change in the solution $\mathbf{x}$ will be proportional to the change in vector $\mathbf{b}$. However, errors in $\mathbf{b}$ in the subspace spanned by the remaining vectors will result in larger effects on $\mathbf{x}$.

2.3.4 Directionality

The above example possibly indicates some sort of directionality in a linear operator, $A : \mathcal{X} \to \mathcal{Y}$ when the singular values are significantly different from each other. For a linear operator that relates input and output spaces, Equation 2.56 may be rewritten as

$$\mathbf{y} = \left(U \Sigma V^T\right)\mathbf{x}$$
$$\underbrace{\left(U^T \mathbf{y}\right)}_{\hat{\mathbf{y}}} = \Sigma \underbrace{\left(V^T \mathbf{x}\right)}_{\hat{\mathbf{x}}} \tag{2.59}$$

so that

$$\hat{y}_i = \sigma_i \hat{x}_i \qquad (2.60)$$

When the basis vectors of the input and output spaces are changed to the singular vectors, this results in simply stretching or shrinking along these vectors. We used Example 2.1 to discuss image and null spaces of a matrix. The following example takes this concept further, expanding on the discussion in Section 2.3.3 for linear operator.

Example 2.3 Significance of Singular Values for Linear Operator

Consider the linear transform

$$\mathbf{y} = \begin{bmatrix} -0.25 & 1.5 \\ -2.25 & 4.25 \end{bmatrix} \mathbf{x} \qquad (2.61)$$

The condition number for the above matrix is 11. The condition number for this matrix is quite good. We will choose this matrix because it is easier to visually demonstrate the significance of singular values for linear operation.

An alternative, *geometric interpretation* of condition number is as follows. Let the vector $\mathbf{x}$ be a unit vector. Let us choose $\mathbf{x}$ randomly in the $\mathcal{X}$ space. If we do this multiple times, the locus of $\mathbf{x}$ will form a unit circle in the $\mathcal{X}$ space. For each of the values of $\mathbf{x}$, we compute $\mathbf{y}$ using the above relation (2.61). The locus of $\mathbf{y}$ will map an ellipse in the $\mathcal{Y}$-space. The ratio of major and minor axes of the ellipse is the condition number. The physical picture is shown in Figure 2.3. The circles denote conditions for major axis of the ellipse, while the diamonds denote conditions for minor axis.

The following code was used to generate Figure 2.3:

```
% Specify A and its svd
A=[-0.2500     1.5000
   -2.2500     4.2500];
[U,S,V]=svd(A);
```

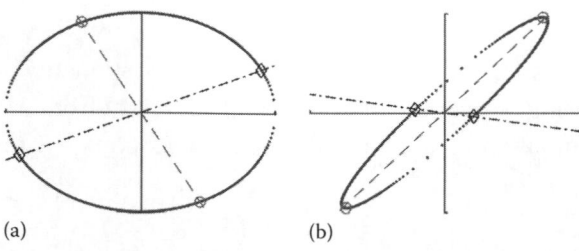

(a)   (b)

FIGURE 2.3 Interpretation of condition number. (a) A circle in the $\mathcal{X}$-space maps to an ellipse on $\mathcal{Y}$-space. (b) Condition number is interpreted as the ratio of major and minor axes of the ellipse.

```
% Generate x randomly on a unit circle
x1=[-1:0.01:1];
x2=sqrt(1-x1.^2);
x=[x1, x1; x2, -x2];
% Plot the original input data
subplot(1,2,1)
plot(x(1,:),x(2,:),'b.'); hold on

% Compute and the output
y=A*x;
subplot(1,2,2);
plot(y(1,:),y(2,:),'b.'); hold on

% Compute the "radius"
ySquare=y.*y;
r=sum(ySquare);
% Minor and major axis directions
% were computed and plotted manually
```

The two circles on the figure correspond to points $(-0.44, 0.898)$ and $(0.44, -0.898)$. Using svd command, the direction of vector $\mathbf{v}_1$ is found to be

$$\mathbf{v}_1 = \begin{bmatrix} -0.443 \\ 0.896 \end{bmatrix}$$

Recall that Equation 2.61 may be written as

$$\underbrace{\left(U^T\mathbf{y}\right)}_{\hat{\mathbf{y}}} = \Sigma\underbrace{\left(V^T\mathbf{x}\right)}_{\hat{\mathbf{x}}} \tag{2.59}$$

The first point therefore corresponds to $\hat{\mathbf{x}} = \begin{bmatrix} 1 & 0 \end{bmatrix}^T$. Consequently, in the $\mathcal{Y}$-space, the corresponding output point is $\hat{y}_1 = \sigma_1$. When translated back to the natural basis, this point is plotted as a circle at $(1.457, 4.807)$. Likewise, the other circle on Figure 2.3 is the point $(-1.457, -4.807)$. Note that these points in the modified basis correspond to

$$\hat{\mathbf{y}} \approx \begin{bmatrix} \sigma_1 \\ 0 \end{bmatrix} \quad \text{and} \quad \hat{\mathbf{y}} \approx \begin{bmatrix} -\sigma_1 \\ 0 \end{bmatrix}$$

respectively.

Generalizing the observations from Example 2.3, let us say that $\sigma_\ell \gg \sigma_{\ell+1}$. Then, span$\{\mathbf{v}_1, \ldots, \mathbf{v}_\ell\}$ corresponds to the direction in the $\mathcal{X}$-space where the result will be *stretched* after linear transformation through matrix A. On the other hand, the vectors that lie in the span$\{\mathbf{v}_{\ell+1}, \ldots, \mathbf{v}_n\}$

get shrunk with this linear transformation through matrix A. A further discussion of the effect of this directionality on system behavior will be presented in Section 5.2.3.

2.4 EIGENVALUES AND EIGENVECTORS

2.4.1 Orientation for This Section

Eigenvalues originated not in linear algebra but in the analysis of rigid-body dynamics, while solving linear ODEs in the late eighteenth and early nineteenth centuries. First year calculus course exposes us to this in the form of eigenvalue problems (or the so-called "Sturm–Liouville problems" for second-order ODEs). The importance of eigenvalues and eigenvectors will be discussed later, in Chapters 3 and 5 for discussing stability and dynamics of ODEs, and in Chapter 8 to introduce stiff ODEs. Needless to say, singular values and eigenvalues are important concepts that we will carry forward in later chapters of this book.

Some of the drawbacks of eigenvalues are that they are applicable to $n \times n$ square matrices only, and eigenvalues of real matrices may be complex. Singular values came much later, perhaps as a way of generalizing the concept of eigenvalues.

With this picture in mind, most linear algebra texts discuss eigenvalues before they discuss singular values. However, I deviate from this narrative because the concepts of singular and eigenvalues need to be tackled on their own merits. I presume that readers are already aware of the concept of eigenvalues. I therefore introduce singular values before eigenvalues because the former presented a natural narrative for the linear operator.

2.4.2 Brief Recap of Definitions

Eigenvalues and eigenvectors were introduced in introductory courses for solving linear ODEs:

$$\frac{d}{dt}\mathbf{y} = A\mathbf{y} \tag{2.62}$$

where $\mathbf{y}$ is an n-dimensional vector. The form of solution to this problem is obtained by comparing it with an equivalent problem in single variable:

$$\frac{dy}{dt} = ay \tag{2.63}$$

The solution for the above problem in single variable is

$$y = c_0 e^{at} \tag{2.64}$$

The ith row for Equation 2.62 is written as

$$y_i' = A_{i1} y_1 + A_{i2} y_2 + \cdots + A_{in} y_n$$

where the right-hand side is linear and differentiation is a linear operator. By analogy, since the superposition principle holds, a solution to Equation 2.62 is

$$\mathbf{y} = e^{\lambda t}\mathbf{c} \tag{2.65}$$

where
 $\mathbf{c}$ is an n-dimensional vector
 λ is a scalar that is yet to be determined

Substituting this in Equation 2.62

$$\frac{d}{dt}\mathbf{y} = Ae^{\lambda t}\mathbf{c} \tag{2.66}$$

Differentiating Equation 2.65

$$\frac{d}{dt}\mathbf{y} = \lambda e^{\lambda t}\mathbf{c}$$

and substituting it in the above equation yields

$$\lambda e^{\lambda t}\mathbf{c} = Ae^{\lambda t}\mathbf{c} \tag{2.67}$$

The scalar $e^{\lambda t}$ is nonzero (except for $\lambda = -\infty$) and may be eliminated. This yields us the following relationship:

$$A\mathbf{c} = \lambda\mathbf{c} \tag{2.68}$$

which is now a linear algebra problem. The above equation is rewritten as

$$(A - \lambda I)\mathbf{c} = \mathbf{0} \tag{2.69}$$

where λ is known as eigenvalue and $\mathbf{c}$ is known as eigenvector. The scalar λ is chosen such that the matrix $(A - \lambda I)$ becomes rank deficient (i.e., rank$(A - \lambda I) < n$). This condition gives us the so-called *characteristic equation*:

$$\det(A - \lambda I) = 0 \tag{2.70}$$

The eigenvector $\mathbf{c}$ is in null space of $(A - \lambda I)$. Let us end this discussion with a summary of key points.

 Definition: For a matrix $A : \mathcal{R}^n \to \mathcal{R}^n$, there exists scalar $\lambda \in \mathcal{C}$ and vector $\mathbf{c} \in \mathcal{C}^n$ such that

$$A\mathbf{c} = \lambda\mathbf{c}$$

where λ is called the eigenvalue and $\mathbf{c}$ is the eigenvector.

Characteristic equation: Equation 2.70 is known as the characteristic equation of matrix A. The above characteristic equation is an nth-order polynomial in λ. Thus, an $n \times n$ matrix will have n eigenvalues. Some of these eigenvalues may be repeated.

Complex eigenvalues: For a real matrix A, eigenvalues may be real or complex. If there are complex eigenvalues, they occur in complex conjugate pairs.

The eigenvector **c** is in the null space of $(A - \lambda I)$. If the n eigenvalues are distinct, the corresponding n eigenvectors are linearly independent. Moreover, if these distinct eigenvalues are real, we have n linearly independent real eigenvectors. This is an important result because n distinct eigenvectors can form basis for the R^n space. However, these eigenvectors may not be orthonormal.

2.4.3 Eigenvalue Decomposition

Eigenvalues and eigenvectors were introduced for solving linear ODEs, $\mathbf{y}' = A\mathbf{y}$. The term is reported to have origins in the German word "eigenwert," which translates (according to Google translate) to "intrinsic." The term eigenvalue or eigenvector may therefore imply "an intrinsic value/vector" that characterizes the matrix. The most important application of eigenvalues and eigenvectors is in the analysis of situations where we need to map a linear transformation of a vector space on itself.

In Section 2.3, physical interpretation of SVD was provided for a linear transformation $\mathbf{y} = A\mathbf{x}$. The mapping was $A: \mathcal{R}^n \rightarrow \mathcal{R}^m$. The input and output vector spaces could be of different dimensions (as seen in Example 2.1) or they could be of same dimension (when $m = n$, such as Example 2.3). Naturally, SVD gave two different coordinate transformations: The orthogonal matrix V as the input basis, the orthogonal matrix U as the output basis, and the diagonal matrix Σ that maps $\hat{x} \in \mathcal{X}$ to $\hat{y} \in \mathcal{Y}$.

In contrast, eigenvalues are of interest when mapping the $\mathcal{R}^n$ vector space on itself. For the ease of this discussion, consider the case where the matrix A has n distinct eigenvalues (this is often a more common case). It can be proved that if the eigenvalues are distinct, the eigenvectors will be linearly independent (proof is left as an exercise). Thus, the eigenvectors can form a basis for coordinate transformation[*]:

$$C = \begin{bmatrix} \mathbf{c}_1 & \mathbf{c}_2 & \cdots & \mathbf{c}_n \end{bmatrix} \tag{2.71}$$

Equivalent to Equation 2.37, change of coordinates to nonorthonormal[†] basis set is represented as

$$\hat{\mathbf{y}} = C^{-1}\mathbf{y} \tag{2.72}$$

---

[*] Standard linear algebra texts often use $A\mathbf{x} = \lambda\mathbf{x}$ to define eigenvalues and eigenvectors. Hence, they represent transformation matrix for eigenvectors as the basis set as $X = \begin{bmatrix} \mathbf{x}_1 & \mathbf{x}_2 & \cdots & \mathbf{x}_n \end{bmatrix}$. Note that this representation and Equation 2.54 are equivalent. Since I have used $\mathbf{x} \in \mathcal{X}$ as input vectors in Section 2.3, I chose to use notation **c** as eigenvectors.

[†] Recall that the eigenvectors may not be orthogonal.

or equivalently

$$\mathbf{y} = C\hat{\mathbf{y}} \qquad (2.73)$$

This can be expanded as

$$\mathbf{y} = \hat{y}_1 \mathbf{c}_1 + \hat{y}_2 \mathbf{c}_2 + \cdots + \hat{y}_n \mathbf{c}_n \qquad (2.74)$$

Premultiplying by matrix A gives us

$$A\mathbf{y} = A\left(\hat{y}_1 \mathbf{c}_1 + \hat{y}_2 \mathbf{c}_2 + \cdots + \hat{y}_n \mathbf{c}_n\right) \qquad (2.75)$$

Using Equation 2.68, this becomes

$$A\mathbf{y} = \left(\lambda_1 \mathbf{c}_1\right)\hat{y}_1 + \left(\lambda_2 \mathbf{c}_2\right)\hat{y}_2 + \cdots + \left(\lambda_n \mathbf{c}_n\right)\hat{y}_n \qquad (2.76)$$

Simple algebraic manipulations yield

$$A\mathbf{y} = \begin{bmatrix} \lambda_1 \mathbf{c}_1 & \lambda_2 \mathbf{c}_2 & \cdots & \lambda_n \mathbf{c}_n \end{bmatrix} \begin{bmatrix} \hat{y}_1 \\ \hat{y}_2 \\ \vdots \\ \hat{y}_n \end{bmatrix} \qquad (2.77)$$

$$A\mathbf{y} = \underbrace{\begin{bmatrix} \mathbf{c}_1 & \mathbf{c}_2 & \cdots & \mathbf{c}_n \end{bmatrix} \begin{bmatrix} \lambda_1 & & & \\ & \lambda_2 & & \\ & & \ddots & \\ & & & \lambda_n \end{bmatrix}}_{C\Lambda} \underbrace{\begin{bmatrix} \hat{y}_1 \\ \hat{y}_2 \\ \vdots \\ \hat{y}_n \end{bmatrix}}_{\hat{y}}$$

Note that the order of matrices C and Λ is important. The product ΛC implies that each element λ_i multiplies the ith row of matrix C, whereas the product $C\Lambda$ implies that each element λ_i multiplies the ith column of matrix C. Thus, the equation *cannot* be written as $A\mathbf{y} = \Lambda C\hat{\mathbf{y}}$.

Substituting (2.72), we get the following important result:

$$\begin{aligned} A\mathbf{y} &= C\Lambda C^{-1}\mathbf{y} \\ A &= C\Lambda C^{-1} \end{aligned} \qquad (2.78)$$

The above equation (2.78) is the key result of *eigenvalue decomposition*.

Diagonalization of a matrix refers to eigenvalue decomposition of matrix A that has n distinct eigenvalues, since the eigenvalue decomposition will result in a diagonal matrix Λ.

When the matrix A is rank deficient, one of its eigenvalues is zero. A zero eigenvalue behaves like any other eigenvalue. There will be an eigenvector associated with the zero eigenvalue. Note that since $(A - \lambda I) = A$, the corresponding eigenvector, **c**, lies in the null space of A. Consider the example

$$A = \begin{bmatrix} 1 & 2 \\ 2 & 4 \end{bmatrix}$$

which we had discussed as an example of rank-deficient matrix. The two eigenvalues are $\lambda_1 = 0$ and $\lambda_2 = 5$. The corresponding eigenvectors are

$$\mathbf{c}_1 = \begin{bmatrix} -2 \\ 1 \end{bmatrix}, \quad \mathbf{c}_2 = \begin{bmatrix} 1 \\ 2 \end{bmatrix}$$

where the first eigenvector, $\mathbf{c}_1$, which corresponds to $\lambda_1 = 0$, is in the null space of A. The above information can be obtained using `eig` command. The readers can verify that the diagonal form

$$\begin{bmatrix} 1 & 2 \\ 2 & 4 \end{bmatrix} = \begin{bmatrix} -2 & 1 \\ 1 & 2 \end{bmatrix} \begin{bmatrix} 0 & 0 \\ 0 & 5 \end{bmatrix} \begin{bmatrix} -2 & 1 \\ 1 & 2 \end{bmatrix}^{-1}$$

is indeed valid.

The above results are also valid when eigenvalues exist in complex conjugate pair.

2.4.4 Applications

The eigenvalue decomposition is the cornerstone of the analysis of linear dynamical systems. It is used in both differential equations

$$\mathbf{y}' = A\mathbf{y} \tag{2.62}$$

and difference equations

$$\mathbf{y}_{k+1} = A\mathbf{y}_k \tag{2.79}$$

In both examples, the matrix A maps the $\mathcal{R}^n$ vector space onto itself.*

First, let us perform a geometric interpretation of eigenvalue decomposition, like that in Example 2.3, before moving to the applications.

---

* Let me repeat the contrast of this case with the examples in Section 2.3. The $n \times n$ square linear operator matrix, for the linear operation $\mathbf{y} = A\mathbf{x}$, mapped a n-dimensional input space to another n-dimensional output space. In case of differential equation, $\mathbf{y}' = A\mathbf{y}$, the matrix A maps the n-dimensional space to itself.

Example 2.4 Significance of Eigenvalues and Eigenvectors

Consider the matrix

$$A = \begin{bmatrix} -0.25 & 1.5 \\ -2.25 & 4.25 \end{bmatrix}$$

from Example 2.3. The eigenvalues and eigenvectors for this matrix can be found using the command `eig` as

$$\lambda_1 = 3.299, \quad \lambda_2 = 0.701$$

$$\mathbf{c}_1 = \begin{bmatrix} 0.3893 \\ 0.9211 \end{bmatrix}, \quad \mathbf{c}_2 = \begin{bmatrix} 0.8446 \\ 0.5354 \end{bmatrix}$$

Figure 2.4 shows the geometric interpretation of eigenvalues and eigenvectors. The dots, which form a unit circle, represent various values of vector $\mathbf{y}$ (as in Example 2.3, each dot represents each of the 101 unit vectors). The plus signs represent corresponding vectors $A\mathbf{y}$. Notice that the two circles are *numerically* the same as that in Figure 2.3; they are plotted on the same axis because they lie on the same vector space. However, the main geometric significance of eigenvalues and eigenvectors is shown as the lines on the plot. The first solid line corresponds to the first eigenvector $\mathbf{c}_1$. When operated by the matrix A, the result is that the vector gets simply *stretched* in the same direction by a factor equal to $\lambda_1 = 3.299$. This is shown by the dashed line in the figure. Likewise, the second thick line is the vector $\mathbf{c}_2$, which gets *shrunk* in the same direction by a factor $\lambda_2 = 0.701$ (this dashed line may not be visible because it is shorter and thinner than $\mathbf{c}_2$).

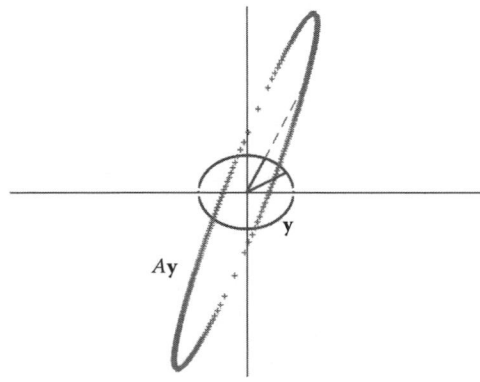

FIGURE 2.4   Geometric interpretation of eigenvalues and eigenvectors. The dots represent $\mathbf{y}$ and the plus symbols represent corresponding $A\mathbf{y}$. The thick solid lines are the eigenvectors $\mathbf{c}_1$ and $\mathbf{c}_2$, whereas the dashed lines are $A\mathbf{c}_1$ *and* $A\mathbf{c}_2$, respectively.

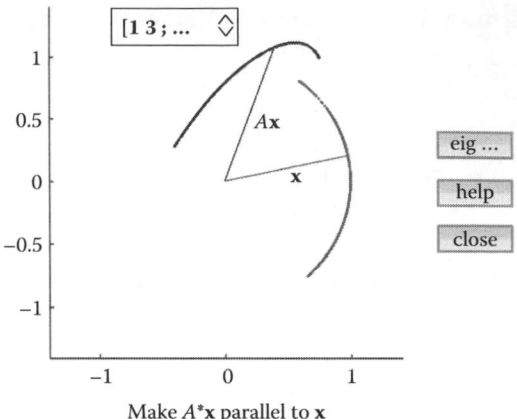

Make $A^*\mathbf{x}$ parallel to $\mathbf{x}$

FIGURE 2.5   Demo window from `eigshow` interactive demonstrator in MATLAB®.

Thus, any vector in the direction of the eigenvector of a matrix gets stretched or shrunk by a factor given by the eigenvalue. If an eigenvalue is negative, the resulting vector $A\mathbf{c}_i$ will be a vector in opposite direction that gets stretched/shrunk by the eigenvalue.

Thus, the physical interpretation is stretching and shrinking of the vector along the direction of an eigenvector by a factor equal to the corresponding eigenvalue. There is an excellent interactive tool in MATLAB called `eigshow` that can be used to exactly understand these concepts. The direction of x can be changed using the mouse, and the vector $A\mathbf{x}$ responds to these changes. Figure 2.5 shows a screenshot of `eigshow` window. By interactively moving the vector $\mathbf{x}$ and tracking how the vector $A\mathbf{x}$ moves, one can understand the concepts discussed in Examples 2.3 and 2.4.

While the above is useful in understanding eigenvalues and eigenvectors, the next example focuses on concepts that will be useful for readers to use them.

Example 2.5   Eigenvalue Decomposition

We will continue with Example 2.4, where we saw that a vector in the direction of $\mathbf{c}_1$ and $\mathbf{c}_2$ is stretched or shrunk by a factor equal to the eigenvalue. This example gives geometric interpretation of how eigenvalue decomposition may be used. Consider the vector

$$\mathbf{y} = \begin{bmatrix} -0.6 \\ -0.8 \end{bmatrix} \Rightarrow A\mathbf{y} = \begin{bmatrix} -1.05 \\ -2.05 \end{bmatrix}$$

These two vectors are represented by solid and dashed line in Figure 2.6, respectively. As shown in the previous section, the two eigenvectors $\mathbf{c}_1$ and $\mathbf{c}_2$ are chosen as

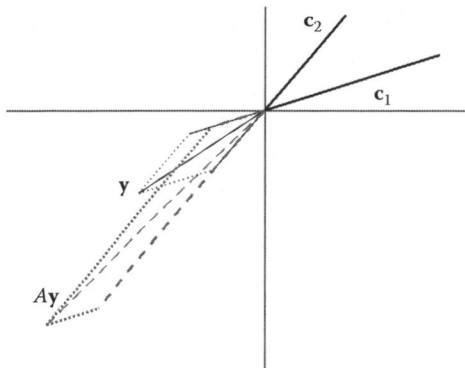

FIGURE 2.6 Significance of eigenvalue decomposition. The thick lines represent eigenvectors c_1 and c_2. The solid line represents the vector $\mathbf{y}$ and the dotted lines "complete the parallelogram." The dashed line represents the vector $A\mathbf{y}$.

basis vectors. The two dotted lines emanating from the point $\mathbf{y}$ complete the parallelogram (e.g., Figure 2.1a). These two solid lines along (and opposite) the direction of eigenvectors represent the projections along the eigenvectors. Since these lines are along the eigenvectors, multiplying with matrix A results simply in expansion of the element $\hat{y}_1$ and shrinking of the element $\hat{y}_2$. These two are represented by the two thick dashed lines in the figure. These thick dashed lines are $A\hat{y}_1$, the coordinate of the solution along first eigenvector, and $A\hat{y}_2$, the coordinate of the solution along the second eigenvector. To get the final solution $A\mathbf{y}$ in the original basis, the thick dotted lines complete the parallelogram to yield the solution $A\mathbf{y}$ displayed in the form of thin dashed line.

This shows the practical application of eigenvalue decomposition. Since any vector that lies along an eigenvector stays along the same eigenvector, when acted upon by the matrix A, the response of a linear system along eigenvectors can be analyzed independently.

To summarize, the steps in computing $A\mathbf{y}$ included the following:

Change of basis to eigenvectors to obtain

$$\hat{\mathbf{y}} = C^{-1}\mathbf{y} = \begin{bmatrix} -0.4236 \\ -0.6223 \end{bmatrix}$$

is represented by thin solid lines in Figure 2.6.

Stretching/shrinking of $\hat{\mathbf{y}}$ by a factor equal to the corresponding eigenvalue gives

$$\begin{bmatrix} \lambda_1(-0.4236) \\ \lambda_2(-0.6223) \end{bmatrix} = \begin{bmatrix} -0.2969 \\ -2.053 \end{bmatrix}$$

where the first element (−0.2969) is the distance along c_1 axis and the second (−2.053) is the distance along c_2 axis. These are represented as the two thick dashed lines in the figure.

Change of basis back to the natural basis

$$C \begin{bmatrix} -0.2969 \\ -2.053 \end{bmatrix} = \begin{bmatrix} -1.05 \\ -2.05 \end{bmatrix}$$

gives us the solution. Geometrically, the resulting vector, $A\mathbf{y}$, is obtained by "completing the parallelogram," which is indicated by thick dotted lines.

The above example showed the case of distinct, real eigenvalues. However, these results are applicable for other situations as well.

2.4.4.1 Similarity Transform

The transformation along the eigenvectors is an example of *similarity transform*. If S is any invertible matrix, then the matrix B obtained as

$$B = S^{-1}AS \tag{2.80}$$

is said to be *similar* to A and the transformation is known as similarity transform. The column vectors of matrix S form the new basis set. The above equation may be written as

$$A = SBS^{-1}$$

Multiplying both sides with eigenvector, $\mathbf{c}$, gives

$$\lambda \mathbf{c} = SBS^{-1}\mathbf{c}$$
$$\lambda \hat{\mathbf{c}} = B\hat{\mathbf{c}}, \quad \hat{\mathbf{c}} = S^{-1}\mathbf{c} \tag{2.81}$$

The eigenvalues of similar matrices are equal, whereas eigenvectors are changed from $\mathbf{c}$ to $S^{-1}\mathbf{c}$. The transformation along eigenvectors as basis set is a special type of similarity transform, since it leads to diagonal (or nearly diagonal) matrix Λ.

2.4.4.1.1 Change of Basis In systems where the matrix multiplication by A maps the vector space on itself, change of basis is represented by similarity transform. This was seen in Example 2.5. Change of basis therefore does not change the eigenvalues of the system.

Change of basis along the eigenvectors is an important practical method for analyzing linear systems.

2.4.4.1.2 Power and Exponent of Similar Matrices Two similar matrices, A and B are related as $A = SBS^{-1}$. We can therefore write A^2 as

$$A^2 = \left(SBS^{-1} \right)\left(SBS^{-1} \right) = SB^2S^{-1}$$

We can multiply the above with matrix A to get higher powers of A. Generalizing,

$$A^i = SB^iS^{-1} \tag{2.82}$$

in a similar manner.

2.4.4.1.3 *Jordan Canonical Form* Jordan canonical form is a special type of similarity transform closely related to eigenvalue decomposition. The idea behind Jordan canonical form is that any matrix A can be written in the following form:

$$A = C\Lambda C^{-1} \tag{2.83}$$

In case of n distinct eigenvalues, the matrix Λ is diagonal, as seen in Equation 2.78. When there are repeated eigenvalues, the eigenvectors corresponding to the repeating eigenvalues may not be linearly independent. If that happens, full diagonalization is not possible. However, it is still possible to decompose the matrix A in the above *Jordan canonical form*. For example, consider a 3×3 matrix whose eigenvalues are $\{\lambda_1, \lambda_2, \lambda_2\}$. It is possible to decompose the matrix A in the following form:

$$A = \begin{bmatrix} \mathbf{c}_1 & \mathbf{c}_2 & \mathbf{c}_3 \end{bmatrix} \begin{bmatrix} \lambda_1 & 0 & 0 \\ 0 & \lambda_2 & 1 \\ 0 & 0 & \lambda_2 \end{bmatrix} \begin{bmatrix} \mathbf{c}_1 & \mathbf{c}_2 & \mathbf{c}_3 \end{bmatrix}^{-1}$$

The above decomposition is nearly diagonal. The scalar "1" appears in the off-diagonal element in each block corresponding to the repeated eigenvalue. The vector $\mathbf{c}_3$ is the "generalized eigenvector" of matrix A.

2.4.4.2 *Linear Differential Equations*
Let us assume that the matrix A is fully diagonalizable. Thus, in case of the differential equation

$$\begin{aligned} \mathbf{y}' &= C\Lambda C^{-1}\mathbf{y} \\ \left(C^{-1}\mathbf{y}\right) &= A\left(C^{-1}\mathbf{y}\right) \end{aligned} \tag{2.84}$$

Transforming the basis to eigenvectors, the above can be written as

$$\hat{\mathbf{y}}' = \Lambda\hat{\mathbf{y}} \tag{2.85}$$

Since Λ is diagonal, each row represents the ith differential equation

$$\hat{y}'_i = \lambda_i \hat{y}_i$$

The solution to this equation, therefore, is

$$\hat{\mathbf{y}} = \underbrace{\begin{bmatrix} e^{\lambda_1 t} & & \\ & \ddots & \\ & & e^{\lambda_n t} \end{bmatrix}}_{e^\Lambda} \hat{\mathbf{y}}(0) \tag{2.86}$$

Transforming it back to the original basis, the above solution is

$$\mathbf{y} = \underbrace{Ce^{\Lambda t}C^{-1}}_{e^{At}}\mathbf{y}(0) \tag{2.87}$$

The *analytical* solution to the ODE (2.62) is given by the above equation. Standard textbooks on linear algebra discuss methods to compute matrix exponent. It is easy to compute for small-size matrices. The MATLAB command expm may be used to compute the matrix exponent numerically. Thus, the evolution of $\mathbf{y}$ at any time t from initial condition $\mathbf{y}_0$ may be calculated in MATLAB as

```
>> yi = expm(A*ti)*y0;
```

where yi is the value $\mathbf{y}_i(t_i)$ calculated at time ti.

The diagonalization from Equation 2.86 also helps us *qualitatively* discuss the stability of linear ODEs. If the eigenvalues are real and negative, the system is asymptotically stable, implying that $\mathbf{y}(t) \to 0$ as $t \to \infty$. If one or more of the eigenvalues is positive, the system is unstable, whereas if an eigenvalue is zero, the system is marginally stable. We will use this property in the discussion of linear stability analysis in Chapter 5 and again in the analysis of nonlinear systems in Chapter 9.

2.4.4.3 Linear Difference Equations

As in the previous discussion on differential equations, let us assume that the matrix A is diagonalizable. The difference equation

$$\mathbf{y}_{k+1} = A\mathbf{y}_k \tag{2.79}$$

may be written as

$$\mathbf{y}_{k+1} = C\Lambda C^{-1}\mathbf{y}_k \tag{2.88}$$

Starting with $k = 0$, we can write

$$\begin{aligned} \mathbf{y}_1 &= C\Lambda C^{-1}\mathbf{y}_0 \\ \mathbf{y}_2 &= C\Lambda C^{-1}\mathbf{y}_1 = C\Lambda C^{-1}\left(C\Lambda C^{-1}\mathbf{y}_0\right) \\ &= C\Lambda^2 C^{-1}\mathbf{y}_1 \\ &\vdots \\ \mathbf{y}_k &= \underbrace{C\Lambda^k C^{-1}}_{A^k}\mathbf{y}_0 \end{aligned} \tag{2.89}$$

Since Λ is a diagonal matrix

$$\mathbf{y} = C \begin{bmatrix} \lambda_1^k & & \\ & \ddots & \\ & & \lambda_n^k \end{bmatrix} C^{-1} \mathbf{y}_0 \tag{2.90}$$

Just as the case of differential equation, the diagonalization from Equation 2.90 also helps us *qualitatively* discuss the stability of linear difference equations. The difference equation (2.79) results in an asymptotically stable behavior if all the eigenvalues are within the unit circle, that is, $|\lambda_i| < 1$, $\forall i$. If one or more of the eigenvalues are outside the unit circle, the system is unstable; whereas eigenvalues on the unit circle implies that the system is marginally stable.

In chemical engineering, difference equations are often used in computer-based process control and logistic maps. We will not discuss difference equations for the rest of this book.

2.5 EPILOGUE

Some of the key concepts in linear algebra were introduced in this chapter. The three types of problems in linear systems of interest to engineers include the following.

First problem involves solving a set of linear equations of the type, $A\mathbf{x} = \mathbf{b}$, given by

$$\underbrace{\begin{bmatrix} a_{11} & a_{12} & \cdots & a_{1n} \\ a_{21} & a_{22} & \cdots & a_{2n} \\ \vdots & \ddots & \ddots & \vdots \\ \vdots & \ddots & \ddots & \vdots \\ a_{m1} & a_{m2} & \cdots & a_{mn} \end{bmatrix}}_{A} \underbrace{\begin{bmatrix} x_1 \\ x_2 \\ \vdots \\ x_n \end{bmatrix}}_{\mathbf{x}} = \underbrace{\begin{bmatrix} b_1 \\ b_2 \\ \vdots \\ \vdots \\ b_m \end{bmatrix}}_{\mathbf{b}} \tag{2.8}$$

The aim is to find the vector $\mathbf{x}$ given the right-hand side, $\mathbf{b}$. The most common example is to solve n linear equations in n unknowns, where the matrix A is an $n \times n$ matrix. However, there are several examples where the number of equations exceeds the number of unknowns. Such examples will be seen in linear least squares problems for parameter estimation (Chapter 10). The MATLAB commands relevant to this type of problems are listed below.

Command	Description	Usage
rank	Compute rank of a matrix	`r=rank(A);`
inv	Compute inverse of a matrix	`x = inv(A)*b;` will solve linear equation as $\mathbf{x} = A^{-1}\mathbf{b}$ if A is full-rank
mldivide	Left division to solve $A\mathbf{x}=\mathbf{b}$ (A is $m \times n$ matrix, and $\mathbf{x}$ is $n \times 1$ and $\mathbf{b}$ is $m \times 1$ vectors)	`x=A\b;` or `x=mldivide(A,b);` solves the linear equation $A\mathbf{x}=\mathbf{b}$, even when A is nonsquare.
lu	LU decomposition of matrix A. Please see Appendix C for details.	`[L,U]=lu(A);` and `[L,U,P]=lu(A);` factorize A into a set of lower/upper triangular matrices such that $A=LU$ or $PA=LU$.

The linear equations problem was cast as a linear combination of column vectors of matrix A:

$$x_1\mathbf{v}_1 + x_2\mathbf{v}_2 + \cdots + x_n\mathbf{v}_n = \mathbf{b} \tag{2.91}$$

where $\mathbf{v}_1, \mathbf{v}_2, \ldots, \mathbf{v}_n$ (see Section 2.2). The various properties of vectors, matrix, and associated vector spaces were discussed. The relevant MATLAB commands are summarized below.

Command	Description	Usage
`null(A)`	Calculate null space of A	`V=null(A);` returns basis vectors of the null space as columns of matrix V.
`orth(A)`	Computes image space of A	`V=orth(A);` returns basis vectors of the image space as columns of matrix V.
`norm(b)`	Compute norm of a vector	`norm(b,i);` computes $(\sum_i \lvert b_i \rvert^p)^{1/p}$ If i is not specified, two-norm is computed.
`norm(A)`	Compute norm of a matrix	Computes norm of a matrix. The two-norm of a matrix is equal to its largest singular value.
`cond`	Condition number of a matrix	`cn=cond(A);` computes the condition number, the ratio of highest to smallest singular values of A.

Example 2.1 used a simple model of blending process to highlight the concepts of null and image spaces, as also to discuss the solution of linear equations. Example 2.2 showed the sensitivity of solution to errors in the data. Condition number was introduced to quantify this sensitivity.

The discussion on null and image spaces led us to SVD (Section 2.3). SVD was introduced in the context of analyzing the second type of problems in linear algebra:

$$\begin{bmatrix} b_1 \\ b_2 \\ \vdots \\ \vdots \\ b_m \end{bmatrix} = \underbrace{\begin{bmatrix} a_{11} & a_{12} & \cdots & a_{1n} \\ a_{21} & a_{22} & \cdots & a_{2n} \\ \vdots & \ddots & \ddots & \vdots \\ \vdots & \ddots & \ddots & \vdots \\ a_{m1} & a_{m2} & \cdots & a_{mn} \end{bmatrix}}_{A} \underbrace{\begin{bmatrix} x_1 \\ x_2 \\ \vdots \\ x_n \end{bmatrix}}_{\mathbf{x}} \tag{2.9}$$

$$\underbrace{\phantom{\begin{bmatrix} b_1 \end{bmatrix}}}_{y}$$

which represented linear mapping between n-dimensional input vector space $\mathcal{X}$ to m-dimensional output vector space $\mathcal{Y}$. Example 2.3 presented geometric interpretation of SVD. With this, one can interpret null and image spaces in context of SVD. This was the second practical context in this chapter: a linear map between an input and an output space.

The third practical context in this chapter is when matrix A represents a linear map between a vector space to itself, most commonly observed in linear differential and difference equations. Section 2.4 and Example 2.5 presented eigenvalue decomposition of a linear system. The physical perspectives of singular value and eigenvalue decompositions were also compared. The following table summarizes the relevant MATLAB commands for the two decompositions and related functions.

Command	Description	Usage
svd(A)	Singular value decomposition	s=null(A); returns ordered singular values of A. [U,S,V]=svd(A); performs SVD such that $A = USV^T$.
eig(A)	Eigenvalues (and eigenvectors) of matrix A	l=eig(A); returns eigenvalues of A. [C,L]=eig(A); returns eigenvectors and eigenvalues of A. Individual columns of V contain the eigenvectors (c1, ..., cn) and the diagonal elements of L contain the corresponding eigenvalues.
jordan(A)	Perform Jordan decomposition	[M,L]=jordan(A); computes the Jordan decomposition, such that $A = MLM^{-1}$. Matrix L is nearly diagonal, with the diagonal elements equal to the eigenvalues of A.
eigshow	SVD and eigenvalue decomposition demonstrator	eigshow provides a GUI-based demonstrator for singular value and eigenvalue decompositions.

This concludes this practical introduction to linear algebra.

EXERCISES

Problem 2.1 (a) Show that linearly independent vectors $v_1, v_2, \ldots, v_n \in \mathcal{R}^n$ span the $\mathcal{R}^n$ space.

(b) Show that Equations 2.18 and 2.19 are equivalent.
Hint: Define the vector

$$\mathbf{v}_i = \begin{bmatrix} \mathbf{v}_{1,i} \\ \mathbf{v}_{2,i} \\ \vdots \\ \mathbf{v}_{n,i} \end{bmatrix}$$

to prove the equivalence.

Problem 2.2 Show that R^n can have at most n linearly independent vectors. In other words, show that vectors $v_1, v_2, \ldots, v_n, v_{n+1} \in \mathcal{R}^n$ are linearly dependent.

Problem 2.3 Show that the row rank and column rank of a matrix are equal.

Problem 2.4 Repeat the steps of Section 2.3.3 (Example 2.2) for solving $Ax = b$, with

$$A = \begin{bmatrix} 1 & 3 \\ 3 & 6.001 \end{bmatrix} \quad \text{and} \quad A = \begin{bmatrix} 1.37 & 1 \\ 0.5 & 0.37 \end{bmatrix}$$

Specifically, find the value of **b** at which the solution is highly sensitive. Change the vector **b** slightly and observe its effect on the solution **x**.

Hint: If you are unable to solve the problem, choose (up to) four different **b** vectors: (i) $b \in \text{span}\{u_1\}$, (ii) $b \in \text{span}\{u_2\}$, (iii) $b \in \text{span}\{v_1\}$, (iv) $b \in \text{span}\{v_2\}$.

Problem 2.5 Show that for an $n \times n$ matrix with distinct eigenvalues, λ_i, the corresponding eigenvectors c_i are linearly independent.

Problem 2.6 Consider the following three matrices:

$$A = \begin{bmatrix} -1.3 & 0.1 \\ 0.2 & -1.4 \end{bmatrix}, \quad B = \begin{bmatrix} -8.05 & -3.95 \\ -7.90 & -4.10 \end{bmatrix}, \quad C = \begin{bmatrix} 4.8 & -6.0 \\ 6.3 & -7.5 \end{bmatrix}$$

Consider the three systems

$$y = Ax, \quad y = Bx, \quad y = Cx$$

Randomly generate 250 data points in domain space, X, using `randn`. Plot the corresponding output in Y-space. Justify the difference in observed response.

Relate the *qualitative* results to the singular values of the three matrices.

Problem 2.7 For the same three matrices in Problem 2.6, choose `y0= [0.7071; 0.7071]`. The solution of the ODE

$$y' = Ay, \quad y(0) = y_0$$

is given by $y = y_0 e^{At}$. This can be computed using `y=y0.*expm(A*t)`. For a range of time values `[0:tMax]`, compute y. Hence, plot y vs. t.

Repeat this for $y' = By, \quad y(0) = y_0$.

Also repeat for $y' = Cy, \quad y(0) = y_0$.

What are the qualitative observations for A, B, and C matrices? Justify the difference in the observed response.

Relate the qualitative results to the eigenvalues and eigenvectors of the three matrices.

CHAPTER 3

Ordinary Differential Equations

Explicit Methods

3.1 GENERAL SETUP

A general ordinary differential equation (ODE), introduced in Chapter 1, is of the form

$$\frac{d\mathbf{y}}{dt} = \mathbf{f}(t, \mathbf{y})$$
$$\mathbf{y}(t_0) = \mathbf{y}_0 \tag{3.1}$$

The *initial condition* at time $t = t_0$ is specified as $\mathbf{y}_0$. In general, $\mathbf{y} \in \mathcal{R}^n$ is an n-dimensional solution variable of interest, and $\mathbf{f}(t, \mathbf{y}) : \mathcal{R}^n \to \mathcal{R}^n$ is a vector-valued function. In this chapter, I first cover numerical techniques to solve an ODE initial value problem (IVP), followed by the application of ODE solution techniques to Process Engineering problems. As discussed in Chapter 2, the convention followed throughout this text is that $\mathbf{y}$ is a $n \times 1$ *column* vector and $\mathbf{f}(\cdot)$ is $n \times 1$ function vector. The independent variable, t, is a scalar.

3.1.1 Some Examples

Examples of systems described by ODEs abound in engineering. For example, the reaction of a species along the length of a plug flow reactor (PFR) is given by an ODE:

$$u \frac{d}{dz} C_A = -k C_A^n \tag{3.2}$$

Since this is a first-order ODE, a single initial condition is required, $C_A(z=0) = C_{A,\,\text{in}}$. Change of temperature of liquid in a well-stirred vessel that is electrically heated is also given by an ODE of the form

$$V\rho c_p \frac{dT}{dt} = Q\rho c_p \left(T_{\text{in}} - T\right) + Q_h A_h \tag{3.3}$$

where
   Q_h is the heat flux from the electric heater
   A_h is the heater surface area

The temperature at the start, $T(t=0) = T_0$ forms the initial condition for this system.

More commonly, we encounter situations where multiple ODEs have to be solved, for example, when there are multiple species in a reactor or a nonisothermal reactor, where energy balance equation models the temperature variation in the reactor. Likewise, there may be multiple heated tanks that need to be modeled, or the height of liquid in the tank may vary.

Equation 3.2 is written in the standard form by dividing throughout by u. If there are multiple species, the mass balance is written for each individual species to obtain

$$\frac{d}{dz} C_A = f_1\left(C_A, C_B, \ldots\right)$$
$$\frac{d}{dz} C_B = f_2\left(C_A, C_B, \ldots\right) \tag{3.4}$$
$$\vdots$$

while the energy balance (for nonisothermal case) may be written as

$$\frac{dT}{dz} = f_{n+1}\left(C_A, C_B, \ldots\right) \tag{3.5}$$

Equations 3.4 and 3.5 may be combined to obtain the set of ODEs:

$$\frac{d}{dz} \underbrace{\begin{bmatrix} C_A \\ C_B \\ \vdots \\ T \end{bmatrix}}_{\mathbf{y}} = \underbrace{\begin{bmatrix} f_1\left(C_A, C_B, \ldots\right) \\ f_2\left(C_A, C_B, \ldots\right) \\ \vdots \\ f_{n+1}\left(C_A, C_B, \ldots\right) \end{bmatrix}}_{\mathbf{f(y)}} \tag{3.6}$$

which is in the standard form (3.1). Since the concentrations and temperature are all specified at the inlet, $\mathbf{y}(z=0) = \mathbf{y}_{\text{in}}$, we have a set of $(n+1)$ ODE-IVP.

Another type of example involves conversion of higher-order ODEs into a set of first-order ODEs. The classical example of this is the mass-spring-damper system. The motion of a body with mass m attached to a spring is given by

$$mx'' = -cx' - kx \tag{3.7}$$

where
   x'' is the acceleration
   $v = x'$ is the velocity

The initial condition involves displacing the mass by a certain distance and releasing it. This is an example of "damped oscillator." The second-order ODE may be written as

$$
\begin{aligned}
x' &= v \\
v' &= -\frac{c}{m}v - \frac{k}{m}x
\end{aligned}
\tag{3.8}
$$

Defining vector $\mathbf{y} = [x \ v]^T$, the above ODE will be written in the standard form*:

$$\mathbf{y}' = \begin{bmatrix} v \\ -\bar{c}v - \bar{k}x \end{bmatrix}, \quad \mathbf{y}(0) = \begin{bmatrix} x_0 \\ 0 \end{bmatrix} \tag{3.9}$$

The solutions of problems of the type (3.6) and (3.9), which give rise to ODE-IVP, are discussed in this chapter.

ODE-BVPs (*boundary value problems*): These form another type of ODEs, where the conditions for various state variables are specified at different points in the domain. A classic example is that of heat conduction in a rod:

$$\frac{d^2T}{dz^2} = -\beta(T - T_a) \tag{3.10}$$

with T_a as the temperature of the ambient. Since this is a second-order ODE, two conditions are required to solve *iy*. If both the conditions are specified at the same location (e.g., $T(0) = T_0$ and $T'(0) = \vartheta_0$), we can convert the second-order ODE into the standard ODE-IVP form (3.1). However, more commonly, the two conditions are specified at either ends of the domain, such as

$$T(0) = T_0, \quad T'(L) = \vartheta \tag{3.11}$$

---

* In fact, this is an example of *linear* system of ODEs: $\mathbf{y}' = A\mathbf{y}$, with $A = \begin{bmatrix} 0 & 1 \\ -\bar{k} & -\bar{c} \end{bmatrix}$.

This leads to ODE-BVP. ODE-BVP cannot directly be solved using methods described in this chapter. They are converted using finite difference approximation to a set of linear or nonlinear equations and are solved. This procedure results in a special structure of the problem. Methods for solving ODE-BVP (as well as other problems that result in similar *numerical structure*) will be discussed in Chapter 7.

Thus, this chapter focuses on *one family* of numerical methods to solve ODE-IVP. An introduction to numerical solution of ODE-IVP and a comparison with numerical integration will be first presented in the remainder of this section. Thereafter, an important family of numerical methods, called *Runge-Kutta (RK) methods*, will be discussed. A majority of discussions in this section will focus on a single-variable problem; extension to multivariable case will be considered thereafter.

3.1.2 Geometric Interpretation

Before proceeding further, I will draw parallels with numerical integration for a problem in a single variable, that is, y is a scalar and f is a scalar function. Numerical integration is covered in Appendix E. If the function $f(\cdot)$ is a function of t only and is independent of y, then numerical integration may be used to "solve" Equation 3.1:

$$\int_{y(a)}^{y(b)} dy = \int_{a}^{b} f(t) dt \tag{3.12}$$

If the integral on the right-hand side is represented by $I = \int_{a}^{b} f(t) dt$, then it is clear that

$$I = y(b) - y(a) \tag{3.13}$$

Thus, $y(b)$ may be considered as simply the integral I, if the initial condition $y(a) = 0$. In contrast, if $f(t, y)$ is a function of the dependent variable y as well, an ODE-IVP is solved.

The geometric interpretation of integration, discussed in Appendix E, is typically well known through the first course in high-school calculus. Integral, if we recall, is the area under a curve: When the function $f(t)$ is plotted against t, the area under the curve between $t = a$ and $t = b$ is the integral I. It bears repeating that in understanding the meaning of integral, the function $f(t)$ is plotted on the Y-axis.

In contrast, when solving ODE-IVP, the solution variable y is plotted on the Y-axis. The initial condition, (a, y_a), forms the starting point and $f(a)$ is the slope at this point. The information about the slope is used to determine the next point, $y(a + h)$. Thus, the geometric interpretation of solving ODE-IVP is to use the information about the slope $f(t)$ to determine how the solution will *propagate* along the y–t plot.

The next question is, what if the function in Equation 3.1 is a function of both t and y? Typically, this is when the phrase "solving an ODE-IVP" is invoked. One way of looking at

ODE-IVP is to consider it as a generalization when $f(\cdot)$ can be a function of both t and y. However, I like to consider ODE-IVP as a more general concept than this.

Consider the following example of ODE-IVP:

$$\frac{dy}{dt} = -t^2 y, \quad y(0) = 1 \tag{3.14}$$

The above equation can be solved analytically, and the initial condition used to obtain

$$y = e^{-t^3/3} \tag{3.15}$$

Note that depending on the initial condition, one gets a *family of curves* when one plots $y(t)$ vs. t. In contrast to integration, ODE-IVP is best understood when the solution, $y(t)$ (and not $f(\cdot)$) is plotted on the Y-axis, as shown in Figure 3.1.

The solution for the above ODE is plotted in Figure 3.1. Unlike integration, we are interested in "tracking" how the solution $y(t)$ evolves with time t, starting at the point (t_0, y_0). The final solution $y(t)$ depends on the initial condition, y_0. For different initial conditions y_0, a family of curves $y = \bar{c}e^{-t^3/3}$ is obtained in the y vs. t space. The solution in the figure is for the initial condition of Equation 3.14.

At $t = 0.5$, the dependent solution variable is $y(0.5) = 0.9592$. The slope of the curve $y(t)$ equals the function value $-t^2 y$, that is, -0.240. Thus, the right-hand side of the ODE (3.14) is nothing but the slope of the curve $y(t)$ at any point on the curve. Solution technique for ODE therefore involves retracing the solid line in Figure 3.1, given the slope $f(t, y)$ at any point in the $t-y$ space. The family of numerical methods discussed in this chapter attempts to use the information of the slope $f(t, y)$ and the projected slopes to predict how $y(t)$ will behave in the future. Another family of numerical methods use information from the past to predict the future $y(t)$.

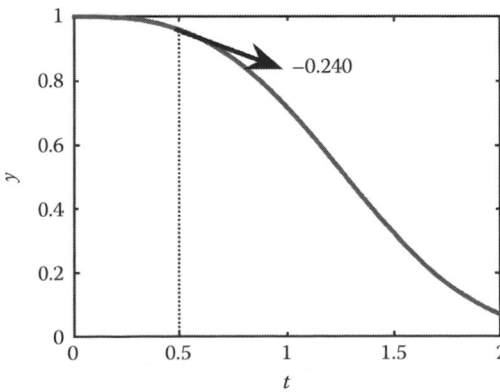

FIGURE 3.1   Solution to the ODE (3.14). The arrow indicates slope of the tangent at $t = 0.5$.

3.1.3 Euler's Explicit Method

The simplest numerical method for evaluating the ODE is Euler's method. Applying the forward difference approximation to y' on the right-hand side of Equation 3.1 yields

$$\frac{y_{i+1} - y_i}{h} = f(t_i, y_i) \tag{3.16}$$

Rearranging the above equation yields Euler's explicit method:

$$y_{i+1} = y_i + hf(t_i, y_i) \tag{3.17}$$

In the above, h is the *step-size* in the independent variable t. The step-size is chosen by users or internally by the algorithm to ensure stability and accuracy of the numerical technique. Starting with the initial values, (t_0, y_0), the above *expression* (3.17) is used recursively to obtain new values y_{i+1} and step forward in time. The code below shows implementation of Euler's explicit method for the ODE-IVP given in (3.14).

Example 3.1 Euler's Explicit Method

Problem: Solve the ODE-IVP given in Equation 3.14 using step-sizes of $h = 0.1, 0.05,$ 0.01 to obtain $y(2)$. Also compare the results with true solution.

    Solution: The code below uses Euler's method with step-size of $h = 0.01$.

```
%% Problem Setup
t0=0; tN=0.5;
y0=1;
h=0.01; N=(tN-t0)/h;
%% Initialize and Solve
t=[0:h:tN]';        % Initialize time vector
y=zeros(N+1,1);
y(1)=y0;            % Initialize solution
for i=1:N
    y(i+1)=y(i)+h*(-t(i)^2*y(i));
end
%% Output and Error
plot(t,y); hold on;
yTrue=exp(-tN^3/3);
err=abs(yTrue-y(end));
```

The result of the above code is shown in Figure 3.2.

    As can be observed from the above figure, the symbols get closer to the true solution as the step-size is decreased. The errors in computing $y(0.5)$ are $e_{0.1} = 0.011$, $e_{0.05} = 0.0057$, and $e_{0.01} = 0.0012$.

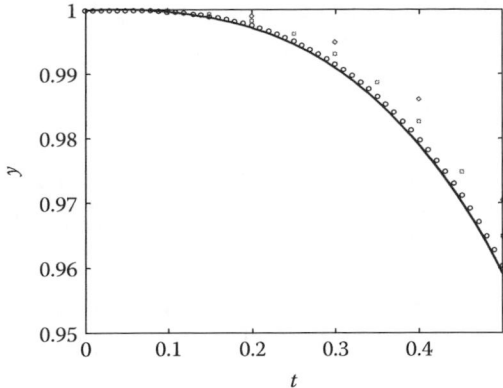

FIGURE 3.2 Comparison of the true solution to ODE-IVP of Equation 3.14 with numerical solutions using Euler's method with $h = 0.1$ (diamonds), $h = 0.05$ (squares), and $h = 0.01$ (circles).

Euler's explicit method can be easily derived from Taylor's series expansion of $y(t)$. Specifically, the expansion of $y(t_i + h)$ around $y(t_i)$ is given by

$$y(t_{i+1}) = y(t_i) + hy'(t_i) + \frac{h^2}{2!} y''(t_i) + \frac{h^3}{3!} y'''(t_i) + \cdots \tag{3.18}$$

Since $y'(t) = f(t, y)$, the above equation may be written as

$$y(t_{i+1}) = y(t_i) + hf(t_i, y_i) + \frac{h^2}{2!} f'(\xi, y(\xi)) \tag{3.19}$$

where $\xi \in [t_i, t_{i+1}]$. The first two terms form Euler's explicit method (3.17), whereas the last term gives the *local truncation error* (LTE) of Euler's explicit method:

$$E_{lte} = \frac{h^2}{2} f'(\xi, y(\xi)) \tag{3.20}$$

where ξ is some point between t_i and t_{i+1}.

The LTE is the error incurred in going from y_i to y_{i+1} numerically, *assuming* the value of y_i was known accurately. However, as shown in the previous example, multiple steps were taken, starting at $t = 0$ and using different step-sizes h to compute the solution $y(t)$ numerically. Any numerical method starting from t_0 and taking multiple steps to reach y_{i+1} will incur errors due to two reasons: error in the value of y_i itself and the error in going from $y_i \rightarrow y_{i+1}$. The net error in using multiple steps of ODE-IVP is known as the *global truncation error* (GTE). The interrelationship between LTE and GTE was introduced in Chapter 1.

The derivation of GTE of Euler's method will be skipped* for brevity. It can be shown that the GTE for Euler's method is

$$E \le \frac{M}{2K}\left(e^{Kt} - 1\right) \cdot h$$

$$K = \max\left\{\frac{\partial f(t,y)}{\partial t}\right\}, \quad M = \max\left\{\frac{df(t,y)}{dt}\right\}$$

(3.21)

The practical significance of Equation 3.21 is that the Euler's explicit method has accuracy of the order h to the power 1, represented in shorthand as $E \sim \mathcal{O}(h^1)$. This was observed in Example 3.1: The error reduced by half when the step-size was halved (from $\varepsilon = 0.011$ for $h = 0.1$ to 0.0057 $h = 0.05$), while the error fell by one order of magnitude (from 0.011 to 0.0012) when the step-size was reduced by one order of magnitude.

The GTE of Euler's method is $\sim \mathcal{O}(h^1)$, while its LTE is $\sim \mathcal{O}(h^2)$. In fact, this is a common trend that will be observed for all ODE solution techniques discussed in this book. If the LTE of the ODE solving method is $\sim \mathcal{O}(h^{n+1})$, its GTE will be $\sim \mathcal{O}(h^n)$. We will call such a method, with GTE $\sim \mathcal{O}(h^n)$, as an nth order method.

3.1.4 Euler's Implicit Method

Euler's explicit method was obtained in Equation 3.16 using the definition of y' as a forward difference formula. Instead, if a backward difference formula is used, then

$$\frac{y_i - y_{i-1}}{h} = f(t_i, y_i)$$

(3.22)

This can be rewritten as

$$y_i - y_{i-1} - hf(t_i, y_i) = 0$$

(3.23)

This equation needs to be solved at each time t_i to obtain the value of y_i. Since the solution y_i is *implicitly* known by solving (nonlinear) equation (3.23). The next example will demonstrate the use of Euler's implicit method.

Example 3.2 Euler's Implicit Method

Problem: Solve the ODE-IVP in Example 3.1 using Euler's implicit method.

Solution: In order to solve the ODE-IVP using Euler's implicit method, the first step is to substitute $f(\cdot)$ in Equation 3.23 to yield

$$y_i - y_{i-1} - h(t_i^2 y_i) = 0$$

---

* Refer to any standard numerical ODE techniques book for the derivation. The derivation of GTE for integration, which is pedagogically simpler, is presented in Appendix E for an interested reader.

Rearranging the above equation yields

$$y_i = \frac{1}{1 + ht_i^2} y_{i-1} \tag{3.24}$$

Knowing the value y_{i-1} and time t_i at which the solution value y_i is desired, the above expression can be used to solve the ODE-IVP using the following code.

```
%% Problem Setup
t0=0; tN=0.5;
y0=1;
h=0.1; N=(tN-t0)/h;
%% Initialize and Solve
y=zeros(N+1,1);
t=[0:h:tN]';      % Initialize time vector
y(1)=y0;          % Initialize solution
for i=2:N+1
    y(i)=y(i-1)/(1+h*t(i)^2);
end
%% Error calculation
yTrue=exp(-tN^3/3);
err=abs(yTrue-y(end));
```

The error in Euler's implicit method follows the same pattern as Euler's explicit method. The figure using Euler's implicit method is similar to that of Figure 3.2 and is skipped. The errors in computing $y(0.5)$ are $e_{0.1} = 0.012$, $e_{0.05} = 0.006$, and $e_{0.01} = 0.0012$.

Like the explicit method, Euler's implicit method also has LTE $\sim \mathcal{O}(h^{n+1})$ and GTE $\sim \mathcal{O}(h^n)$.

Before moving further, one point of implementation detail needs to be mentioned. Since Euler's method is rather simple, a comparison of codes in Examples 3.1 and 3.2 will be useful. One of the common errors among beginners with computational methods is related to "bookkeeping." Comparing with the two Euler's codes, the `for` loop is executed starting at $i=2$ (which corresponds to $t=1.h$). Note that according to Equation 3.23 for the implicit method, *past* information is used to determine the current value, y_i. Thus, `y(i)` is computed using `y(i-1)` in Example 3.2. On the other hand, in the explicit method, current information (i.e., t_i and y_i) is used to determine the future value y_{i+1}. Thus, the for loop starts at $i=1$, and the *current* information `y(i)` is used to generate the future values `y(i+1)`.

Although Euler's implicit method code could very well be written for `i=1:N`, I will try to keep the MATLAB® codes consistent with the equations shown in the overall derivation.

3.1.5 Stability and Step-Size

Convergence is an important property of a numerical technique. An ODE solution technique is said to be convergent if the numerical solution approaches the true solution value as the step-size $h \to 0$. In other words, $\lim\limits_{h \to 0} |y_n - y(t_n)| = 0$.

Stability is another important property. A numerical solution is said to be stable if the numerical solution y_n remains bounded as $n \to \infty$. Typically, there is a range of step-sizes, h, for which the numerical method for solving ODE is stable.*

3.1.5.1 Stability of Euler's Explicit Method

Euler's explicit method has a limit on the maximum value of step-size h that can be used; choosing a step-size larger than this value will result in an unstable solution. Consider the recursive expression (3.17) used in Euler's method

For the problem in Example 3.1, substituting the value of initial conditions in $f(t_0, y_0)$ yields

$$y_1 = \left(1 - h t_0^2\right) y_0$$

Further

$$y_2 = \left(1 - h t_1^2\right) y_1 = \left(1 - h t_1^2\right)\left(1 - h t_0^2\right) y_0$$

$$\vdots$$

$$y_n = \prod_{i=0}^{n-1} \left(1 - h t_i^2\right) y_0$$

The above equation is guaranteed to be stable if the following condition is satisfied:

$$\left|1 - h t^2\right| \leq 1$$

Since both h and t are positive, the above condition implies

$$h \leq \frac{2}{t^2} \tag{3.25}$$

With the end-time in Example 3.1 chosen as 10 s, run the code again. With $h = 0.1$, the value of $y(10)$ is obtained as $\sim 10^{17}$ because Euler's method is unstable. On the other hand, Euler's method correctly gives the value of $y(10) = 0$ with step-size of $h = 0.01$.

---

* Since all the numerical methods for solving ODE have their errors $\sim \mathcal{O}(h^n), n \geq 1$, these numerical techniques will be convergent if they are stable. Hence, there is a concept of *zero-stability*, which implies stability of the numerical technique when $h \to 0$.

The stability analysis can be formalized when $f(\cdot)$ is an explicit function of y only. As seen in case studies in Section 3.5 and other chapters, this is true of a majority of chemical engineering problems of interest. Before proceeding to the general case, consider the case when $f(y)$ is linear, that is, the ODE-IVP is

$$y' = -\lambda y \tag{3.26}$$

The analytical solution for this ODE is $y(t) = y_0 e^{-\lambda t}$, which is stable because $y(t)$ is bounded as $t \to \infty$. Substituting $f(y)$ in Equation 3.17

$$
\begin{aligned}
y_i &= y_i - h\lambda y_{i-1} = (1 - h\lambda) y_{i-1} \\
&= -h\lambda (-h\lambda y_{i-2}) \\
&\quad \vdots \\
\therefore y_i &= (-h\lambda)^i y_0
\end{aligned}
\tag{3.27}
$$

Using the above equation, the relation for y_{i-1} can be substituted as

$$
\begin{aligned}
y_i &= (1 - h\lambda)\left[(1 - h\lambda) y_{i-2}\right] \\
&= (1 - h\lambda)^2 y_{i-2}
\end{aligned}
\tag{3.28}
$$

Continuing this, it is easy to see that

$$y_i = (1 - h\lambda)^i y_0 \tag{3.29}$$

This is a convergent series if the term $(1 - h\lambda)$ lies between -1 and 1. In other words, Euler's method is stable if

$$\left|1 - \lambda h\right| \leq 1 \tag{3.30}$$

Analogous to Equation 3.25, the condition for stability is given by

$$h \leq \frac{2}{\lambda} \tag{3.31}$$

3.1.5.2 Error and Stability of Euler's Implicit Method

While a detailed analysis of error and stability will be skipped in this section, I draw your attention to the fact that the error in Euler's implicit method was similar to that obtained in the explicit method. Both implicit and explicit Euler's methods have a GTE $\sim \mathcal{O}(h^1)$.

Consider the stability of Euler's implicit scheme. Since both h and t_i are positive, the term $\left(1 + ht_i^2\right)^{-1}$ from Equation 3.24 is positive and less than 1. Therefore, Euler's implicit method is stable for any choice of h. Indeed, when Example 3.2 is run for $h = 0.1$ and $t_N = 10$, the implicit method is still stable. When the step-size is further increased to $h = 1$, the implicit method is still stable. Same results for stability of Euler's implicit method are obtained for the linear ODE case, $y' = -\lambda y$ as well. Any positive value of the step-size h will result in stable solution.

In contrast to the explicit methods, implicit methods are often "globally stable" for any range of values of step-size h. Even when the numerical methods are not globally stable, their stability margins are significantly broader than the corresponding explicit methods of equivalent accuracy. This shall be discussed further in Section II of this book. Specifically, Chapter 8 is devoted to implicit ODE solving techniques for a variety of problems.

In addition to determining the step-size, stability analysis has significant practical ramifications on the choice of numerical method for solving ODEs. This will be demonstrated toward the end of this chapter, in Section 3.5.4 (please see Example 3.15 as well). In case of ODEs in multiple variables, if the values of λ_i determine the step-size. The "speed" of evolution of the ODE is governed by the smallest eigenvalue (slowest mode), whereas the step-size is limited by the largest eigenvalue. If the ratio of the two is very large, an explicit ODE solver may not converge. For such systems, an implicit ODE solver is required to ensure good speed of convergence.

3.1.6 Multivariable ODE

Next, a multivariable case is considered, with $y \in \mathcal{R}^n$ as an n-dimensional vector. The reader is reminded that for the entire book, an n-dimensional vector will be a column vector (unless specified otherwise), that is, it will be an $n \times 1$ vector. Recall the discussion in Chapter 2, $\mathbf{f}: \mathcal{R}^n \rightarrow \mathcal{R}^n$, because it maps the $\mathcal{Y}$-space onto itself. Thus, $\mathbf{f}(t, \mathbf{y})$ is an $n \times 1$ vector of nonlinear functions of t and $\mathbf{y}$. With these definitions, extension to a multivariate case is straightforward:

$$\mathbf{y}_{i+1} = \mathbf{y}_i + \mathbf{f}\left(t_i, \mathbf{y}_i\right) \tag{3.32}$$

For a linear case, the ODE-IVP may be written as

$$\mathbf{y}' = -A\mathbf{y} \tag{3.33}$$

Hereafter, the boldface will be dropped for convenience.* Euler's method is written as

$$y_i = y_{i-1} - hAy_{i-1} = \left(I - Ah\right)y_{i-1} \tag{3.34}$$

---

* With some notable exceptions such as interpolation and integration, most of the methods discussed in this book are, in general, applicable to multivariable cases. The derivations will be done for single-variable case. Extension to multivariable cases will be clear from the context.

where I is an $n \times n$ identity matrix. Note that in a multivariable case, the order of multiplication is important. Continuing on the same lines

$$y_i = (I - Ah)^i y_0$$

From Chapter 2, we know that the above equation is stable if eigenvalues of the matrix $(I - Ah)$ are in the unit circle. If eigenvalues of A are $\{\lambda_1, \ldots, \lambda_n\}$, then the eigenvalues of $(I - Ah)$ are $\{(1 - \lambda_1 h), \ldots, (1 - \lambda_n h)\}$. Thus, following same arguments as before, the choice of h should be such that all eigenvalues of Ah are positive and less than 2, which is equivalent to saying that the step-size h should be chosen such that

$$0 \le h \le \frac{2}{|\lambda_i|}, \quad \forall i \tag{3.35}$$

3.1.6.1 Nonlinear Case
For a nonlinear ODE, it is possible to *estimate* the step-size required for stability by linearizing the system at the current point. Thus, defining

$$\lambda = -\left. \frac{\partial f}{\partial y} \right|_{t_i, y_i} \tag{3.36}$$

the original ODE (3.1) can be written as

$$\frac{d}{dt} \bar{y} \approx -\lambda \bar{y}$$

Thus, the choice of h for stability is given by

$$h \le \frac{2}{\left(-\left. \frac{\partial f}{\partial y} \right|_{t_i, y_i} \right)} \tag{3.37}$$

Likewise, in a multivariable case, the matrix A is replaced by the Jacobian

$$J = \begin{bmatrix} \dfrac{\partial f_1}{\partial y_1} & \dfrac{\partial f_1}{\partial y_2} & \cdots & \dfrac{\partial f_1}{\partial y_n} \\ \dfrac{\partial f_2}{\partial y_1} & \dfrac{\partial f_2}{\partial y_2} & \cdots & \dfrac{\partial f_2}{\partial y_n} \\ \vdots & \vdots & \ddots & \vdots \\ \dfrac{\partial f_n}{\partial y_1} & \dfrac{\partial f_n}{\partial y_2} & \cdots & \dfrac{\partial f_n}{\partial y_n} \end{bmatrix} \tag{3.38}$$

and the eigenvalues of $[J]$ are of interest in determining the step-size.

This concludes our rather long introduction to ODE-IVP.

3.2 SECOND-ORDER METHODS: A JOURNEY THROUGH THE WOODS

Euler's method is the simplest numerical technique to solve an ODE-IVP. The order of accuracy, given by the GTE, is of the order $\mathcal{O}(h^1)$. This makes Euler's method fairly inaccurate. As we witnessed in the example in the previous section, reducing the step-size by a factor improved the accuracy linearly by the same factor. This section introduces higher-order ODE methods for solving ODEs, which are more accurate than Euler's methods.

3.2.1 Some History

Euler's method was described by Leonhard Euler in ca. 1770s. An improvement to Euler's method was proposed by Karl Heun in late nineteenth century. The idea draws inspiration from the trapezoidal rule in integration. To compute y_{i+1}, Euler's explicit method uses the slope at the ith point, $f(t_i, y_i)$; Euler's implicit method uses the slope at $(i + 1)$th point, $f(t_{i+1}, y_{i+1})$. The accuracy can be improved if the average of the two slopes is used. Heun proposed that instead of using an average of the slopes at the two points, one could use the slope at i and the "projected slope" at $(i + 1)$. The projected point at $(i + 1)$ is

$$y_i + hf(t_i, y_i) \tag{3.39}$$

the slope at this point is

$$f\left(t_{i+1}, y_i + hf(t_i, y_i)\right) \tag{3.40}$$

and the improved Euler's method is given by

$$y_{i+1} = y_i + \frac{h}{2}\left(f(t_i, y_i) + f\left(t_{i+1}, y_i + hf(t_i, y_i)\right)\right) \tag{3.41}$$

This Heun's method is more accurate than Euler's method. Specifically, its GTE is $\sim \mathcal{O}(h^2)$, that is, it is a second-order accurate method. This shows that it is possible to use more information to improve the accuracy of numerical ODE solution methods. Due to the importance of differential equations in engineering and science, ODE solution techniques have received a lot of attention in the twentieth century. Various families of methods have been developed. This chapter will focus on explicit RK family of methods. RK methods use multiple points between i and $(i + 1)$ for projecting the slopes and use an average of slopes at these projected points. These methods are of varying degrees of accuracy and are abbreviated as RK-n, where n represents the order of GTE $\sim \mathcal{O}(h^n)$. Second-order RK methods (i.e., methods with GTE $\sim \mathcal{O}(h^2)$) are considered in this section for pedagogical purposes. Thereafter, extension to higher-order methods is discussed in Section 3.3.3.

Analysis of the improved Euler's method and interpretation in context of RK-2 method will be discussed in Section 3.2.2, whereas interpreting the improved Euler's method in the context of other methods is presented in Section 3.2.5.

3.2.2 Runge-Kutta (RK-2) Methods

The solution using RK-2 method marches forward using the following equation:

$$y_{i+1} = y_i + h\left(w_1 k_1 + w_2 k_2\right) \tag{3.42}$$

where

$$\begin{aligned} k_1 &= f\left(t_i, y_i\right) \\ k_2 &= f\left(t_i + ph, y_i + qh \cdot k_1\right) \end{aligned} \tag{3.43}$$

Notice that the *slope* $f(t_i, y_i)$ from Equation 3.17 is replaced by a weighted sum, $S = w_1 k_1 + w_2 k_2$, where the values k_1 and k_2 may be interpreted as slopes calculated at two different points. RK-2 methods involve function evaluations with (at least) two intermediate points. These function evaluations are represented as k_1 and k_2 in Equation 3.42.

Comparing with Equation 3.41, one can see that the improved Euler's method uses $p = q = 1$. As we will see presently, RK-2 is not a single method but a family of methods. The derivation of the formula and error analysis is discussed next.

3.2.2.1 Derivation for RK-2 Methods

Let y_i represent the current value of the dependent variable at time t_i. Starting with this value as initial condition, let $\overline{y}\left(t_{i+1}\right)$ represent the true solution. Using the Taylor's series expansion for the true solution around the current value y_i

$$\overline{y}_{i+1} = y_i + h\left.\frac{dy}{dt}\right|_i + \frac{h^2}{2!}\left.\frac{d^2 y}{dt^2}\right|_i + \frac{h^3}{3!}\left.\frac{d^3 y}{dt^3}\right|_i + \cdots \tag{3.44}$$

In the above equation, note that

$$\frac{dy}{dt} = f\left(t, y\right) \quad \text{and} \quad \frac{d^2 y}{dt^2} = \frac{df}{dt} = \frac{\partial f}{\partial t} + \frac{\partial f}{\partial y}\frac{dy}{dt}$$

Using this, Equation 3.44 may be written as

$$\overline{y}_{i+1} = y_i + hf^{(i)} + \frac{h^2}{2}\left(f_t^{(i)} + f_y^{(i)}f^{(i)}\right) + \frac{h^3}{6}\left.\frac{d^3 y}{dt^3}\right|^{(i)} + \cdots \tag{3.45}$$

Here, we have used the shorthand notation:

$$f^{(i)} \triangleq f\left(t_i, y_i\right), \quad f_t^{(i)} \triangleq \frac{\partial}{\partial t}f\left(t_i, y_i\right), \quad f_y^{(i)} \triangleq \frac{\partial}{\partial y}f\left(t_i, y_i\right)$$

Likewise, we will use Taylor's series expansion for Equation 3.43:

$$k_2 = f(t_i, y_i) + ph \frac{\partial}{\partial t} f(t_i, y_i) + qh \cdot k_1 \frac{\partial}{\partial y} f(t_i, y_i) + h^2 T \tag{3.46}$$

In the above expression, we have represented the terms in $\sigma(h^2)$ and higher as T. Rearranging the above and substituting in Equation 3.42, we get

$$y_{i+1} = y_i + hw_1 f^{(i)} + hw_2 \left[f^{(i)} + phf_t^{(i)} + qhf^{(i)} f_y^{(i)} + h^2 T \right] \tag{3.47}$$

$$y_{i+1} = y_i + h(w_1 + w_2) f^{(i)} + h^2 (w_2 p) f_t^{(i)} + h^2 (w_2 q) f_y^{(i)} + h^3 w_2 T \tag{3.48}$$

Comparing Equations 3.45 and 3.48, we get the following relations:

$$w_1 + w_2 = 1$$
$$w_2 p = w_2 q = \frac{1}{2} \tag{3.49}$$

$$\bar{y}_{i+1} - y_{i+1} = h^3 \left[\frac{1}{6} (y''')^{(i)} - w_2 T \right] \tag{3.50}$$

Equation 3.49 gives the relationship for choosing p, q, w_1, and w_2, whereas (3.50) shows that the *LTE* of RK-2 method is $\sim \mathcal{O}(h^3)$. Specifically, using the mean value theorem, it can be shown that the error in RK-2 methods is

$$\varepsilon_{RK2} = \frac{-h^3}{12} y'''(\xi, y(\xi)), \quad \xi \in [t_i, t_{i+1}] \tag{3.51}$$

3.2.2.2 Heun's Method

There are four unknowns and three equations in (3.49). Thus, one of the parameters can be chosen independently. As discussed earlier, RK-2 Heun's method uses $p = 1$. This gives $q = 1, w_1 = 0.5, w_2 = 0.5$. Thus, Heun's method may be written as

$$y_{i+1} = y_i + \frac{h}{2}(k_1 + k_2)$$
$$k_1 = f(t_i, y_i) \tag{3.52}$$
$$k_2 = f((t_i + h), (y_i + hk_1))$$

The GTE of an RK-2 method, such as Heun's, is $\mathcal{O}(h^2)$. Thus, when the step-size, h, is reduced by a factor of 10, the error reduces approximately by a factor of 100. Consequently, RK-2 methods are more accurate than Euler's explicit method, as demonstrated in the next example.

Example 3.3 RK-2 Heun's Method

Problem: Solve the problem of Example 3.1 using Heun's method in MATLAB.

Solution: The code for RK-2 Heun's method is given below. We make one improvement in our code compared to Example 3.1. We define a function myFun, which is used into Heun's solver to compute $f(t, y)$. MATLAB's *anonymous function* is used in this example (see Appendix A for a brief on MATLAB functions and scripts).

```
% ODE-IVP Using RK-2 Heun's Method
% Solve ODE of the type: y' = f(t,y)
%% Problem Setup
t0=0; tN=0.5;
y0=1;
h=0.01; N=(tN-t0)/h;
myFun=@(t,y) -t^2*y;
%% Initialize and Solve
t=[0:h:tN]';        % Initialize time vector
y=zeros(N+1,1);
y(1)=y0;            % Initialize solution
for i=1:N
    k1=myFun(t(i),y(i));
    y2=y(i)+h*k1;
    k2=myFun(t(i)+h,y2);
    y(i+1)=y(i) + h/2*(k1+k2);
end
```

The underlined command above defines the function myFun as an anonymous function. The two inputs, *t* and *y*, are defined by the function handle @(t,y): The function has two input arguments *t* and *y* and returns a single output $-t^2y$. Note that while *t* is a scalar, it is possible to have **y** as a vector (although in this example, *y* is scalar as well).

The above code is written for single-variable Heun's method with step-size $h = 0.5$. We can change the first line of the code and execute it again for a different step-size. Figure 3.3 compares the results obtained using two step-sizes: $h = 0.5$ and $h = 0.1$. It is clear from the figure that there are errors in the numerical solution of the ODE and that these errors are greater for the larger step-size of $h = 0.5$. The errors in computing $y(0.5)$ using the two step-sizes are $e_{0.5} = 0.022$ and $e_{0.1} = 8.2 \times 10^{-4}$.

Each step of Heun's method requires function $f(t, y)$ to be evaluated twice. Thus, with $h = 0.1$, there are 10 function evaluations required for obtaining $y(0.5)$. On the other hand, Euler's explicit method requires one function evaluation per step, so the same number of function evaluations are required when step-size of $h = 0.05$ is used in Euler's method. The corresponding error computed in Example 3.1 was 5.7×10^{-3}. Comparing with the

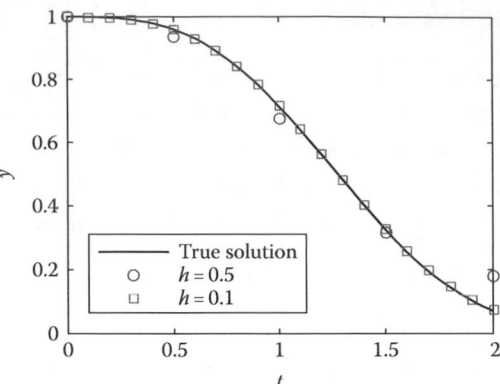

FIGURE 3.3   Comparison of Heun's method using $h = 0.5$ (circles) and $h = 0.1$ (squares) with the true solution (solid line). Smaller value of h results in higher accuracy.

error in Heun's method for $h = 0.1$ (Example 3.2) shows the power of using a higher-order method: The error is nearly one order of magnitude lower for the same number of function evaluations.

The geometric interpretation of Heun's method is as follows. The slope at (t_i, y_i) of Euler's method is replaced by a weighted sum: $h(k_1 + k_2)/2$, geometrically, an average of slopes k_1 and k_2. The former is nothing but the slope at (t_i, y_i). This slope is used to project a point at t_{i+1}. The value k_2 is the slope calculated at this projected point. Thus, Heun's method progresses using an average of the slope at (t_i, y_i) and a projection at t_{i+1} using the slope k_1.

3.2.2.3 Other RK-2 Methods

The derivation in Section 3.2.2.1 shows that *a family of RK-2 methods* can be obtained by choosing one of the parameters, such as the parameter p in the formulation. Another example of RK-2 methods include the *midpoint method* $(p = 0.5)$, which is given by

$$y^{i+1} = y^{(i)} + hk_2$$
$$k_1 = f\left(t^{(i)}, y^{(i)}\right) \tag{3.53}$$
$$k_2 = f\left(\left(t^{(i)} + \frac{h}{2}\right), \left(y^{(i)} + \frac{h}{2}k_1\right)\right)$$

Here, k_2 is the slope at the projected "midpoint" between $t^{(i)}$ and $t^{(i+1)}$. Projected value of y at that point is $(y^{(i)} + h/2\,k_1)$.

Likewise, the so-called *Ralston's method* is derived to minimize truncation error. In this method

$$p = q = 3/4, \quad w_1 = 1/3 \quad \text{and} \quad w_2 = 2/3 \tag{3.54}$$

John Butcher designed a convenient way to represent the coefficients of an RK method, now called the *Butcher tableau*. A simplified tableau for RK-2 is represented as shown below:

$$
\begin{array}{c|cc}
0 & & \\
p & q & \\
\hline
 & w_1 & w_2
\end{array}
$$

The bottom-most row contains weights of the (two) slopes, whereas the left-most row contains the parameter p. The Butcher's tableau will be revisited later in this section for general RK methods.

3.2.3 Step-Size Halving: Error Estimate for RK-2

The LTE for RK-2 methods was derived in the previous section and is shown in Equation 3.51. It is usually not possible to compute error, ε, mainly because the value of $f''(\xi, y(\xi))$ cannot be calculated easily. Since the true solution is not known, calculating ε does not seem to be straightforward. Hence, one needs to *estimate* what the truncation error is likely to be. This section is devoted to discussing ways to estimate the truncation error and use this information to improve the numerical ODE-IVP scheme. Although the information in this section is discussed for RK-2 methods, the principles are more general and applicable to other numerical schemes as well.

Consider again the second-order RK method for the ODE-IVP:

$$
y' = f(t, y), \quad y(t_i) = y_i \tag{3.55}
$$

starting at the point (t_i, y_i). LTE implies the error in numerically computing y_{i+1}, starting from this point, without accounting for the error accumulated in computing y_i itself. As before, if $\bar{y}_{i+1}$ represents the true solution of the ODE_IVP (3.55), then

$$
\bar{y}_{i+1} = y_{i+1} + ch^3 \tag{3.56}
$$

where, $c \sim -f'''(\xi)/12$ is the factor representing the error as per Equation 3.51. The error, ch^3, can be estimated if there is an alternate way to compute y_{i+1}. The first way to do so is to compute y_{i+1} with a smaller step-size. Alternatively, y_{i+1} may be computed using a method with different level of accuracy (such as RK-3 method, which has LTE of $\mathcal{O}(h^4)$).

Consider the case where y_{i+1} is also calculated in two steps with step-size of $h/2$. The same RK-2 method is employed. The first application of RK-2 method yields

$$
\bar{y}_{i+\frac{1}{2}} = y_{i+\frac{1}{2}} + c_1 \left(\frac{h}{2}\right)^3 \tag{3.57}
$$

In the second application, $\bar{y}_{i+\frac{1}{2}}$ is not known. Thus, the net error includes the LTE in the second application, as well as the effect of the first error:

$$\bar{y}_{i+1} \approx y_{i+1} + c_1 \left(\frac{h}{2}\right)^3 + c_2 \left(\frac{h}{2}\right)^3 \tag{3.58}$$

Note the use of approximation sign above. If the function $f(t,y)$ in Equation 3.55 is independent of y, then the above relationship is exact. Subtracting Equation 3.56 from Equation 3.58 and representing the difference between the solutions using two steps and a single step as $\hat{\Delta}_{i+1}$

$$0 = \hat{\Delta}_{i+1} + h^3 \left(\frac{c_1}{2^3} + \frac{c_2}{2^3} - c\right), \quad \hat{\Delta}_{i+1} = y_{i+1}\left[\frac{h}{2}\right] - y_{i+1}[h] \tag{3.59}$$

In the above, while $c \sim -f'''(\xi)/12$

$$c_1 \sim -f'''(\xi_1)/12 \left(with\, \xi_1 \in \left[t_i, t_{i+\frac{1}{2}}\right]\right) \quad and \quad c_2 \sim -f'''(\xi_2)/12 \left(with\, \xi_2 \in \left[t_{i+\frac{1}{2}}, t_{i+1}\right]\right)$$

If the step-size is reasonably small, then we can assume $c \approx c_1 \approx c_2$. Thus

$$0 \approx \hat{\Delta}_{i+1} - ch^3 \left(1 - \frac{1}{2^2}\right) \tag{3.60}$$

$$E_{i+1} \sim ch^3 \approx \hat{\Delta}_{i+1} \left(\frac{2^2}{2^2 - 1}\right) \tag{3.61}$$

This result can be generalized to any numerical scheme. Although the LTE was used in the above derivation for ease of explanation, the result is more appropriately defined in terms of GTE. If the *GTE* of a numerical scheme is $\sim \mathcal{O}(h^n)$, then the error may be estimated as

$$E_i \sim \hat{\Delta}_i \left(\frac{2^n}{2^n - 1}\right), \quad \hat{\Delta}_i = y_{i+1}\left[\frac{h}{2}\right] - y_{i+1}[h] \tag{3.62}$$

Example 3.4 Error Estimation for Heun's Method

Problem: Estimate the error in using Heun's method for the problem in Example 3.3 and compare it with the actual error.

Solution: In this problem, the code from Example 3.3 will be used to estimate the error using Equation 3.62 for $h = 0.1$.

First, the code is executed with the values of h=0.1; tN=0.1. This gives the solution $y_{0.1}[h]$. The result is stored in variable yOneStep. The code is executed again with the value of h=0.05, keeping tN the same. The solution obtained at $t = 0.1$ using this method is the value $y_{0.1}\left[\dfrac{h}{2}\right] \equiv$ yTwoStep. Next, the difference between the values calculated using two steps and a single step gives $\widehat{\Delta}_{0.1}$:

```
>> Delta=yTwoStep-yOneStep;
>> errEstimate=4*Delta/3
errEstimate =z
    1.6673e-04
```

Recall that the true value of $y(t) = e^{-t3/3}$. Thus, the *actual* error in computing $y_{0.1}[h]$ is

$$y_{\text{true}}(0.1) - y_{0.1}[h] = 1.6672 \times 10^{-4}$$

Notice that the error estimated in the above example is close to the true error. It should be remembered that the difference $\widehat{\Delta}_i$, as per Equation 3.83, estimates the error in the *single-step* implementation of RK-2 method. The next example demonstrates how the value of $\widehat{\Delta}_i$ can be used to obtain a more accurate approximation of the ODE solution.

3.2.4 Richardson's Extrapolation

Let me use the following example to demonstrate Richardson's extrapolation method to improve the accuracy of a numerical technique.

Example 3.5 Richardson's Extrapolation for Heun's Method

Problem: Use the values of $y_{0.1}[h], y_{0.1}\left[\dfrac{h}{2}\right]$, and $\widehat{\Delta}_{0.1}$ to get an improved solution for the ODE-IVP from Example 3.4.

Solution: According to Equation 3.61, $ch^3 \sim \dfrac{4}{3}\widehat{\Delta}_{0.1}$. Substituting this in Equation 3.56, an improved approximation of the ODE-IVP solution can be obtained as

$$\overline{y}_{0.1} \approx y_{0.1}[h] + \frac{4}{3}\widehat{\Delta}_{0.1}$$

Continuing where we left off in Example 3.4, an improved solution can be obtained as

```
>> yImproved=yOneStep+4*Delta/3
yImproved =
    0.9997
>> errNew=abs(yTrue-yImproved)
errNew =
    1.2156e-08
```

Note that the error in calculating $\bar{y}_{0.1}$ decreased significantly. This method of combining results from different step-sizes to obtain an improved numerical method is known as *Richardson's extrapolation*.

So, does using Richardson's extrapolation mean that there is no error in the improved numerical solution? As one might guess it, the answer to that question is no. It is important to notice that in Equation 3.61, we wrote that the error was approximately proportional to $\hat{\Delta}_{i+1}$. Richardson's extrapolation will serve to improve the order of accuracy, not eliminate the error.

Consider a numerical technique has $GTE \sim \mathcal{O}\left(h^n\right)$. As earlier, $y_t[h]$ will be used to represent the numerical solution obtained using step-size h at time t. Note that in addition to the leading error term, there are higher-order terms that have to be considered as well:

$$\bar{y}_t = y_t\left[h\right] + ch^n + dh^{n+1} \tag{3.63}$$

Likewise, a similar expression can be written with step-size $\dfrac{h}{2}$:

$$\bar{y}_t = y_t\left[\frac{h}{2}\right] + c\left(\frac{h}{2}\right)^n + d\left(\frac{h}{2}\right)^{n+1} \tag{3.64}$$

Multiplying the above equation by 2^n and subtracting

$$\left(2^n - 1\right)\bar{y}_t = 2^n y_t\left[\frac{h}{2}\right] - y_t\left[h\right] - \frac{d}{2}h^{n+1} \tag{3.65}$$

$$\bar{y}_t = \underbrace{\frac{2^n y_t\left[\dfrac{h}{2}\right] - y_t\left[h\right]}{2^n - 1}}_{y_{\text{improved}}} - \underbrace{\frac{d}{2\left(2^n - 1\right)}h^{n+1}}_{E \sim \mathcal{O}\left(h^{n+1}\right)} \tag{3.66}$$

Consider application of the above formula to Heun's method. Since $n = 2$ for Heun's method

$$y_{\text{improved}} = \frac{4y_t\left[\dfrac{h}{2}\right] - y_t\left[h\right]}{3}$$

$$= \frac{4}{3}\left(y_t\left[\frac{h}{2}\right] - y_t\left[h\right]\right) + y_t\left[h\right]$$

$$= y_t\left[h\right] + \frac{4}{3}\hat{\Delta}_t$$

This is the same formula that was used in Example 3.5. Richardson's extrapolation improves the accuracy of the numerical method to the next remaining term in the truncated series.

NOTE: RICHARDSON'S EXTRAPOLATION AND ERROR ESTIMATION

Consider the effort required in calculating $y_{0.1}$ using Richardson's extrapolation in Example 3.5. The function $f(t, y)$ is evaluated twice in computing $y_{0.1}[h]$ and four times in computing $y_{0.1}\left[\dfrac{h}{2}\right]$. Thus, six function evaluations are required for obtaining a GTE of $\mathcal{O}(h^3)$. In contrast, the best RK-3 method gives the same GTE with only three function evaluations. Thus Richardson's extrapolation is useful only in those cases where both error estimation and improvement in numerical scheme are simultaneously needed. It finds use in numerical differentiation and numerical integration.

Embedded RK method, discussed in the next section, is a preferred method for online error estimation for ODE-IVP in most commercial solvers.

Richardson's extrapolation is not just limited to ODE solution techniques. It may be used for any numerical method. In Problem 3.2, you'll use Richardson's extrapolation for trapezoidal rule in numerical integration and for numerical differentiation in Problem 3.3. Using a numerical method with accuracy $\sim \mathcal{O}(h^n)$, let χ_h be the numerical solution using step-size h and $\chi_{h/2}$ be the numerical solution using step-size $h/2$. The Richardson's extrapolation formula gives

$$\chi_{true} = \underbrace{\frac{2^n \chi_{h/2} - \chi_h}{2^n - 1}}_{\chi_{improved}} - \underbrace{\frac{d}{2(2^n - 1)} h^{n+1}}_{E \sim \mathcal{O}(h^{n+1})} \tag{3.67}$$

A final note is that if the numerical technique has error terms given by

$$\bar{\chi} = \chi_{num} + ch^n + dh^{n+2}$$

then the error from Richardson's extrapolation improves to $\mathcal{O}(h^{n+2})$.

3.2.5 Other Second-Order Methods (*)

Note: This section is included for completion and may be skipped without affecting the continuity of the discussion.

3.2.5.1 Trapezoidal Rule: An Implicit Second-Order Method

The improved Euler's method takes an average between slopes at ith point and projected slope at $(i + 1)$th point. Instead, if the actual slope at the endpoint were used, we will get an implicit trapezoidal rule:

$$y_{i+1} = y_i + \frac{h}{2}\left(f(t_i, y_i) + f(t_{i+1}, y_{i+1})\right) \tag{3.68}$$

This is an implicit, second-order method. It will be discussed again in Chapter 8, where implicit ODE methods and methods for solving differential algebraic equations (DAEs) are presented. More generally, it falls in the *Adams-Moulton family of methods* and may be said to be second-order AM-2 method. In RK methods, the projections on the "future" points are used; the previously computed points, $y_{i-1}, y_{i-2}, \ldots$, are ignored. The Adams-Moulton method fits a polynomial on the past points to compute

$$y_{i+1} = y_i + h\left(af_{i+1} + \left(b_1 f_i + b_2 f_{i-1} + \cdots + b_n f_{i-n+1}\right)\right)$$

Since the right-hand side depends on y_{i+1}, this is an implicit family of methods

3.2.5.2 Second-Order Adams-Bashforth Methods

The idea of Adams-Bashforth methods is like the Adams-Moulton method, except that they are explicit. Thus, $a = 0$ and Adams-Bashforth method is given by

$$y_{i+1} = y_i + h\left(b_1 f_i + b_2 f_{i-1} + \cdots + b_n f_{i-n+1}\right)$$

The second-order AB-2 method is given by

$$y_{i+1} = y_i + \frac{3h}{2} f_i - \frac{h}{2} f_{i-1} \tag{3.69}$$

Note that AB-1 method is nothing but Euler's method.

3.2.5.3 Predictor-Corrector Methods

The improved Euler's method was originally proposed by Wilhelm Heun as a "predictor-corrector" method. The explicit Euler's method

$$y_{i+1}^{\text{pred}} = y_i + hf\left(t_i, y_i\right)$$

is used to predict the next point, and a higher-order method

$$y_{i+1} = y_i + \frac{h}{2}\left(f\left(t_i, y_i\right) + f\left(t_{i+1}, y_{i+1}^{\text{pred}}\right)\right)$$

is used to improve this prediction. In general, AB_{n-1} is used as a predictor and AM_n is used as a corrector. This is done not merely to improve the accuracy of the numerical method (AB_n can instead be used if accuracy was the only concern); the predictor-corrector methods have improved stability properties than explicit AB methods but are still explicit, unlike AM methods.

MATLAB solver `ode115` uses Adam's predictor-corrector methods of varying orders, from $\mathcal{O}(h^1)$ to $\mathcal{O}(h^{15})$. This is an *explicit* solver.

3.2.5.4 Backward Differentiation Formulae

The idea of backward differentiation formulae (BDFs) is to use the past information much the same way as AM methods. However, instead of fitting polynomial to past values of f_{i-j}, the past values of the solution are used instead to improve the order of accuracy:

$$y_{i+1} = \left(c_1 y_i + c_2 y_{i-1} + \cdots + c_n y_{i-n+1} \right) + ahf \left(t_{i+1}, y_{i+1} \right)$$

The second-order BDF formula is written as

$$y_{i+1} = \frac{4}{3} y_i - \frac{1}{3} y_{i-1} + \frac{2h}{3} f \left(t_{i+1}, y_{i+1} \right) \tag{3.70}$$

MATLAB solver `ode15s` uses first- to fifth-order BDF method. It is the second most popular solver in MATLAB and is useful for solving stiff ODE-IVP problems, as well as some types of DAE problems. The need for a stiff solver like `ode15s` is introduced in a case study in Section 3.5.4 and the implicit methods are discussed in more detail in Chapter 8.

3.3 HIGHER-ORDER RUNGE-KUTTA METHODS

The previous section focused on second-order methods. The RK methods are popular for solving ODE-IVP because higher-order methods with better accuracy are available; they are relatively easy to code being explicit methods; and they allow adaptive step-sizing, where the step-size h is varied to meet the error tolerances. Higher-order RK methods will be discussed in this section.

3.3.1 Explicit Runge-Kutta Methods: Generalization

The key idea behind RK methods is to replace $f(t, y)$ for computation of y_{i+1} with a *weighted slope* calculated at multiple points, that is

$$y_{i+1} = y_i + h \cdot S \left(t_i, y_i \right) \tag{3.71}$$

where $S(.)$ is a term that can be explicitly calculated from known quantities t_i and y_i. Higher-order RK methods include more terms based on the projection of $f(t, y)$. A generic RK method that involves m-terms may be written as

$$y_{i+1} = y_i + \left(w_1 k_1 + w_2 k_2 + \cdots + w_m k_m \right) \tag{3.72}$$

where

$$\begin{aligned}
k_1 &= f \left(t_i, y_i \right) \\
k_2 &= f \left(\left(t_i + p_2 h \right), \left(y_i + q_{21} h k_1 \right) \right) \\
k_3 &= f \left(\left(t_i + p_3 h \right), \left(y_i + q_{31} h k_1 + q_{32} h k_2 \right) \right) \\
&\cdots
\end{aligned} \tag{3.73}$$

The first term, k_1, is just the slope at ith point; the second term, k_2, involves coefficients p_2 and q_{21}. Thus, k_2 is a slope computed at some projected point between i and $(i+1)$. The terms in subsequent stages use the slopes, k_j, computed so far. A compact way of representing an RK method with m terms is

$$y_{i+1} = y_i + \sum_{j=1}^{m} w_j k_j \qquad (3.74)$$

$$k_j = f\left(\left(t_i + p_j h\right), y_i + \left(h \sum_{\ell=1}^{j-1} q_{j\ell} k_\ell\right)\right) \qquad (3.75)$$

The Butcher's tableau is a convenient way of representing all the weights for RK and similar family of methods. The general Butcher's tableau is written as

p_1	q_{11}	$\cdots$	$\cdots$		
p_2	q_{21}	q_{22}	$\cdots$		
$\vdots$	$\vdots$	$\vdots$	$\ddots$	$\ddots$	$\vdots$
p_m	q_{m1}	q_{m2}	$\cdots$	$q_{m,m-1}$	$q_{m,m}$
	w_1	w_2	$\cdots$	w_{m-2}	w_m

or the shorthand notation

$$\begin{array}{c|c} p & Q \\ \hline & w \end{array}$$

Note that in RK family of methods, the first row is zero, since the first term k_1 is simply the slope $f(t_i, y_i)$. For explicit methods, all the diagonal elements and superdiagonal elements in Q are zero. This is because, as seen in Equation 3.75, the summation is only up to $(j-1)$. Thus, k_2 depends on h, t_i, y_i, k_1 but not on itself and subsequent terms; k_3 depends on h, t_i, y_i, k_1, k_2 but not on itself and subsequent terms; and so on.

The Butcher's Tableau for *explicit* RK methods using m terms is given by

0					
p_2	q_{21}				
$\vdots$	$\vdots$				
p_m	q_{m1}	q_{m2}	$\cdots$	$q_{m,m-1}$	
	w	w	$\cdots$	w_{m-1}	w_m

Note that all the diagonal elements, $q_{11}, q_{22}, \ldots$ are zero, making these explicit ODE methods.

NOTE: SOME INTERESTING INFORMATION

The Butcher tableau became a standard way of representing coefficients of numerical ODE solvers in the last few decades, after John Butcher published his book on differential equations. The Butcher tableau for explicit RK methods has a strictly lower triangular structure, with diagonal and superdiagonal elements as zero.

However, there exists a class of *implicit* RK methods that have the diagonal elements in the matrix Q as nonzero. Methods that have nonzero superdiagonal elements may not be categorized as RK methods.

The so-called Lobatto and Radau methods are fully implicit methods, where the entire Q matrix may contain nonzero elements. These are not RK-type methods but instead have roots in quadrature and collocation formulae. Interestingly, these methods were proposed *before* RK methods. In fact, Reheul Lobatto died in 1866, 1 year before Martin Kutta was born! It is only recently that Lobatto and Radau methods are expressed in terms of the Butcher tableau.

The explicit RK methods with m stages are of the form of Equations 3.74 and 3.75. The number of stages m is nothing but the number of terms k_i that are used in solving the ODE-IVP. The Butcher's tableau for explicit RK methods is such that its first row is zero ($p_1 = 0, q_{1\ell} = 0$) and that Q matrix is strictly lower triangular (i.e., all the diagonal elements $q_{\ell\ell}$ and superdiagonal elements are zero). On the other hand, the order of an RK method is the order of GTE; thus, the nth order RK method has GTE $\sim \mathcal{O}(h^n)$. One interesting result for all explicit RK methods is that the order of the RK method does not exceed the number of stages, that is, $n \leq m$. This means that the RK method with four terms can at most be $\mathcal{O}(h^4)$ accurate. There is an upper limit on the order of the RK method with m terms. However, it is possible to choose the weights p, Q, and w such that the order of accuracy is lower. The standard RK-3 method is written as

$$y_{i+1} = y_i + \frac{h}{6}\left(k_1 + 4k_2 + k_3\right) \tag{3.76}$$

where

$$k_1 = f(t_i, y_i), \quad k_2 = f\left(t_i + \frac{h}{2}, y_i + \frac{hk_1}{2}\right), \quad k_3 = f(t_i + h, y_i - hk_1 + 2hk_2)$$

The Butcher tableau for this RK-3 method is

0			
0.5	0.5		
1	−1	2	
	$\frac{1}{6}$	$\frac{4}{6}$	$\frac{1}{6}$

This method has LTE of $\sim \mathcal{O}\left(h^4\right)$ and a GTE of $\sim \mathcal{O}\left(h^3\right)$. Just as we had seen in RK-2 methods, it is possible to choose different values of weights p, Q, and w to obtain various RK-3 methods involving three stages. The following example presents a MATLAB code for using standard RK-3 for solving ODE-IVP.

Example 3.6 Standard RK-3 Method

Problem: Use the standard RK-3 method of Equation 3.76 to solve the ODE of Example 3.3.

Solution: The code for Heun's method from Example 3.3 is modified as below to implement the standard RK-3 method. As before, the anonymous function is used in the code below.

```
% ODE-IVP Using Standard RK-3 Method
% Solve ODE of the type: y' = f(t,y)
%% Problem Setup
t0=0;      tN=0.5;
h=0.01;    N=(t3.t0)/h;
y0=1;
myFun=@(t,y) -t^2*y;
%% Initialize and Solve
t=[0:h:tN]';      % Initialize time vector
y=zeros(N+1,1);
y(1)=y0;          % Initialize solution
for i=1:N
    k1=myFun(t(i),y(i));
    t2=t(i)+h/2;
    y2=y(i)+h/2*k1;
    k2=myFun(t2,y2);
    t3=t(i)+h;
    y3=y(i)-h*k1+2*h*k2;
    k3=myFun(t3,y3);
    y(i+1)=y(i) + h/6*(k1+4*k2+k3);
end
%% True solution and error
yTrue=exp(-t.^3/3);
err=abs(yTrue-y);
disp([h,err(end)]);
```

The above code uses RK-3 method to solve the ODE-IVP. The following results are obtained when the code is run with various step-sizes.

h	err
0.1	6.6908e-06
0.05	8.3830e-07
0.01	6.6685e-09

When the step-size is halved from 0.1 to 0.05, the error reduces by a factor of 8; whereas, when the step-size is reduced by one order of magnitude, the error reduces by three orders of magnitude. This is because RK-3 is a $\mathcal{O}\left(h^3\right)$ method.

It should be noted that the *actual* error will also depend on $f''''(\xi)$ terms. So, the actual change in error may not be exactly $(h_1/h_2)^n$; this is just an indicator of how the numerical method behaves in a qualitative (and semiquantitative) way.

Similarly, higher-order RK methods have also been derived. Fourth-order RK methods will be discussed in a separate Section 3.3.3 due to their popularity in ODE solving.

3.3.2 Error Estimation and Embedded RK Methods

I mentioned in the previous section that two ways of estimating errors in RK methods are to use step-size halving (Section 3.2.3) or to use two different RK methods of differing accuracy. The latter method is discussed here.

Let us consider the RK-2 midpoint method

$$y_{i+1}^{RK2} = y_i + hk_2 \tag{3.77}$$

$$k_1 = f\left(t_i, y_i\right), \quad k_2 = f\left(t_i + \frac{h}{2}, y_i + \frac{h}{2}k_1\right)$$

is used to solve the ODE-IVP. The solution is related to the true value as

$$\bar{y}_{i+1} = y_{i+1}^{RK2}\left[h\right] + ch^3 \tag{3.56}$$

since the LTE of RK-2 method is $\sim \mathcal{O}\left(h^3\right)$. Third-order RK-3 methods have a greater accuracy than RK-2 methods. The solution using RK-3 method (employing three function evaluations[*]) was represented as

$$y_{i+1}^{RK3} = y_i + h\left(w_1 k_1 + w_2 k_2 + w_3 k_3\right)$$

and the numerical solution of the RK-3 method is related to the true solution as

$$\bar{y}_{i+1} = y_{i+1}^{RK3}\left[h\right] + dh^4 \tag{3.78}$$

Note that unlike Examples 3.3 and 3.6, we are considering LTE in the above equations. The difference between the two is given by

$$0 = \underbrace{y_{i+1}^{RK3}\left[h\right] - y_{i+1}^{RK2}\left[h\right]}_{\Delta_{i+1}} + dh^4 - ch^3$$

---

[*] An RK-3 method requires at least three function evaluations. It is possible to derive RK-3 methods with more than three function evaluations as well. I have kept this discussion simple with three function evaluations only, though the arguments are generalized to any higher-order method.

Thus, noting that the leading error term is ch^3, the above equation is approximated as

$$ch^3 \approx \Delta_{i+1} \equiv E_{i+1} \tag{3.79}$$

Thus, the difference between the single application of RK-3 and RK-2 methods provides an estimate of the error for the RK-2 method. This is demonstrated in the next example, where the error in using RK-2 midpoint method will be estimated. We will use the standard RK-3 method of Example 3.6:

$$y_{i+1}^{RK3} = y_i + \frac{h}{6}\left(k_1 + 4k_2 + k_3\right) \tag{3.76}$$

where

$$k_1 = f\left(t_i, y_i\right), \quad k_2 = f\left(t_i + \frac{h}{2}, y_i + \frac{hk_1}{2}\right), \quad k_3 = f\left(t_i + h, y_i - hk_1 + 2hk_2\right)$$

The following example uses the RK-2 and RK-3 methods together to estimate the *LTE* in RK-2.

Example 3.7 Error Estimation for RK-2

Problem: To estimate the error in RK-2 method using single implementation of RK-3.

Solution: The code for Heun's method from Example 3.3 is modified as below. The term yMP calculates y_{i+1} using RK-2 midpoint method, whereas the term z calculates y_{i+1}^{RK3} using Equation 3.88. The difference is captured in the variable delta. The code for single use of RK-2 midpoint method along with error estimate is given below.

```
% Define problem conditions
h = 0.1;
t0 = 0;
y0 = 1;
% Calculate intermediate points
k1 = -t0*t0*y0;
t2 = t0+h/2;
y2 = y0+h/2*k1;
k2 = -t2*t2*y2;
t3 = t0+h;
y3 = y0-h*k1+2*h*k2;
k3 = -t3*t3*y3;
% Midpoint method solution
yMP = y0+h*k2;
% Error estimate
```

```
z   = y0 + h/6*(k1+4*k2+k3);
del = abs(z-yMP);
err = abs(exp(-h^3/3) - yMP);
```

The above code is executed for various values of *h* and the results are tabulated below:

```
h       del          err
0.02    6.667e-7     6.667e-7
0.1     8.325e-5     8.328e-5
0.5     9.115e-3     9.561e-3
```

It is clear from the above table that Δ_1 is a reasonable approximation of local error E_1.

Thus, using RK methods of different orders to estimate the numerical error is the first key idea. Notice an important factor in the above example: The terms k_i used in the RK-2 and RK-3 methods are the same. In fact, we could write the RK-2 method as

$$y_{i+1}^{\text{RK2}} = y_i + h\left(0k_1 + 1k_2 + 0k_3\right)$$

to represent that the two methods use the same terms with different weights. The *extended Butcher tableau* is written as

$$
\begin{array}{c|ccc}
0 & & & \\
0.5 & 0.5 & & \\
1 & -1 & 2 & \\
\hline
& 0 & 1 & 1 \\
& \dfrac{1}{6} & \dfrac{4}{6} & \dfrac{1}{6}
\end{array}
$$

Generalizing this, we can write it as

$$
\begin{array}{c|c}
p & Q \\
\hline
& w^{\text{RK}n} \\
& w^{\text{RK}(n+1)}
\end{array}
$$

The above is the second key idea behind *embedded RK* methods.

The third key idea is that error is proportional to the *n*th power of *h* for an RK-*n* method. If the RK-*n* method is used with two different step-sizes, *h* and h_{new}, then as per Equation 3.61

$$\frac{E_{\text{new}}}{E} = \frac{\left(h_{\text{new}}\right)^{n+1}}{\left(h\right)^{n+1}} \tag{3.80}$$

where the LTE of the method is $\mathcal{O}\left(h^{n+1}\right)$. Recall that in Equation 3.79, the RK-2 method has LTE $\sim \mathcal{O}\left(h^3\right)$, the RK-3 method has LTE $\sim \mathcal{O}\left(h^4\right)$ and their difference, $\Delta \sim ch^3$. This difference, Δ, between the solutions from two different RK methods is approximately equal to the error E of the lower order RK method, as seen in Equation 3.79.

Let us say that the desired accuracy of the numerical method was $E_{new} = \varepsilon_{tol}$. Then, from Equation 3.80

$$h_{new} \sim h\left(\frac{\varepsilon_{tol}}{\Delta}\right)^{1/n} \tag{3.81}$$

is designed to give the error tolerance desired from the ODE solver.

Example 3.8 Step-Size Adaptation in RK-2 Method

Here, let us combine the results from Example 3.7 to improve the error performance of RK-2 midpoint method using Equation 3.81.

The value of y_{MP}, z and their difference Δ was computed in Example 3.7 for $h = 0.5$ as $y_{MP} = 0.9688$, $z = 0.9596$, $|\Delta| = 9.115e-3$. Let us say that the desired error tolerance was $\varepsilon_{tol} = 10^{-5}$. From Example 3.7, we also know that the error for $h = 0.02$ was 6.7×10^{-7} and that using $h = 0.1$ was 8.3×10^{-5}. So, the step-size that gives error of 10^{-5} would lie somewhere between these two values of step-sizes. Using Equation 3.81, we get

$$h_{new} = 0.5\left(\frac{10^{-5}}{9.115 \times 10^{-3}}\right)^{1/3}$$

which yields us $h_{new} = 0.0516$. Let us now run RK-2 midpoint method with this step-size. To do so, the code in Example 3.7 was executed again with h=0.0516. With no other change in the code:

```
err =
    1.1448e-05
```

This indicates that Equation 3.81 may be used to determine the step-size that gives approximately the desired error.

3.3.2.1 MATLAB® Solver ode23

MATLAB has a variety of solvers in their ODE suite. MATLAB has a solver ode23 that uses a combination of second- and third-order RK methods with adaptive step-sizing.

It is an explicit RK solver with four stages. The Butcher tableau for this *Bogacki-Shampine* method is

0				
0.5	0.5			
0.75	0	0.75		
1	2/9	1/3	4/9	
	2/9	1/3	4/9	
	7/24	1/4	1/3	1/8

(3.82)

The interested reader is referred to the paper by Shampine and coworkers for more information on this and other solvers in MATLAB. The `ode23` solver has another interesting property called "*first same as last*" (FSAL for short). The coefficients used for calculation of k_4 are the same as the weights w_{RKn}. In other words, the projected value of $k_4(t_i, y_i)$ calculated at the ith step is the same as $f(t_{i+1}, y_{i+1})$. Thus, one does not need to calculate k_1 at each step; k_1 at the $(i+1)$th step equals the value of k_4 calculated at the previous step.

The adaptive step-size RK solvers work in a similar manner as described in this section. The step-size that was determined from the previous step is used to compute the solution y_{i+1} at the next step using the first RK method (from extended Butcher's tableau). Using the second RK method from the extended Butcher's tableau, the value Δ (which is $\Delta = z - y_{i+1}$) is computed. The new step-size is computed using Equation 3.81. This forms the step-size for the next step. The same procedure is repeated at each point in computing the trajectory of $y(t)$.

The procedure is slightly different for the first step. The method starts with some (possibly arbitrary) estimate of the step-size and uses Equation 3.81 to iteratively determine an appropriate step-size. Once determined, this step-size is used in computing y_1; thereafter the method proceeds as explained before.

Writing a code for adaptive step-size RK method will be skipped in this chapter.

3.3.3 The Workhorse: Fourth-Order Runge-Kutta

In this chapter, I have presented first-, second-, and third-order RK methods. The Bogacki-Shampine method was a four-stage RK method. With different values of weights, RK-2 and RK-3 methods were obtained. The best order of accuracy for a four-stage RK method is $\sim \mathcal{O}(h^4)$. The overall equation for a single step using four intermediate points is given by

$$y_{i+1} = y_i + h\left(w_1 k_1 + w_2 k_2 + w_3 k_3 + w_4 k_4\right)$$

(3.83)

where

$$k_1 = f(t_i, y_i), k_m = f\left((t_i + p_m h), \left(y_i + h \sum_{j=1}^{m-1} q_{mj} k_j\right)\right) \qquad (3.84)$$

For all explicit RK methods, the order of the RK method does not exceed the number of stages. The highest order of accuracy possible for $m = 2$ stages is $n = 2$; for $m = 3$ stages is $n = 3$; and for $m = 4$ stages is $n = 4$. This means that RK method with four terms can at most be $\mathcal{O}(h^4)$ accurate; it is not possible to have $\mathcal{O}(h^5)$ or higher accuracy with four stages. However, the interesting fact is that the highest order of accuracy possible for any $m = 5$ stage RK method is still $\mathcal{O}(h^4)$; the minimum number of terms required for $\mathcal{O}(h^5)$ RK-5 method is $m = 6$. This is an interesting result; it also happens to be the reason for popularity of RK-4 methods. The popularity of RK-4 methods is because including the fifth term k_5 does not improve the order of accuracy of RK method compared to the best four-stage RK method.

3.3.3.1 Classical RK-4 Method(s)

The classical RK-4 method is given by the following formula:

$$y_{i+1} = y_i + \frac{h}{6}(k_1 + 2k_2 + 2k_3 + k_4) + \mathcal{O}(h^5) \qquad (3.85)$$

where

$$k_2 = f\left(t_i + \frac{h}{2}, y_i + \frac{h}{2}k_1\right), \quad k_3 = f\left(t_i + \frac{h}{2}, y_i + \frac{h}{2}k_2\right), \quad k_4 = f(t_i + h, y_i + hk_3)$$

Thus, the Butcher tableau for RK-4 classical method is given by

0				
$1/2$	$1/2$			
$1/2$	0	$1/2$		
1	0	0	1	
	$1/6$	$1/3$	$1/3$	$1/6$

The classical RK-4 method uses four intermediate stages and yields a method that has a global accuracy of $\mathcal{O}\left(h^4\right)$.

3.3.3.2 Kutta's 3/8th Rule RK-4 Method

Several other RK-4 methods can also be derived. Another popular method is that from Kutta, proposed in the same paper as the classical RK-4 method. This method is similar to Simpson's 3/8th rule for integration (see Appendix E for discussion on integration formulae), if the function f is function of t only. The formula is given by

$$y_{i+1} = y_i + \frac{h}{8}\left(k_1 + 3k_2 + 3k_3 + k_4\right) + \mathcal{O}\left(h^5\right) \tag{3.86}$$

The Butcher's tableau for this method is

0				
$1/3$	$1/3$			
$2/3$	$-1/3$	1		
1	1	-1	1	
	$1/8$	$3/8$	$3/8$	$1/8$

Since the previous *method* is called *classical RK-4*, this RK-4 method is often called 3/8th rule RK-4 method.

The classical RK-4 method was described originally in their seminal work by Carl Runge and Martin Kutta. The method is rather elegant and easy to implement. When numerical computing started becoming popular in the first half of the twentieth century, RK-4 became an ODE-IVP method of choice. An additional term did not give any advantage because the best accuracy from a five-stage RK method still remains $\mathcal{O}\left(h^4\right)$. Later, when computational speeds improved, the $\mathcal{O}\left(h^4\right)$ method with *adaptive step-sizing* provided sufficiently high accuracy that RK-4 remained the most popular explicit ODE-IVP technique.

3.4 MATLAB® ODE45 SOLVER: OPTIONS AND PARAMETERIZATION

MATLAB's default ODE-IVP solver is `ode45`, which uses Dormand Prince (RK-DP) method. This uses a combination of fifth- and fourth-order methods. The fifth-order method gives the value of the differential variable, y_{i+1}, at the next time t_{i+1}. Based on the discussion in the previous section, it is clear that the minimum number of intermediate stages required for a fifth-order RK method is six. Since this is an embedded RK method, the same function evaluations will be used by the fourth-order RK method as well.

The fourth-order method is used for error control. The extended Butcher tableau for this method is given by

0							
1/5	1/5						
3/10	3/40	9/40					
4/5	44/45	−56/15	32/9				
8/9	19372/6561	−25360/2187	64448/6561	−212/729			
1	9017/3168	−355/33	46732/5247	49/176	−5103/18656		
1	35/384	0	500/1113	125/192	−2187/6784	11/84	
	35/384	0	500/1113	125/192	−2187/6784	11/84	
	5179/57600	0	7571/16695	393/640	−92097/33920	187/2100	1/40

The first set of weights give the fifth-order method. Thus, the next value of y_{i+1} is given by

$$y_{i+1} = y_i + h\left(\frac{35}{384}k_1 + \frac{500}{1113}k_3 + \frac{125}{192}k_4 - \frac{2187}{6784}k_5 + \frac{11}{84}k_6\right) \tag{3.87}$$

Since this is an embedded RK method, the same function evaluations are used in calculation of the fourth-order solution. Additionally, since k_7 is not used in computing y_{i+1} as per Equation 3.87, it is computed as follows:

$$\bar{\zeta}_{i+1}^{(7)} = y_i + h\left(\frac{35}{384}k_1 + \frac{500}{1113}k_3 + \frac{125}{192}k_4 - \frac{2187}{6784}k_5 + \frac{11}{84}k_6\right)$$
$$k_7 = f\left(t_{i+1}, \bar{\zeta}_{i+1}^{(7)}\right) \tag{3.88}$$

The fourth-order embedded method gives

$$\bar{y}_{i+1}^{(RK4)} = y_i + h\left(\frac{5179}{57600}k_1 + \frac{7571}{16695}k_3 + \frac{393}{640}k_4 - \frac{92097}{339200}k_5 + \frac{187}{2100}k_6 + \frac{1}{40}k_7\right) \tag{3.89}$$

The difference

$$\Delta = y_{i+1} - \bar{y}_{i+1}^{(RK4)} \tag{3.90}$$

provides error estimate.

Additionally, RK-DP method is a *first same as last* (FSAL) method. Notice that the coefficients for computing k_7 on the last row of q weights are the same as the coefficients used in Equation 3.87. In other words, since the coefficients used in Equation 3.87 to compute y_{i+1} are the same as the coefficients used for computing k_7 as per Equation 3.88, the function value k_7 computed in the current step is the same as the first function evaluation to compute k_1 in the next step. Thus, one additional function evaluation is avoided due to the FSAL property. Thus, at each step, because k_1 is nothing, but k_7 of the previous step, six function evaluations (to compute $k_2, \ldots, k_7$) are required for each step of RK-DP method.

The case study Section 3.5 will extensively use ode45 solver for solving ODE-IVP. Before that, we end this section with the following brief example that shows application of ode45. All the ODE methods in MATLAB follow the same format. They require a function that takes scalar-valued t and vector **y** as input argument and return $f(t, y)$ as the output. Function **f** should be of the same dimension as the vector **y**. The syntax for ode45 is

```
[tSol,ySol]=ode45(@(t,y) myFun(t,y), tSpan, y0);
```

where tSol is the time vector at which the solutions are obtained, ySol are the solutions $y(t)$, tSpan gives the time span for which the solution is desired, and y0 is the initial condition at time tSpan(1). The matrix ySol has as many rows as the number of time-points in tSol and as many columns as the dimension of vector **y**.

The next example demonstrates use of ode45 for a 1D problem.

Example 3.9 Using ode45 for Solving Example 3.3

The initial condition and function was already defined in earlier examples as

```
% Conditions for solving the ODE
t0=0;    tN=0.5;
y0=1;
myFun=@(t,y)  -t^2*y;
```

Note that since the solution is desired in the range $[t_0, t_N]$, this forms the vector tSpan. Thus, the remainder of the code consists of the following single line (and comment):

```
% Solving the ODE
[tSol,ySol]=ode45(@(t,y) myFun(t,y), [t0,tN], y0);
```

This is all that is required to solve the ODE-IVP using ode45 solver. The solver takes step-size of approximately ~0.0125 to meet the desired tolerance.

3.5 CASE STUDIES AND EXAMPLES

The concepts presented in this chapter will now be applied for solving practical problems of interest to chemical engineers.

3.5.1 An Ideal PFR

A PFR is an idealized reactor, which is used to carry out the following general reaction:

$$A \rightarrow \text{Products}$$

The inlet consists of the reactant introduced at a concentration C_{A0}. If the inlet volumetric flowrate is Q, the velocity is given by $u = Q/A_{cs}$, where A_{cs} is the area of cross-section of the PFR. The molar flowrate is $F_A = QC_A$. The model for a PFR at steady state is given by

$$\frac{d}{dz}(uC_A) = R_A \tag{3.91}$$

Consider the case where the reaction rate is given by the Langmuir-Hinshelwood model. If there is no change in volume due to reaction, u is constant and the overall model for the system is written as

$$u\frac{dC_A}{dz} = -\frac{kC_A}{\sqrt{1 + K_r C_A^2}}, \quad C_{A0} = 1\,\text{mol/m}^3 \tag{3.92}$$

The kinetic parameters for the system are

$$k = 2\,\text{s}^{-1}, \quad K_r = 1\,\text{mol}^2/\text{m}^6$$

First, this problem will be solved using ode45 solver. Thereafter, the problem will be recast as a design problem for the PFR and solved using numerical integration. This example will thus also be used to illustrate the relationship between integration and ODE-IVP. In Chapter 5, extensions of the PFR will be undertaken as case studies.

3.5.1.1 Simulation of PFR as ODE-IVP

It is necessary to espouse good MATLAB programming practices right at the outset. Toward this goal, the code will be written in a modular fashion, keeping it flexible for further extensions and utilization of MATLAB's functionalities to maintain a consistent coding structure. An important aspect of good programming practices that I evangelize includes separation of problem definition and solution. I will define a single "*driver*" script (which is executed by the user) that provides problem definitions and handles overall project execution and uses separate function(s) as required. The function file, insofar as possible, may avoid defining parameters of the problem. The parameters are defined in a separate file (for larger projects) or in the driver script (smaller problems such as the current example). These parameters are passed to the solver function through use of parameterized anonymous functions in MATLAB. The following example illustrates this method as well.

Example 3.10 Conversion in PFR

Problem: For the PFR described above, use ode45 to obtain concentration C_A along the length of a 0.5 m long PFR. Also compute the conversion X_A. The inlet flowrate is $Q = 0.1$ m³/s and cross-sectional area is $A_{cs} = 0.25$ m².

Solution: The first task is to consider making the function file that will be passed to ode45; this function will take z and Ca as inputs and returns dC. The computation of dC is as per Equation 3.92. The values of rate parameters, k and K, as well as inlet velocity u, form the parameters for the PFR. This function will be named pfrModelFun.

The inlet concentration C_{A0} forms the initial condition. It is, in general, not a model parameter. The model parameters, initial conditions, and solution parameters will be defined in the driver script. In earlier versions of MATLAB, due to the lack of other options, model parameters were passed to the function using global variables. This practice is discouraged; a parameter structure will be passed to the function instead.

Once the ODE is solved, a separate section computes and plots the final results. The conversion is defined as follows:

$$X_A = \frac{F_{A0} - F_A}{F_{A0}} = \frac{Q(C_{A0} - C_A)}{Q C_{A0}}$$

The core of the function pfrModelFun consists of the following lines:

```
rxnRate=k*Ca/sqrt(1+Kr*Ca^2);   % Reaction rate
dC=-1/u*rxnRate;   % Overall model equation
```

In addition to the above, the parameters, k, Kr, and u need to be obtained from modelParam structure. This structure will be defined in the driver script pfrLHDriver and passed using anonymous function definition: @(z,Ca) pfrModelFun(z,Ca,modelParam).

The resulting **model function** pfrModelFun.m is given below:

```
function dC=pfrModelFun(z,Ca)
% To compute dC/dz for a PFR for a
%     single Langmuir-Hinshelwood rate
% Model Parameters
k=modelParam.k;
Kr=modelParam.Kr;
u=modelParam.u0;
% Calculation of PFR model equation
rxnRate=k*Ca/sqrt(1+Kr*Ca^2);
dC=-1/u*rxnRate;
```

The following is the **driver script** `pfrLHDriver.m`:

```
% Driver script for simulation of a PFR
% with single reaction using LH kinetics
%% Model Parameters
modelParam.k =2;
modelParam.Kr=1;
% Reactor conditions
Acs=0.25;
L=0.5;
modelParam.u0=0.1/Acs;
modelParam.C0=1;
%% Simulating the PFR
[Z,CA]=ode45(@(z,Ca) pfrModelFun(z,Ca,modelParam), ...
    [0 L],modelParam.C0);
%% Computing and plotting results
XA=(modelParam.C0-CA)/modelParam.C0;
plot(Z,XA);
xlabel('Length (m)'); ylabel('conversion');
```

Figure 3.4 is a plot of conversion vs. length of the PFR. The conversion increases with the length of the PFR. Conversion of 75% is attained at PFR length of 0.32 m. Conversely, PFR volume of 0.08 m³ is required for attaining 75% conversion.

3.5.1.2 Numerical Integration for PFR Design

The model equation of PFR (Equation 3.92) can be rewritten in the form of design equation that is familiar to chemical engineers:

$$\frac{d\left(Q/A_{cs}\right)C_A}{dz} \equiv \frac{dF_A}{dV} = -\frac{kC_A}{\sqrt{1+K_r C_A^2}} \tag{3.93}$$

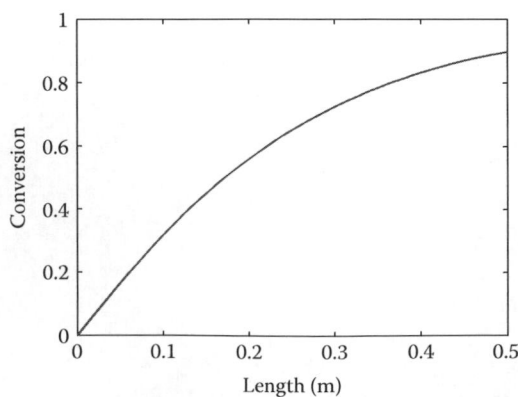

FIGURE 3.4   Conversion as a function of PFR length.

From the definition of conversion

$$F_A = F_{A0}(1 - X_A), \quad C_A = C_{A0}(1 - X_A)$$

Thus, the design equation is written as

$$F_{A0}\left(-\frac{dX}{dV}\right) = -\frac{kC_{A0}(1 - X_A)}{\sqrt{1 + K_r C_{A0}^2 (1 - X_A)^2}}$$

Rearranging, the PFR design equation is given by

$$V = \int_0^{X_{pfr}} \underbrace{\frac{F_{A0}}{k'}\left(\frac{\sqrt{1 + K_r'(1 - X_A)^2}}{(1 - X_A)}\right)}_{fval} dX, \quad k' = kC_{A0}, \quad K_r' = K_r C_{A0}^2 \qquad (3.94)$$

This integration can be solved using trapezoidal rule (MATLAB command `trapz`), as described in Appendix E.

Example 3.11 Design of a PFR

Problem: Solve the PFR design equation to obtain reactor volume for 75% conversion.

Solution: The following code was used for PFR design. The initial part of the code remains similar to the PFR driver. The remainder of the code is given below.

```
%% Design Parameters
kPrime=modelParam.k*CA0;
KrPrime=modelParam.Kr*CA0^2;
FA0=(u0*Acs)*CA0;
%% Design of PFR
Xpfr=0.75;
XALL=[0:0.01:Xpfr];
RALL=kPrime*(1-XALL)./sqrt(1+KrPrime*(1-XALL).^2);
fval=FA0./RALL;
% PFR Volume
V=trapz(XALL,fval);
```

The resulting value of volume is given below:

```
>> V =
    0.0798
```

Numerical integration requires splitting of the integration domain into several intervals. In this example, I chose the interval width as $h = 0.01$. The function value `fval` can be calculated as $\left(F_{A0}\right) \dfrac{1}{R\left(X_A\right)}$. This can be conveniently done using the powerful vector operations of MATLAB (see Appendix A for a primer). The function `trapz` uses the trapezoidal rule (also see Appendix E) to compute the PFR volume. The result of 0.0798 m³ is close to the approximate value of 0.07 m³ obtained in the previous section.

3.5.1.3 Comparison of ODE-IVP with Integration

The PFR example was intended to illustrate a couple of points regarding ODE-IVP and integration. When the function $f(\cdot)$ depends only on the independent variable, the two can be said to be equivalent. However, the example above demonstrates the relationship between the two very well.

First, when we asked the question: "what is the conversion achieved at a given length (volume) of PFR," I chose ODE-IVP. This was because ODE-IVP naturally tracks how the dependent variable (C_A or conversion) varies with PFR volume (or length). On the other hand, when we asked the design question: "what volume (or length) is required to achieve certain conversion," I could use Figure 3.4 to obtain PFR volume that yielded 75% conversion. Alternatively, it was convenient to rearrange the ODE in the form of $\dfrac{dV}{dX_A} = \left(F_{A0}\right)\dfrac{1}{R\left(X_A\right)}$, making the desired value of PFR volume as a target variable needed to achieve desired X_{pfr}. The geometric interpretation of this design procedure becomes clear when $\left(F_{A0}\right)\dfrac{1}{R\left(X_A\right)}$ is plotted against X_A in Figure 3.5. The PFR design volume is nothing but the area under this curve. Note that in case of ODE-IVP, the evolution of solution X_A vs. volume is plotted; whereas in numerical integration, function value `fval`$=\left(F_{A0}\right)\dfrac{1}{R\left(X_A\right)}$ is plotted vs. X_A (Figure 3.5), with the volume given by the shaded area under the `fval` curve.

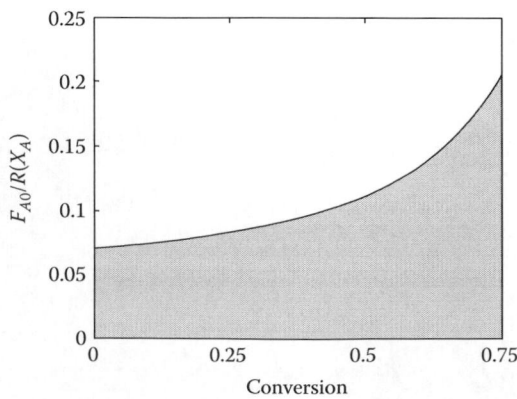

FIGURE 3.5   Plot of `fval` vs. conversion required for design of a PFR volume.

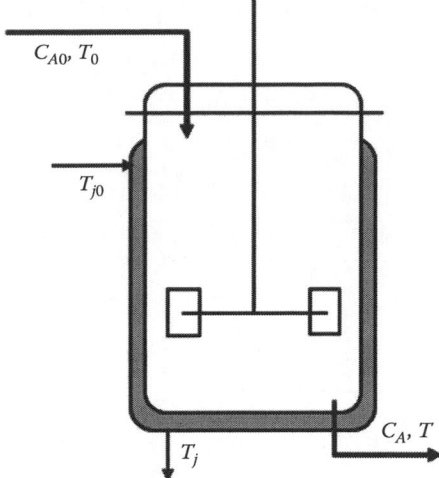

FIGURE 3.6   Schematic of a jacketed CSTR.

3.5.2 Multiple Steady States: Nonisothermal CSTR

Another standard problem in chemical engineering is that of a continuous stirred tank reactor (CSTR) with exothermic reaction. The heat generated on reaction is removed from the system using a cooling jacket. The system is shown in Figure 3.6.

3.5.2.1 Model and Problem Setup

The overall model for the CSTR includes mass balance and energy balances for the reactor and cooling jacket. The resulting model equations are given below:

$$\frac{dC_A}{dt} = \frac{F}{V}\left(C_{A0} - C_A\right) - r_A$$

$$V\rho c_p \frac{dT}{dt} = F\rho c_p \left(T_0 - T\right) + \left(-\Delta H\right)r_A V - UA\left(T - T_j\right) \tag{3.95}$$

$$V_j \rho_j c_j \frac{dT_j}{dt} = F_j \rho_j c_j \left(T_{j0} - T_j\right) + UA\left(T - T_j\right)$$

with the reaction rate given by $r_A = k_0 \exp\left(-\dfrac{E}{RT}\right)C_A$. The values of various parameters are given in Table 3.1.

This problem is solved in the following example.

TABLE 3.1   Model Parameters and Operating Conditions for Jacketed CSTR

CSTR Inlet	Cooling Jacket	Other Parameters
$F = 25$ L/s	$F_j = 5$ L/s	$C_{A0} = 4$ mol/L
$V = 250$ L	$V_j = 40$ L	$k_0 = 800$ s^{-1}
$\rho = 1000$ kg/m^3	$\rho_j = 800$ kg/m^3	$E/R = 4500$ K
$c_p = 2500$ J/kg·K	$c_p = 5000$ J/kg·K	$(-\Delta H) = 250$ kJ/mol
$T_0 = 350$ K	$T_{j0} = 25°$C	$UA = 20$ kW/K

Example 3.12 Jacketed CSTR

Problem: Solve the transient nonlinear CSTR problem for inlet temperature 350 K. Initially, at $t = 0$, the concentration within the CSTR is 0.1 mol/L, the CSTR temperature is 600 K and cooling jacket temperature is 500 K. Plot how the three variables change with time.

Solution: In order to solve the above problem, the solution vector is defined as

$$
\mathbf{y} = \begin{bmatrix} C_A \\ T \\ T_j \end{bmatrix}
$$

The model equations are written in the standard form $\mathbf{y}' = f(t, \mathbf{y})$, where both y and dy are column vectors* of size 3×1. All the parameters need to be converted to SI units.

The core part of the code consists of the following model description:

```
dy(1,1)=F/V*(C0-Ca)-rxnRate;
dy(2,1)=F/V*(T0-T)+DH*rxnRate/(rhoCp)-hXfer/(V*rhoCp);
dy(3,1)=Fj/Vj*(Tj0-Tj)+hXfer/(VRhoC_j);
```

Prior to this, the following terms need to be defined: `rxnRate` (rate of reaction), `hXfer` (heat transfer term $UA(T - T_j)$), `rhoCp` (ρc_p), and `VRhoC_j` ($V_j \rho_j c_j$). This example continues after a brief interlude.

First, I will describe a "traditional" way of writing the MATLAB function that is passed to the `ode45` solver. In earlier versions of MATLAB, if one needed to pass a large number of parameters (such as the parameters in Table 3.1), doing so through function arguments was inconvenient and error-prone. Hence, the model parameters required in a specific function were defined within the same function, whereas the parameters required in multiple functions were shared using `global` variables. The function, `cstrFunOld.m` exemplifies this traditional method.

```
function dy = cstrFunOld(t,y)
global C0 T0 Tj0
%% Parameters
F=25/1000; % m^3/s
V=0.25;    % m^3
<other parameters also defined here>
```

---

* Keeping with the convention followed throughout this book, unless otherwise stated, all vectors are defined as $n \times 1$ column vectors.

```
%% Key variables
Ca=y(1); T=y(2); Tj=y(3);
rxnRate=k0*exp(-E/T)*Ca;
hXfer=UA*(T-Tj);
rhoCp=rho*Cp;
VRhoC_j=Vj*rhoj*cj;
%% Model equations
dy(1,1)=F/V*(C0-Ca)-rxnRate;
dy(2,1)=F/V*(T0-T)+DH*rxnRate/(rhoCp)-hXfer/(V*rhoCp);
dy(3,1)=Fj/Vj*(Tj0-Tj)+hXfer/(VRhoC_j);
```

The above code maintains separation of parameter definition and model execution functions: The main part of the code starts at the second section (demarcated by `%% Key variables`), prior to which all the parameters are defined. The inlet conditions are shared with the driver function using `global` variables. However, a few years ago (perhaps around 2010), anonymous functions and structure arrays were introduced in MATLAB. My experience working with cross-functional teams in the industry has led me to embrace these as powerful tools for writing better MATLAB programs. I personally evangelize this "modern" method of programming, as shown below.

3.5.2.2 Simulation of Transient CSTR

With their recent introduction in MATLAB, structure arrays form a more convenient means to pass the model parameters to a function. The function argument variables have a local *scope* within the function. This is a recommended programming practice followed in all programming languages.

The core of the code remains nearly unchanged; the model parameters are now accessed through structure fields. For example, calculation of reaction rate is modified as

```
rxnRate=par.k0 * exp(-par.E/T)*Ca;
```

Here, `par` is a structure that is passed to the function, `cstrFun`, as an argument. The fields of the variable, `par.k0` and `par.E` represent the preexponential factor and activation energy of the reaction, respectively. Similarly, the next line is rewritten as

```
hXfer=par.UA*(T-Tj);
```

The overall function file `cstrFun.m` is written as below:

Example 3.12 (Continued)

```
function dy = cstrFun(t,y,par)
%% Inlet and current conditions
C0=par.C0; T0=par.T0; Tj0=par.Tj0;
Ca=y(1);   T=y(2);     Tj=y(3);
```

```
%% Key Variables
rxnRate=par.k0 * exp(-par.E/T)*Ca;
hXfer=par.UA*(T-Tj);
rhoCp=par.rho*par.cp;
VRhoC_j=par.Vj*par.rhoj*par.cj;
tau=par.V/par.F;      % Residence time
tauJ=par.Vj/par.Fj;
%% Model Equations
dy(1,1)=(C0-Ca)/tau-rxnRate;
dy(2,1)=(T0-T)/tau+par.DH*rxnRate/rhoCp-hXfer/(par.V*rhoCp);
dy(3,1)=(Tj0-Tj)/tauJ+hXfer/(VRhoC_j);
```

Comparing with the earlier code (`cstrFunOld`), the parameters are not defined within the function. Instead, they will be defined in the driver script (`cstrDriver.m`) and passed to the function through `par` structure. `cstrDriver` script is given below:

```
% Simulation of a jacketed CSTR with
%        exothermic first-order reaction
%% Define model parameters (SI units)
modelPar.F=25/1000; % CSTR parameters
modelPar.V=0.25;
modelPar.rho=1000;
modelPar.cp=2500;
modelPar.Fj=5/1000; % Cooling jacket
modelPar.Vj=40/1000;
modelPar.rhoj=800;
modelPar.cj=5000;
modelPar.k0=800;    % Rate constants
modelPar.E=4500;
modelPar.DH=250000;
modelPar.UA=20000;
%% Define inlet and initial conditions
modelPar.C0=4000;
modelPar.T0=350;
modelPar.Tj0=298;
y0 = [100; 600; 500];
tSpan = [0 50];
%% Solving and plotting
[T,Y]=ode15s(@(t,y) cstrFun(t,y,modelPar),tSpan,y0);
subplot(2,1,1);
plot(T,Y(:,1));
xlabel('time (s)'); ylabel('C_A (mol/m^3)')
subplot(2,1,2);
plot(T,Y(:,2:3));
xlabel('time (s)'); ylabel('T, T_j (K)')
```

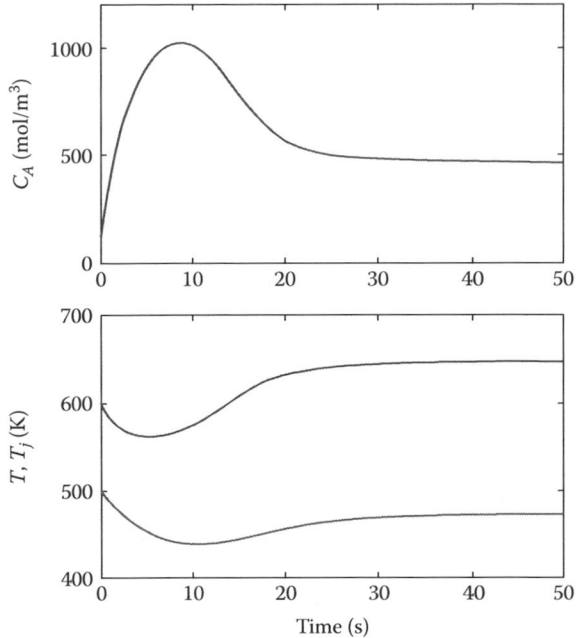

FIGURE 3.7   Transient simulations of jacketed CSTR.

The transient simulation results are shown in Figure 3.7. This example illustrates simulation of a jacketed CSTR. At the end of 50 s, the three modeled variables have the values, $C_A = 461.0$, $T = 647.6$, $T_j = 472.6$.

3.5.2.3 Step Change in Inlet Temperature

At $t = 50$ s, the inlet temperature of the CSTR is changed to 298 K. The following example shows the simulation of CSTR from 50 to 100 s.

Example 3.13 Step Change in Inlet Temperature

Problem: Starting with the conditions of Example 3.12 at the end of 50 s, simulate the system with step change in T_0 to 298 K.

Solution: After the last line of the previous code, the following is added:

```
%% Simulation for next 500s with step change in T0
tSpan=[50, 200];
y50=Y(end,:);
modelPar.T0=298;
[T2,Y2]=ode15s(@(t,y) cstrFun(t,y,modelPar),tSpan,y50);
subplot(2,1,1);
plot(T2,Y2(:,1));
ylabel('C_A (mol/m^3)')
subplot(2,1,2);
plot(T2,Y2(:,2:3));
xlabel('time (s)'); ylabel('T, T_j (K)')
```

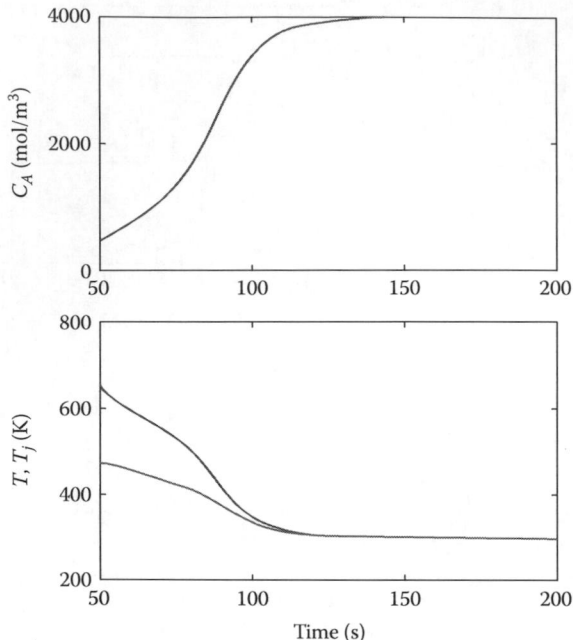

FIGURE 3.8   Transient CSTR response after T_0 is reduced to 298 K.

Figure 3.8 shows the time evolution of the modeled variables for 150 s after step-down in the inlet temperature. The outlet concentration increases and the temperatures drop rapidly after the step change. The outlet concentration is now close to the inlet concentration of 4 mol/L, and almost no reaction is taking place in the system. At the end of 200 s, the values of the corresponding variables are $C_A = 3991$, $T = 298.8$, $T_j = 298.4$.

This example shows steady state multiplicity. When the inlet temperature was $T_0 = 350$ K, the system was in a steady state with high conversion. When the inlet temperature was reduced to $T_0 = 298$ K, the system transitioned to another steady state with low conversion. This issue of steady state multiplicity will be discussed in more detail in Section II of this book.

3.5.3 Hybrid System: Two-Tank with Heater

Consider the system consisting of two heated tanks in series, as shown in Figure 3.9. The two tanks are connected with tubes at the bottom and at a certain height H. Electric heaters are used to heat up the fluid in the tanks. This is an example of so-called *hybrid system*, which consists of continuous variables as well as switching behavior. In this example, there is a switching on/off of the flow in the intermediate connecting tube. If the height of liquid in a tank is greater than H, then the fluid flows through this connecting tube. When the liquid levels are below H, fluids flow through the bottom pipe only.

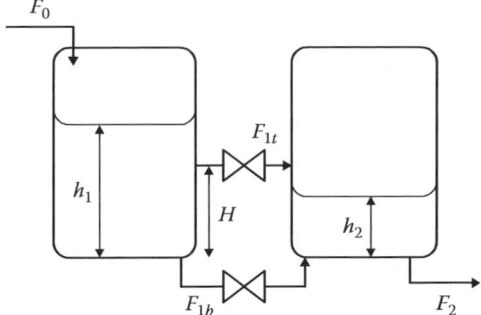

FIGURE 3.9 Schematic of two tanks in series.

The model for the system is given below:

$$A_1 \frac{d}{dt} h_1 = F_0 - F_{1t} - F_{1b}$$

$$A_2 \frac{d}{dt} h_2 = F_{1t} + F_{1b} - F_2$$

$$\frac{d}{dt} T_1 = \frac{F_0}{A_1 h_1} (T_0 - T_1) + \frac{Q_1}{A_1 h_1 \rho c_p}$$

$$\frac{d}{dt} T_2 = (F_{1t} + F_{1b})(T_1 - T_2) + \frac{Q_2}{A_2 h_2 \rho c_p}$$

(3.96)

The first two equations are material balance on the two tanks. The last two equations come from energy balance. These equations are obtained after rearranging the original energy balance equation, where the term on the left-hand side is $\left(A_i h_i \rho c_p T_i \right)' = \rho c_p A_i \left(h_i \cdot T_i' + T_i \cdot h_i' \right)$. The last term is simplified using the appropriate material balance to obtain the equations given above.

The flowrates in the above equations are given by

$$F_{1b} = c_1 \sqrt{h_1 - h_2}, \quad F_2 = c_2 \sqrt{h_2}$$

(3.97)

and

$$F_{1t} = \begin{cases} 0 & h_1 \le H \\ c_1 \sqrt{h_1 - H} & h_1 > H, h_2 \le H \\ c_1 \sqrt{h_1 - h_2} & h_2 > H \end{cases}$$

(3.98)

The parameters are as follows:

Areas of the tanks: $A_1 = A_2 = 0.25$ m^3

Inlet conditions: $F_0 = 0.4$ m^3/min, $T_0 = 25°$C

Valve coefficients: $c_1 = c_2 = 0.6$ m$^{2.5}$/min

Other conditions: $\rho c_p = 1$ cal/cm^3, $Q_1 = Q_2 = 100$ kW

Height of intermediate pipe: $H = 0.5$ m

The condition for calculation of F_{1t} can be handled in a couple of ways. One way is to use if-then statements. Another, more elegant method is to note that first expression in Equation 3.98 (i.e., when $h_1 \le H$) can be written as $F_{1t} = c_1 \sqrt{H - H}$:

$$
F_{1t} = \begin{cases} c_1 \sqrt{H - H} & h_1 \le H \\ c_1 \sqrt{h_1 - H} & h_1 > H, h_2 \le H \\ c_1 \sqrt{h_1 - h_2} & h_2 > H \end{cases} \tag{3.98}
$$

When $h_2 > H$, the second value inside the square root is h_2; else it is H. Likewise, the first value inside the square root is H when $h_1 \le H$ and equal to h_1 otherwise. Thus, the above equation may be written as

$$
F_{1t} = c_1 \sqrt{\bar{h}_1 - \bar{h}_2}, \quad \bar{h}_i = \max(h_i, H) \tag{3.99}
$$

The forthcoming example illustrates the use of this conditional statement to evaluate the flowrate and hence solve the model of two-tank heater system. Unlike previous example that used SI units, I will use minutes as the unit for time and KJ for energy for convenience.

Example 3.14 Hybrid Two-Tank Heater System

Problem: Solve the two-tank heater system described above. Initially, both the tanks contain 100 liters of water at 25 °C.

Solution: First, all the quantities are converted to consistent units. Height will be in meters and temperature in °C. The parameters requiring unit conversion are

$$
\rho c_p = 4180 \, \text{kJ/kg}, \quad Q_1 = Q_2 = 100 \, \text{kJ/s} \times 60 \, \text{s/min} = 6000 \, \text{kJ/min}
$$

The computation of F_{1t} will be done using Equation 3.99 as

```
del=max(H,h1)-max(H,h2);
F1t=par.c1*sqrt(del);
```

The first line defines del as the term within the square root, which is then used in computing F1t in the next line. The MATLAB function heated2TankFun.m is

```
function dy=heated2TankFun(t,y,par)
% Model for hybrid two-tank system with heater
```

```
%% Getting variables and computing terms
h1=y(1); h2=y(2); T1=y(3); T2=y(4);
% Flow rates
F1b=par.c1*sqrt(h1-h2);
F2=par.c2*sqrt(h2);
del=max(par.H,h1)-max(par.H,h2);
F1t=par.c1*sqrt(del);
% Heating rate
heater1=par.Q1/(par.A1*h1*par.rhoCp);
heater2=par.Q2/(par.A2*h2*par.rhoCp);
%% Model equations
dy=zeros(4,1);
dy(1)=(par.F0-F1b-F1t)/par.A1;
dy(2)=(F1b+F1t-F2)/par.A2;
dy(3)=par.F0/(par.A1*h1)*(par.T0-T1)+heater1;
dy(4)=(F1b+F1t)*(T1-T2)+heater2;
```

The corresponding driver script `heated2TankDriver.m` is given below:

```
% Drive script: heated2TankDriver.m
% Simulation of hybrid two-tank system with heater
%% Model parameters
modelPar.A1=0.25;   modelPar.A2=0.25;
modelPar.F0=0.4;    modelPar.T0=25;
modelPar.c1=0.6;    modelPar.c2=0.6;
modelPar.Q1=6000;   modelPar.Q2=6000;
modelPar.rhoCp=4180;
modelPar.H=0.5;
%% Initial conditions
h1init=0.4;
h2init=0.35;
y0=[h1init; h2init; 25; 25];
tSpan=[0 10];
%% Solving and plotting
[T,Y]=ode45(@(t,y) heated2TankFun(t,y,modelPar),...
    tSpan, y0);
subplot(2,1,1);
plot(T,Y(:,1:2));
ylabel('height (m)')
subplot(2,1,2);
plot(T,Y(:,3:4));
xlabel('time (s)'); ylabel('T (deg C)')
```

The results are shown in Figure 3.10. Since the two liquid heights are fairly close to each other initially, the net flow out of the second tank is greater than the flow in. Hence, the level first falls, before increasing again.

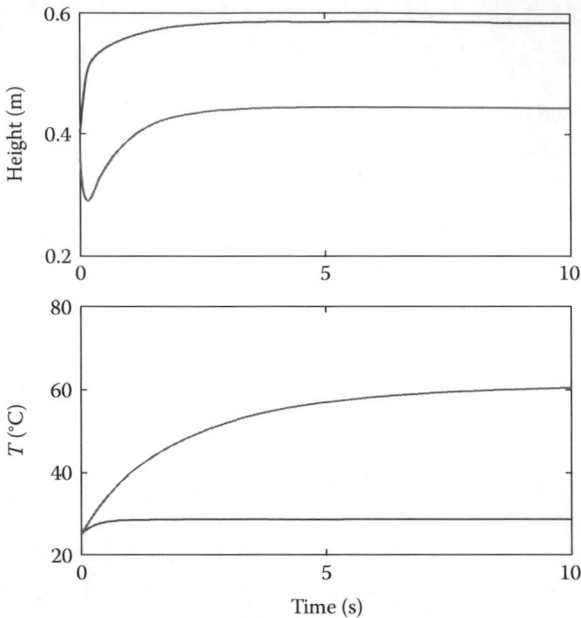

FIGURE 3.10 Transient results in terms of height and temperature of water in the two tanks.

The next figure (Figure 3.11) shows the behavior of the system at higher flowrate of 0.48 m³/min. At this higher flowrate, water levels in both the tanks increase above the intermediate pipe. Note that at steady state, $F_0 = (F_{1b} + F_{1t}) = F_2$ so that the volume of water in the two tanks does not change.

3.5.4 Chemostat: Preview into "Stiff" System

A chemostat is a continuous stirred biological reactor, where a growth medium is continuously added to the reactor containing unconverted substrates, metabolic products, and microorganisms. The inlet to the chemostat usually contains only the substrate(s); it typically does not contain microorganisms. The microorganisms consume the substrate to grow, and in the process, make useful products. Since the chemostat is essentially a CSTR, the product stream from the reactor consists of the same components as the chemostat itself. The concentrations of all the important species: substrate(s), intermediates (if any), metabolites and products, and biomass are modeled as a transient ODE system. The growth rate of the microorganisms can often be complex. The simplest model that captures the essential features of microorganism grown is given by Monod. The growth rate is given by

$$r_g = \left(\mu^{\max} \frac{[S]}{K_S + [S]} \right)[X] \tag{3.100}$$

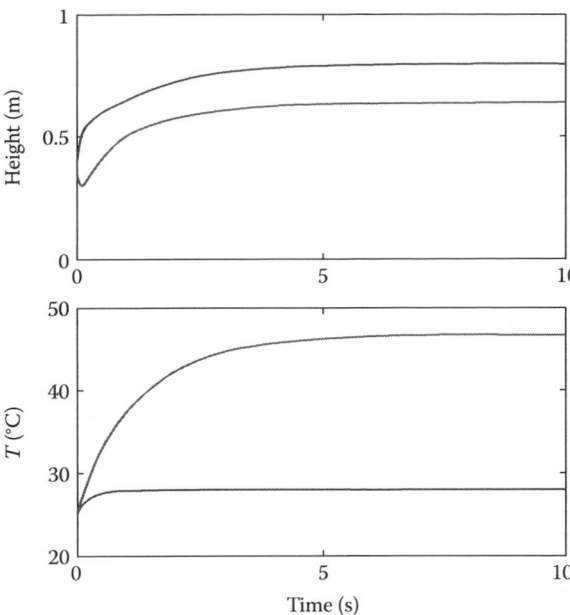

FIGURE 3.11   Transient results for a higher inlet flowrate of 480 L/min.

where $[S]$ and $[X]$ are substrate and biomass concentrations, respectively. The rate constant μ^{max} is the maximum growth rate and K_S is the saturation constant. At large substrate concentration, when $[S] \gg K_S$, the growth rate is zero-order in substrate concentration; whereas when $[S]$ falls substantially, the growth rate becomes first-order in substrate concentration.

The model equations are given by the following ODEs:

$$\frac{dS}{dt} = D(S_f - S) - r_g$$
$$\frac{dX}{dt} = -DX + r_g Y_{xs} \qquad\qquad (3.101)$$
$$\frac{dP}{dt} = -DP + r_g Y_{ps}$$

The rate constants are given as $\mu^{max} = 0.5$, $K_S = 0.25$, $Y_{xs} = 0.75$, and $Y_{ps} = 0.65$.

The next example demonstrates simulation of chemostat.

Example 3.15 Chemostat: Bioreactor with Monod Kinetics Model

Problem: Solve the chemostat model given by Equation 3.101 for feed concentration of $S_f = 5$ g/mL and dilution rate of $D = 0.1$ h^{-1}. The chemostat initially has 5 g/mL of the substrate, and inoculant is added to that initial $[X] = 0.02$ g/mL.

Solution: This is a straightforward implementation in ode45, similar to the earlier examples. The solution consists of the following *driver* script:

```
% driverMonod: Modeling a Chemostat with Monod kinetics
%% Reactor parameters
monodParam.mu=0.5;
monodParam.K=0.25;
monodParam.Yxs=0.75;
monodParam.Yps=0.65;
monodParam.Sf=5;
monodParam.D=0.1;
% Initial conditions
Y0=[5.0; 0.02; 0];
%% Solving and plotting
[tS,YS]=ode45(@(t,y) funMonod(t,y,monodParam),[0 30],Y0);
plot(tS,YS);
```

The second-last line uses ode45 to calculate the solutions tS and YS. The function, funMonod, passed to the ODE solver is given below:

```
function dy=funMonod(t,y,monodParam)
mu=monodParam.mu;      K=monodParam.K;
Yxs=monodParam.Yxs;    Yps=monodParam.Yps;
D=monodParam.D;        Sf=monodParam.Sf;
%% Variables to solve for
S=y(1); X=y(2); P=y(3);
%% Model equations
rg= mu*S/(K+S)*X;
dy(1,1)=D*(Sf-S) - rg;
dy(2,1)=-D*X + Yxs*rg;
dy(3,1)=-D*P + Yps*rg;
end
```

The solver takes more than a hundred steps to reach the solution. On MATLAB 2016a version, tS is a 109-length vector, whereas the size of YS is 109 × 3. Figure 3.12 shows the transient response of the chemostat for this problem.

This example also follows the same pattern as the previous ones. It is a good programming practice to declare all the model parameters at one location. As the code grows more complex for large projects, this way of modularizing the codes proves very useful. As a good coding practice in the industry, we often went a step further. The parameters related to the *microbial growth kinetics* would be located in one structure (say monodParam), which include $\mu^{max}, K_S, Y_{xs}, Y_{ps}$. On the other hand, the chemostat *operating parameters*—that is, D and S_f—would be included in another structure (say reactorParam). While this demarcation would be an "overkill" for this problem, it becomes useful in large projects where modularity is a key.

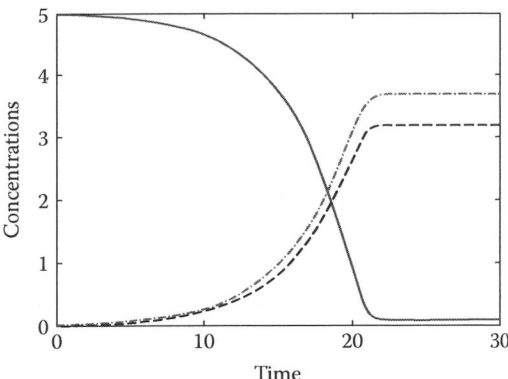

FIGURE 3.12   Transient response of a chemostat.

For the parameter values of Example 3.15, the chemostat response is smooth. Initially, at low biomass concentrations, the substrate concentration falls gradually. After around 10 h, there is a significant reduction in substrate concentration as the higher biomass concentration results in faster consumption of the substrate. As substrate concentration falls, the system behavior would transition from an initial zero-order response in [S] to eventually first-order in [S]. Finally, the system reaches steady state.

Now consider the case when the saturation constant, K_S, in Equation 3.100 is reduced to a value of 0.005 g/mL. The fourth line in `driverMonod.m` is changed, and the code is run again. For this value of saturation constant, the `ode45` solver becomes unstable. The results are shown in Figure 3.13. The initial "zero-order response" is similar to that seen earlier, in Figure 3.12. However, when the substrate concentration drops, the reaction rate becomes first-order in [S], resulting in a rapid decrease in reaction rate that the explicit `ode45` solver is unable to capture. This instability of the explicit solver is not very different from the

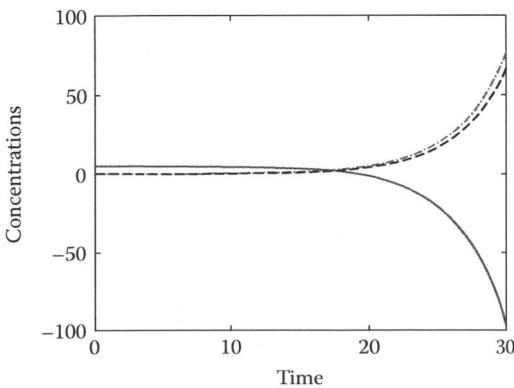

FIGURE 3.13   Transient chemostat response for $K_S = 0.005$ g/mL using `ode45`. The solver becomes unstable due to the *stiff* nature of the ODE.

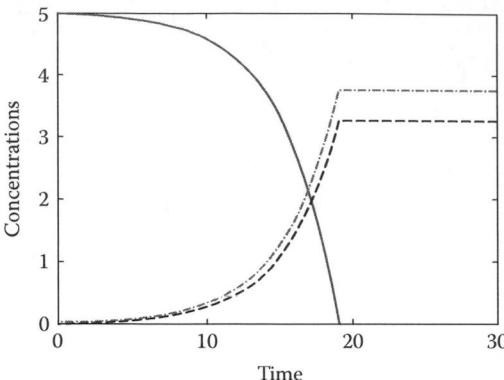

FIGURE 3.14 Transient chemostat response for $K_S = 0.005$ g/mL using ode15s. A stable solution is obtained as ode15s is a stiff solver.

one discussed in Section 3.1.5, where the stability of Euler's explicit and implicit methods were compared.

STIFF SYSTEM

A stiff system refers to a system of ODEs where one or more components of the system have a significantly different time scale of response compared to other components.

In the case of the chemostat, at low values of substrate concentration, the growth rate is significantly lower when the reaction is first-order in [S] compared to the reaction rate being independent of [S] (i.e., zero-order in [S]). The transition from zero- to first-order happens when the substrate concentration is of the same order of magnitude as K_S. When the value of K_S is very low, the system undergoes a very rapid reduction in the growth rate, r_g. Consequently, the rate at which substrate concentration decays falls precipitously. This can be seen in Figure 3.14.

Figure 3.14 is obtained using ode15s, which is a *stiff ODE solver* provided in MATLAB. The solver ode15s uses an implicit backward difference formula to solve the ODE. As can be seen from this example, ode15s is the solver of choice if the system of ODEs is stiff, or when ode45 does not work.

Comparing Figures 3.12 and 3.14, one can see that the former results in smoother curves for [S], [X], and [P] than ode15s. This is because ode45 is a higher-order solver and is a better solver that gives an excellent trade-off between accuracy and speed of computation. As described in Section 3.4, it uses an embedded RK4 method. As a rule of thumb, which is suggested by MATLAB and corroborated by experience of several expert users, ode45 should be the first solver of choice for a generic ODE problem. If the problem is stiff and ode45 does not work, ode15s usually provides the next best option.

A further discussion on ode15s will be provided in Chapter 8 when discussing implicit ODE solver and solution of DAEs. With this note, we end the discussion in this chapter.

3.6 EPILOGUE

The problem of solving ODE-IVP was introduced in Section 3.1. Several types of example problems can be solved as ODE-IVP. Higher-order ODEs can be converted to a set of first-order ODEs, or the problem to solve may consist of multiple simultaneous ODEs. For ODE-IVP, the initial conditions are specified at a single point in the solution domain. All the components are written together as a solution vector, **y**. The strategies to obtain ODE-IVP in the standard form

$$\frac{d\mathbf{y}}{dt} = \mathbf{f}(t,\mathbf{y})$$
$$\mathbf{y}(t_0) = \mathbf{y}_0$$

(3.1)

were discussed briefly in Section 3.1.1 and case studies provided in Section 3.5.

Second-order ODE-IVP methods were discussed in Section 3.2. This chapter focuses on explicit RK family of methods, which are discussed at length. The stability problems with explicit RK methods were discussed, and Section 3.5.4 demonstrated the practical implications on a simulation problem. Section 3.2.5 provided a brief peek into some other important ODE-IVP methods. These classes of methods are available in the ODE suite of MATLAB.

Section 3.3 extended the discussion to higher-order RK methods. Adaptive step-size methods were then discussed. All the methods in MATLAB ODE suite are adaptive step-size methods. These methods were referred to at various points in the chapter, most specifically in Sections 3.2.5 and 3.4.

The practical aspect of solving ODE-IVP in MATLAB is summarized now. For many problems of interest to chemical engineers, explicit ODE-IVP methods will prove sufficient. The MATLAB solver `ode45` is the primary go-to solver for these cases. While other explicit solvers (such as `ode23` and `ode115`) are available, `ode45` provides very high accuracy at good computational speed.

There are several other ODE-IVP problems that are difficult to solve. Such stiff ODEs will be discussed in Chapter 8. If `ode45` fails to converge, the stiff solver `ode15s` would be the next choice. This variable-order solver is typically less accurate and may not be used as the first solver in MATLAB. The MATLAB help documentation is an excellent resource on *implementing* an ODE solver. However, choosing which ODE solver is suitable requires some understanding of how the ODE-IVP are solved, as discussed in this book.

If your problem requires you to use a different solver than `ode45` and `ode15s`, reading Chapter 8 after this chapter will provide a better background to make an informed choice.

EXERCISES

Problem 3.1 Solve Example 3.1 using RK-2 midpoint method from Equation 3.53 and RK-2 Ralston's method from Equation 3.54.

Problem 3.2 Use Richardson's extrapolation for numerical integration.

Problem 3.3 Use Richardson's extrapolation for numerical differentiation.

Problem 3.4 Solve Example 3.3 using the classical RK-4 method from Section 3.3.3.

Problem 3.5 Solve the mass-spring-damper problem, $mx'' = -cx' - kx$, using `ode45`. Choose appropriate values of c and k. The initial condition is that the system is moved to a location $x = 0.2$ and released without any velocity $x'(0) = 0$. Plot how location and velocity vary with time.

Problem 3.6 Repeat the above Problem 3.5 using your own RK-2 Heun's method code. For this purpose, you may write your own code from scratch or build upon Heun's code from Example 3.3.

Problem 3.7 Repeat the above problem using RK-3 and RK-4 methods.

CHAPTER 4

Partial Differential Equations in Time

4.1 GENERAL SETUP

Chapter 3 covered systems defined by ordinary differential equations, where the dependent variable varied either in time or in a single spatial dimension. There are several examples where the dependent variables vary in time and space or in multiple spatial dimensions. Such models are represented by partial differential equations (PDEs). A general first-order PDE in two independent variables is given by

$$\left(\phi, \frac{\partial\phi}{\partial x_1}, \frac{\partial\phi}{\partial x_2}, x_1, x_2\right) = 0 \tag{4.1}$$

where
  ϕ is the dependent variable
  x_1 and x_2 are the two independent variables (e.g., x_1 could be time t, and x_2 could be axial coordinate z; or x_1 and x_2 could be axial and radial coordinates z and r)

If the above function is nonlinear in the dependent variable or its derivatives, then the overall PDE is nonlinear.

A generic second-order PDE contains one or more of the second-order terms, $\partial^2\phi/\partial x_1^2$, $\partial^2\phi/\partial x_2^2$, and/or $\partial^2\phi/\partial x_1\partial x_2$ as well.

We usually encounter either first-order or second-order PDEs in (chemical) engineering. Higher-order PDEs are not common. These PDEs are further of a specific type called *quasilinear PDEs*. These may be written in general as

$$A\frac{\partial^2\phi}{\partial x_1^2} + B\frac{\partial^2\phi}{\partial x_1\partial x_2} + C\frac{\partial^2\phi}{\partial x_2^2} + D\frac{\partial\phi}{\partial x_1} + E\frac{\partial\phi}{\partial x_2} + F = 0 \tag{4.2}$$

The coefficients A to F may themselves be functions of x_1, x_2, and/or ϕ. It would be clear from the description above that the PDE (4.2) is *second-order PDE* if one or more of A, B, and C are nonzero. If A, B, and C are all zero, this is a *first-order PDE*. Thus, the order of PDE is the highest order of differentiation of ϕ in the overall equation.

The above PDE needs to have initial or boundary conditions. The number of boundary conditions required in a particular dimension depends on the order of the PDE in that dimension. For example, if A is nonzero, two boundary conditions are required in x_1; if A is zero, then single initial condition is sufficient in x_1.

4.1.1 Classification of PDEs

Homogeneous PDE: The PDE is said to be *homogeneous* if $F = 0$.

Linear vs. Nonlinear PDE: The above PDE is linear if A to E are all functions of independent variables only, whereas F is a linear function of ϕ (i.e., $F = F_0 + F_1 \phi$). The coefficients A to F_1 should not depend on ϕ. For example, the PDE

$$\frac{\partial \phi}{\partial t} + t^2 \frac{\partial \phi}{\partial x} = \sin(t) \tag{4.3}$$

is linear, because the coefficient $E = t^2$ and the *source term* $F = \sin(t)$ are not a function of the dependent variable, ϕ. On the other hand

$$\frac{\partial \phi}{\partial t} + \phi \frac{\partial \phi}{\partial x} = \sin(t) \tag{4.4}$$

is nonlinear, because the coefficient $E = \phi$ depends on the dependent variable, ϕ. Additionally, the following PDE

$$\frac{\partial \phi}{\partial t} + \left(\frac{\partial \phi}{\partial x} \right)^2 = G(t,x) \tag{4.5}$$

is also nonlinear. It should be noted that Equation 4.5 is a first-order PDE since the highest order of differentiation with respect to t and x is first order. While Equations 4.3 and 4.4 are quasilinear PDEs, Equation 4.5 is not.

4.1.2 Brief History of Second-Order PDEs

We are going to concern ourselves with PDEs of the type given in Equation 4.2 only, since these are the most common types of PDEs in chemical engineering. Partial differentiation was used in calculus even before PDEs were identified as key mathematical tools. Leibnitz and l'Hospital used partial differentiation in the late seventeenth century. Following Leonhard Euler's formalization of "infinitesimal calculus," the role of PDEs in describing physical laws for balance of mass, energy, momentum, etc., started getting discovered. In 1752, d'Alembert introduced the 1D wave equation:

$$\phi_{tt} = v^2 \phi_{xx}$$

to model vibrating strings.* A decade later, Euler generalized this to 2D and 3D wave equation. The Laplace equation

$$\nabla^2 \phi = \phi_{xx} + \phi_{yy} = 0$$

was introduced to model Newton's gravitational potential. Siméon Poisson modified the Laplace equation to model Gauss's law for gravity in differential form. This equation is now well known as the Poisson equation and is written in a general form as $\nabla^2 \phi = \rho$. Finally, in 1822, Fourier introduced the transient heat equation:

$$\phi_t = k\phi_{xx}$$

for modeling heat flow.

The wave equation, Laplace/Poisson equations, and heat equation are considered prototypical examples of three main types of PDEs of importance to engineers, namely, hyperbolic PDEs, elliptic PDEs, and parabolic PDEs, respectively. The next section discusses classification of first- and second-order PDEs and their implication for numerical solutions.

4.1.3 Classification of Second-Order PDEs and Practical Implications

This classification of PDEs will prove important for the selection of appropriate numerical techniques. The classification of PDEs as elliptic, hyperbolic, or parabolic is based on the value of

$$\mathbb{D} = B^2 - 4AC \tag{4.6}$$

The PDE is said to be elliptic if $\mathbb{D} < 0$, hyperbolic if $\mathbb{D} > 0$, and parabolic if $\mathbb{D} = 0$. More important than the definition is the physical significance of the terms. These terms are perhaps motivated from the equivalence with corresponding curves in conics: ellipse, hyperbola, and parabola. The implication of these characteristics for solving the PDEs is discussed here.

4.1.3.1 Elliptic PDE

Elliptic PDEs describe systems that have reached a steady state and where diffusion is a dominant phenomenon (or at least where diffusion significantly affects the independent variable). Just as ellipse is a smooth curve, elliptic PDEs describe dependent variables that vary smoothly over space. If the conditions at the boundaries undergo an abrupt step change, the *diffusive terms* will tend to smoothen out its effect in the domain. For example, the following elliptic PDE describes temperature distribution at steady state with heat conduction as the dominant phenomenon:

$$0 = k\nabla^2 T + \mathcal{S}(T) \tag{4.7}$$

---

* Here I have used shorthand notation ϕ_{tt} to represent $\dfrac{\partial^2 \phi}{\partial t^2}$, and so on.

which for Cartesian coordinate in 2D is represented as

$$0 = k\left(\frac{\partial^2 T}{\partial x^2} + \frac{\partial^2 T}{\partial y^2}\right) + \mathcal{S}(T) \tag{4.8}$$

In the above equation, $B = 0$, $A = C = k$; thus $\mathbb{D} < 0$ and the PDE is classified as elliptic. Thus, an elliptic PDE in two dimensions has the diffusive term dominant in both the dimensions. Since the differential equations are second order, two boundary conditions are needed in both X- and Y-coordinates. For example, if Equation 4.8 governs temperature in a rectangular plate, boundary conditions are specified for each of the four edges of the plate. The key feature of elliptic PDEs is that the temperature at any point in the domain will be governed by the conditions at all the four boundaries of the plate.

Elliptic PDEs will not be covered in this chapter. The reader may refer to Chapter 7 for discussion on solving elliptic PDEs.

4.1.3.2 Hyperbolic PDE

Hyperbolic PDEs are used to describe *propagation* of waves or *convection* of material or energy in transport problem. In conics, a hyperbola is an open curve with two disconnected branches. By analogy, hyperbolic PDEs can be used to model those systems that are characterized by propagation of an abrupt or step change at boundary along the domain; they also model systems where this propagation or convection results in discontinuities, such as in shock waves.

The 1D wave equation introduced earlier is an example of hyperbolic PDE widely studied in introductory engineering math courses:

$$\phi_{tt} = v^2 \phi_{xx} \tag{4.9}$$

Writing this equation in standard form shows $\mathbb{D} > 0$, classifying it as a hyperbolic PDE.

Let us introduce the following coordinate transformation:

$$\xi = x + vt, \quad \eta = x - vt \tag{4.10}$$

With this transformation, one can verify that Equation 4.9 reduces to

$$\frac{\partial^2 \phi}{\partial \xi \partial \eta} = 0 \tag{4.11}$$

and the solution to the wave equation is given by

$$\phi = f_1(\xi) + f_2(\eta) \tag{4.12}$$

$$\phi(x,t) = f_1(x + vt) + f_2(x - vt) \tag{4.13}$$

The curves $x+vt=c_1$ and $x-vt=c_2$ are called characteristic curves. What does this physically mean? Consider a curve $x-vt=1$. The second term in Equation 4.13 becomes $f_2(1)$ and thus is a constant. Thus, the value of f_2 is constant along the second characteristic. Likewise, the value of $f_1(\cdot)$ is constant along the first characteristic.

The key behavior of a hyperbolic PDE relevant to numerical solution techniques is that the information at the boundary is propagated along the two characteristic curves.

A similar quantitative response is observed in some types of first-order PDEs, as discussed next.

4.1.3.3 First-Order Hyperbolic PDEs

Let us consider another analysis of the 1D wave equation described above. Let us define two new variables:

$$\theta = \frac{\partial \phi}{\partial t}, \quad \psi = v \frac{\partial \phi}{\partial x} \tag{4.14}$$

Clearly, one can see that taking partial differentiation ψ_t yields

$$\frac{\partial \psi}{\partial t} = v \frac{\partial \theta}{\partial x} \tag{4.15}$$

Likewise, partial differential $\theta_t = \phi_{tt}$, which using Equation 4.9 can be written as

$$\frac{\partial \theta}{\partial t} = v \frac{\partial \psi}{\partial x} \tag{4.16}$$

Thus, the wave equation (4.9) can be converted into the following matrix form:

$$\frac{\partial}{\partial t} \begin{bmatrix} \theta \\ \psi \end{bmatrix} - \begin{bmatrix} 0 & v \\ v & 0 \end{bmatrix} \frac{\partial}{\partial x} \begin{bmatrix} \theta \\ \psi \end{bmatrix} = 0 \tag{4.17}$$

This equation is in the form $\Phi_t - A\Phi_x = 0$. Using the concepts of eigenvalues discussed in Chapter 2, it is easy to diagonalize the above equation. Noting that the eigenvalues of the above matrix are $\pm v$ and the eigenvectors are $\mathbf{v}_1 = \begin{bmatrix} 1/\sqrt{2} & 1/\sqrt{2} \end{bmatrix}^T$ and $\mathbf{v}_2 = \begin{bmatrix} -1/\sqrt{2} & 1/\sqrt{2} \end{bmatrix}^T$, we can change to coordinate axes to $\begin{pmatrix} \mathbf{v}_1 & \mathbf{v}_2 \end{pmatrix}$. Indeed, it is easy to do this coordinate transformation. Recall from Chapter 2 that the physical interpretation of transforming to $\begin{pmatrix} \mathbf{v}_1 & \mathbf{v}_2 \end{pmatrix}$ implies that we define the following:

$$\vartheta = \frac{1}{\sqrt{2}} (\theta + \psi), \quad \varpi = \frac{1}{\sqrt{2}} (\theta - \psi) \tag{4.18}$$

It is easy to verify that Equations 4.15 and 4.16 get converted to the following *decoupled set of first-order PDEs*:

$$\frac{\partial \vartheta}{\partial t} - v\frac{\partial \vartheta}{\partial x} = 0 \tag{4.19}$$

$$\frac{\partial \varpi}{\partial t} + v\frac{\partial \varpi}{\partial x} = 0 \tag{4.20}$$

The second-order PDE is converted into a set of two first-order ODEs. One would recognize that the first term is the time derivative term, whereas the second is a *convection* term. Thus, balance equations involving convection form a set of hyperbolic PDEs.

The curves $(x + vt)$ and $(x - vt)$ form the characteristic curves for PDEs (4.19) and (4.20), respectively. The key numerical feature of hyperbolic PDEs is the convection or propagation along the characteristics influences the solution at any point in the domain.

4.1.3.4 Parabolic PDE

Parabolic PDEs are used to solve initial value problems (IVPs) where the system is diffusion dominated in one spatial dimension. Parabola is a smooth open curve, which represents transition between an ellipse (closed smooth curve) and hyperbola (disconnected open curves). In the same manner, a parabolic PDE is used to describe systems that propagate in one direction, while the dependent variable has a smooth behavior due to a diffusive component along the other dimension. A transient heat equation

$$\frac{\partial T}{\partial t} = \alpha\frac{\partial^2 T}{\partial z^2} + S(T) \tag{4.21}$$

is a prototypical example of parabolic PDE. This is an IVP in one dimension (t) and a boundary value problem in the other dimension (z).

The key numerical feature of parabolic PDEs is the propagation of the solution along one direction, whereas the solution at any point in the domain is influenced by both boundaries in the other (z) direction.

4.1.4 Initial and Boundary Conditions

Initial conditions: They are most often specified in the form of the value of the variable ϕ defined at a certain initial time for the entire domain in other independent variables.

In case of first-order hyperbolic PDE, one boundary condition is required for each of the dependent variables, whereas two boundary conditions are required for parabolic PDEs. There are three general types of boundary conditions as follows.

Dirichlet condition: The value of independent variable itself is specified at one boundary:

$$\phi(t, z = z_0) = \phi_0 \tag{4.22}$$

For example, temperature at one end or concentration entering a reactor may be specified.

Neumann condition: Here, the derivative of independent variable is specified at one boundary:

$$\frac{\partial \phi}{\partial z}\bigg|_{t,z_1} = b \tag{4.23}$$

For example, heat flux is zero at insulated boundary or the mass flux may be specified.

Mixed condition: Here, the boundary condition includes both the value of the variable and its derivative at a boundary:

$$\mathfrak{g}\left(\phi(t,z_1), \frac{\partial \phi}{\partial z}\bigg|_{t,z_1}\right) = 0 \tag{4.24}$$

Danckwerts boundary condition is an example of mixed boundary condition. Another example of mixed boundary condition is when the heat flux at boundary equals the rate of heat exchange with the surroundings by convection. The mixed boundary condition often reduces to the form

$$k\frac{\partial \phi}{\partial z}\bigg|_{t,z_1} = \mathfrak{g}\big(\phi(t,z_1)\big) \tag{4.25}$$

4.2 A BRIEF OVERVIEW OF NUMERICAL METHODS

Various methods have been developed for solving PDEs. This section provides a brief overview of some important classes of methods. The methods discussed here are only indicative; this is not intended to be a comprehensive or accurate review of numerical PDE-solving techniques. I will use the example of a plug flow reactor (PFR) to motivate the idea.

4.2.1 Finite Difference

The first class of methods for solving PDEs is finite difference techniques. The domain is divided into multiple intervals. The differentiation formulae (see Appendix B) are used to approximate numerical derivatives. Thus, the spatial derivative at any location, i, is written using backward difference formula as

$$u\frac{\partial C}{\partial z}\bigg|_i = u_i \frac{C_i - C_{i-1}}{h}$$

or using the central difference formula as

$$u\frac{\partial C}{\partial z}\bigg|_i = u_i \frac{C_{i+1} - C_{i-1}}{2h}$$

The solutions are obtained at the individual discrete points in the domain. In a transient PFR, the concentration will also vary with time. The time derivative may be written as

$$\frac{\partial C}{\partial t} = \frac{C_{p,i} - C_{p-1,i}}{\Delta_t}$$

giving rise to *implicit methods*.

Explicit in time finite difference method is obtained if the time derivative is written as

$$\frac{\partial C}{\partial t} = \frac{C_{p+1,i} - C_{p,i}}{\Delta_t}$$

4.2.2 Method of Lines

Method of lines (MoL) is like that method described above, except that finite difference is used *only* for the spatial domain. The finite difference in space

$$\left.\frac{\partial C}{\partial z}\right|_i = u_i \frac{C_i - C_{i-1}}{h}, \quad \text{or} \quad u\left.\frac{\partial C}{\partial z}\right|_i = u_i \frac{C_{i+1} - C_{i-1}}{2h}$$

converts the original PDE into a set of ODEs. The solution vector then consists of the problem variable at discrete points in the spatial domain:

$$\phi = \begin{bmatrix} C_1 & C_2 & \cdots & C_{n+1} \end{bmatrix}^T$$

and the ODEs are written in the standard for $\phi' = f(\phi)$. These may then be solved using ODE solution techniques mentioned in Chapter 3.

> The finite difference and MoL techniques will be discussed in detail in this book. This chapter will use MoL to solve parabolic and hyperbolic PDEs by converting them into a set of ODEs. Hence, the title of this chapter indicates the solution of *transient PDEs*.

4.2.3 Finite Volume Methods

Finite volume methods (FVM) are a different class of PDE solution techniques. Instead of discretizing the domain into various intervals and writing the differential equation in terms of variables at the nodes, the domain is discretized in *finite volumes*. The original equation is written in the so-called conservative form. Each finite volume behaves as a physical material volume in the domain, with the balance equation applicable to the volume.

The PFR also consists of multiple finite volumes in series. The flux term was derived after letting the finite volume shrink to $\delta z \to 0$ as

$$\frac{\partial}{\partial z}(uC) = \lim_{\delta z \to 0} \frac{(uC)_{\text{out}} - (uC)_{\text{in}}}{\delta z}$$

The idea of FVM is to model instead the finite volume itself; the flux terms give inlet and outlet from the boundaries of the volume and the transient and reaction source terms pertain to the reacting material in the volume itself.

Thus, in principle, FVM applied to a PFR may be considered as multiple CSTRs in series. FVM are commonly used in computational fluid dynamics (CFD) software. FVM will not be discussed further in the rest of this book.

4.2.4 Finite Element Methods

Finite element methods (FEM) use a slightly different approach for approximating the solution after discretizing the domain. Instead of expressing the differential term as a finite difference or solving the equation over a finite volume, the solution is expressed as a combination of basis functions:

$$C = \sum c_i \psi_i$$

where
ψ_i denote basis functions
c_i denote the coefficients

One of the common methods to choose the basis functions is that the functions ψ_i have value 1 at the ith node and zero at all other nodes. The solution C is the linear combination of the basis functions over the entire domain. Once the basis functions are chosen and discretization is fixed, solution of a PDE using FEM mainly involves finding the coefficients c_i.

FEM became popular in solid mechanics. However, applications of FEM have now been found in a wide variety of problems. FEM continue to be popular in solid mechanics and structural problems; they are also finding growing use in multiphysics problems where various types of physics are involved in solving a single problem.

Again, FEM are beyond the scope of this book. It is also mentioned briefly to contrast it with our method of choice: finite difference in space.

4.3 HYPERBOLIC PDE: CONVECTIVE SYSTEMS

Hyperbolic PDEs are most commonly encountered in chemical engineering for transient simulations of convective systems. The conservation equations are of the form

$$\frac{\partial \phi}{\partial t} + \frac{\partial}{\partial x} f_\phi = S(\phi) \tag{4.26}$$

The above equation is first-order hyperbolic PDE when the term f_ϕ is the convective flux. The convective or advective component is given by

$$f_\phi = u\phi \tag{4.27}$$

where u is the velocity. For example, in continuity equation, which is an overall mass balance, $\phi = \rho$, the average density and $S(\phi) = 0$ (except for nuclear reactions). Furthermore, for the sake of initial discussion, we will assume that u is constant. Thus, we are initially interested in hyperbolic PDEs of the type

$$\frac{\partial \phi}{\partial t} + u \frac{\partial \phi}{\partial x} = \mathcal{S}(\phi) \tag{4.28}$$

At a later stage, we will relax this condition and allow u to vary.

The example that we will use in this discussion is that of a PFR. Species conservation equation for a PFR involves conservation of moles or mass of a species. In case of the former, $\phi = C_A$ where C_A is the concentration of reacting species, A. The source term $\mathcal{S}$ captures creation or disappearance of the species due to reaction. Thus, the species balance for a PFR yields the following (set of) first-order hyperbolic PDEs:

$$\frac{\partial C_A}{\partial t} + \frac{\partial}{\partial x}(uC_A) = r_A(\mathbf{C}) \tag{4.29}$$

In several cases, the velocity u is constant* (or nearly constant). In such cases, the PFR is governed by the following PDE:

$$\frac{\partial C_A}{\partial t} + u \frac{\partial C_A}{\partial x} = r_A(\mathbf{C}) \tag{4.30}$$

We will now discuss various numerical methods for solving hyperbolic PDE (4.28).

4.3.1 Finite Differences in Space and Time[†]

The simplest procedure to solve a first-order hyperbolic PDE is to use finite difference in time and space. Let us choose Δ_t as the step-size in the t-direction and Δ_x as the step-size in the x-direction. The spatial domain is of a finite length L so that the number of spatial divisions is $n = L/\Delta_x$. The discretization procedure will convert the PDE into a set of algebraic equations or expressions, which will be used to march forward in time. As described in Chapter 2, we may use either explicit or implicit method in time for marching forward.

4.3.1.1 Upwind Difference in Space

An *explicit* $\mathcal{O}(\Delta_t)$ finite difference in time is the simplest choice for discretizing the PDE. We start with initial conditions at $t = 0$ and march forward in time. We are nominally free to choose the finite difference scheme in space. Let us choose first-order backward difference formula for discretizing in space. The convective term is then written as

$$\left[\frac{\partial \phi}{\partial x}\right]_{p,i} = \frac{\phi_{p,i} - \phi_{p,i-1}}{\Delta_x} \tag{4.31}$$

---

* This is true for liquid-phase reactions and isothermal gas-phase reactions. The velocity changes in nonisothermal gas-phase reactions, and/or in gas-phase reactions with significant change in the number of moles due to reaction.

[†] The methods discussed in this section are useful for understanding some numerical basis behind finite difference techniques for solving hyperbolic PDEs. A reader interested in numerical methods may continue reading this section, whereas a reader interested in process simulation may skip this section without loss of continuity.

where p is the time index such that $t = p\Delta_t$ and i is the spatial index such that $x = i\Delta_x$. The source term is calculated at time $p\Delta_t$ and location $i\Delta_x$ and is represented as $\mathcal{S}_{p,i} \triangleq \mathcal{S}(p\Delta_t, i\Delta_x)$. Forward difference in time yields the following $\mathcal{O}(\Delta_t^1)$ accurate approximation:

$$\frac{\partial \phi}{\partial t} = \frac{\phi_{p+1,i} - \phi_{p,i}}{\Delta_t} \tag{4.32}$$

Substituting in Equation 4.28, the difference equation is given by

$$\phi_{p+1,i} = \phi_{p,i} - \frac{u\Delta_t}{\Delta_x}\left[\phi_{p,i} - \phi_{p,i-1}\right] + \Delta_t \mathcal{S}_{p,i} \tag{4.33}$$

which is an explicit expression to calculate $\phi_{p+1,i}$ and march forward in time.

The *local stability* of the above equation can be analyzed using the so-called *von Neumann stability analysis*. It is local because it ignores the effect of boundary conditions and is derived for homogeneous equation (i.e., without the source term). The derivation of stability conditions is relatively straightforward, but will be skipped since it is beyond the scope of this book. The backward difference formula is stable if velocity u is positive and the following Courant condition is satisfied:

$$\left|u\right|\frac{\Delta t}{\Delta x} \le 1 \tag{4.34}$$

When the flow is from right to left, that is, when the velocity is negative, backward difference formula is unstable and the *forward difference formula* should be used.

Thus, the *upwind* difference formula is given by

$$\phi_{p+1,i} = \begin{cases} \phi_{p,i} - \dfrac{u\Delta_t}{\Delta_x}\left[\phi_{p,i} - \phi_{p,i-1}\right] + \Delta_t \mathcal{S}_{p,i}, & \text{if } u > 0 \\[2mm] \phi_{p,i} - \dfrac{u\Delta_t}{\Delta_x}\left[\phi_{p,i+1} - \phi_{p,i}\right] + \Delta_t \mathcal{S}_{p,i}, & \text{if } u < 0 \end{cases} \tag{4.35}$$

Figure 4.1 shows the stencil of dependence for the upwind difference scheme. The square represents the point $(p+1, i)$ in the T–X space where the solution needs to be found. The solid line and arrow connect points used to compute the spatial and temporal derivatives, respectively. The filled circles denote points where the values are known. The stencil is shown for all methods discussed in this section.

Let us consider the physical interpretation of the Courant condition. When $u > 0$, the material flows from left to right. The conditions at any spatial location are therefore influenced by the node that is upstream of itself. Thus, using a backward difference formula retains this physical condition that ϕ_i is influenced by preceding nodes ϕ_{i-1}. Same can be observed for opposite flow direction also. Furthermore, the Courant condition can be written as

$$\left|u\right| \le \frac{\Delta x}{\Delta t}$$

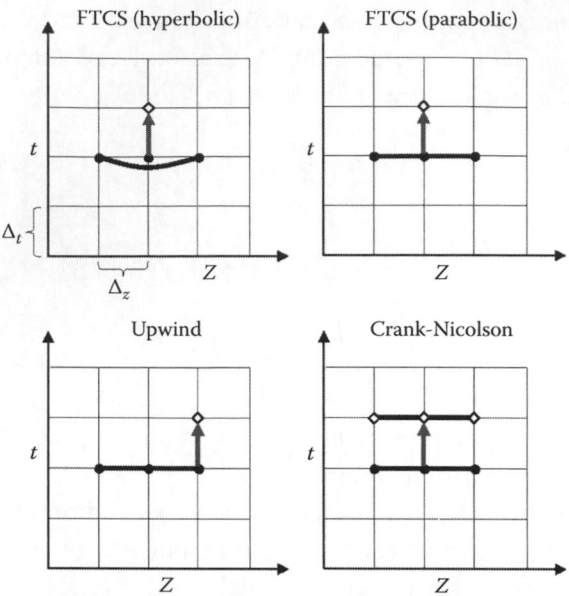

FIGURE 4.1   Schematic diagram of various finite difference schemes. Thin lines represent the computing grid, with the thin dashed line indicating points in the future. Solid circles are known points at current time and empty diamonds represent unknown points to be calculated. The thick solid lines link points connected by finite difference in space; arrows link points connected by finite difference in time.

Since u is the physical speed of propagation, the numerical rate $u_{num} = \Delta x/\Delta t$ should be greater than the speed u to stably capture the effects of convection. Thus, the spatial differencing should maintain *direction* of information propagation and the grid in time domain (i.e., Δt) should be brought closer than the rate at which information convects.

The forward in time, first-order upwind differencing method discussed here is not accurate enough to be useful for practical examples. The discussion here, however, will serve as a useful reminder for future analysis.

4.3.1.2 Forward in Time Central in Space (FTCS) Differencing

The accuracy of the numerical technique may be improved using a central difference formula for the spatial derivative:

$$\left[\frac{\partial \phi}{\partial x}\right]_{p,i} = \frac{\phi_{p,i+1} - \phi_{p,i-1}}{2\Delta_x} \tag{4.36}$$

Substituting this in Equation 4.28 yields the following formula:

$$\phi_{p+1,i} = \phi_{p,i} - \frac{u\Delta_t}{2\Delta_x}\left[\phi_{p,i+1} - \phi_{p,i-1}\right] + \Delta_t \mathcal{S}_{p,i} \tag{4.37}$$

The above is an explicit relation. Unfortunately, the FTCS method is *unstable* for any choice of Δ_t. The proof can be arrived at using the von Neumann stability analysis. This implies that any error in the numerical solution of $\phi_{p,i}$ will grow unbounded and the FTCS method may not be used for first-order hyperbolic PDEs.

4.3.1.3 Lax-Friedrichs Scheme

Lax and Friedrichs proposed a simple way to make the FTCS method stable. They replaced the term $\phi_{p,i}$ on the right-hand side of Equation 4.37 with an average of the values at its neighboring nodes: $\phi_{p,i} \rightarrow (\phi_{p,i+1} + \phi_{p,i-1})/2$. Thus, the updated equation using Lax-Friedrichs method is

$$\phi_{p+1,i} = \frac{1}{2}\left(\phi_{p,i+1} + \phi_{p,i-1}\right) - \frac{u\Delta_t}{2\Delta_x}\left[\phi_{p,i+1} - \phi_{p,i-1}\right] + \Delta_t \mathcal{S}_{p,i} \tag{4.38}$$

The von Neumann stability analysis results in the same Courant stability condition:

$$|u|\frac{\Delta t}{\Delta x} \le 1 \tag{4.34}$$

Let us analyze Lax-Friedrichs scheme further by rearranging Equation 4.38 as

$$\phi_{p+1,i} - \phi_{p,i} = \frac{1}{2}\left(\phi_{p,i+1} - 2\phi_{p,i} + \phi_{p,i-1}\right) - \frac{u\Delta_t}{2\Delta_x}\left[\phi_{p,i+1} - \phi_{p,i-1}\right] + \Delta_t \mathcal{S}_{p,i} \tag{4.39}$$

Dividing by Δ_t

$$\frac{\phi_{p+1,i} - \phi_{p,i}}{\Delta_t} = \frac{\Delta_x^2}{2\Delta_t}\frac{\left(\phi_{p,i+1} - 2\phi_{p,i} + \phi_{p,i-1}\right)}{\Delta_x^2} - \frac{u\Delta_t}{2\Delta_x}\left[\phi_{p,i+1} - \phi_{p,i-1}\right] + \mathcal{S}_{p,i} \tag{4.40}$$

Clearly, the first term above is an additional term, which is a finite difference approximation of

$$\frac{\Delta_x^2}{2\Delta_t}\frac{\left(\phi_{p,i+1} - 2\phi_{p,i} + \phi_{p,i-1}\right)}{\Delta_x^2} \equiv \frac{\Delta_x^2}{2\Delta_t}\left[\frac{\partial^2 \phi}{\partial x^2}\right]_{p,i} \tag{4.41}$$

Thus, Lax-Friedrichs scheme makes the FTCS method stable by introducing a *numerical diffusion* term, described in Equation 4.41.

4.3.1.4 Higher-Order Methods

Accuracy in time can be improved using higher-order differencing in time.

The leapfrogging method uses central difference formula to approximate the time derivative as well. Recall that the central difference formula is $\phi'_p = \left(\phi_{p+1} - \phi_{p-1}\right)/\Delta_t$. The update equation can be derived in a straightforward manner and is left as an exercise. The leapfrogging method is stable for a range of Δ_t values given by the Courant condition. However, the problem is that the time derivative does not include values at time instance p. Thus, the alternate mesh points are decoupled, due to which the errors at even- and odd-numbered nodes evolve independently of each other. Consequently, this method often does not give acceptable results.

Lax-Wendroff method, which is a combination of Lax-Friedrichs and leapfrogging methods, is another alternative that is $\mathcal{O}\left(\Delta_t^2\right)$ accurate in time. It uses the Lax-Friedrichs scheme with a half step to determine values $\phi_{p+0.5,i\pm0.5}$. These two values are then used in leapfrogging method, which is again implemented with half-steps. Consequently, this is a two-step explicit formula. While the method is good in theory, it faces several issues when handling stiff systems or systems with highly nonlinear source term, $\mathcal{S}(\phi)$.

Higher-order upwind method: While the above two methods provided higher-order differencing in time, it sometimes also becomes necessary to use higher-order differencing in space. For example, three-point upwind difference equation can be used to improve spatial accuracy of upwind method. This will be considered in Section 4.3.3 for "method of lines." The derivation of forward-in-time three-point upwind method is also left as an exercise.

In summary, I have discussed several methods for using finite differences in time and space. I have personally found explicit-in-time finite differences inadequate (poor accuracy and stability). Crank-Nicolson (Section 4.3.2) and MoL (Section 4.3.3) are my preferred numerical methods for solving practical problems involving PDEs.

4.3.2 Crank-Nicolson: Second-Order Implicit Method

Chapter 3 discussed implicit and explicit methods for solving ODEs, where it was shown that implicit (Euler's) method is globally stable. The same principle works for hyperbolic PDEs as well. Using finite difference in time, Equation 4.28 is written as

$$\frac{\phi_{p+1,i} - \phi_{p,i}}{\Delta_t} = -u\left[\frac{\partial\phi}{\partial x}\right]_{p+1,i} + \mathcal{S}\left(\phi_{p+1,i}\right) \tag{4.42}$$

Note that both $\partial\phi/\partial x$ and $\mathcal{S}(\phi)$ are computed at time location $(p+1)$. A central difference formula may be used for $\partial\phi/\partial x$ and the first-order fully implicit formula may be derived. The derivation is left as an exercise.

The above first-order implicit method is not very useful due to its low accuracy. Crank and Nicolson, in the 1940s, derived a second-order implicit formula. Although they derived it for a general parabolic PDE, it is applicable to hyperbolic PDEs as well. The time differencing is based on the trapezoidal rule (see Appendix D for numerical integration and Chapter 8 for ODEs). The idea is similar to the one discussed with Heun's method in Chapter 3: A higher-order accurate method is obtained by taking the average of function values at $t=p\Delta_t$ and $t=(p+1)\Delta_t$. Specifically, the right-hand side of Equation 4.42 is replaced by an average at the two points. This results in the following implicit formula for *Crank-Nicolson method*:

$$\frac{\phi_{p+1,i} - \phi_{p,i}}{\Delta_t} = -\frac{u}{2}\left(\left[\frac{\partial\phi}{\partial x}\right]_{p,i} + \left[\frac{\partial\phi}{\partial x}\right]_{p+1,i}\right) + \frac{1}{2}\left(\mathcal{S}_{p,i} + \mathcal{S}_{p+1,i}\right) \tag{4.43}$$

$$\frac{\phi_{p+1,i} - \phi_{p,i}}{\Delta_t} + \frac{u}{2}\left(\frac{\phi_{p,i+1} - \phi_{p,i-1}}{2\Delta_x} + \frac{\phi_{p+1,i+1} - \phi_{p+1,i-1}}{2\Delta_x}\right) - \frac{1}{2}\left(\mathcal{S}_{p,i} + \mathcal{S}_{p+1,i}\right) = 0 \qquad (4.44)$$

The above is a nonlinear equation, which is written for each spatial location, $1 \leq i \leq n$. All the quantities at time-coordinate p are known and the quantities at $p+1$ are to be computed. The above is a set of nonlinear equations. I have not yet discussed methods for solving nonlinear equations. Hence, I will defer numerical solution of Crank-Nicolson method until later.

4.3.2.1 Preview of Numerical Solution

Equation 4.44 does not represent a single equation, but rather a set of nonlinear equations for all spatial locations at the next time, $(p + 1)$. We define a vector of solution variables

$$\Phi_{p+1} = \begin{bmatrix} \phi_{p+1,1} & \phi_{p+1,2} & \cdots & \phi_{p+1,n} \end{bmatrix}^T \qquad (4.45)$$

with the convention followed in this book of using *column vectors*. The above set of equations (4.44) is written in standard vector form as

$$\mathbf{G}\left(\Phi_{p+1}\right) = \mathbf{0} \qquad (4.46)$$

where $\mathbf{G}$ is a vector of functions, with Equation 4.44 representing the ith row. Various methods for nonlinear algebraic equations are discussed in Chapter 6. An example demonstrating the use of Crank-Nicolson method is deferred to Chapter 8 in Section II of this book.

4.3.3 Solution Using Method of Lines

Since the spatial domain is finite, the discretization in space discussed in Sections 4.3.1 and 4.3.2 are of sufficient accuracy. In the previous chapter, methods for solving transient ODEs with high accuracy were discussed. MoL combines spatial discretization with higher-order transient ODE solvers to solve transient PDEs. The core idea behind MoL is to convert the PDE into a set of ODE-IVP by discretizing in space.

As discussed in Section 4.3.1, the spatial domain is discretized* in n intervals, with $h = L/n$. If the central difference formula is used in space, the PDE (4.28) get converted to a set of n coupled ODE-IVP:

$$\frac{d\phi_i}{dt} = -u\frac{\phi_{i+1} - \phi_{i-1}}{2\Delta_x} + \mathcal{S}\left(\phi_i\right) \qquad (4.47)$$

---

* I use h to represent step-size in discretization. In the previous section, since discretization in both space and time was discussed, I had used Δ_x instead. I have used the symbols h and Δ_x replaceably to represent step-size. The notations will be clear from the context they are used.

Here, we have only retained index for space and dropped the time-index since the equation is now converted to an ODE in time. If we define the vector

$$\Phi = \begin{bmatrix} \phi_1 & \phi_2 & \cdots & \phi_n \end{bmatrix}^T \tag{4.45}$$

then we will get a set of n ODE-IVP in the standard form. As before, we are not limited to central difference in space; any *appropriate* finite difference approximation (see Appendix B for details on numerical differentiation) may be used as well.

In the remainder of this section, we will use the example of an isothermal PFR to compare central difference, first-order upwind and second-order upwind methods.

4.3.3.1 MoL with Central Difference in Space

The transient mass balance on a PFR results in a first-order hyperbolic PDE. The PFR model with a first-order irreversible reaction without volume change is

$$\frac{\partial C_A}{\partial t} + u \frac{\partial C_A}{\partial z} = -kC_A$$

$$C_A(t=0) = C_{A0}, \quad C_A(t,z=0) = C_{A,in} \tag{4.48}$$

The numerical solution of the PDE needs to be benchmarked with a known solution. For this, we use the steady state model for the PFR:

$$\frac{dC_A}{dz} = -\frac{k}{u} C_A \tag{4.49}$$

which can be solved analytically to obtain

$$C_A = C_{A,in} \exp\left(-\frac{kz}{u}\right) \tag{4.50}$$

The numerical solution will involve discretizing Equation 4.48 in space, followed by solving the resulting system of ODEs using an appropriate ODE solver (such as `ode45` or `ode15s`). The next example shows the use of central difference formula in space for the PFR example.

Example 4.1 Transient PFR Using Central Difference in Space

Problem: Solve the species material balance for a PFR described by the PDE (4.48) using MoL. The inlet velocity is $u = 0.4$ m/min, $k = 0.2$ min^{-1}, length of the PFR is 0.5 m, and inlet concentration is 1 mol/L. Assume the initial concentration in the PFR is same as 1 mol/L as well.

Solution: Equation 4.48 is discretized in *n* equal intervals. Since inlet conditions are specified, we do not need to solve for the inlet node using an ODE solver. Hence, for the sake of convenience, the inlet node is denoted as "0," the internal nodes go from 1 to $(n-1)$, and the end node is denoted as *n*th node. The equations for the internal nodes after discretization are

$$\frac{d}{dt}C_i = -u_i \frac{C_{i+i} - C_{i-1}}{2h} - kC_i, \quad i = 2 \text{ to } (n-1)$$

$$\frac{d}{dt}C_i = -u_i \frac{C_{i+i} - C_{in}}{2h} - kC_i, \quad i = 1$$

(4.51)

Since this is an IVP in spatial domain as well, boundary condition is not specified at the end node. The domain equation itself is used at the end node, such as by using a backward difference formula:

$$\frac{d}{dt}C_i = -u_i \frac{C_i - C_{i-1}}{h} - kC_i, \quad i = n$$

(4.52)

The vector of dependent variables that is to be solved for in the ODE solver is

$$Y = \begin{bmatrix} C_1 & C_2 & \cdots & C_n \end{bmatrix}^T$$

Function file for ODE solver: The first step is to write the function file, `pfrMOLLinFun.m`. As described in the Chapter 3, this function takes time and *Y* as input arguments are returns *dY/dt* as the output, with variables `Y` and `dY` being $n \times 1$ column vectors. The core of this code involves the calculation of reaction term (`r`) and convection term (`convec`):

```
r = k*Y(i);                          is calculated for all values of i
convec=u0/(2*h)*(Y(i+1)-Y(i-1));     for i = 2 to (n-1)
convec=u0/h*(Y(i+1)-C0);             for i = 1
convec=u0/h*(Y(i)-Y(i-1));           for i = n
```

In the above, `h=L/n`. The function also requires the following variables to be defined:

```
k, u0, C0, L, n
```

I have followed a consistent and modular style of coding, wherein all the parameters related to reactor operation are defined once in the main driver script and passed on as function arguments. Recall that the function information is provided to the solver `ode45` using the following anonymous function definition:

```
@(t,Y) pfrMOLLinFun(t,Y,modelParam)
```

where the parameters mentioned above are provided to the function using modelParam structure.

The complete function file pfrMOLLinFun.m is given below:

```
function dY=pfrMOLLinFun(t,Y,par)
% Function file for transient PFR problem
% to be solved using method of lines
%% Get the parameters
k=par.k;
C0=par.C0;
u=par.u0;
L=par.L;
n=par.n;
h=L/n;
%% Model equations
dY=zeros(n,1);
for i=1:n
    r=k*Y(i);
    if i==1
        convec=u/(2*h) * (Y(i+1)-C0);
    elseif (i==n)
        convec=u/h * (Y(i)-Y(i-1));
    else
        convec=u/(2*h) * (Y(i+1)-Y(i-1));
    end
    dY(i)=-convec-r;
end
```

Driver script to solve the problem: As described in Chapter 1, the driver script consists of three parts: defining problem parameters (the list was given above), solving the ODE, and result output. The ODEs will be solved using ode45 in MATLAB®, with the prescribed initial conditions. For results, the final steady state profile of C_A vs. z is plotted as figure(1) and the transients in outlet concentration plotted as figure(2).

```
% Driver script for simulating transient PFR
% using Method of Lines (central difference)
%% Model Parameters
modelParam.k=0.2;
modelParam.u0=0.4;    u0=modelParam.u0;
modelParam.C0=1;      C0=modelParam.C0;
modelParam.L=0.5;

%% Initializing and simulating the PFR
n=100;
modelParam.n=n;
h=modelParam.L/n;
```

```
Y0=ones(n,1)*C0;
[T,CA]=ode45(@(t,Y) pfrMOLLinFun(t,Y,modelParam), ...
    [0 10],Y0);
```

```
%% Plotting results
% Steady state plot
C_ss=[C0, CA(end,:)];      % Results: steady state
Z=[0:h:h*n];               % Location
CModel=C0*exp(-modelParam.k*Z/u0);
figure(1)
plot(Z,C_ss,'-',Z,CModel,'.');
xlabel('Length (m)'); ylabel('C_A (mol/L)');

% Exit concentration vs. time
figure(2);
plot(T,CA(:,end));
xlabel('time (min)'); ylabel('C_A (mol/m^3)');
```

The results at steady state are plotted in Figure 4.2. Symbols represent the analytical solution for comparison, whereas the solid line represents the numerical solution. There is clearly instability observed in the numerical solution. The numerical solution oscillates around the true solution.

This example demonstrates the unsuitability of central difference formula in MoL solution with explicit transient ODE solver for convection-dominated systems.

4.3.3.2 MoL with Upwind Difference in Space
In Section 4.2.1, the central and upwind differencing schemes were compared. Since ode45 is a variable-step fourth-order explicit ODE solver, those *actual* stability results do not apply, though the observed *trend* is still applicable. Analogously in MoL, upwind differencing in

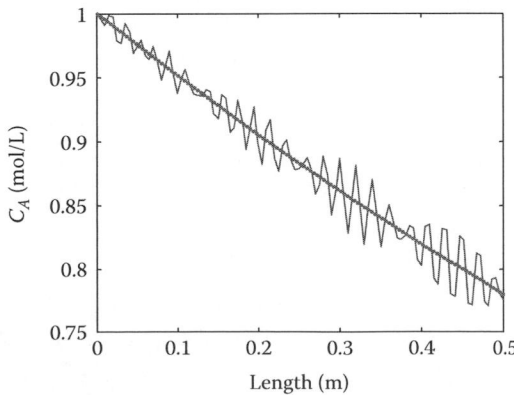

FIGURE 4.2 Numerical results with MoL employing central differences in space (solid line) compared with true solution at steady state (symbols).

space will provide better stability behavior when coupled with ode45. When the flow is in the positive z-direction, upwind implies a backward difference formula. Thus

$$\frac{d}{dt}C_i = -u_i \frac{C_i - C_{i-1}}{h} - kC_i, \quad \forall i \quad \text{with } C_0 \leftarrow C_{A,\text{in}} \tag{4.53}$$

The two-point backward difference formula is only $\mathcal{O}\left(h^1\right)$ accurate, but is easy to implement. If a greater accuracy is desired, a higher-order formula is needed for special discretization to improve accuracy. A three-point backward difference formula

$$\frac{d}{dt}C_i = -u_i \frac{3C_i - 4C_{i-1} + C_{i-2}}{2h} - kC_i, \quad i = 2 \text{ to } n \quad \text{with } C_0 \leftarrow C_{A,\text{in}} \tag{4.54}$$

can be used, with an $\mathcal{O}\left(h^2\right)$ accuracy.

The most convenient option for the first internal node ($i = 1$) is to use a standard two-point backward difference formula:

$$\frac{d}{dt}C_i = -u_i \frac{C_i - C_{A,\text{in}}}{h} - kC_i, \quad i = 1 \tag{4.55}$$

While this is simple to use and does not adversely affect the stability of upwind method, its disadvantage is that it is only $\mathcal{O}\left(h\right)$ accurate. Since the leading error is governed by the least accurate method, using a two-point formula makes the overall scheme less accurate.

NOTE ON SECOND-ORDER UPWIND

My personal experience with higher-order upwind differencing has been significantly more positive than what the naïve view of theory would suggest. The statement made above, that the overall accuracy of numerical method is governed by the worst-case error, is indeed correct. However, the error and stability analyses we performed were *local*. While Equation 4.55 indeed is $\mathcal{O}\left(h^1\right)$ accurate, the boundary condition $C_{A,in}$ is assumed to be known exactly without any error. On the other hand, all the terms on the right-hand side of Equation 4.54 may have numerical errors.

Consequently, in my experience, the performance of MoL employing three-point upwind formula (4.54) for internal nodes combined with the two-point formula (4.55) at the first node has been excellent for a large fraction of practical simulation problems.

The next example presents MoL with upwind differencing in space, for the conditions in Example 4.1. First, the standard two-point backward difference formula will be used. A comparison between central difference (Example 4.1) and upwind difference (Example 4.2) will demonstrate the suitability of the latter for hyperbolic PDEs. Thereafter, the PFR problem will be solved for a greater residence time (i.e., with lower inlet velocity) to compare the two- and three-point upwind differences.

Example 4.2 Transient PFR with Backward Difference in Space

Problem: Solve Example 4.1 using upwind difference for spatial discretization.

Solution: The problem will be solved in two parts. First, let us compare two-point upwind with central difference in space.

Two-point formula: The driver script remains the same as Example 4.1. In the PFR function, the first section remains the same; only the second section (named "`Model Equations`") is modified as below:

```
%% Model equations
dY=zeros(n,1);
for i=1:n
    r=k*Y(i);
    if i==1
        convec=u/h*(Y(i)-C0);
    else
        convec=u/h*(Y(i)-Y(i-1));
    end
    dY(i)=-convec-r;
end
```

The rest of the code remains the same. The simulations are run for a time span that is long enough for the system to reach steady state. The concentration C_A (obtained at the last time instance) is plotted against spatial location. Figure 4.3 shows that the numerical technique is stable and the numerical solution closely matches the true solution at steady state.

Comparison with three-point formula: The two-point upwind method gives good performance for the conditions above. The conversion from the PFR is less than 25%. Conversion can be increased by increasing the residence time (increase the length or decrease the inlet velocity). The operating condition for the PFR is changed to inlet

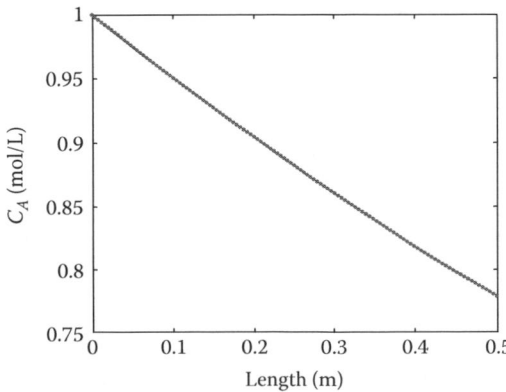

FIGURE 4.3   MoL with backward differences in space (solid line) compared with true solution at steady state (symbols). The line may not be clearly visible since the symbols overlap closely.

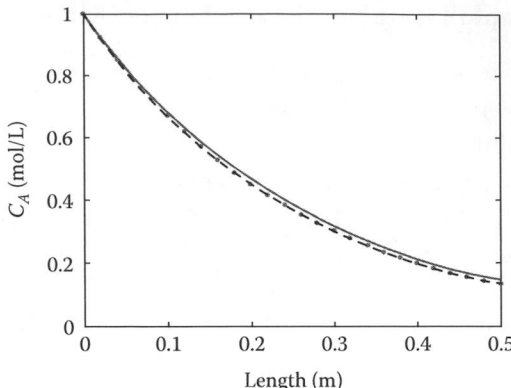

FIGURE 4.4   Comparison of two-point (solid line) and three-point (dashed line) backward difference formula solutions at steady state.

velocity of $u_0 = 0.05$ and the number of axial nodes is reduced to n=25. The simulation time span is increased to [0, 25], which is long enough to reach steady state. Steady state results are plotted in Figure 4.4 as a solid line. Clearly, the numerical solution differs from the analytical one by ~2%–3%.

The accuracy can be improved by using the three-point upwind method. The code for three-point upwind method is left as an *exercise*. The dashed line in Figure 4.4 shows that the three-point upwind difference is more accurate, as can be expected from the fact that it is $\mathcal{O}\left(h^2\right)$ accurate. Transient evolution of the concentration at the PFR exit is plotted in Figure 4.5 as solid (two-point upwind) and dashed (three-point upwind) lines. The two methods show some difference as we approach steady state. Based on comparison with steady state solution in Figure 4.4, the three-point results are more accurate.

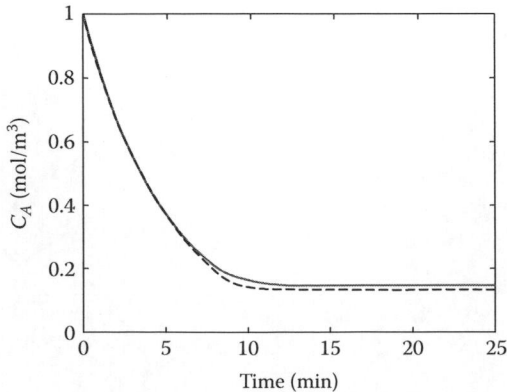

FIGURE 4.5   Comparison of transient results for two- and three-point backward difference formulae.

4.3.4 Numerical Diffusion

Another key issue in the finite difference methods is the so-called numerical diffusion. This was discussed for the Lax-Friedrichs scheme, where the modification to improve stability of central difference was shown to be equivalent to introducing diffusion. The reality is that numerical diffusion is introduced by all the finite difference methods.

I will demonstrate this using the following example of a tracer injection. Let us consider the flow of material through a pipe. In the absence of axial diffusion (i.e., for an ideal plug flow), any packet of fluid spends its residence time $\tau = L/u$ within the pipe. Initially, pure water flows through the pipe (initial condition is $C(0) = 0$). At certain time, $t = 0^+$, a dye is injected into the inlet at an inlet concentration of 1 mol/L. The flow of this tracer through the pipe can be modeled as a plug flow system with no reaction:

$$\frac{\partial C}{\partial t} + u\frac{\partial C}{\partial z} = 0, \quad C(t, z = 0) = C_{\text{in}}, \quad C(t = 0) = 0 \tag{4.56}$$

Let the inlet velocity be 0.1 m/min and the PFR length as 0.5 m. Based on the physics of the system, we expect that at 1 min after the start of injection, the tracer (at 1 mol/min) to cover the first 0.1 m, whereas the remainder contains "original" dye-free fluid packets that have not yet been pushed out of the PFR. This tracer moves as a step through the PFR as time progresses until $t = 0.5/0.1$ min, after which the entire PFR contains water with dye.

The above problem can be solved using the code from Example 4.2, with k=0. Simulation results obtained using upwind method with a very large number of nodes, n=10000, is shown by solid lines in Figure 4.6. This is close to the true solution. The dotted lines represent n=1000, whereas dashed lines represent the original number of nodes n=100. All these cases do not show perfectly vertical lines. Clearly, we observe that the tracer has "diffused" into adjoining packets, leading to a smooth transition from 1 to 0 mol/L. The amount of diffusion is negligible with n=10000 (solid lines), the effect is obvious with n=1000

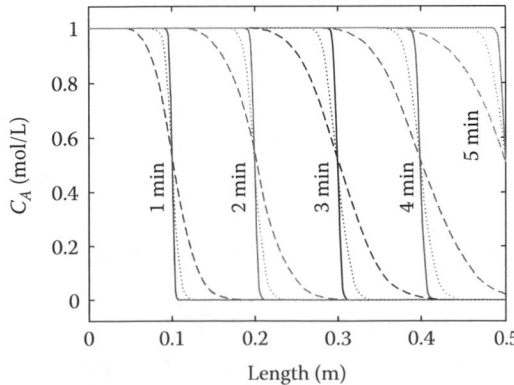

FIGURE 4.6   Simulation of tracer injected in a pipe at various times using 10,000 nodes (solid line), 100 nodes (dashed lines), and 1,000 nodes (dotted lines).

(dotted lines) and prominent with n=100 (dashed lines). The cause of this observed behavior is *numerical diffusion*.

Consider the upwind difference in space. Although we wrote $\phi_i' = (\phi_i - \phi_{i-1})/h$, we have neglected the higher-order terms. In fact, the approximation using Taylor's series is

$$C_{A,(i-1)} = C_{A,i} - h \left. \frac{\partial C_A}{\partial z} \right|_i + \frac{h^2}{2!} \left. \frac{\partial^2 C_A}{\partial z^2} \right|_i - \frac{h^3}{3!} \left. \frac{\partial^3 C_A}{\partial z^3} \right|_i + \cdots \tag{4.57}$$

Therefore

$$\frac{C_{A,i} - C_{A,(i-1)}}{h} = \left. \frac{\partial C_A}{\partial z} \right|_i - \frac{h}{2} \left. \frac{\partial^2 C_A}{\partial z^2} \right|_i + \frac{h^2}{6} \left. \frac{\partial^3 C_A}{\partial z^3} \right|_i - \cdots \tag{4.58}$$

Substituting in the upwind difference–based MoL from Equation 4.53

$$\frac{d}{dt} C_i = -u_i \left[\left. \frac{\partial C_A}{\partial z} \right|_i - \frac{h}{2} \left. \frac{\partial^2 C_A}{\partial z^2} \right|_i + \frac{h^2}{6} \left. \frac{\partial^3 C_A}{\partial z^3} \right|_i - \cdots \right] \tag{4.59}$$

The leading error in the higher-order terms is

$$u_i \frac{h}{2} \left. \frac{\partial^2 C_A}{\partial z^2} \right|_{\xi \in [i-1, i]} \tag{4.60}$$

This is indeed the *numerical diffusion*, with the diffusion coefficient

$$\mathcal{D}_{\text{numerical}} = 0.5 h\, u_i \tag{4.61}$$

The numerical diffusion decreases with decreasing step-size, h, as was evident in Figure 4.6.

The reaction source term in transient PFR predominates and masks the effects of numerical diffusion. This is also true for the majority of problems of interest to chemical engineers. Thus, first- or second-order upwind methods are sufficiently accurate with reasonable step-size, h. However, in some problems involving tracking of a moving front, high resolution numerical methods must be used. Such problems are not considered in this book and a discussion of high-resolution hyperbolic PDE methods is beyond the scope of this text.

4.4 PARABOLIC PDE: DIFFUSIVE SYSTEMS

Parabolic PDEs result most often when simulating processes with diffusive component in one direction. Common examples of such systems include transient heat conduction in a solid body, problems involving axial convection and radial diffusion, transient reactor with axial diffusion, etc. The discussion in the preceding section can be extended to transient equations with diffusion in spatial dimension. Numerically, diffusion introduces second-order differential in space, resulting in parabolic PDEs. Transient diffusion (or conduction) equations

$$\frac{\partial \phi}{\partial t} = \frac{\partial}{\partial z}\left(\mathcal{D}\frac{\partial \phi}{\partial z}\right) + \mathcal{S}(\phi) \tag{4.62}$$

or transient convection-diffusion problems

$$\frac{\partial \phi}{\partial t} + \frac{\partial}{\partial z}(u\phi) = \frac{\partial}{\partial z}\left(\mathcal{D}\frac{\partial \phi}{\partial z}\right) + \mathcal{S}(\phi) \tag{4.63}$$

represent general model forms for parabolic PDEs.

The velocity u and the coefficient $\mathcal{D}$ may be constant, leading to quasilinear parabolic PDE:

$$\frac{\partial \phi}{\partial t} + u\frac{\partial \phi}{\partial z} = \mathcal{D}\frac{\partial^2 \phi}{\partial z^2} + \mathcal{S}(\phi) \tag{4.64}$$

Packed-bed reactor with axial mixing: The PFR model of the previous section assumed fluid elements to move as "plugs" and axial back-mixing was ignored. A packed bed reactor with axial mixing will be used as an example of parabolic PDE in this section:

$$\frac{\partial C_A}{\partial t} + u\frac{\partial C_A}{\partial z} = \mathcal{D}_e \frac{\partial^2 C_A}{\partial z^2} - kC_A \tag{4.65}$$

The initial condition remains the same as in the previous (PFR) case:

$$C_A(t=0,z) = C_0 \tag{4.66}$$

Two boundary conditions are required. The inlet boundary condition may be concentration specified at the inlet, whereas a no-flux condition may be used at the outlet:

$$C_A(t,z=0) = C_{in}, \quad \left.\frac{\partial C_A}{\partial z}\right|_{t,z=L} = 0 \tag{4.67}$$

The above two represent Dirichlet and Neumann boundary conditions, respectively.

The previous example was solved in dimensional quantities. For this example, let us use a nondimensional formulation. Let us define variable $\phi = C_A/C_{in}$ as the dimensionless concentration and $\zeta = z/L$ as the dimensionless length. Using the residence time as the key time scale in the reactor, we define $\tau = t/(L/u)$. This yields

$$C_{in}\frac{u}{L}\frac{\partial \phi}{\partial \tau} + \frac{uC_{in}}{L}\frac{\partial \phi}{\partial \zeta} = \frac{\mathcal{D}_e C_{in}}{L^2}\frac{\partial^2 \phi}{\partial \zeta^2} - kC_{in}\phi \tag{4.68}$$

Rearranging

$$\frac{\partial \phi}{\partial \tau} + \frac{\partial \phi}{\partial \zeta} = \frac{1}{\text{Pe}} \frac{\partial^2 \phi}{\partial \zeta^2} - \text{Da} \cdot \phi$$

$$\text{where} \quad \text{Pe} = \frac{uL}{\mathcal{D}_e}, \quad \text{Da} = \frac{kL}{u} \tag{4.69}$$

are Péclet and Damköhler numbers, respectively. Péclet number is the ratio of convective to diffusive transport rates and the Damköhler number* is the ratio of reaction to convective transport rates. The higher the Péclet number, the greater is the contribution of the convective transport compared to axial diffusion. The reaction time scale for an nth order reaction is

$$\tau_{rxn} = \frac{r(C_A)}{C_A} = k_n C_A^{n-1} \tag{4.70}$$

In summary, Equation 4.69 is the parabolic PDE of interest:

$$\frac{\partial \phi}{\partial \tau} + \frac{\partial \phi}{\partial \zeta} = \frac{1}{\text{Pe}} \frac{\partial^2 \phi}{\partial \zeta^2} - \text{Da} \cdot \phi \tag{4.69}$$

subject to initial and boundary conditions

$$\phi(\tau = 0, \zeta) = \phi_0 \tag{4.71}$$

$$[\phi]_{\zeta=0} = 1, \quad \left.\frac{\partial \phi}{\partial \xi}\right|_{\zeta=1} = 0 \tag{4.72}$$

The next subsections discuss numerical methods to solve parabolic PDEs.

A SIDENOTE ON TREATMENT IN THIS SECTION

Most books on numerical techniques provide a detailed treatment of parabolic PDEs of the form of Equation 4.62, where the convective term is neglected. The general convective-diffusive equation (4.64) is often treated separately, if at all. The presence of convective term affects the nature of solution, and hence the choice of numerical scheme. However, following the discussion in Section 4.3, we are equipped to handle the general equation (4.64). I will therefore present a unified treatment in this section, so that the techniques discussed herein are equally applicable to the transient diffusive PDE (4.62).

4.4.1 Finite Difference in Space and Time

The forward in time and central in space (FTCS) method can be used to discretize PDE (4.64). Although the FTCS method was unstable for hyperbolic PDEs, the presence of the diffusive term, $\mathcal{D}\phi_{zz}$, makes the FTCS method stable for parabolic PDEs. Recall that the Lax-Friedrichs method stabilized FTCS for hyperbolic PDEs by introducing a

---

* More precisely, this is known as Damköhler number of the first kind, Da_I. The Damköhler number of the second kind is the ratio of reaction to *transverse* mass transport rates. Mass transfer in this example is *axial* mass transfer.

numerical diffusion term. The presence of a physical diffusion in parabolic PDE makes the original FTCS method conditionally stable.

As before, let us choose Δ_t as the step-size in the t-direction, Δ_z as the step-size in the z-direction, and $n = L/\Delta_z$ as the number of spatial divisions. The central difference formula for the diffusive term is

$$\left[\frac{\partial^2 \phi}{\partial z^2} \right]_{p,i} = \frac{\phi_{p,i+1} - 2\phi_{p,i} + \phi_{p,i-1}}{\Delta_z^2} \tag{4.73}$$

The forward difference in time and central difference for convective term remain the same as before. Substituting the finite differences in Equation 4.64 and rearranging yields the FTCS update:

$$\phi_{p+1,i} = \phi_{p,i} - \frac{u\Delta_t}{2\Delta_z}\left[\phi_{p,i+1} - \phi_{p,i-1} \right] + \frac{D\Delta_t}{\Delta_z^2}\left[\phi_{p,i+1} - 2\phi_{p,i} + \phi_{p,i-1} \right] + \Delta_t S_{p,i} \tag{4.74}$$

Comparing the above equation with Equation 4.37, we see that the diffusive term is additional. The diffusive term, $\dfrac{D\Delta_t}{\Delta_z^2}\left[\phi_{p,i+1} - 2\phi_{p,i} + \phi_{p,i-1} \right]$, makes the FTCS stable for parabolic PDEs.

The condition for stability of FTCS is

$$\frac{D\Delta_t}{\Delta_x^2} < \frac{1}{2}, \quad \frac{u^2\Delta_t}{2D} < 1 \tag{4.75}$$

The first condition implies Δ_t should be chosen such that the grid is sufficient to capture the physical diffusive effects governed by the coefficient D. The second condition ensures that the diffusive component at the local grid is dominant over the convective component. Note that for a purely diffusive system, the second condition is trivially satisfied since $u = 0$.

Leapfrogging and *Lax-Wendroff* methods, which are both $\mathcal{O}\left(\Delta_t^2 \right)$ accurate, can be adapted for parabolic PDEs as well. They follow the same principles as elaborated before.

4.4.2 Crank-Nicolson Method

The implicit Crank-Nicolson method, described in Section 4.3.2, is applicable for parabolic PDEs as well. Crank-Nicolson method is unconditionally stable* for both hyperbolic and parabolic PDEs. In other words, there is no limit on the step-size Δ_t to ensure stability. The reader can verify that the Crank-Nicolson method for Equation 4.64 is

$$\frac{\phi_{p+1,i} - \phi_{p,i}}{\Delta_t} + \frac{u}{2}\left(\frac{\phi_{p,i+1} - \phi_{p,i-1}}{2\Delta_x} + \frac{\phi_{p+1,i+1} - \phi_{p+1,i-1}}{2\Delta_x} \right) - \frac{1}{2}\left(S_{p,i} + S_{p+1,i} \right)$$

$$- \frac{D}{2}\left(\frac{\phi_{p,i+1} - 2\phi_{p,i} + \phi_{p,i-1}}{\Delta_x^2} + \frac{\phi_{p+1,i+1} - 2\phi_{p+1,i} + \phi_{p+1,i-1}}{\Delta_x^2} \right) = 0 \tag{4.76}$$

---

* It should be noted that these stability conditions are derived with constant source term. In practice, the stability conditions may be violated due to strongly nonlinear source terms. Also, these conditions are derived locally, implying that the effect of boundary conditions or strongly varying coefficients $\left(u \text{ or } D \right)$ is not covered in these derivations.

The above is a nonlinear equation, which is written for each spatial location, $1 \leq i \leq n$. The quantities at time $p + 1$ should be computed simultaneously using an appropriate nonlinear equation solver. Chapters 6 and 7 will discuss techniques that will equip us to solve Equation 4.76 and the solution will be discussed in Chapter 8.

4.4.3 Method of Lines Using MATLAB® ODE Solvers

PDF (4.64) is discretized in the spatial domain using central difference formula and solved using MoL. This works like the method discussed in Section 4.3.3, with an added benefit that due to the axial diffusion term, the central difference is likely to be stable. I will demonstrate this method using the reactor with axial mixing, modeled in Equation 4.69, with initial conditions, (4.71) and boundary conditions, (4.72).

4.4.3.1 MoL with Central Difference in Space

As before, the solution vector is defined as

$$Y = \begin{bmatrix} \phi_1 & \phi_2 & \cdots & \phi_n \end{bmatrix}^T \tag{4.77}$$

and the step-size, $h = L/n$. The discretized equations for the internal nodes ($i = 2$ to $n-1$) are

$$\frac{d}{d\tau}\phi_i = -\frac{\phi_{i+1} - \phi_{i-1}}{2h} + \frac{\phi_{i+1} - 2\phi_i + \phi_{i-1}}{Pe_i\, h^2} - Da_i \cdot \phi_i \tag{4.78}$$

The nondimensional inlet boundary condition yields the equation for the first node ($i = 1$):

$$\frac{d}{d\tau}\phi_i = -\frac{\phi_{i+1} - 1}{2h} + \frac{\phi_{i+1} - 2\phi_i + 1}{Pe_i\, h^2} - Da_i \cdot \phi_i, \quad i = 1 \tag{4.79}$$

Neumann or mixed boundary conditions can be handled in a couple of ways:

1. Use backward difference formula for the boundary node, n.

2. Use the so-called *ghost-point* method. This will be our method of choice.

The above domain equation for the end node ($i = n$) introduces a variable at a new point, ϕ_{n+1}. This point is not within the domain that is being solved, but can be eliminated using the exit boundary condition. The Neumann boundary condition at exit yields

$$\frac{\partial \phi_n}{\partial z} = \frac{\phi_{n+1} - \phi_{n-1}}{2h} = 0 \tag{4.80}$$

Substituting the value of ϕ_{n+1} into the equation above, the following ODE is obtained for $i = n$:

$$\frac{d}{d\tau}\phi_i = \frac{-2\phi_i + 2\phi_{i-1}}{Pe_i\, h^2} - Da_i \cdot \phi_i, \quad i = n \tag{4.81}$$

As an example, we will modify the PFR conditions from Example 4.1. The residence time is $t_{res} = 10$ and Damköhler number is Da = 2. Considering the effective diffusivity, $\mathcal{D}_e = 5 \times 10^{-3}$ m²/min, which is at the higher end of the spectrum, for small molecules, the value of Pe = 1. For this condition, the model at steady state is

$$\frac{d^2\phi}{d\xi^2} - \frac{\partial\phi}{\partial\xi} - 2\phi = 0 \tag{4.82}$$

Noting that the roots of the equation

$$\left(D^2 - D - 2\right)\phi = 0 \tag{4.83}$$

are (2, −1), the steady state solution for the system can be obtained as

$$\phi = \frac{e^{2\xi} + 2e^{3-\xi}}{1 + 2e^3} \tag{4.84}$$

The next example shows a numerical solution of the axial dispersion model using MoL for these conditions.

Example 4.3 Reactor with Axial Dispersion

Problem: Obtain transient solutions for a reactor with axial dispersion for Da = 2, Pe = 1, and residence time of 10 minutes.

Solution: The solution follows on similar lines as the PFR problem, except that being a parabolic PDE, central difference formula can be comfortably used. The function file is given below:

```
function dY=axialDiffLinFun(t,Y,par)
% Function for transient reactor w/ axial mixing
% to be solved using method of lines
%% Get the parameters
n=par.n;
Pe=par.Pe;
Da=par.Da;
%% Model equations
h=1/n;
dY=zeros(n,1);
for i=1:n
    r=Da*Y(i);
    if i==1
        convec=(Y(i+1)-1)/(2*h);
        diffu =(Y(i+1)-2*Y(i)+1)/(Pe*h^2);
```

```
    elseif (i==n)
        convec=0;
        diffu =(-2*Y(i)+2*Y(i-1))/(Pe*h^2);
    else
        convec=(Y(i+1)-Y(i-1))/(2*h);
        diffu =(Y(i+1)-2*Y(i)+Y(i-1))/(Pe*h^2);
    end
    dY(i)=-convec+diffu-r;
end
```

The driver script is like the ones we have used before, and hence will be skipped for brevity. The steady state analytical solution is computed as

```
CModel=(exp(2*Z)+2*exp(3-Z))/(1+2*exp(3));
```

and plotted as symbols for comparison. The numerical solution at various times ($\tau = [0.02,\ 0.1,\ 0.2]$, which corresponds to 0.2, 1, and 2 min) is plotted in Figure 4.7. The dashed line is the steady state numerical solution and the symbols are the analytical solution. Clearly, the numerical solution closely matches the analytical solution at steady state, even with $n=25$ axial nodes.

As we did in the previous section, we can run the simulations with initial concentration as $\phi=0$ and no reaction condition of $Da = 0$. This will simulate tracer experiment, where a dye is injected in a reactor and its fate is tracked. Two different conditions are simulated, when $Pe = 5$ (solid lines) and when $Pe = 50$ (dashed lines). For a system that is convection dominant, $Pe \to \infty$. As axial mixing increases, $\mathcal{D}$ increases and, consequently, Péclet number decreases. Thus, the former case ($Pe = 5$) is further away from the PFR than the latter ($Pe = 50$).

Figure 4.8 shows the transient evolution of step tracer in a reactor with axial diffusion. Although at steady state, both cases show uniform concentration of $\phi(\xi) = 1$, the transients

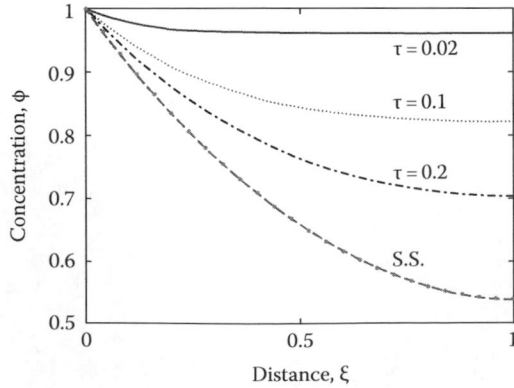

FIGURE 4.7 Concentration profile within a packed bed reactor with axial dispersion and first-order reaction at various times. Symbols represent analytical solution at steady state.

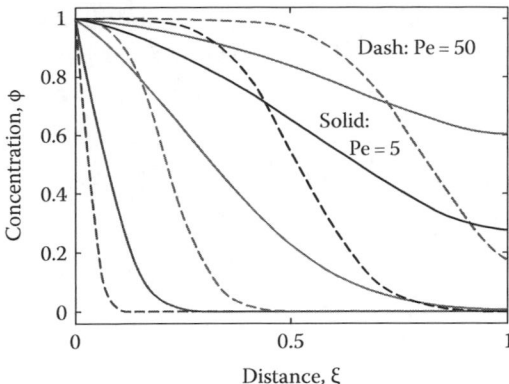

FIGURE 4.8 Simulation of tracer injected in a reactor with axial dispersion at various times. Solid lines represent Pe = 5, and dashed lines represent Pe = 50.

are significantly different. For a larger Pe value, we see a diffuse tracer front traveling in the reactor, whereas the tracer is highly diffused at the lower Pe value.

4.4.4 Methods to Improve Stability

The method of choice discussed in this chapter for parabolic PDEs is the MoL with central difference in spatial domain. The resulting set of ODEs can be solved using MATLAB's `ode45` solver. This provides a good trade-off between ease of implementation, stability, and accuracy. If the problem is stiff, the implicit solver `ode15s` may be used instead. Likewise, MoL using `ode45` may be too slow, in which case again `ode15s` may be used instead.

In some problems, the velocity may vary significantly with time and/or space. Sometimes, the above methods do not provide stable results for such problems. Recall that the stability condition for explicit methods require us to discretize such that the diffusive term is dominant compared to the convective terms. If this condition is not met, a solution using standard MoL may show unstable behavior (similar to oscillations in Figure 4.2 or instability observed in Chapter 3). In such a case, one may resort to a mixture of central difference formula for diffusive term and upwind difference for convective term.

If these methods fail, as is possible in some highly stiff problems, fully implicit methods may be attempted. The Crank-Nicolson method, usually, provides a good starting point for such cases. If all else fails, specialized solvers available at online repositories may be used. Unfortunately, for such cases, it may be pragmatic to use alternative languages such as FORTRAN or C++, where well-tested numerical libraries are available.

4.5 CASE STUDIES AND EXAMPLES

4.5.1 Nonisothermal Plug Flow Reactor

First, a recap of isothermal PFR with first-order reaction was considered in Example 4.2. The model for the PFR was given by Equation 4.30. The PDE was converted into a set of ODEs using MoL. Since this was a hyperbolic PDE, the convective term was approximated using upwind differencing.

In this case study, we will build upon the isothermal PFR model to consider a nonisothermal liquid-phase PFR with complex reaction kinetics. The velocity may be assumed constant for the liquid-phase system. The PFR material balance remains the same as before:

$$\frac{\partial C_A}{\partial t} + u \frac{\partial C_A}{\partial z} = r_A(C_A) \tag{4.30}$$

The reaction rate is given by

$$r_A = -\frac{kC_A}{K_r + C_A}$$

$$k = k_0 \exp\left(-\frac{E}{RT}\right), \quad K_r = K_0 \exp\left(-\frac{\Delta E_r}{RT}\right) \tag{4.85}$$

The transient energy balance for the PFR is given by

$$\frac{\partial T}{\partial t} + u \frac{\partial T}{\partial z} = \frac{(-\Delta H_r) r}{\rho c_p} \tag{4.86}$$

The model for nonisothermal PFR consists of two hyperbolic PDEs, (4.30) and (4.86), subject to Dirichlet boundary condition at the inlet (inlet concentration is specified). The initial conditions will be assumed to be the same as at the inlet. The parameters are given in Example 4.4 itself.

Now, we are solving for two variables, C_A and T. The above two PDEs are discretized using the upwind method. The spatial domain is discretized into n intervals and the two variables are specified at each of the nodes. The variable to be solved for will therefore be

$$\Phi = \begin{bmatrix} C_{A1} \\ T_1 \\ C_{A2} \\ T_2 \\ \vdots \\ \vdots \\ C_{An} \\ T_n \end{bmatrix}$$

Note the arrangement of variables in the column vector Φ. We first list out both the variables at node-1, followed by both variables at node-2, and so on. The above structure of Φ is a personal preference, due to convenience of this organization of the dependent variables. Note that when Equations 4.30 and 4.86 are discretized, the concentration and temperature depend on the values at the same ith node and the values at its neighboring nodes. Thus, I find the above structuring more convenient to use.

MATLAB provides a useful command `reshape`, to reshape matrices. It is convenient to convert the above matrix into the following form using `reshape` command:

$$\bar{\Phi} = \begin{bmatrix} C_{A1} & C_{A2} & \cdots & C_{An} \\ T_1 & T_2 & \cdots & T_n \end{bmatrix} \tag{4.87}$$

The MATLAB command to do this is

```
YBar=reshape(Y,2,n);
```

The individual columns of Y are reshaped into a $2 \times n$ matrix. Reshape can also be used to convert the above matrix back into the original form:

```
Y=reshape(YBar,2*n,1);
```

One needs to be careful in using this command. The elements in the first column are "read" in a sequence, followed by the second column, and so on.

The transient adiabatic PFR problem is solved in the example below.

Example 4.4 Nonisothermal PFR

Problem: Consider a liquid-phase reaction takes place in a nonisothermal PFR. The model equations are given by Equations 4.30 and 4.86. Temperature and concentration at the inlet are 450 K and 1 mol/m^3, respectively.

The reaction rate is given by Equation 4.85. The activation energy is $E = 60$ kJ/mol and $\Delta E_r = -10$ kJ/mol. The rate constants are specified at the temperature $T_1 = 450$ K as $k_1 = 2$ and $K_{r,1} = 1$. The values at any temperature T are given by

$$k = k_1 \exp\left[-\frac{E}{R}\left(\frac{1}{T} - \frac{1}{T_1}\right)\right], \quad K_r = K_{r1} \exp\left[-\frac{\Delta E_r}{R}\left(\frac{1}{T} - \frac{1}{T_1}\right)\right]$$

The heat of reaction is $(-\Delta H) = 100$ kJ/mol and fluid density and the specific heat capacity is $\rho c_p = 800$ J/mol K.

Solve the PFR problem for $L = 2$ m and $u_0 = 0.4$ m/min.

Solution: I will reuse the function file from Example 4.2 as a starting point.

1. *Reaction rate calculation*: As a first step, we will make the function more modular by using a separate function to calculate the reaction rate. Since reaction rate is dependent on only the *local* temperature and concentration, the function to calculate the reaction rate is given below:

```
function r=rxnRate(T,C,par)
% Function computes local rate of reaction
```

```
% given temperature and concentration
Rgas=par.Rgas;        % Gas constant
% Rate constants
k1=par.k1;
E=par.E;
T1=par.T1;
K1=par.K1;
DEr=par.DEr;
k =k1*exp(-E/Rgas*(1/T-1/T1));
Kr=K1*exp(-DEr/Rgas*(1/T-1/T1));
% Compute the reaction rate
r=k*C/sqrt(1+Kr*C^2);
end
```

The advantage of a modular function is that the overall PFR solver can be reused for a variety of different conditions. For example, if simulating a PFR for a different process, the core of PFR solver does not change; only the reaction rate and process parameters change. The former is captured in the reaction rate function, whereas the latter are provided in the driver script.

2. *Function for PFR balances*: Next, let us write the function for PFR balance equations. The function arguments will be time t and solution vector Y. First step is to extract concentration and temperature from Y. Indeed, this can be done by using CA=Y(1:2:end); and T=Y(2:2:end);.

However, I will show a more convenient and flexible way to do so using reshape:

```
YBar=reshape(Y,2,n);
CA=YBar(1,:);
T=YBar(end,:);
```

Again, there is nothing special or different about using reshape. It is a personal choice based on my experience. For example, if the system contained nsp species (instead of a single species), the second line will change to

```
CA=YBar(1:nsp,:);
```

At this stage, I should mention that using the colon-index notations is also easy, but I find reshape method more elegant.

The discretized equation will contain two terms: the discretized convective term and the reaction term. We will use upwind method for the former, while the latter is computed by invoking the function rxnRate that we created earlier. This is computed for each of the *n* nodes in a for loop. This forms the core "%% Model Equations" section of the function file, where we will compute both dC and dT. The vector dY will be extracted from these two using reshape again.

The function file is given below:

```
function dY=pfrAdiabfun(t,Y,par)
% Function file for transient adiabatic PFR
% solved using method of lines
%% Get the parameters
n=par.n;          % Number of nodes
L=par.L;
h=L/n;

u=par.u0;         % Velocity remains constant
C0=par.Cin;       % Inlet concentration
T0=par.Tin;       % Inlet temperature
rCp=par.rCp;
DH=par.DH;

%% Variables
Y=reshape(Y,2,n);
CA=Y(1,:);
T=Y(2,:);
dC=zeros(n,1);
dT=zeros(n,1);

%% Model equations
for i=1:n
    % Reaction rate
    r=rxnRate(T(i),CA(i),par);
    % Convection term
    if i==1
        convecC=u/h*(CA(i)-C0);
        convecT=u/h*(T(i)-T0);
    else
        convecC=u/h*(CA(i)-CA(i-1));
        convecT=u/h*(T(i)-T(i-1));
    end

    % Model equations
    dC(i)=-convecC-r;
    dT(i)=-convecT+(-DH)*r/rCp;
end
%% Return dY as vector
dY=reshape([dC';dT'],2*n,1);
```

Recall the caution I had expressed in using reshape. Note the structure of matrix $\bar{\Phi}$ from Equation 4.87: The first row contains concentration and the second row contains temperature. Since *dC* and *dT* were defined as column vectors,* the form of matrix $\bar{\Phi}$ can be obtained using matrix $\bar{\Phi}$.

This completes our description of the adiabatic PFR function file.

3. *Driver script*: As before, the driver script is in three sections: parameter definition, problem solving, and postprocessing for displaying the results. The script is given below. First, the PFR parameters are defined and stored in model-Param structure. The solution is initialized with uniform conditions C_{A0} and T_0 in the PFR. Note the use of reshape to convert from $\Phi \leftrightarrow [C_A; T]$. Transient and steady state results are plotted in the final section:

```
% Driver script for simulation of a PFR
% with single reaction & complex kinetics
%% Model Parameters
modelParam.k1=0.2;
modelParam.K1=1;
modelParam.T1=450;
modelParam.E=60*1000;
modelParam.DEr=-10*1000;
modelParam.DH=-100*1000;
modelParam.rCp=800;
modelParam.Rgas=8.314;
% Reactor conditions
modelParam.L=2;
modelParam.u0=0.4;
modelParam.Cin=1;
modelParam.Tin=450;
%% Simulating the PFR
n=100;
modelParam.n=n;
Y0=[ones(1,n)*modelParam.Cin;
    ones(1,n)*modelParam.Tin];
Y0=reshape(Y0,2*n,1);
[tSol,Ysol]=ode15s(@(t,Y) pfrAdiabFun(t,Y,modelParam), ...
    [0:0.1:10],Y0);


%% Displaying results
% Extract values for plotting
CA=Ysol(:,1:2:end);
T=Ysol(:,2:2:end);
XA=(modelParam.Cin-CA)/modelParam.Cin;
h=modelParam.L/n;
Z=[0:n]*h;
```

---

* In fact, in this example, it may even be more convenient to define dC and dT as row vectors. However, I have found that beginners often find it confusing when different conventions are used for different examples. Hence, it is more for the pedagogical reason that I have stuck to the convention of defining vectors as column vectors, except when there is a good reason to do otherwise.

```
% Steady state conversion vs. location
figure(1); plot(Z,XA(end,:));
xlabel('Location (m)'); ylabel('Conversion');

% Plot outlet C and T vs. time
figure(2); plot(tSol,CA(:,end));
xlabel('Time (min)'); ylabel('Outlet C_A (mol/m^3)');
figure(3); plot(tSol,T(:,end));
xlabel('Time (min)'); ylabel('Outlet T (K)');

% Plot profiles at various times
figure(4);
subplot(2,1,1); plot(Z,T([101,6,11,16,21],:));
ylabel('Temperature');
subplot(2,1,2); plot(Z,CA([101,6,11,16,21],:));
xlabel('Location (m)');
ylabel('C_A (mol/m^3)');
```

The results are shown in Figures 4.9 and 4.10, which will be discussed presently.

Grid independence study is an important step that must be performed to ensure that the numerical procedure gives an acceptable solution. In examples until this point, we had the analytical solution to compare. However, the PFR model in this example cannot be solved analytically. As we have seen multiple times before, decreasing the step-size improves the accuracy. Grid independence study verifies the adequacy of the chosen grid by comparing the results with two different grid sizes. A rule of thumb that we employ is that the solution is taken as grid-independent if *halving* the step-size (i.e., doubling the number of nodes) causes no further change in the solution. Figure 4.9 shows an example of grid-independence study. For this example, we may consider the solutions from $n = 100$ to be grid independent (thick line in the figure), because doubling the number of nodes to 200 did not cause a significant change in the axial profile of conversion at steady state.

Figure 4.10 shows the temporal evolution of concentration and temperature in the PFR. Initially, the temperature and concentration are uniform. As reaction takes place, eventually

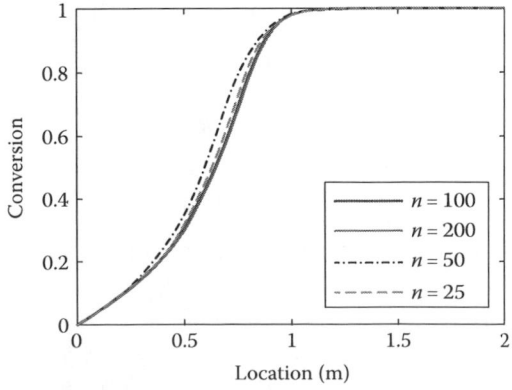

FIGURE 4.9   Conversion vs. axial location at steady state for different spatial discretization.

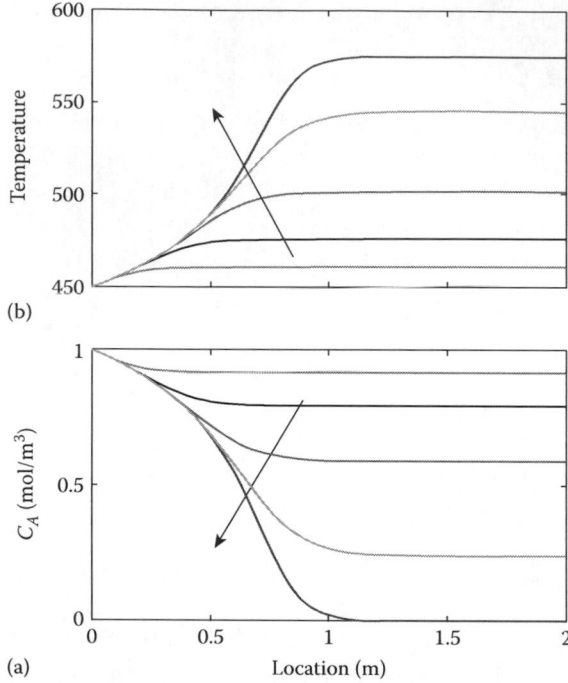

FIGURE 4.10   Temperature (a) and axial profiles of concentration (b) at various times from start (0.5, 1.0, 1.5, 2 min, and steady state). The arrow marks increasing time, t.

the steady state is reached. The arrow shows increasing time. As time progresses, the conversion and temperature both increase smoothly to reach the steady state value.

4.5.2 Packed Bed Reactor with Multiple Reactions

Let us now consider an example of a packed bed reactor with multiple species and reactions. Unlike the example in Section 4.4, a dimensional form of the equation will be used in this example. To keep things simple, a constant fluid velocity is assumed. This assumption is valid for liquid-phase systems and highly dilute gas-phase systems (e.g., catalyst for exhaust emissions control). For a general gas-phase system, this assumption is not valid. Example of a gas-phase system (with varying velocity) will be considered in a case study in Section 9.5.3. The following reactions take place in the reactor:

$$
\begin{aligned}
A &\rightarrow P+Q, \quad 0.05C_A \\
A &\rightarrow P+R, \quad 0.005C_A C_P
\end{aligned}
\tag{4.88}
$$

There are five species in the system: A, P, Q, R, and inerts. The model for a packed bed reactor is given by

$$
\frac{\partial C_k}{\partial t} + u\frac{\partial C_k}{\partial z} = \mathcal{D}_{\text{eff}}\frac{\partial^2 C_k}{\partial z^2} + \sum_j v_{kj}r_j
\tag{4.89}
$$

where k represents the five species and j sums over the two reactions. The value of $\mathcal{D}_{eff}$ can be computed from Taylor's correlation as $\mathcal{D}_{eff} = 3.57\, u_0 d\sqrt{f}$. The boundary condition at the inlet is the Dirichlet condition; inlet concentrations of all the species are known. The reactor inlet consists of the reactant A and inerts. A boundary condition is not implemented at the outlet; instead, backward difference formula is used at the boundary node and the diffusive term is neglected. The initial condition is that the reactor contains only inerts. Thus, the initial and inlet boundary conditions are given by

$$
\begin{aligned}
C(0,z) &= \begin{bmatrix} 0 & 0 & 0 & 0 & C_{I0} \end{bmatrix}^T \\
C(t,z=0) &= \begin{bmatrix} C_{A,\text{in}} & 0 & 0 & 0 & C_{I,\text{in}} \end{bmatrix}^T
\end{aligned}
\tag{4.90}
$$

We will use MoL to solve the PDE. The spatial domain is discretized in n intervals, with step-size $h = L/n$. Central difference formula will be used for convective as well as diffusive terms, which are given by

$$
\left[\frac{\partial C_k}{\partial z} \right]_i = \frac{C_{k,i+1} - C_{k,i-1}}{2h}, \quad \left[\frac{\partial^2 C_k}{\partial z^2} \right]_i = \frac{C_{k,i+1} - 2C_{k,i} + C_{k,i-1}}{h^2}
$$

The solution vector consists of the five concentrations at each of the n locations:

$$
\bar{\mathbf{Y}} = \begin{bmatrix} C_{A1} & C_{A2} & \cdots & C_{An} \\ \vdots & \vdots & \vdots & \vdots \\ C_{I1} & C_{I2} & \cdots & C_{In} \end{bmatrix}
\tag{4.91}
$$

It is convenient to express $\bar{\mathbf{Y}}$ in the above form. However, the ODE solver, `ode45`, requires $\mathbf{Y}$ as a column vector.[*] The interconversion between the two can be easily done as

$\bar{\mathbf{Y}} =$ `reshape(Y, nSpecies,n);`

The rate of reaction will be calculated as a 2×1 vector:

$$
\mathbf{r} = \begin{bmatrix} k_1 C_A \\ k_2 C_A C_P \end{bmatrix}
\tag{4.92}
$$

and the axial dispersion coefficient is computed as $\mathcal{D}_{eff} = 0.013\,\text{m}^2/\text{s}$. The next example demonstrates this problem.

---

[*] MATLAB experts know that this is not strictly true. However, if you were one, you wouldn't be reading this book. So, let us assume this statement is true!

Example 4.5 Isothermal Packed Bed Reactor with Multiple Reactions

Problem: Solve the isothermal packed bed reactor problem described above. The PBR has a diameter of 25 cm and a length of 10 m, and inlet flowrate is 10 L/s. The inlet feed contains $C_A = 5$ and $C_I = 45$ mol/m³.

Solution: This will be solved using MoL, as described above. The reaction rates will be calculated using a separate function:

```
function r=reactionRate(C,par)
%% Calculate reaction rate
r(1,1)=par.k(1)*C(1);
r(2,1)=par.k(2)*C(1)*C(2);
end
```

Writing the main function file: The main function file consists of a code to compute $\left[\dfrac{dC_k}{dt}\right]_i$. As per Equation 4.89, this will consist of three parts: convective term, `conv-Term`; axial diffusive term, `diffTerm`; and reaction term, `rxnTerm`. The stoichiometric matrix for this example is defined as

$$v = \begin{bmatrix} -1 & -1 \\ 1 & -1 \\ 1 & 0 \\ 0 & 1 \\ 0 & 0 \end{bmatrix}$$

The powerful matrix operations of MATLAB can be used to compute the reaction term for all five species at any location simultaneously, because

$$\sum_j v_{kj} r_j = \begin{bmatrix} -1 & -1 \\ 1 & -1 \\ 1 & 0 \\ 0 & 1 \\ 0 & 0 \end{bmatrix} \underbrace{\begin{bmatrix} k_1 C_A \\ k_2 C_A C_P \end{bmatrix}}_{r} \tag{4.93}$$

Equipped with the above information, the function for computing the overall model can be written as follows:

```
function dY=packBedFun(t,Y,par)
%% Get parameters and variables
nsp=par.nSpecies;
n=par.n;    h=par.h;
```

```
Y=reshape(Y,nsp,n);
dY=zeros(nsp,n);
CIn=par.CIn;
nuStoic=par.stoicCoef;
%% Model equations
for i=1:n
    C=Y(:,i);
    rxnTerm=nuStoic*reactionRate(C,par);
    if (i==1)
        diffTerm=par.DCoef*(Y(:,i+1)-2*C+CIn)/h^2;
        convTerm=par.u0*(Y(:,i+1)-CIn)/(2*h);
    elseif(i==n)
        diffTerm=0;
        convTerm=par.u0*(C-Y(:,i-1))/h;
    else
        diffTerm=par.DCoef*(Y(:,i+1)-2*C+Y(:,i-1))/h^2;
        convTerm=par.u0*(Y(:,i+1)-Y(:,i-1))/(2*h);
    end
    dY(:,i)=-convTerm + diffTerm + rxnTerm;
end
dY=reshape(dY,n*nsp,1);
```

The use of `reshape` command to interconvert $\mathbf{Y} \Leftrightarrow \bar{\mathbf{Y}}$ is underlined in the code above. Let us investigate the term `convTerm` (for internal nodes). The spatial derivative is

$$\frac{\partial}{\partial z}\begin{bmatrix} C_A \\ C_P \\ \vdots \\ C_I \end{bmatrix} = \begin{bmatrix} \dfrac{C_{A,i+1}-C_{A,i-1}}{2h} \\ \dfrac{C_{P,i+1}-C_{P,i-1}}{2h} \\ \vdots \\ \dfrac{C_{I,i+1}-C_{I,i-1}}{2h} \end{bmatrix} = \frac{1}{2h}\left(\begin{bmatrix} C_{A,i+1} \\ C_{P,i+1} \\ \vdots \\ C_{I,i+1} \end{bmatrix} - \begin{bmatrix} C_{A,i-1} \\ C_{P,i-1} \\ \vdots \\ C_{I,i-1} \end{bmatrix} \right)$$

Based on the definition of $\bar{\mathbf{Y}}$ as per Equation 4.91, the first term in the brackets is `Y(:,i+1)` and the second term is `Y(:,i-1)`. Hence, the convection term was written as

```
convTerm = par.u0*(Y(:,i + 1) - Y(:,i - 1))/(2*h);
```

Likewise, `diffTerm` follows the same arguments. Both these yield a 5×1 column vector. The reaction source term, $\sum_j \nu_{kj} r_j$, as per Equation 4.93, is computed for all five species in `rxnTerm`, which also yields a 5×1 column vector.[*]

---

[*] This demonstrates how the power of MATLAB matrix operations is utilized fully to write highly efficient and readable codes. Recall my insistence on defining all vectors as column vectors, unless the structure of a problem or storage of "historical" data suggests otherwise.

Driver script: The driver script is straightforward and follows the same pattern as before. The parameters, inlet boundary values, and initial conditions are specified. The MATLAB command `repmat` is used to obtain the entire initial condition vector `Y0`. The ODE is solved using `ode45` for a time span of `[0 100]`. The *grid independence* was verified by running the code for $n = 25$ nodes and comparing the results with $n = 50$. The closeness of the two suggests that $n = 25$ gives a solution with sufficient accuracy. The code with this value of n is given below:

```
% Driver script for simulation of a
% packed bed reactor with multiple species
%% Model parameters
Q=0.01;
d=0.25;
Acs=pi/4*d^2;
u0=Q/Acs;          modelParam.u0=u0;
L=10;              modelParam.L=L;
DCoef=0.013;       modelParam.DCoef=DCoef;
% Reaction conditions
modelParam.nSpecies=5;
modelParam.k=[0.05; 0.005];
modelParam.stoicCoef=[-1 -1; 1 -1; 1 0; 0 1; 0 0];
% Inlet conditions
CaIn=5.0;   CiIn=45.0;
CIn=[CaIn; 0; 0; 0; CiIn];
modelParam.CIn=CIn;

%% Initialization and solution
n=25;          modelParam.n=n;
h=L/n;         modelParam.h=h;
C0=[0;0;0;0;CiIn];
Y0=repmat(C0,n,1);
[tSol,YSol]=ode45(@(t,Y) packBedFun(t,Y,modelParam),...
    [0 100], Y0);

%% Plotting results
figure(1)
Z=0:h:L;
C_All=[CIn', YSol(end,:)];
C_All=reshape(C_All,5,n+1);
plot(Z,C_All(1:4,:));
```

Figure 4.11 shows the axial concentration profiles for the four species at steady state. The profiles show expected behavior.

The code above will be used to further analyze the behavior of a packed bed reactor. Figure 4.12 shows the temporal variations in the concentrations of the four species at the exit. The residence time in the reactor is 49 s. The concentrations stay at their initial condition

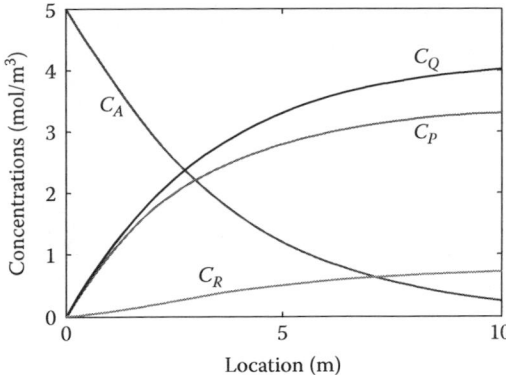

FIGURE 4.11 Axial concentration profiles of the four species at steady state.

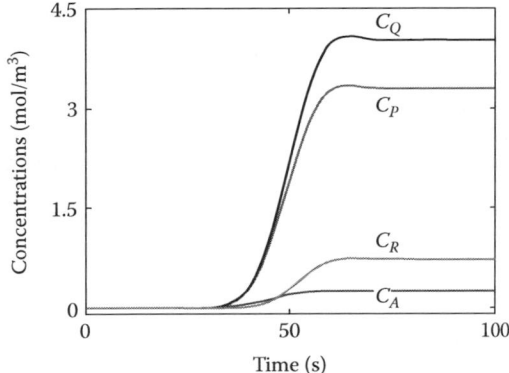

FIGURE 4.12 Transient profiles of the four species (except the inert) at the exit. The inlet conditions start affecting the exit only after some time, which depends on the residence time in the reactor.

for the initial 40 s. The effect of inlet conditions only start being observed at the exit after a sufficient time has elapsed. The results seem to indicate that convection is predominant under the operating conditions considered in this example. In the absence of the diffusive term (and numerical diffusion), the exit concentrations would be affected only after 49 s. However, the diffusive term makes the overall response smoother.

The importance of the diffusive term is further verified by comparing the results of the packed bed reactor with a PFR (dashed line) in Figure 4.13. At the nominal value of $D_{eff} = 0.013$, the two curves are fairly close to each other. Since PFR represents the case with no axial diffusion, one could conclude that the convection is dominant and diffusion plays only a minor role in this packed bed reactor example. The dash-dot line in Figure 4.13 represents an order of magnitude greater value of $\mathcal{D}_{eff}$. Clearly, this increase in axial diffusion results in a decrease in conversion.

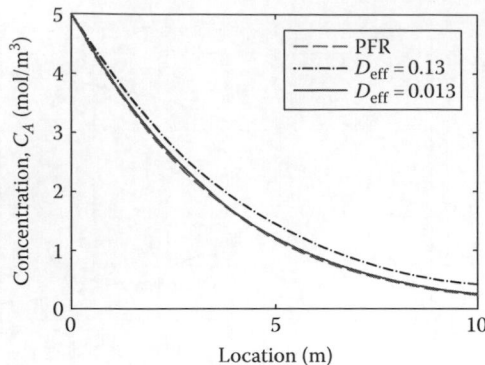

FIGURE 4.13   Axial profile of C_A for packed bed reactor with $D_{eff}=0.013$ (solid line) and results from steady state PFR simulations (dashed line) for comparison. The profile for a much higher value of $D_{eff}=0.13$ (dash-dot line) is also shown.

4.5.3 Steady Graetz Problem: Parabolic PDE in Two Spatial Dimensions
4.5.3.1 Heat Transfer in Fluid Flowing through a Tube
Graetz problem is a standard problem in heat transfer in a fluid flowing through a tube, where the fluid gets progressively heated due to higher boundary temperature or a constant boundary flux. This is an example of steady state parabolic PDE. The primary mode of transport in the axial direction is through convection while that in the radial direction is diffusion. The steady state heat equation is given by

$$u \frac{\partial T}{\partial z} = \frac{\alpha}{r} \frac{\partial}{\partial r}\left(r \frac{\partial T}{\partial r} \right) \tag{4.94}$$

where
   u is the velocity
   α is the thermal diffusivity

The inlet temperature at $z=0$ is specified. The symmetry boundary condition is applicable at the center, whereas either the temperature (Dirichlet) or the heat flux (Neumann) is specified at the wall. We will consider the problem where boundary temperature is specified:

$$\begin{aligned} T(z=0,r) &= T_0 \\ \left[\frac{\partial T}{\partial r} \right]_{r=0} &= 0, \quad T(r=R) = T_b \end{aligned} \tag{4.95}$$

Furthermore, parabolic velocity profile is assumed to exist in the tube:

$$u(r) = u_{max}\left(1 - \left(\frac{r}{R}\right)^2 \right) \tag{4.96}$$

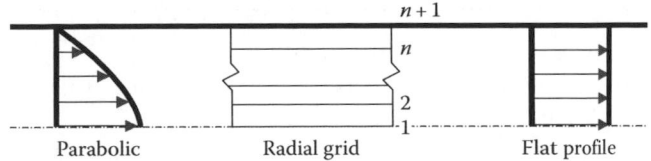

FIGURE 4.14   Velocity profile and radial grid used in Graetz problem.

It needs to be emphasized (for readers without heat transfer background) that for a tube, R is a given value (radius) while r is a variable that changes from 0 at the center to R at the wall.

We have simulated transient parabolic PDEs before in this chapter. The PDE (4.94) is exactly of the same form, the only difference being that both independent variables are spatial. The same tools employing MoL can be applied to this example as well. We will discretize using finite differences in radial direction. The velocity profile and computational grid are schematically shown in Figure 4.14. The location at the axis of symmetry ($r = 0$) is 1. Since the temperature at the final node is specified, we do not need to solve for the final node. Discretizing the PDE for radian nodes 1 to n yields the following set of ODEs in axial direction:

$$\frac{dT}{dz} = \frac{\alpha}{u_j}\left[\frac{T_{j+1} - 2T_j + T_{j-1}}{h^2} + \frac{1}{r_j}\frac{T_{j+1} - T_{j-1}}{2h}\right], \quad j = 2 \text{ to } (n-1)$$

$$\frac{dT}{dz} = \frac{\alpha}{u_j}\left[\frac{T_b - 2T_j + T_{j-1}}{h^2} + \frac{1}{r_j}\frac{T_b - T_{j-1}}{2h}\right], \quad j = n \tag{4.97}$$

where T_b is the boundary temperature specified. The symmetry boundary condition at the first node results in the condition $T_2 - T_0 = 0$, yielding the following ODE:

$$\frac{dT}{dz} = \frac{\alpha}{u_j}\left[\frac{2T_{j+1} - 2T_j}{h^2}\right], \quad j = 1 \tag{4.98}$$

Note that since $\partial T/\partial r$ is zero at the boundary, the second term drops out of the above equation. The following example shows the simulation of fluid flowing in a tube with hot walls. This is the Graetz problem with constant boundary temperature.

Example 4.6 Constant Wall Temperature Graetz Problem with Parabolic Velocity

Problem: Solve the constant wall temperature Graetz problem of Equation 4.95, with a fully developed parabolic velocity profile Equation 4.96 and $u_{max} = 0.5$ m/s. The tube is 2 m long and its diameter is 10 cm. Thermal diffusivity of the fluid is $\alpha = 10^{-4}$ m²/s.

Solution: This is a standard parabolic PDE, where we use MoL. Discretizing in radial direction leads to a set of ODEs given by Equations 4.97 and 4.98. The following function file, which follows the same pattern as previous examples, is used:

```
function dTdz=graetzFun(z,T,par)
% Function to compute dT/dz for
% heat transfer in a heated tube
r=par.r;          % Radial coordinate
u=par.u;          % Radial velocity profile
n=par.n;          % Nbr of radial nodes
h=par.h;          % Radial step-size
alpha=par.alpha;
Tb=par.Tb;
% Model equations
dTdz=zeros(n,1);
for j=1:n
    if (j==1)
        term1=(2*T(j+1)-2*T(j))/h^2;
        term2=0;
    elseif (j==n)
        term1=(Tb-2*T(j)+T(j-1))/h^2;
        term2=(Tb-T(j-1))/(2*h*r(j));
    else
        term1=(T(j+1)-2*T(j)+T(j-1))/h^2;
        term2=(T(j+1)-T(j-1))/(2*h*r(j));
    end
    dTdz(j)=alpha/u(j)*(term1+term2);
end
```

Although obvious to intermediate and expert users, it bears repeating that the overall methodology and structure of the file remains exactly as before. The following driver script is used to solve the Graetz problem:

```
% Driver script for Graetz Problem:
% Heat transfer in a tube with hot wall
%% Model parameters and conditions
L=2.0;            % 2 m long tube
R=0.05;           % 10 cm tube dia
n=100;    h=R/n;
r=[0:n]'*h;       % Radial coordinate
modelPar.alpha=0.0001;

% Inlet and boundary conditions
modelPar.Tin=300;
modelPar.Tb=400;
uMax=0.5;
u=uMax*(1-(r/R).^2);      % Parabolic
% u=uMax/2*ones(n+1,1); % Plug-flow
```

```
modelPar.L=L;    modelPar.R=R;
modelPar.n=n;    modelPar.h=h;
modelPar.r=r;
modelPar.u=u;
```

```
%% Set up and solve the PDE
T0=ones(n,1)*modelPar.Tin;
opts=odeset('relTol',1e-12,'absTol',1e-10);
[Zsol,Tsol]=ode15s(@(z,T) graetzFun(z,T,modelPar),...
    [0, L], T0, opts);
```

```
%% Plot results and post-process
% Include wall temperature in Tsol
nAxial=length(Zsol);
Twall=modelPar.Tb*ones(nAxial,1);
Tsol=[Tsol,Twall];

% Temperature vs. location
figure(1);
plot(Zsol,Tsol(:,[1,26,51,76,86]));
xlabel('Axial location, z (m)');
ylabel('Temperature (K)');
% Temperature contours
figure(2)
subplot(2,1,1)
contourf(Zsol,r,Tsol');
colorbar
xlabel('Axial location, z (m)');
```

Figure 4.15 shows the temperature profile vs. axial distance for four different radial locations in the tube. The final temperature is higher and the temperature increases more rapidly as one goes closer to the hot wall.

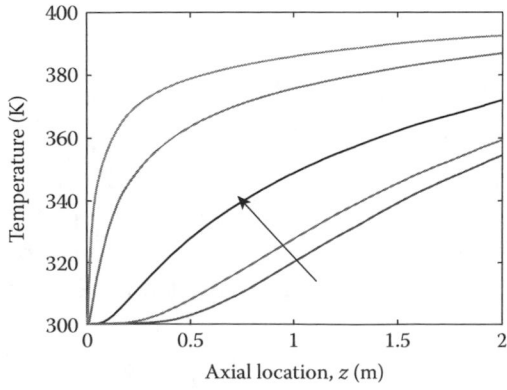

FIGURE 4.15   Axial temperature profiles at various locations: $r = 1.25, 2.50, 3.75$, and 4.25 cm from the central symmetry axis. The arrow indicates increasing radial coordinate.

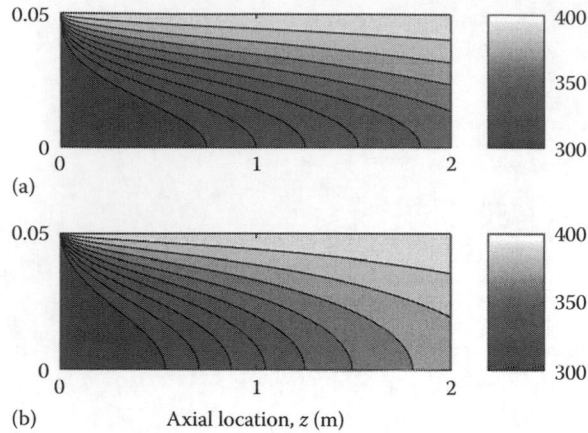

FIGURE 4.16   Contours of temperature for parabolic (a) velocity profile and plug flow (b).

4.5.3.2 Effect of Velocity Profile

The above problem was solved with parabolic velocity profile. This is an appropriate assumption for a fully developed laminar flow, which is what one expects in practical situations. As a numerical exercise, it is worthwhile to compare the results with a flat velocity profile (such as the one we assume under plug flow conditions). Figure 4.16 shows velocity contours for the two cases: Darker regions indicate lower temperature and lighter regions indicate higher temperatures. The flat velocity profile can be imposed in Example 4.6 by uncommenting the underlined expression in the code:

```
u=uMax/2*ones(n+1,1);    % Plug-flow
```

Recall that the average velocity in a tube is $u_{avg} = u_{max}/2$. Comparing the contour maps, the warm region "diffuses" into the tube faster with plug flow than parabolic velocity profile. This is expected since the Nusselt number (Nu) for the two cases is 5.78 (plug) and 3.66 (parabolic).

4.5.3.3 Calculation of Nusselt Number

I mentioned above that the *asymptotic* value of Nu for a parabolic velocity profile is 3.66. The value of Nu is high when the thermal boundary layer is developing and settles to the asymptotic value when the thermal boundary layer is developed. Standard heat transfer or transport phenomena books show these correlations (Dittus-Boelter or Sieder-Tate correlations are quite popular in UG chemical and mechanical engineering textbooks). Let us numerically arrive at the Nu profile.

Recall that Nu is defined as

$$Nu = \frac{hD}{k} \tag{4.99}$$

The heat flux at the wall is given by

$$k\left[\frac{\partial T}{\partial r}\right]_R$$

This is the rate at which energy enters the tube through a unit area of the wall. At any axial location, if $\langle T \rangle$ is the average temperature, the convective term for heat transfer to the fluid is given by

$$h\left(T_b - \langle T \rangle\right)$$

Since at steady state these two are equal *at any axial location*, we get

$$\mathrm{Nu} = D\frac{h}{k} = 2R\frac{\left[\dfrac{\partial T}{\partial r}\right]_R}{\left(T_b - \langle T \rangle\right)} \tag{4.100}$$

Note that this definition is valid for each axial location.

The term $\partial T/\partial r$ can be obtained using three-point backward difference formula:

$$\left[\frac{\partial T}{\partial r}\right]_R = \frac{3T_{n+1} - 4T_n + T_{n-1}}{2h} \tag{4.101}$$

The average temperature $\langle T \rangle$ is the so-called mixing-cup temperature:

$$\langle T \rangle_z = \frac{\displaystyle\int_0^R T(z,r) \cdot ru(r)\,dr}{\displaystyle\int_0^R ru(r)\,dr} \tag{4.102}$$

Since we have already precomputed vectors r and u, the denominator is calculated using trapz. Likewise, we will use trapz to calculate the numerator at each axial location. The code for calculating Nu profile is given below.

Example 4.7 Nusselt Number Profile for Example 4.6

The code for calculating Nu using Equation 4.100 is given below. This is a continuation of the code from Example 4.6.

```
% Calculate mixing-cup T
Tavg=zeros(nAxial,1);
```

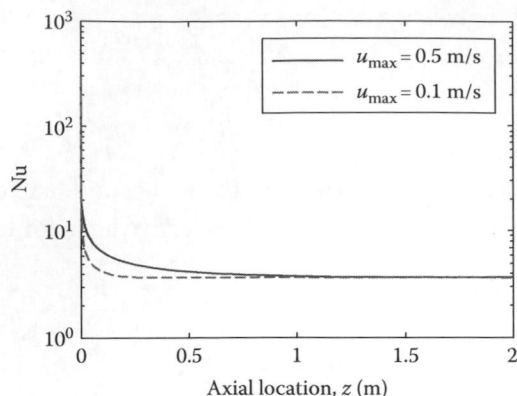

FIGURE 4.17   Nu profile for parabolic velocity profile for two values of u_{max}.

```
denom=trapz(r,u.*r);
for i=1:nAxial
    Tradial=Tsol(i,:);
    numer=trapz(r,u.*r.*Tradial');
    Tavg(i)=numer/denom;
end
% Calculate dT/dr at wall
dTdr=zeros(nAxial,1);
for i=1:nAxial
    dTdr(i)=(3*Tb-4*Tsol(i,n)+Tsol(i,n-1))/(2*h);
end
% Calculate and plot Nusselt Nbr
Nu=2*R* dTdr ./ (modelPar.Tb-Tavg);
figure(3)
semilogy(Zsol,Nu)
```

Figure 4.17 shows the Nu profile for fully developed velocity in a tube. Note the logarithmic scale on Y-axis. The value of Nu settles at its asymptotic value when the temperature profile is fully developed. Hence, at a lower velocity of 0.1 m/s, the asymptotic value of Nu is reached very rapidly.

The other example of Graetz problem is when the flux is specified at the boundary. This is left as an exercise for the reader.

4.6 EPILOGUE

Numerical techniques for solving transient PDEs were the focus of this chapter. An introduction to PDEs was presented in Section 4.1. Most of the distributed parameter problems of interest to chemical engineers are expressed in the form of quasilinear PDEs of the first and second order. The historic classification of these PDEs as elliptic, hyperbolic, and parabolic was discussed. Elliptic PDEs arise in multidimensional steady state balances of systems where diffusion terms play an important or predominant role. Hyperbolic PDEs describe convection-dominated systems. The solution propagates without damping along the characteristic curves. Parabolic PDEs, on the other hand, lie somewhere between elliptic and hyperbolic; like hyperbolic systems, the solution propagates along the first independent variable (usually time), but like elliptic systems, the solution at any point in the domain is influenced by both the boundaries.

Numerical methods based on finite difference approximation and MoL were discussed in Section 4.3 for hyperbolic PDEs and in Section 4.4 for parabolic PDEs. Methods that are explicit in time are attractive because they are computationally less intensive than fully implicit methods. MoL was the predominant approach discussed in this chapter to solve the hyperbolic and parabolic PDEs. It involves discretizing the PDEs in space to convert them into a series of ODEs in time. ODE solution techniques from the previous chapter may then be used for solving the resulting problem.

The physical nature of the hyperbolic and parabolic PDEs determines the choice of numerical method to solve the problem. I showed that central difference formula for discretizing the convection term can result in an unstable behavior of the solution of PDEs. Thus, upwind method (either first or higher order) was suggested for spatial discretization in hyperbolic PDEs.

Section 4.3.4 also discussed the issue of numerical diffusion in hyperbolic PDEs.

The diffusion term in parabolic PDEs results in a stabilizing effect, allowing one to use the central difference formula for discretizing in the spatial domain.

The Crank-Nicolson method was introduced as a fully implicit in-time numerical solution technique based on finite differences (Sections 4.3.2 and 4.4.2). However, this method was not discussed further. The actual solution using Crank-Nicolson method will be discussed in Chapter 8, in Section II of this book.

In summary, MoL is the method of choice for solving hyperbolic and parabolic PDEs. MoL allows us to use higher-order formulae for integrating in time. The nature of the two systems dictates the use of upwind method and central difference for discretization in hyperbolic and parabolic PDEs, respectively. The MATLAB ode45 and ode15s are the preferred solvers for solving the discretized equations.

EXERCISES

Problem 4.1 (a) Derive the leapfrogging scheme for first-order hyperbolic PDE (4.28) using a central difference in time and central difference in space.

(b) Derive forward-in-time second-order upwind method for first-order hyperbolic PDE (4.28) with $u > 0$. Use forward difference approximation for ϕ_t and a three-point backward difference formula for ϕ_x.

(c) How does the upwind method derived in (b) above change for $u < 0$?

(d) Derive the fully implicit formula for hyperbolic PDE using first-order approximation in time and central difference in space. Extend it to parabolic PDEs as well.

Problem 4.2 Solve the PFR with first-order reaction (see Example 4.1) using various explicit forward difference in time and upwind in space discretization from Section 4.3.1.

Problem 4.3 Complete the problem in Example 4.2 by writing the code for MoL with three-point upwind method. Verify results shown in Figures 4.4 and 4.5.

Problem 4.4 Convert the general parabolic PDE (4.64) into a set of ODEs using MoL. Use central difference formula for the diffusive term and upwind method for the convective term.

Problem 4.5 The adiabatic PFR problem in Section 4.5.1 was solved ignoring the axial diffusion. If the axial diffusion was relevant, this will result in parabolic PDE.
(a) If $\mathcal{D}$ is the diffusivity and λ is the thermal conductivity, then write down the mass and energy balance equations for a reactor with axial diffusion.
(b) Solve the above problem using MoL. The Péclet number for heat and mass transfer are $Pe_H = uL/\alpha = 1$ and $Pe_M = uL/\mathcal{D} = 2.5$, respectively. Note that $\alpha = \lambda/\rho c_p$ is the thermal diffusivity. No-flux boundary conditions may be chosen at the exit, that is, $\left.\dfrac{\partial C}{\partial z}\right|_{z=L} = 0$ and $\left.\dfrac{\partial T}{\partial z}\right|_{z=L} = 0$.

Problem 4.6 Solve the Graetz problem with flux specified at the boundary. The model equation and boundary conditions are given by

$$u\frac{\partial T}{\partial z} = \frac{\alpha}{r}\frac{\partial}{\partial r}\left(r\frac{\partial T}{\partial r}\right) \tag{4.94}$$

$$T(z = 0, r) = T_0$$
$$\left[\frac{\partial T}{\partial r}\right]_{r=0} = 0, \quad \left[\frac{\partial T}{\partial r}\right]_{r=R} = \backslash beta \tag{4.103}$$

$$u(r) = u_{max}\left(1 - \left(\frac{r}{R}\right)^2\right) \tag{4.96}$$

Problem 4.7 Compute the Nu profiles for the above problem and compare with the constant wall temperature case.

CHAPTER 5

Section Wrap-Up

Simulation and Analysis

Several examples were covered as case studies in this part of the book. Chapter 2 laid the linear systems basics, which we will use in this chapter to bring together the contents in this part. The next two chapters focused on process simulation examples, where the model was intended to capture how the solutions evolve in time.

Chapter 3 covered examples that can be cast into an ordinary differential equation–initial value problem (ODE-IVP) of the type

$$\frac{d\mathbf{y}}{dt} = \mathbf{f}(t,\mathbf{y}) \tag{5.1}$$

$$\mathbf{y}(t_0) = \mathbf{y}_0 \tag{5.2}$$

These included transient simulations, where $\mathbf{y} \in \mathcal{R}^n$ is a solution vector that varied in time, as well as cases where the solution variable $\mathbf{y}$ varied along a single spatial direction. The following case studies were discussed:

- A steady state plug flow reactor (PFR)

- Batch and continuous stirred reactors for exothermic reactions and biological reactions

- A two-tank heater, which presented an example of a hybrid* system

---

* A hybrid system is governed by continuous dynamics and binary (or integer) switching variables. In the two-tank heater case study, the flow between two tanks was turned on or off based on the height of the liquid in the two tanks. In systems engineering parlance, such systems are known as *autonomous* hybrid systems. In contrast, the other type of hybrid systems involves ones where an inlet may be switched on or off (binary decision).

Chapter 4 extended our analysis to systems that vary in more than one dimensions (space and time, or two spatial dimensions). Such systems* are governed by partial differential equations (PDEs). The chapter focused on hyperbolic and parabolic PDEs. Such systems arise due to material, energy, or momentum balance equations and are represented as

$$\frac{\partial \phi}{\partial t} + \frac{\partial}{\partial z}(u\phi) = \frac{\partial}{\partial z}\left(\Gamma \frac{\partial \phi}{\partial z}\right) + \mathcal{S}(\phi) \tag{5.3}$$

The first term on the right-hand side is the diffusive term. The PDE is said to be parabolic if this term is included in the model, whereas it is hyperbolic if it can be ignored. The method of lines approach used finite difference approximation to discretize the z-dimension and convert the PDE (5.3) into a set of ordinary differential equations (ODEs) of the form Equation 5.1. With such a conversion

$$\mathbf{y} \triangleq \begin{bmatrix} \phi_1 \\ \phi_2 \\ \vdots \\ \phi_n \end{bmatrix} \tag{5.4}$$

where ϕ_i represents the variable ϕ at the ith spatial node. A good balance of case studies was provided in Chapter 4 to cover the most important systems:

- Transient PFR (an example of first-order hyperbolic PDEs)

- Transient packed bed reactor (an example of parabolic PDEs)

- Steady state heat transfer in a heated tube (parabolic PDE in two spatial dimensions)

This chapter builds upon the understanding gained in the previous two chapters for simulation and linear analysis of process systems. The examples are meant to highlight a rather broad spectrum of problems. Often, textbooks or courses deal with examples that can be bucketed into "ODE-IVP" or "parabolic PDE" or "hyperbolic-PDE" molds. However, in simulating practical systems, such distinctions are not so clearly defined. The case studies in this chapter are intended to break these definitions. The intention here is to show that *concepts* in the previous chapters are very much applicable, it is just a matter of breaking these molds from our minds.

---

* Process variables in such systems vary with both time and space. Hence, systems governed by PDEs are referred to as *distributed parameter systems*. On the other hand, systems defined by ODEs have their properties evolve in one direction (time for transient systems, or one spatial dimension for examples such as PFR), and are therefore termed as *lumped parameter systems*. This distinction is not particularly useful for *simulating* the processes, but is sometimes useful in analysis and control.

The core concepts that I will build on in this chapter include eigenvalues, eigenvectors, and decomposition (Chapter 2); explicit and stiff ODE solving using `ode45` and `ode15s` (Chapter 3); and finite difference discretization using central or upwind differencing (Chapter 4).*

5.1 BINARY DISTILLATION COLUMN: STAGED ODE MODEL

Distillation is an important unit operation in chemical engineering. A continuous distillation column is used for the separation of two or more miscible close-boiling fluids. Figure 5.1 shows schematic of a distillation column. Distillation is an example of *staged* equilibrium operation. Mole fraction of a component is a key process variable, which varies along the height of the column. The condenser at the top of the column forms the first stage, whereas the reboiler at the bottom of the column forms the Nth stage. The feed is introduced at the feed tray, which is somewhere between the bottom and top stages. In a standard binary distillation column, the light component is obtained at the top and the heavier component at the bottom.

5.1.1 Model Description

The mole fraction of the lighter component will be the process state variable.

Figure 5.1 also shows representation of a single equilibrium stage in the column. The liquid and vapor are in equilibrium, and they flow from one stage to the next, and the process is thus modeled as a sequence of equilibrium stages. The figure also shows a single ith stage in the column. The mole fractions of the lighter component in the liquid and gas phase on this stage are represented as x_i and y_i, respectively. These are assumed to be in equilibrium. The molar flowrate of liquid and vapor streams exiting the stage are L_i and V_i, respectively.

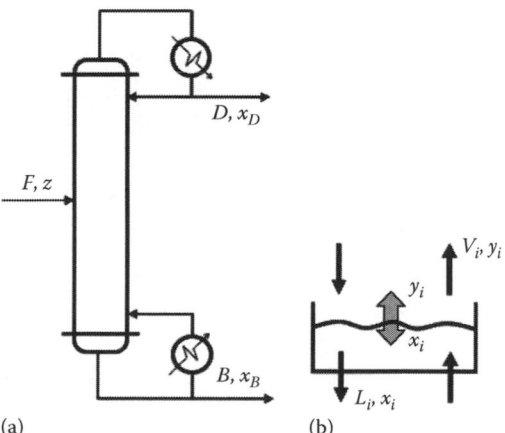

FIGURE 5.1　Schematic of (a) a distillation column and (b) a single equilibrium stage in the column.

---

* In each chapter, we spent some time discussing theory, and then solving problems that used these concepts to provide efficient solutions. I would like to emphasize that the definitions such as stiff vs. nonstiff or hyperbolic vs. parabolic are not critical. What matters more is an *understanding* about how these categories affect our choice of numerical techniques.

The distillate is drawn from the condenser at the top of the column. The distillate flow-rate is D and the mole fraction is x_D. The bottoms is drawn from the reboiler at a flowrate of B and mole fraction x_B. The feed enters the feed tray as a saturated liquid, with a flowrate F and mole fraction z.

The liquid holdup on each stage is assumed constant. Thus, the overall material balance requires the following condition to be satisfied:

$$F = D + B$$

A couple more assumptions are made to simplify the model. The first one is that of constant relative volatility, which is defined as

$$\alpha = \frac{y/(1-y)}{x/(1-x)} \tag{5.5}$$

Larger the value of α, easier it is to separate the two components. Constant relative volatility is a reasonable assumption for mixtures that are close to ideal. Thus, the liquid- and vapor-phase compositions are related to each other as

$$y_i = \frac{\alpha x_i}{1 + (\alpha - 1) x_i} \tag{5.6}$$

The second assumption is that of a constant molar flowrate. In other words, L_i and V_i are constant in the stripping (below the feed stage) and rectification (above the feed stage) sections.

The molar balance on the light component (see schematic in Figure 5.1) on each tray yields

$$M_i \frac{dx_i}{dt} = V_{i+1} y_{i+1} + L_{i-1} x_{i-1} - V_i y_i - L_i x_i \tag{5.7}$$

The model for the feed tray (see Equation 5.12) will include feed term as well.

One of the assumptions was the molar flowrate remains constant. Thus, the vapor-phase flowrate remains constant at V within the entire column. Likewise, within the rectification section ($i > i_{\text{feed}}$), the liquid molar flowrate is constant at L. However, feed gets added as liquid on the feed tray; so, in the stripping section ($i \leq i_{\text{feed}}$), $L_i = L + F$.

The liquid stream that enters the reboiler is split into two streams, bottoms and a boil-up stream. Thus

$$L + F = V + B$$

Likewise, the vapor that enters condenser is split into distillate and reflux stream:

$$V = L + D$$

Thus, in the rectification section

$$L_i = L, \quad V_i = V \tag{5.8}$$

and in the stripping section

$$L_i = L_s, \quad V_i = V, \quad \text{where } L_s = L + F \tag{5.9}$$

5.1.2 Model Equations and Simulation

Final model equations for all stages can now be summarized as

$$\text{Condenser: } M_C \frac{dx_1}{dt} = Vy_2 - Dx_1 - Lx_1 \tag{5.10}$$

$$\text{Rectification stage: } M \frac{dx_i}{dt} = V\left(y_{i+1} - y_i\right) + L\left(x_{i-1} - x_i\right) \tag{5.11}$$

$$\text{Feed stage: } M \frac{dx_i}{dt} = V\left(y_{i+1} - y_i\right) + Lx_{i-1} + Fz_f - L_s x_i \tag{5.12}$$

$$\text{Stripping stages: } M \frac{dx_i}{dt} = V\left(y_{i+1} - y_i\right) + L_s\left(x_{i-1} - x_i\right) \tag{5.13}$$

$$\text{Reboiler: } M_R \frac{dx_N}{dt} = L_s x_{N-1} - Vy_N - Bx_N \tag{5.14}$$

The following example simulates the binary distillation column for an easy separation process with $\alpha = 4$. The distillation column has five stages, one each in the rectification and stripping sections.

Example 5.1 Binary Distillation with Five Stages

Problem: Starting with an initial mole fraction of 0.5, simulate a binary distillation column with five stages for the conditions given in the following table:

Feed: $F = 25$ mol/s	Distillate: $D = 12.5$ mol/s	Tray holdup: $M = 200$ mol
$x_f = 0.5$	Vapor: $V = 30$ mol/s	Holdup: $M_C = M_R = 2$ kmol

Solution: We will simulate the system in the so-called "D-V mode." Here, the distillate and vapor flowrates are specified. Since there are only two degrees of freedom, the other flowrates are computed from these two. Other modes are the L-V mode, or where reflux ratio (RR = L/D) is specified along with one bottoms parameter.

The readers who have followed the previous chapters will realize that the binary distillation function file is written in the same manner as before:

```
function dx = binaryDistFun(t,x, par)
% Five-stage binary distillation column
%% Distillation Parameters
alpha=par.alpha;
M=[ par.Mc;
    par.M*ones(3,1);
    par.Mr];    % All liquid holdup

V=par.V;  L=par.L;
D=par.D;  B=par.B;
F=par.F;  xf=par.xf;
Ls = L+F;

%% Binary distillation model
y = (alpha*x) ./ (1 + (alpha-1)*x);

dx(1,1) = -L*x(1) - D*x(1) + V*y(2);
dx(2,1) = L*(x(1)-x(2)) + V*(y(3)-y(2));
dx(3,1) = L*x(2) + F*xf - Ls*x(3) + V*(y(4)-y(3));
dx(4,1) = Ls*(x(3)-x(4)) + V*(y(5)-y(4));
dx(5,1) = Ls*x(4) - B*x(5) - V*y(5);
dx = dx ./ M;
```

The driver code for the distillation column is given below. Again, the driver script is self-explanatory. The above code as well as the driver script are hard-coded for five stages. Expanding it to a larger number of stages is left as an exercise. Since $\alpha = 4$, this is a rather easy separation.

```
% Transient binary distillation simulation
%% Inlet and design conditions
modelPar.xf=0.5;
modelPar.M=200;
modelPar.Mc=2000;
modelPar.Mr=2000;

F=25;
D=12.5;    % D-V mode: D is given
V=30.0;    % D-V mode: V is given
L=V-D;
B=F-D;
modelPar.F=F;
modelPar.B=B; modelPar.D=D;
modelPar.V=V; modelPar.L=L;
modelPar.alpha=4;
```

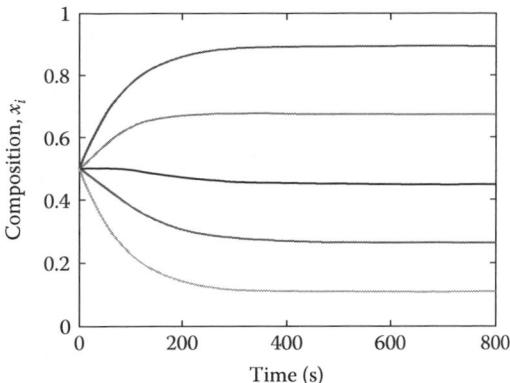

FIGURE 5.2   Transient results for the data given in the table.

```
%% Initial condition and time
tSpan = [0, 2000];   % Time span
x0 = 0.5*ones(5,1); % Initial conditions
[tSol,xSol]=ode45(@(t,x) binaryDistFun(t,x,modelPar),...
    tSpan, x0);

%% Plotting results
figure(1)
plot(tSol,xSol);
figure(2)
plot(xSol(end,:)); hold on;

R=L/D;
disp(['Reflux ratio: ',num2str(R), ...
    '=> xd: ',num2str(xSol(end,1))]);
```

The transient simulation results are shown in Figure 5.2. Starting at the initial conditions, the compositions settle at their steady state values in about 10 min. At simulation conditions, the distillate at 12.5 mol/s contains 89.2% by moles of lighter component. The reflux ratio is 1.4.

5.1.3 Effect of Parameters: Reflux Ratio and Relative Volatility

Textbooks on mass transfer and separations focus on reflux ratio (RR = L/D) as a key variable for designing a distillation column. Keeping D constant at 12.5 mol/s, increasing the reflux ratio increases the liquid reflux flowrate L as well as the reboiler vapor flowrate V. The steady state results for three different values of V are shown in Figure 5.3. The value of reflux ratio is indicated in the figure for comparison. The three conditions correspond to V = 30.0, 37.5, and 112.5 mol/s, respectively. Increasing the reflux ratio clearly shows an increase in the purity of the distillate and bottom. This comes at the cost of a higher amount of energy input. Vaporizing a higher amount of fluid in the reboiler increases reboiler duty, and therefore, energy is required. Additionally, for the same external flowrate, the internal

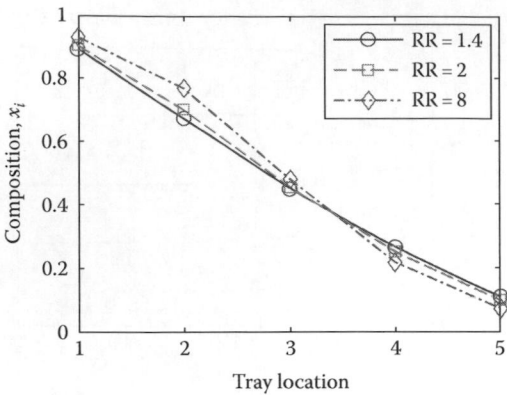

FIGURE 5.3   Steady state profile of compositions at various trays for three different conditions.

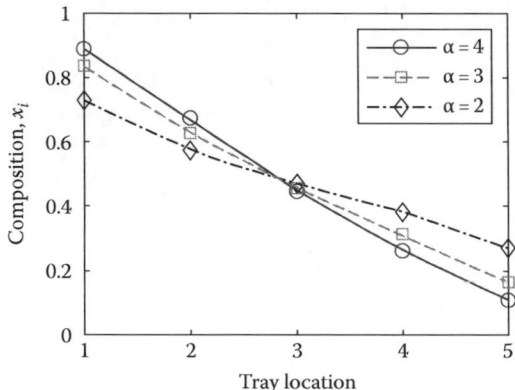

FIGURE 5.4   Effect of relative volatility on separation efficiency of binary distillation.

flowrates (L, V, and L_s) are significantly higher to achieve the same separation. This will require a larger distillation column.

Alternatively, a distillation column with a larger number of trays may be used to achieve a higher amount of separation. This is left as an exercise.

Relative volatility is an indicator of how easy or difficult separation through distillation will be. Two close-boiling liquids will have relative volatility closer to 1 and are difficult to separate. On the other hand, mixtures with large α are easier to separate. Figure 5.4 shows the performance of a distillation column for three different values of relative volatility. As relative volatility decreases, the purity of the two components also decreases. This indicates that separation of liquids with small values of α would require a larger number of trays or higher reflux or both.

5.2 STABILITY ANALYSIS FOR LINEAR SYSTEMS

Analysis of nonlinear systems is a vast and important field. A nonlinear system displays a very rich repertoire of interesting dynamics. We will defer nonlinear analysis and bifurcation to Chapter 9 in Section II of this book. The first step toward the analysis of nonlinear systems in multiple dimensions is linear stability analysis. In this section, the need for a linear stability analysis will be first motivated using the chemostat example. Qualitatively different types of dynamical behavior of linear systems will be reviewed in subsequent subsections in this section.

As I shall argue in this section that analysis of linear systems itself is an important field. The analysis in this section will build upon linear algebra concepts discussed in Chapter 2, specifically eigenvalues and eigenvectors and vector subspace. It will be useful to review those concepts to get a better understanding of linear system analysis.

Even to an undergraduate student, this topic is not entirely new. We have been introduced to the topic of stability analysis in the Process Dynamics and Control course. One may recall the discussion that a stable system has all its poles in the left-half plane. The same requirement for closed-loop systems gave rise to the Routh-Hurwitz criterion for stability. These criteria were introduced for analysis in the Laplace domain; similar criteria will be generalized and discussed in the time domain, applied to a (multidimensional) system* of ODEs $\mathbf{y}' = A\mathbf{y}$.

5.2.1 Motivation: Linear Stability Analysis of a Chemostat

A detailed nonlinear and bifurcation analysis of a chemostat will be discussed in Section 9.2. Here, we will motivate linear stability analysis using the chemostat example from Section 3.5.4. Recall that one of the steady states for the chemostat was

$$S_{ss} = 0.091, \quad X_{SS} = 3.68, \quad P_{SS} = 3.19 \tag{5.15}$$

Since S and X do not depend on P, we will ignore P from further discussion. Recall from Chapter 3 that starting at an arbitrary initial condition, the system attained steady state and reached the above values. This indicates that the above steady state is a *stable* steady state.

The stability of a system can be analyzed in the vicinity of the steady state by linearizing the nonlinear model at that steady state. A function $\mathbf{f}(\mathbf{y})$ can be linearized around steady state $\mathbf{y}_{SS}$ using multivariate Taylor's series expansion:

$$\mathbf{f}(\mathbf{y}) \approx \mathbf{f}(\mathbf{y}_{SS}) + \nabla \mathbf{f}(\mathbf{y}_{SS})(\mathbf{y} - \mathbf{y}_{SS}) \tag{5.16}$$

where $J = \nabla \mathbf{f}(\mathbf{y}_{SS})$ is the Jacobian. Only the first-order term is retained for linearization. By definition, at the steady state solution, $\mathbf{f}(\mathbf{y}_{SS}) = \mathbf{0}$. Thus, the linearized model is given by

$$\frac{d}{dt}\overline{\mathbf{y}} \approx \mathbf{J}\overline{\mathbf{y}} \tag{5.17}$$

---

* The discussion in control texts is for a controlled system $\mathbf{y}' = A\mathbf{y} + B\mathbf{u}$. Here, we are only analyzing the "autonomous" part of the system, without the input manipulation. The stability *concepts*, though, are general.

where $\bar{\mathbf{y}} = \mathbf{y} - \mathbf{y}_{ss}$ represents deviation from the steady state value. In the case of the chemostat, the Jacobian is calculated as

$$J = \begin{bmatrix} \dfrac{\partial f_1}{\partial S} & \dfrac{\partial f_1}{\partial X} \\ \dfrac{\partial f_2}{\partial S} & \dfrac{\partial f_2}{\partial X} \end{bmatrix} \tag{5.18}$$

Since the model equations for the system are written as

$$\begin{bmatrix} f_1 \\ f_2 \end{bmatrix} = \begin{bmatrix} D(S_f - S) - \dfrac{\mu SX}{K+S} \\ DX - \dfrac{\mu SX}{K+S} Y_{xs} \end{bmatrix} \tag{5.19}$$

the partial derivatives can be calculated analytically to yield the Jacobian:

$$J = \begin{bmatrix} -D - \dfrac{\mu XK}{(K+S)^2} & -\dfrac{\mu S}{K+S} \\ \dfrac{\mu XK}{(K+S)^2} Y_{xs} & -D + \dfrac{\mu S}{K+S} Y_{xs} \end{bmatrix} \tag{5.20}$$

The Jacobian value at this steady state is

$$J = \begin{bmatrix} -4.06 & -0.1333 \\ 2.97 & 0 \end{bmatrix}$$

The eigenvalues and eigenvectors of the system are computed using the `eig` command. The eigenvalues are $\lambda_1 = -3.96$ and $\lambda_2 = -0.1$, and the corresponding eigenvectors are

$$v_1 = \begin{bmatrix} -0.8 \\ 0.6 \end{bmatrix}, \quad v_2 = \begin{bmatrix} 0.0337 \\ -0.9994 \end{bmatrix}$$

Eigenvalue decomposition was discussed in Chapter 2. Any matrix with distinct eigenvalues can be written in the form $J = V \Lambda V^{-1}$, where V is a matrix whose n columns are composed of n eigenvectors of the matrix. Coordinate transformation was also discussed in

Chapter 2. Change of coordinates to the above eigenvectors can be represented as matrix multiplication:

$$\mathbf{y} = [\mathbf{v}_1 \ \mathbf{v}_2]\hat{\mathbf{y}} \tag{5.21}$$

so that $\mathbf{y} = [v_1 \quad v_2]\hat{\mathbf{y}}$, where $\hat{\mathbf{y}}$ is the same vector in the transformed coordinate system. Therefore

$$V\hat{\mathbf{y}}' = JV\hat{\mathbf{y}}$$
$$\hat{\mathbf{y}}' = \underbrace{V^{-1}JV}_{\Lambda}\hat{\mathbf{y}} \tag{5.22}$$

Since Λ is a diagonal vector, we can write the two ODEs as

$$\frac{d}{dt}\hat{y}_i = \lambda_i \hat{y}_i \Rightarrow \hat{y}_i = e^{\lambda_i t} \tag{5.23}$$

The above is stable iff the real part of the eigenvalues, $re(\lambda_i) < 0$. In vector form

$$\hat{\mathbf{y}} = e^{\Lambda t} \Rightarrow \mathbf{y} = Ve^{\Lambda t}V^{-1}$$

Stability condition for $y' = \mathbf{J}y$: Thus, the condition for stability of a linear ODE is that the real part of all eigenvalues of the matrix $\mathbf{J}$ should be less than zero. When the eigenvalues of the Jacobian are plotted on a complex plane, all the eigenvalues should lie in the left-half plane for the system to be linearly stable.

Example 5.2 Stability of Linear System

```
mu=0.5;       K=0.25;
Yxs=0.75;
Sf=5;         D=0.1;
% First steady state
S=0.0909;    X=3.6818;
a1=mu*S/(K+S);
b1=mu*X*K/(K+S)^2;
% Linearization and analysis
J=[-D-b1,    -a1;
     b1*Yxs, -D+a1*Yxs];
[V, Lambda]=eig(J);
% Verify Stability
x0=[1;1];
f=@(t,y) J*y;
[tSol,xSol]=ode45(f,[0 100],x0);
```

One can check that the linear system is indeed stable, as evidenced by xSol reaching the origin at the end time. The values of Jacobian and its eigenvalues and eigenvectors are

$$J = \begin{bmatrix} -4.06 & -0.1333 \\ 2.97 & 0 \end{bmatrix}, \quad \lambda_i = \{-3.96, -0.1\} \tag{5.24}$$

$$v_1 = \begin{bmatrix} -0.8 \\ 0.6 \end{bmatrix}, \quad v_2 = \begin{bmatrix} 0.0337 \\ -0.9994 \end{bmatrix} \tag{5.25}$$

5.2.1.1 Phase Portrait at the Steady State

A phase plane is a plane with the two state variables plotted as two axes. Each curve on the plot represents a trajectory that the system takes starting at various locations in the plane. A collection of such curves that defines the dynamic response of the system constitutes the phase portrait. Figure 5.5 shows the phase portrait for the linear bioreactor example. Circles denote the various initial conditions in the 2D space. Starting at any point, the system responds rapidly along the first eigenvector v_1, whereas the response along the second eigenvector v_2 is more sluggish. This gives rise to the phase portrait as seen in Figure 5.5. Eigenvalues and eigenvectors can be used to qualitatively analyze the dynamic response of a system, which will be discussed in this section.

5.2.1.2 Trivial Steady State and Analysis

The above steady state is not the only steady state of the system. If the bioreactor does not initially contain any biomass, no reaction can take place since the reactor input also does not contain any microorganisms. Thus, with $X(0) = 0$, the exit concentrations are $X = 0$ and $S = S_f$. This can also be derived from the chemostat model:

$$S' = D(S_f - S) - \frac{\mu S X}{K + S} \tag{5.26}$$

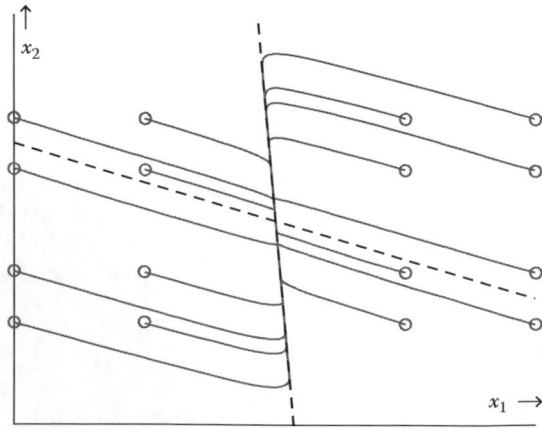

FIGURE 5.5 Phase-plane plot for linear bioreactor from Example 5.2. The two dashed lines represent the two eigenvectors.

$$X' = -DX + \frac{\mu SX}{K+S} Y_{xs} \tag{5.27}$$

At steady state, $S' = X' = 0$; the second equation

$$0 = X\left(-D + \frac{\mu S}{K+S} Y_{xs}\right) \tag{5.28}$$

has two solutions. The first one was mentioned in (5.15) above, whereas $X = 0$ (and hence $S = S_f$) is the second solution. This second solution may be called trivial solution.

Starting from a condition $X = 0$, we always reach the trivial solution. However, if the simulations are started with a very small positive value (say $X = 10^{-3}$), the system still reaches the steady state denoted in (5.15). This behavior can be further understood by calculating the Jacobian at the trivial steady state. Specifically

$$J = \begin{bmatrix} -0.1 & -0.476 \\ 0 & 0.257 \end{bmatrix}$$

the two eigenvalues are $\lambda_1 = -0.1$ and $\lambda_2 = 0.257$, and the corresponding eigenvectors are

$$v_1 = \begin{bmatrix} 1 \\ 0 \end{bmatrix}, \quad v_2 = \begin{bmatrix} -0.8 \\ 0.6 \end{bmatrix}$$

With a change of coordinate axes to the eigenvectors, the first eigenvalue represents a stable response. The first eigenvector is the *Stable Subspace*: Starting at any arbitrary point along v_1, the system asymptotically approaches the steady state. Since the first eigenvector is the S-axis, an initial condition $[S_0 \quad 0]^T$, where S_0 is any arbitrary value, will result in a stable response. This is exactly what we observed with a nonlinear system in Chapter 3, where $X(0) = 0$ led to the trivial solution. The system linearized at the trivial steady state can be analyzed in a manner similar to Example 5.2, with steady state values, S=5 and X=0.

In summary, the eigenvalues of the Jacobian matrix determine the *local stability* of a system. For an $n \times n$ Jacobian matrix, if all n eigenvalues have negative real parts, then the system is stable in the vicinity of that steady state. If only m eigenvalues have negative real parts (with $m < n$), the m-dimensional subspace spanned by the corresponding eigenvectors is the stable subspace, whereas the remaining $(n-m)$-dimensional subspace is the unstable subspace of the system. The stability of a linear system is further analyzed in this section.

5.2.2 Eigenvalues, Stability, and Dynamics

Any nonlinear system can be analyzed in the vicinity of a steady state by linearizing it and computing its eigenvalues to predict is dynamic behavior near the steady state. Additionally, linear system analysis provides an extremely powerful tool to analyze a general n-dimensional system. The preceding discussion not only motivates the need to better understand linear dynamics but also provides a template to do so.

The case of real and distinct eigenvalues was discussed in the preceding section. Consider the ODE

$$\mathbf{y}' = A\mathbf{y} \tag{5.29}$$

Coordinate transformation of the form

$$\hat{\mathbf{y}}' = \underbrace{V^{-1}AV}_{\Lambda}\hat{\mathbf{y}} \tag{5.22}$$

converted the original ODE into a set of decoupled ODEs. Based on this diagonalization, we argued that eigenvalues in the left-half plane (i.e., eigenvalues with negative real parts) are stable. This result is not limited to distinct eigenvalues but is applicable for all linear dynamical systems. The proof of the above result is skipped since this is beyond the focus of this book.

Nonetheless, let us take the discussion further using 2D examples in the rest of this section. We will first take the case of real and distinct eigenvalues, and then extend it to other general cases.

5.2.2.1 Dynamics When Eigenvalues Are Real and Distinct

Solution for linear differential equations is often a part of first-year undergraduate calculus course. Let us relate the results obtained for matrix system in Equations 5.22 and 5.23 to the more familiar form. For a 2×2 matrix, A, the solution derived in Equation 5.22 was

$$\hat{\mathbf{y}} = \begin{bmatrix} c_1 e^{\lambda_1 t} \\ c_2 e^{\lambda_2 t} \end{bmatrix} \tag{5.30}$$

The solution, $\mathbf{y}$, is obtained by changing the coordinate system back to the original coordinates. From Equation 5.21

$$\mathbf{y} = \begin{bmatrix} \mathbf{v}_1 & \mathbf{v}_2 \end{bmatrix} \begin{bmatrix} c_1 e^{\lambda_1 t} \\ c_2 e^{\lambda_2 t} \end{bmatrix} \tag{5.31}$$
$$\mathbf{y} = c_1 \mathbf{v}_1 e^{\lambda_1 t} + c_2 \mathbf{v}_2 e^{\lambda_2 t}$$

which is a familiar result from an introductory calculus course. The constants of integration are obtained by substituting the initial condition:

$$\mathbf{y}_0 = c_1 \mathbf{v}_1 + c_2 \mathbf{v}_2 = \begin{bmatrix} \mathbf{v}_1 & \mathbf{v}_2 \end{bmatrix} \begin{bmatrix} c_1 \\ c_2 \end{bmatrix} \tag{5.32}$$

The implication of the above result is interesting. If the initial condition $\mathbf{y}_0$ lies on the first eigenvector, $\mathbf{y}_0 = c_1 \mathbf{v}_1$ and $c_2 = 0$.

The expression (5.32) also clarifies the geometric interpretation of results discussed in previous subsection, which is summarized below:

1. If both λ_1 and λ_2 are negative, the system is stable, starting at any point in the state space.

2. If both λ_1 and λ_2 are positive, the system is unstable at any point in the state space.

3. If λ_1 is negative and λ_2 is positive, then the system is a saddle node, with the first eigenvector as stable subspace and the second eigenvector as unstable subspace. In other words, if the initial condition is $\mathbf{y}_0 = \alpha \mathbf{v}_1$, then the system is stable. This is because as per (5.32), the solution is $\mathbf{y}(t) = \alpha \mathbf{v}_1 e^{\lambda_1 t}$; since $\lambda_1 < 0$, the system is stable.

   The dynamics for this case is interesting. If $\mathbf{y}_0$ is not in the stable subspace, the final solution is represented as

$$\mathbf{y}(t) = c_1 \mathbf{v}_1 e^{\lambda_1 t} + c_2 \mathbf{v}_1 e^{\lambda_1 t}$$

   The dynamics *along* $\mathbf{v}_1$ are convergent and stabilize after sufficient time, whereas dynamics in the direction of $\mathbf{v}_2$ are divergent.

4. The final case is when λ_1 is negative and $\lambda_2 = 0$. The system is stable along the first eigenvector, whereas it remains unchanged along $\mathbf{v}_2$.

   As a first example, consider

$$\mathbf{y}' = \underbrace{\begin{bmatrix} -4 & 1 \\ 2 & -3 \end{bmatrix}}_{A} \mathbf{y} \tag{5.33}$$

The two eigenvalues are $\lambda = \{-5, -2\}$, and the two eigenvectors are

$$\mathbf{v}_1 = \begin{bmatrix} 1 \\ -1 \end{bmatrix}, \quad \mathbf{v}_2 = \begin{bmatrix} 1 \\ 2 \end{bmatrix}$$

Figure 5.6 shows the transient response of the ODE for two different initial conditions. The first is starting from the first eigenvector. Specifically, $\mathbf{y}_0 = \mathbf{v}_1$. The system will evolve along this eigenvector, with its response governed by the first eigenvalue; that is

$$\mathbf{y}(t) = \begin{bmatrix} e^{-5t} \\ -e^{-5t} \end{bmatrix}$$

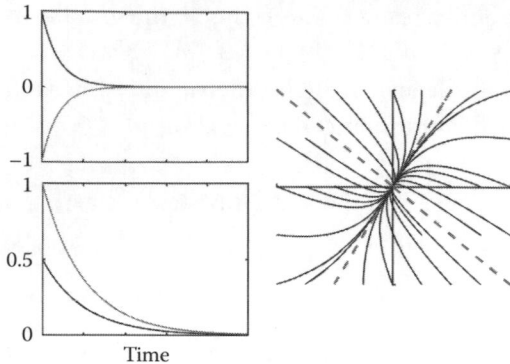

FIGURE 5.6 Transient response and phase portrait for linear ODE with `A=[-4, 1; 2, -3]`. The two eigenvectors are `v1=[1; -1]` and `v2=[1; 2]`.

In contrast, when $\mathbf{y}_0 = 0.5\mathbf{v}_2$, the system response will be given by

$$\mathbf{y}(t) = \begin{bmatrix} 0.5e^{-2t} \\ e^{-2t} \end{bmatrix}$$

These two are shown on the left panels of Figure 5.6. Clearly, the system responds faster if the initial condition is along the eigenvector $\mathbf{v}_1$, whereas the response is slower if the initial condition is along the eigenvector $\mathbf{v}_2$. Thus, the more negative eigenvector is the *fast mode* of the system and the less negative eigenvector is the *slow mode* of the system.

The right-hand panel of Figure 5.6 shows a phase-planeplot (also known as phase portrait). This is a 2D plot of state variable y_1 vs. y_2. It shows how the system responds starting at various points in the state space. Specifically, with various initial conditions, `ode45` is used to find the variation of $\mathbf{y}$ with time; the locus of points y_1 vs. y_2 on the phase plane gives the overall phase portrait of the system. The phase portrait charts out the dynamic response of the system in the relevant regions in the two main dimensions. The dashed lines on the phase portrait represents the system starting along either of the two eigenvectors. Since $\mathbf{v}_1 = [1 \ -1]^T$ is the fast eigenmode, the system response along $\mathbf{v}_1$ is faster than the response along $\mathbf{v}_2$. The trajectories therefore approach $\mathbf{v}_2$ and then move along it to reach the origin.

Figures 5.7 and 5.8 show the phase portraits of the linear system when the eigenvalues are real and distinct. The eigenvalues are mentioned in the figure captions, and the eigenvectors are indicated as dashed thick lines. When the two eigenvalues are negative and close to each other (Figure 5.7a), the trajectories move uniformly in the state space. When the two eigenvalues are an order of magnitude different (as in panels b and c), the trajectories move rapidly along the fast eigenvector to reach the slow eigenvector; after this, the trajectories move along the slow eigenvectors. When the eigenvalue is increased to 0, the trajectories move only along the first eigenvector and remain unchanged along the second eigenvector.

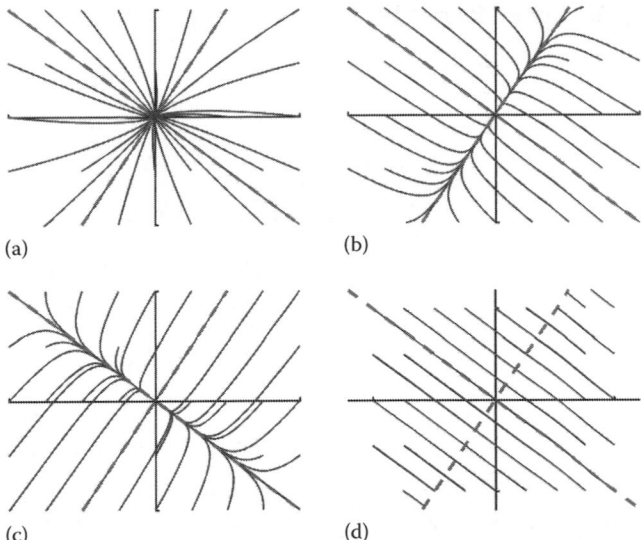

FIGURE 5.7   Phase portrait for various 2D linear systems with two distinct real eigenvalues: (a) a stable node with ($\lambda_1 = -5, \lambda_2 = -4$), (b) a stable node with ($\lambda_1 = -5, \lambda_2 = -0.5$), (c) a stable node with ($\lambda_1 = -0.5, \lambda_2 = -5$), and (d) a case when one eigenvalue is 0. The two eigenvectors are v1=[1; -1] and v2=[1; 2].

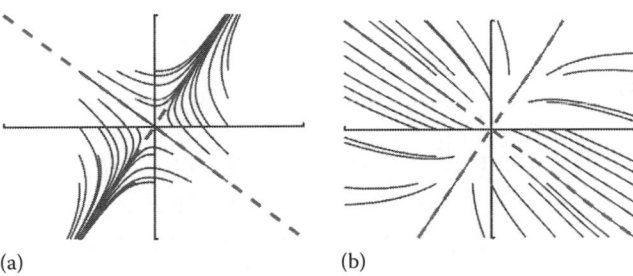

FIGURE 5.8   Phase portrait for various unstable 2D linear systems with two distinct real eigenvalues: (a) an unstable saddle node ($\lambda_1 = -5, \lambda_2 = 2$) and (b) an unstable node ($\lambda_1 = 5, \lambda_2 = 2$). The two eigenvectors are v1=[1; -1] and v2=[1; 2].

When one of the eigenvalues becomes positive (see Figure 5.8a), the system behaves as a *saddle* node. The first eigenvector is the stable subspace, whereas the second eigenvector is unstable. The overall system is unstable. If the initial condition is along the stable eigenvector, the system asymptotically goes to origin. Starting from other points in the state space, the system reaches the unstable eigenvector and moves along that eigenvector to instability. When the other eigenvalue is positive as well (Figure 5.8b), the system becomes unstable in the entire 2D space. The linear systems of ODEs for which the results in Figures 5.7 and 5.8

are generated are given below. The subscripts represent the phase portrait responses (*a*) through (*f*):

$$A_{(a)} = \begin{bmatrix} -4.667 & 0.333 \\ 0.667 & -4.333 \end{bmatrix} \quad A_{(b)} = \begin{bmatrix} -3.5 & 1.5 \\ 3 & -2 \end{bmatrix}$$

$$A_{(c)} = \begin{bmatrix} -2 & -1.5 \\ 3 & -3.5 \end{bmatrix} \quad A_{(d)} = \begin{bmatrix} -3.333 & 1.667 \\ 3.333 & -1.667 \end{bmatrix}$$

$$A_{(e)} = \begin{bmatrix} -2.667 & 2.333 \\ 4.667 & -0.333 \end{bmatrix} \quad A_{(f)} = \begin{bmatrix} 4 & -1 \\ -2 & 3 \end{bmatrix}$$

One can easily verify using the `eig` command that the above matrices represent the ODE systems $\mathbf{y}' = A_{(\cdot)}\mathbf{y}$ for Figures 5.7 and 5.8.

The phase portrait when eigenvalues are complex is shown in Figure 5.9. Complex eigenvalues always occur in complex conjugate pairs. Since

$$e^{(a+ib)t} = e^{at}\left(\cos(bt) + i\sin(bt)\right)$$

the real part of the eigenvalues determine stability and the imaginary part results in oscillatory behavior. Thus, if the real part of the eigenvalues $re(\lambda_i)$ is negative, the system exhibits

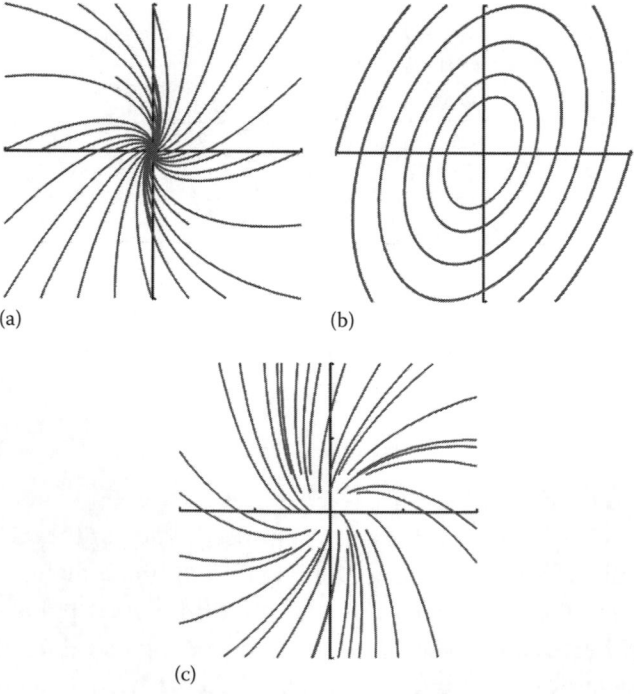

(a) (b)

(c)

FIGURE 5.9 Various examples when eigenvalues are a complex conjugate pair: (a) a stable spiral when $re(\lambda_i) < 0$, (b) a center when λ_i are imaginary, and (c) an unstable spiral when $re(\lambda_i) > 0$.

stable spiral in phase portrait, if re(λ_i) is positive, the system exhibits *unstable spiral*, and if the eigenvalues are purely imaginary, the phase portrait consists of concentric closed cyclic trajectories. This is known as *center*. This behavior is discussed in the next example.

5.2.2.2 An Example

A classical example is a mass-spring system. The motion of a mass suspended from an ideal spring that is displaced from its natural position is given by

$$mx'' + kx = 0 \tag{5.34}$$

Consider displacement (x) and velocity (v) as the two state variables:

$$\frac{d}{dt}\begin{bmatrix} x \\ v \end{bmatrix} = \begin{bmatrix} 0 & 1 \\ -k/m & 0 \end{bmatrix}\begin{bmatrix} x \\ v \end{bmatrix} \tag{5.35}$$

The eigenvalues of the above matrix are $\lambda_i = \pm i\sqrt{k/m}$. Thus, the mass-spring system is qualitatively represented by the phase portrait in Figure 5.9b. The mass at rest in its equilibrium position, $x = 0$, continues to remain at rest at that position. If the mass is displaced from its equilibrium position, the spring makes the mass return to its original position. However, in the absence of damping, the mass will keep oscillating around its equilibrium position. If the original displacement of mass was greater, the mass will keep oscillating with a larger amplitude. The phase portrait would therefore be one of the various closed cycles, which depends on the original displacement of the mass from its equilibrium position.

5.2.2.3 Summary

Figure 5.10 summarizes the qualitative responses of the system for various eigenvalues denoted in the central figure on a complex plane. Eigenvalues on a real axis result in stable

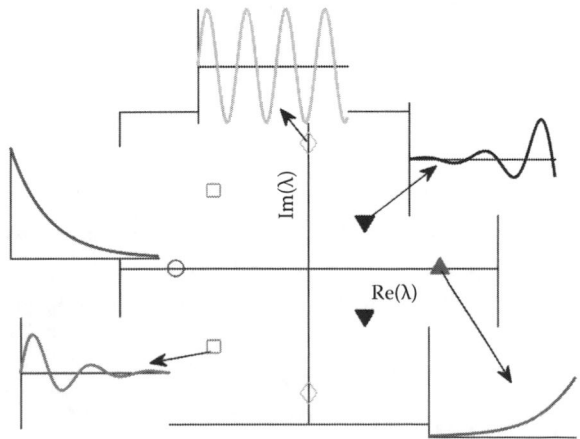

FIGURE 5.10   Eigenvalues and qualitative response of the systems.

(left-half plane) or unstable (right-half plane) responses without oscillations. Eigenvalues on the imaginary axis, on the other hand, result in oscillatory responses, with the amplitude of oscillations remaining constant. Finally, complex eigenvalues (with nonzero real and imaginary components) result in oscillatory responses that stabilize or grow, based on the real part.

5.2.3 Transient Growth in Stable Linear Systems

The discussion on the dynamics of a linear system of ODEs

$$\mathbf{y}' = A\mathbf{y}$$

has so far focused on the effect of eigenvalues on the system dynamic behavior. The eigenvalues determine the *asymptotic* or long-time behavior of the system. Eigenvalues with negative real parts indicate stable response. When the two eigenvalues are close to each other, the system does not "favor" any direction (see Figure 5.7a). However, when the ratio of eigenvalues increases, the system response shows "faster" and "slower" modes (see Figure 5.7b and c). The larger the magnitude of negative eigenvalue, the faster is the system response along the corresponding eigenvector. This contrasting behavior along fast and slow modes becomes especially significant as the ratio of the eigenvalues increases.

While the previous discussion focused on eigenvalues, the importance of eigenvectors will be highlighted in this subsection.

5.2.3.1 Defining Normal and Nonnormal Matrices

Before proceeding further, let us define what is meant by a *normal matrix*. A matrix A is said to be normal if it commutes with its (conjugate) transpose. For a real-valued matrix A, the condition for normality is

$$AA^T = A^T A \tag{5.36}$$

An important property that follows from this definition is that the eigenvectors of a normal matrix are orthogonal to each other. Conversely, a nonnormal matrix is one where the eigenvectors are not orthogonal. Geometric interpretation is that a highly nonnormal matrix will have its eigenvectors further away from being orthogonal (or at a closer angle to each other).

There are several ways to quantify nonnormality. The one definition that is aligned to the discussion in this section is adapted from the one presented by Ruhe.* When a matrix is *normal*, the eigenvalues and singular values are equal. Thus, if $|\lambda_i|$ and σ_i represent an ordered set of eigenvalues (based on modulus) and singular values, respectively, then

$$\max_i \left| \sigma_i - |\lambda_i| \right| \tag{5.37}$$

---

* Ruhe, A., On the closeness of eigenvalues and singular values for almost normal matrices, *Linear Algebra Applications*, **11** (1975), 87–94.

is one measure of nonnormality. Another measure is based on condition number. The condition number of a *normal matrix* is equal to the ratio of its largest and smallest (by modulus) eigenvalues. Thus, the relative values of condition number and the ratio of eigenvalues, that is

$$\mu = \frac{\sigma_1}{\sigma_n}, \quad \bar{\mu} = \frac{|\lambda_{max}|}{|\lambda_{min}|} \tag{5.38}$$

quantifies the nonnormality of a matrix. A normal matrix has $\mu / \bar{\mu} = 1$; an increase in this ratio indicates greater departure from normality.

5.2.3.2 Analysis of Nonnormal Systems

Now that the normal and nonnormal systems have been defined, let us shift the focus on using the simple 2×2 example to demonstrate the salient features of a system that displays nonnormal dynamical behavior. Such an analysis has been used to study weather patterns, instabilities in combustion systems, etc., where both the long-term system stability as well as short-term transients are equally important. Short-term surges in the energy of the system may result in severe weather conditions or high-energy failure of devices.

Transient response of the original system

$$\mathbf{y}' = \underbrace{\begin{bmatrix} -4 & 1 \\ 2 & -3 \end{bmatrix}}_{A} \mathbf{y} \tag{5.33}$$

was shown in Figure 5.6. The matrix is nonnormal; however, the two eigenvectors are close to being orthogonal. Hence, the condition number

```
>> cond(A)
ans =
    2.6180
```

is close to the ratio of its eigenvalues, 2.5. Thus, the system is only slightly nonnormal. Thus, starting at an initial condition, the magnitude of dependent state variables y_1, y_2 falls monotonically to origin.

Now consider another matrix

$$A_1 = \begin{bmatrix} 4 & -9 \\ 6 & -11 \end{bmatrix}$$

which has the same eigenvalues ($\lambda = \{-5, \ -2\}$ as the original matrix, A. The eigenvectors of A_1 are given as

$$\mathbf{w}_1 = \begin{bmatrix} 1 \\ 1 \end{bmatrix}, \quad \mathbf{w}_2 = \begin{bmatrix} 1.5 \\ 1 \end{bmatrix}$$

The eigenvectors of matrix A_1 make a small angle ($\mathbf{w}_1$ and $\mathbf{w}_2$ are 11.3° apart). The condition number of this matrix

```
>> cond(A1)
ans =
    25.3606
```

is an order of magnitude greater than the ratio of its eigenvalues. The condition number is a metric that indicates how sensitive a system is to a change in the input parameters. The asymptotic (long-term) response of the system is stable, and no eigenvector is significantly faster or slower than the other. However, the closeness of the eigenvectors affects the short-term transients of the system.

The transient response of the system for various initial conditions is shown in Figure 5.11. The two panels on the left show the response of the system vs. time starting at four different initial conditions:

$$\mathbf{y}(0) = \begin{bmatrix} 0.8 \\ 0.6 \end{bmatrix}, \quad \mathbf{y}(0) = \begin{bmatrix} 0.8 \\ -0.6 \end{bmatrix}, \quad \mathbf{y}(0) = \begin{bmatrix} 0.2 \\ 0.98 \end{bmatrix}, \quad \mathbf{y}(0) = \begin{bmatrix} 0.2 \\ -0.98 \end{bmatrix}$$

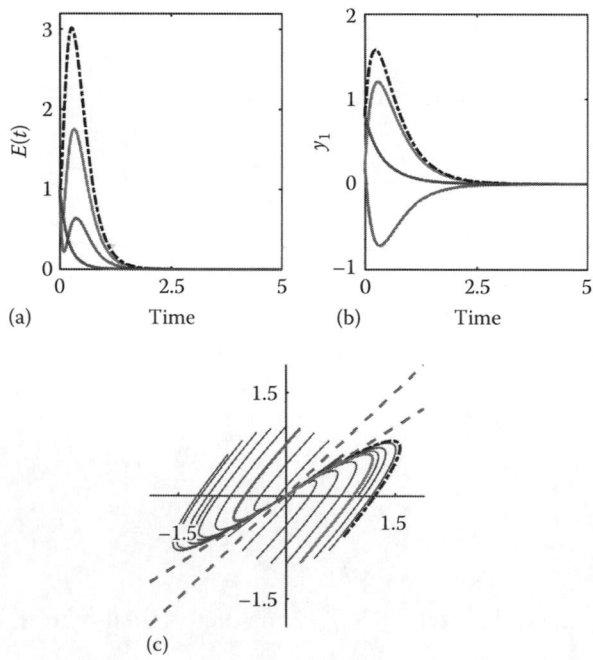

FIGURE 5.11 Demonstration of transient growth in a nonnormal system: (a) energy $E(t)$ vs. time, (b) transient response of $y_1(t)$ vs. time, and (c) phase portrait. The four thick lines in the phase portrait correspond to the four lines in the two left transient panels.

Figure 5.11a shows the change in energy of the system with time. For the purpose of this example, energy $E(t)$ is defined as

$$E(t) = \frac{\mathbf{y}^T\mathbf{y}}{\mathbf{y}_0^T\mathbf{y}_0} = \frac{y_1^2(t) + y_2^2(t)}{y_1^2(0) + y_2^2(0)} \tag{5.39}$$

The first initial condition, $\mathbf{y}_0 = [0.8 \quad 0.6]^T$, starts between the two eigenvectors (see phase portrait in Figure 5.11c, where the dashed lines represent eigenvectors). The value of $E(t)$ as well as $y_1(t)$ and $y_2(t)$ decrease monotonically. This behavior seems qualitatively similar to that observed in nearly normal systems.

The transient response starting at the second initial condition is shown as dash-dot lines in all the three panels. A fairly large surge in energy is observed, with the maximum value of $E(t)$ exceeding 3. Likewise, the maximum value of $y_1(t)$ attained is 1.575. Some initial conditions show large transient growth, whereas some other initial conditions do not. This is an important feature of transient response of nonnormal systems: For certain initial conditions, a large surge in short-term transients of the system is observed before the system settles down asymptotically to its equilibrium point (origin).

The phase portrait of the nonnormal system (Figure 5.11c) is exactly as expected. The initial response of the system is along the fast eigenvector, $\mathbf{w}_1$, and eventually approaches and moves parallel to the slower eigenvector, $\mathbf{w}_2 = [1.5 \quad 1]^T$. Since the two eigenvectors form an acute angle, any trajectory starting from outside this region gets stretched along $\mathbf{w}_1$, before returning to the origin along $\mathbf{w}_2$.

The transient growth will be even higher if either the two eigenvectors are brought closer to each other or if the ratio of eigenvalues is increased. For example, with

$$A_2 = \begin{bmatrix} 8.5 & -13.5 \\ 9 & -14 \end{bmatrix}$$

the transient growth in $E(t)$ is significantly higher. Note that the eigenvectors of A_2 are the same $(\mathbf{w}_1, \mathbf{w}_2)$ as those of A_1, whereas the eigenvalues are $\lambda = \{-5, \quad -0.5\}$. Since the ratio of eigenvalues is much greater than in the previous case, a larger transient growth is observed. Note that the condition number for this matrix is $\mu = 212$. Phase-plane analysis and transient growth for this system is left as an exercise.

Although the rules for phase-plane analysis remain the same for highly nonnormal systems, the *consequence* of skewed eigenvectors manifests as a large transient growth.

5.3 COMBINED PARABOLIC PDE WITH ODE-IVP: POLYMER CURING

This is an example that is relevant to the curing of thermosetting polymers or natural rubber. The process, in the familiar case of rubber, is known as vulcanization. In either case, the polymer or rubber sap is mixed with cross-linking agents (e.g., sulfur, in case of rubber), promoters, and additives. The mixture is put in a mold and heated to a predetermined temperature. Reactions take place at these elevated temperatures, which result in a cross-linking of the polymeric chains. The mixture solidifies and gets its properties based on the degree of cross-linking that happens during the process. Indeed, the actual process involves tens or hundreds of reactions and is complicated. Moreover, geometries required of these materials (such as tyre) are also highly complex. What we will do in this case study is to define a simplified problem that is motivated by this important process. We will retain just a couple of reactions and simulate a 1D slab geometry as a representative system.

Consider a 2 cm long slab, which is originally at room temperature. The two ends of the slab are heated using a constant temperature source, which results in a rise in slab temperature due to conduction. At elevated temperature, the following reactions take place:

$$P \xrightarrow{k_1} C \xrightarrow{k_2} D$$

where
  P is the original polymer
  C is the cross-linked polymer
  D is dead-ended polymer

A specific range of C and D are needed for the product to have desired properties.

The simplified geometry is a 1D slab with its ends kept at 180°C. Since the material is a semisolid material in a mold, the temperature effects are modeled by a 1D transient heat equation:

$$\frac{\partial T}{\partial t} = \alpha \frac{\partial^2 T}{\partial z^2} + S(T, \mathbf{C}) \tag{5.40}$$

where $S(T, \mathbf{C})$ is the source term due to reactions. We will neglect reaction heat effects. The two reactions take place at each location in the slab. Since there is no transport of material within the solid slab, the equations governing the concentration are

$$\frac{\partial}{\partial t} C_P = -r_1$$
$$r_1 = k_{10} \exp\left(-\frac{E_1}{RT}\right) C_P \tag{5.41}$$

$$\frac{\partial}{\partial t} C_C = r_1 - r_2$$
$$r_2 = k_{20} \exp\left(-\frac{E_2}{RT}\right) C_C \tag{5.42}$$

$$\frac{\partial}{\partial t} C_D = r_2 \tag{5.43}$$

The initial concentration of P in the mixture is 0.5 mol/L; there is no C or D initially. The system is initially at room temperature (25°C), when the heating is started. Dirichlet boundary conditions are imposed at the two ends of the slab since the temperature is specified. The next example shows the methodology.

Example 5.3 Heat Transfer and Reactions of Thermosetting Polymer

Problem: Solve the example of thermosetting polymer slab, which is cured by exposing it to elevated temperature. The slab is 2 cm wide and thermal diffusivity is 5×10^{-8} m²/s. The rate constants are given by

$$k_1 = 5\times10^8 \text{ min}^{-1}, \quad \frac{E_1}{R} = 9500\,\text{K}, \quad k_2 = 10^{11}\,\text{min}^{-1}, \quad \frac{E_2}{R} = 13,500$$

Compute the temperature profile and how the cross-link and dead-end concentrations vary within the slab with time. The slab is heated for 75 min with the two ends held at 180°C and allowed to cool under ambient conditions.

Solution: We will discretize the parabolic PDE (5.40):

$$\frac{dT_i}{dt} = \alpha \frac{\left(T_{i+1} - 2T_i + T_{i-1}\right)}{h^2} + S\left(T_i, C_i\right) \tag{5.44}$$

Equations 5.41 through 5.43 are written for each node:

$$\frac{d}{dt}C_{X,i} = R_{X,i} \tag{5.45}$$

where $X \equiv P, C, D$ represents the three species. Since the boundary temperatures are known, we will write the above equations for only the $(n-1)$ internal nodes. We define the solution variable as

$$\Phi = \begin{bmatrix} T_1 & C_{P1} & C_{C1} & C_{D1} & \cdots & T_{n-1} & C_{P,n-1} & C_{C,n-1} & C_{D,n-1} \end{bmatrix}^T \tag{5.46}$$

and write the function file that computes $d\Phi$ as below:

```
function dPhi=curingFun(t,Phi,par)
% Model for simplified polymer curing example
% Combination of parabolic PDE for energy
%          and ODE for material balances
%% Model parameters
alpha=par.alpha;
Tb1=par.Tb1;
Tb2=par.Tb2;
n=par.n;
h=par.h;
diffPar=alpha/h^2;
```

```
%% Model variables
Phi=reshape(Phi,4,n-1);
T=Phi(1,:);
C=Phi(2:end,:);

%% Model equations
dT=zeros(1,n-1);
dC=zeros(3,n-1);
for i=1:n-1
    r=rxnRate(T(i),C(:,i),par);

    % Energy balance
    if (i==1)
        diffCoef=T(i+1)-2*T(i)+Tb1;
    elseif (i==n-1)
        diffCoef=Tb2-2*T(i)+T(i-1);
    else
        diffCoef=T(i+1)-2*T(i)+T(i-1);
    end
    dT(i)=diffPar*diffCoef;
    % Material balance
    dC(:,i)=par.nuStoich * r;
end
dPhi=reshape([dT;dC],(n-1)*4,1);
end
```

Calculation of the reaction rate involves two reactions. The reaction term for each species is given by

$$R_X = \sum_j v_{X,j} r_j \tag{5.47}$$

where $v_{X,j}$ is the stoichiometric coefficient for species X in jth reaction. The matrix v is a 3×2 matrix, whereas the reaction rate vector is a 2×1 vector. The stoichiometric matrix is

$$v_{X,j} = \begin{bmatrix} -1 & 0 \\ 1 & -1 \\ 0 & 1 \end{bmatrix} \tag{5.48}$$

The reaction rate is computed in the function given below:

```
function r=rxnRate(T,C,par)
k0=par.k0;
E=par.E;
k=k0 .* exp(-E/T);
```

```
% First reaction
r(1,1)=k(1)*C(1);        % k1*C_P
% Second reaction
r(2,1)=k(2)*C(2);        % k2*C_C
end
```

Note that the main challenge in this problem was converting the given set of PDE and ODEs into a standard form after discretization. Once this was achieved as Equations 5.40 through 5.43, the rest followed the same procedure as we have done before. On the same lines, the driver script also follows the same structure as earlier problems. Since the solution vector is in the form of Equation 5.46, the initial conditions need to be handled appropriately.

We will split the driver solution into two parts. The first part is slab heating, and the second part is cooling down. The ODE solver, ode45, is used for a time span of [0:75]. We use colon notation in time span because the solution is desired at every 1 min interval. When the solution is obtained, the last row of the solution vector contains the values at time 75 min.

Thereafter, to simulate cooling in atmosphere, the solver is called again with a time span of [75:90] and with the initial condition on the second call being the value at the final time of the first call. The driver script is given below:

```
% Simulation of "simplified" polymer curing
% Combination of parabolic PDE for energy
%          and ODE for material balances
%% Model parameters
L=0.02;
n=50;
h=L/n;
slabParam.L=L;
slabParam.n=n;         slabParam.h=h;
% Energy balance parameters
Tb1=180+273;
Tb2=180+273;
slabParam.Tb1=Tb1;
slabParam.Tb2=Tb2;
slabParam.alpha=5e-8*60;
% Reaction parameters
slabParam.k0= [5e8, 1e11];
slabParam.E = [9500, 13500];
slabParam.nuStoich=[-1, 0;
    1, -1;
    0, 1];
%% Initial conditions and simulation
T0=25+273;
C0=[0.5;0;0];
Y0=repmat([T0;C0],n-1,1);
```

```
% Part-1: Heating up
tSpan=0:75;
[tSol,YSol]=ode15s(@(t,y) curingFun(t,y,slabParam),...
    tSpan,Y0);
Y0=YSol(end,:);    % Solution at t=75

% Part-2: Cool-down
tSpan=75:120;
slabParam.Tb1=25+273;
slabParam.Tb2=25+273;
[tSol2,YSol2]=ode15s(@(t,y) curingFun(t,y,slabParam),...
    tSpan,Y0);
tSol=[tSol;tSol2(2:end)];
YSol=[YSol;YSol2(2:end,:)];

%% Plotting results
figure(1)
Z=[0:n]*h;
T1=YSol([6,11,21,41,81],1:4:end);
T1=[Tb1*ones(5,1), T1, Tb2*ones(5,1)];
plot(Z,T1);

MidT=YSol(:,101);
MidC=YSol(:,102:104);
figure(2);
subplot(2,1,1); plot(tSol,MidT);
subplot(2,1,2); plot(tSol,MidC(:,2)./midC(:,1));
```

Figure 5.12 shows the temperature profile along the length of the slab at various times (5, 10, 20, 40, and 80 min). The last line occurs 5 min after the start of the cool-down

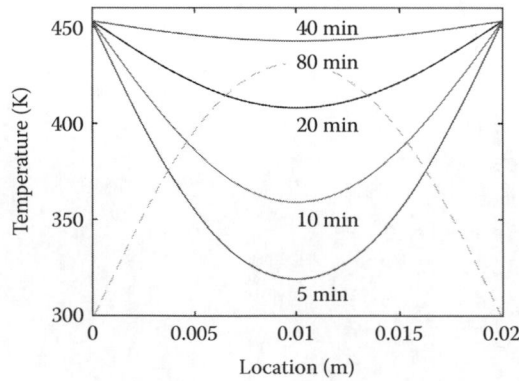

FIGURE 5.12   Evolution of spatial temperature profile in the 1D slab at various times. Solid lines represent heating phase and dashed line 5 min into the cooling phase.

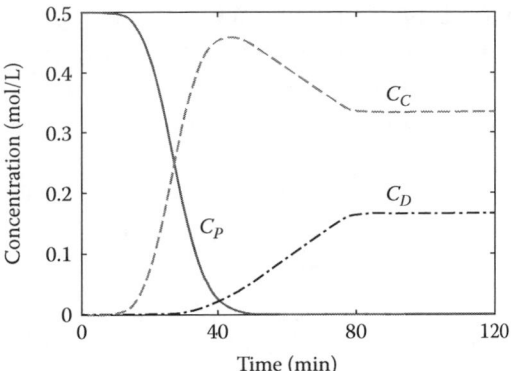

FIGURE 5.13   Concentration of the three species at the center of the slab during the 2 h process.

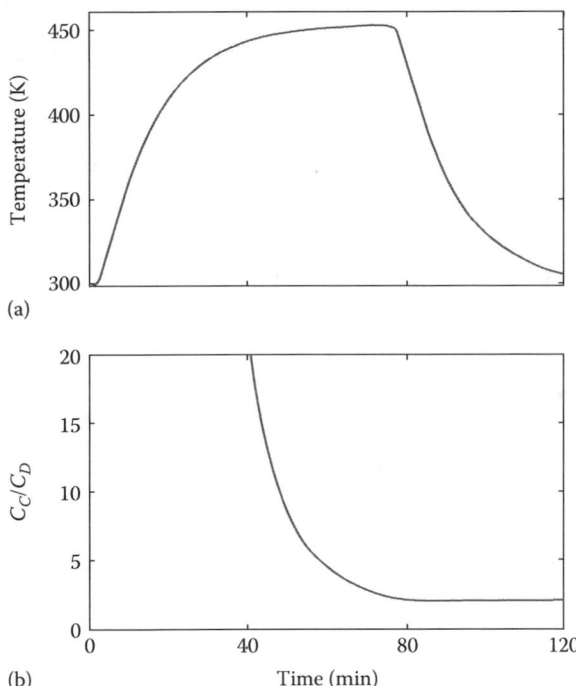

FIGURE 5.14   Variation of (a) temperature and (b) ratio of crosslink and dead-end concentrations (C_C/C_D) at the center of the slab as a function of time.

period and is shown as dashed line. Due to low thermal diffusivity, significant temperature gradients are present during the process.

Figures 5.13 and 5.14 show temporal variations in several key variables at the center of the slab. Figure 5.13 indicates that the cross-link concentration, C_C goes through maxima. Moreover, although significant temperature gradients exist in the slab, the reactions stop nearly 6 min into the cooling period, although the center temperature is still quite high (Figure 5.14 top panel). This is attributed to the fact that the reaction rate falls significantly due to large activation energy. Finally, the bottom panel of Figure 5.14 shows that at the end of the process, the two desirable conditions of curing are met: complete conversion of the original polymer ($C_p \approx 0$) and a good crosslink:dead-end ratio.

A real polymer curing process is more complex, with tens or hundreds of reactions. Moreover, heat effects due to reaction may also be important. The aim of this case study was to introduce students to a different type of problem, where parabolic PDE and ODE-IVP are solved simultaneously in process simulation. The intent of this exercise is to demonstrate that a complex problem is easily solvable using tools in this book, once we break it down to a standard form.

5.4 TIME-VARYING INLET CONDITIONS AND PROCESS DISTURBANCES

The examples considered so far in this book considered constant values of inlet. For example, we simulated examples where the inlet flowrate or temperature was constant or there was a single step change. In this section, I will present examples where the inlet conditions or process parameters change with time. We will take the example of a biological reactor with Monod kinetics, which was discussed in detail in Section 3.5.4.

The model and parameters for simulating a continuous stirred bioreactor summarized from Chapter 3 (along with parameters) are given below:

$$\frac{dS}{dt} = D\left(S_f - S\right) - r_g$$

$$\frac{dX}{dt} = -DX + r_g Y_{xs} \tag{5.49}$$

$$\frac{dP}{dt} = -DP + r_g Y_{ps}$$

$$\text{where} \quad r_g = \left(\mu^{max} \frac{[S]}{K_S + [S]}\right)[X] \tag{5.50}$$

The rate constants are given as $\mu^{max} = 0.5$, $K_S = 0.25$, $Y_{xs} = 0.75$, and $Y_{ps} = 0.65$.

5.4.1 Chemostat with Time-Varying Inlet Flowrate

Incorporating a time-varying inlet flowrate is straightforward. However, it can sometimes be confusing if the flowrate varies in a complex manner. We will tackle this problem in a couple of steps. First, we consider the flowrate to vary in a sinusoid manner around the nominal value. Next, we will consider the case where the flowrate is varied as a series of steps and ramps. The nominal value of dilution rate $D = Q/V = 0.3$ h^{-1}; the dilution rate will vary ±20% around this value.

Example 5.4 Chemostat with Sinusoidal Input

Problem: Solve the chemostat problem with sinusoid input, where the dilution rate varies by ±20% round the nominal value of 0.2 h^{-1}.

   Solution: The driver script remains the same as before. The frequency of the sinusoid is defined in a `monodParam` structure.

   The function file used in the calculation of $d\mathbf{Y}/dt$ (for a constant dilution rate) was discussed in Section 3.5.4. The fourth line in that function was

```
D=monodParam.D;
```

It is easy to see that simulating sinusoid input simply means replacing that line with a sine function that calculates

$$D = D_0 \left[1 + 0.2 \sin\left(2\pi f t \right) \right] \tag{5.51}$$

where f is the frequency of the sinusoid. Indeed, simulating a sinusoidal input is as simple as replacing the original assignment for D with the expression (5.51).

   However, to set ourselves up for the next example, we will define a separate function, `varyingInlet`, to do this calculation. It is easy to see that this function is

```
function D=varyingInlet(t,y,par)
    D0=par.D;
    f=par.freq;
    D=D0*(1+0.2*sin(2*pi*f*t));
end
```

This function will be used within a `monodFun.m` function:

```
function dy=monodFun(t,y,monodParam)
%% System parameters
mu=monodParam.mu;
K=monodParam.K;
Yxs=monodParam.Yxs;
Yps=monodParam.Yps;
Sf=monodParam.Sf;
D=varyingInlet(t,y,monodParam);

%% Variables to solve for
S=y(1);
X=y(2);
P=y(3);
%% Calculating the model
rg=mu*S/(K+S)*X;
dy=zeros(3,1);
```

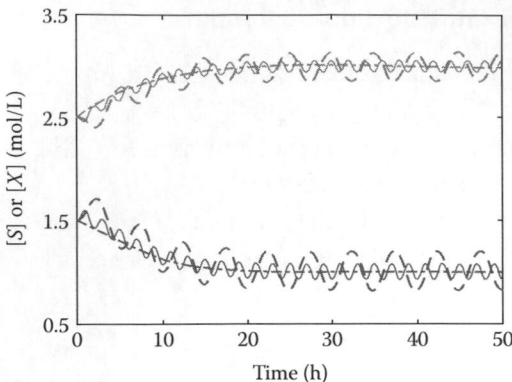

FIGURE 5.15 Simulation of a chemostat with ±20% sinusoid input for a frequency of 0.5 Hz (solid line) and 0.2 Hz (dashed line). The results with constant (nonsinusoidal) input are shown as dash-dot line.

```
dy(1)=D*(Sf-S)  -  rg;
dy(2)=-D*X  +  Yxs*rg;
dy(3)=-D*P  +  Yps*rg;
```

The underlined line is where we inserted the calculation of dilution rate, D. Figure 5.15 shows the transient behavior of chemostat with sinusoidal input, with the dash-dot line representing the case with constant D for comparison. The solid line indicates a 0.5 h^{-1} input sinusoid, whereas the dashed line indicates a slightly lower frequency of 0.2 h^{-1}. The output oscillates in both the cases. The output oscillates with a greater amplitude when the frequency of sinusoid is lower.

The above example demonstrates how a sinusoidal input can be used in simulations. The sinusoid can be replaced by a square wave using the following command instead:

```
D=D0*(1+0.2*square(2*pi*f*t));
```

Note again how straightforward it is to include periodic input signals in simulations. The comparison between sinusoidal and square-wave inputs with a frequency of 0.2 h^{-1} is shown in Figure 5.16.

In my experience teaching MATLAB®, students have often found it easy to incorporate a periodic input trajectory, as shown in Example 5.4. However, using a more complex input trajectory was more challenging. The next example will show how easy it is to incorporate complex input trajectories as well. Let us modify the above problem as follows for pedagogical reasons: We wish to decrease the dilution rate to 0.1 h^{-1} for improving the product yield. However, due to the fragile nature of the bioorganisms, we are unable to do so rapidly. Hence, we will hold the dilution rate at 0.3 h^{-1} for one hour, then ramp down to 0.25 h^{-1} in the next hour, hold it at 0.25 h^{-1} for another hour, ramp down to 0.2 h^{-1} in the next hour, and so on until we reach 0.1 h^{-1}.

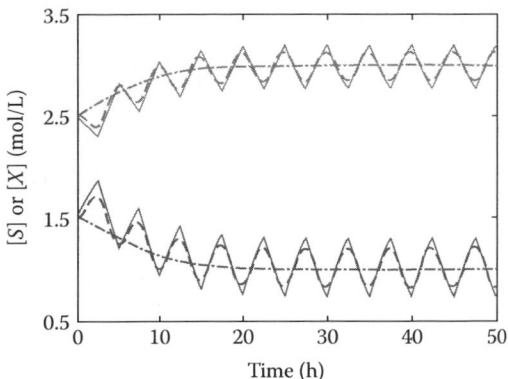

FIGURE 5.16   Simulation of chemostat using sinusoid and square waves at the same frequency of 0.2 Hz.

Example 5.5 Chemostat with Gradual Ramping Down of Input

Problem: Repeat the chemostat problem with the dilution rate ramped down gradually from 0.3 to 0.1 h^{-1} in four ramp-down sequences.

Solution: In this example, we have retained the same driver script and chemostat function as before. The only change will be in the function `varyingInlet`:

```
function D=varyingInlet(t,y,par)
% Dilution ramped down in a series of four
% ramp-down-and-hold sequence
if (t<1)
    D=0.3;
elseif(t<2)
    D=0.3-0.05*(t-1);
elseif(t<3)
    D=0.25;
elseif(t<4)
    D=0.25-0.05*(t-3);
elseif(t<5)
    D=0.2;
elseif(t<6)
    D=0.2-0.05*(t-5);
elseif(t<7)
    D=0.15;
elseif(t<8)
    D=0.15-0.05*(t-7);
else
    D=0.1;
end
```

The chemostat is simulated, and the results are shown in Figure 5.17, with the top panel showing the output concentration and bottom panel showing input dilution rate.

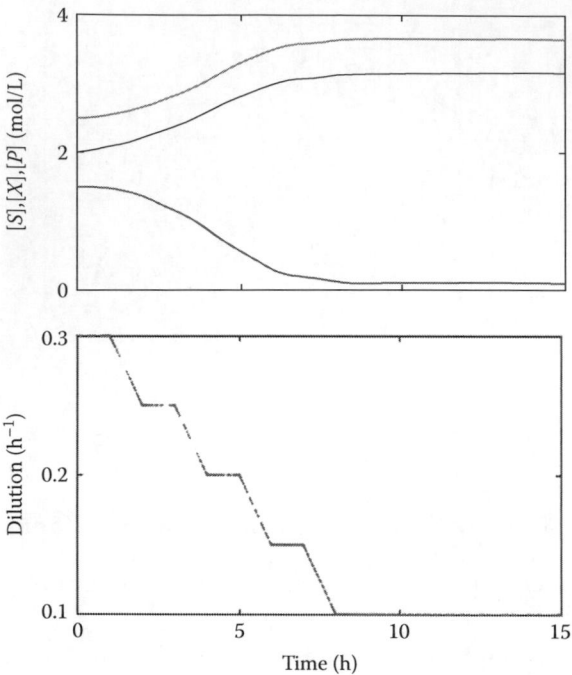

FIGURE 5.17 Chemostat with dilution rate ramped down in a sequence of ramp-down-and-hold. Starting at 0.3 h⁻¹, the dilution rate is held for an hour followed by ramp down at the rate of 0.05 for the next hour.

5.4.2 Zero-Order Hold Reconstruction in Digital Control

Modern advanced computer control systems are *digital* controllers, in contrast to conventional *analog* controllers. In a digital controller, the process variables are measured from the system (and control actions injected) at *discrete* intervals of time. The chemostat is a *continuous-time* system, which is governed by a set of ODEs of the type

$$\frac{d\mathbf{y}}{dt} = \mathbf{f}(t, \mathbf{y})$$
(5.1)

The time interval at which measurements are available in a discrete-time system is known as sampling interval. We will represent sampling interval as Δt_s. Let us say that the chemostat was monitored at the rate of $\Delta t_s = 0.2$ h, then information is available at times $t = 0$, 0.2, 0.4, …. In discrete time, this is represented at measurement intervals k, which is an integer. We represent these values as $\mathbf{y}(k)$. In other words, the values at times mentioned above are represented as $\mathbf{y}(0)$, $\mathbf{y}(1)$, $\mathbf{y}(2)$, … and the equivalent *discrete-time model* is given by

$$\mathbf{y}(k+1) = f_\Delta(\mathbf{y}(k))$$
(5.52)

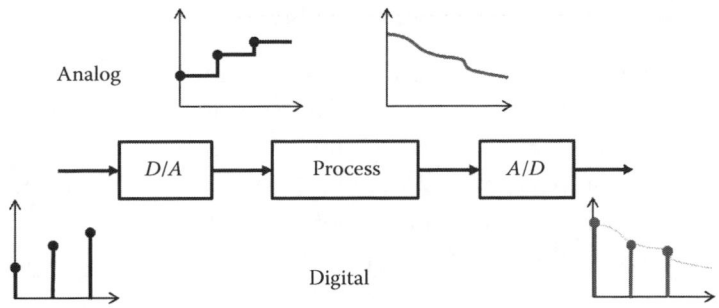

FIGURE 5.18   Block diagram of the chemostat process with digital-to-analog reconstruction at the input and analog-to-digital sampling of the output.

Let us say that the dilution rate was a manipulated variable, then the discrete-time model with manipulated control input is given by

$$\mathbf{y}(k+1) = f_\Delta\big(\mathbf{y}(k), v(k)\big) \tag{5.53}$$

The process of converting a discrete-time signal $v(k)$ to a continuous-time input $D \equiv v$ is known as *reconstruction*, whereas the process of obtaining $f_\Delta(\cdot)$ of Equation 5.53 from $f(\cdot)$ of Equation 5.1 is known as *model discretization*.

I will use an example to describe both reconstruction (using *zero-order hold*) and discretization. Figure 5.18 shows block diagram for this procedure.

For the sake of this example, let us consider that the controller uses some algorithm to determine the following input sequence (which is repeated three times).

Zero-Order Hold (ZOH) is a method for reconstructing a continuous-time input from a discrete-time signal, wherein the value of the input is held constant for the duration of the time interval. In other words, for the data given in the Table 5.1, the dilution rate is kept at $D = 0.3$ for 0–0.2 h, then at $D = 0.28$ for 0.2–0.4 h, then at $D = 0.26$ for 0.4–0.6 h, and so on. The mathematical representation of ZOH is

$$v = v_k, \quad \text{for } k\Delta t_s < t \le (k+1)\Delta t_s \tag{5.54}$$

At time $k\Delta t_s$, the value of state variables is $\mathbf{y}(k)$. Based on the ZOH reconstruction, the value of v is kept constant in the interval. We can solve the ODE-IVP starting at $t = k\Delta t_s$ and $\mathbf{y} = \mathbf{y}(k)$. This will yield us behavior of the system at future time instances. The system is sampled at $t = (k+1)\Delta t_s$, until which time, the input is held constant. Therefore, the *discrete-time* model is nothing but the ODE-IVP solution at time $(k+1)\Delta t_s$, starting with $\mathbf{y} = \mathbf{y}(k)$ at $t = k\Delta t_s$.

TABLE 5.1   Discrete-Time Signal for a Chemostat

k	0	0.25	0.5	0.75	1	1.0	1.25	1.5	1.75	2	2.25	2.5	2.75	3.0
$v(k)$	.3	.28	.26	.27	.3	.28	.24	.21	.20	.20	.18	.19	.17	.18

Example 5.6 Chemostat with Discrete-Time Inputs

Problem: Solve the chemostat problem with discrete-time inputs given in Table 5.1.

Solution: In this example, the chemostat function, monodFun remains the same as the one from Section 3.5.4. The driver script is adequately modified for the discrete-time system to (i) incorporate ZOH to get input signal for the chemostat from the discrete-time signal from Table 5.1, (ii) solve ODE for a single time-step Δt_s, and (iii) *sample* the output at the end of the time-step. The driver script is given below:

```
% Simulates a discrete-time Chemostat with
% zero-order-hold inlet flowrate
%% Reactor parameters
monodParam.mu=0.5;
monodParam.K=0.25;
monodParam.Sf=5;
monodParam.Yxs=0.75;
monodParam.Yps=0.65;

%% Initial conditions and flowrate
S0=1.5;
X0=2.5;
P0=2.0;
monodParam.D=0.3;
monodParam.ts=0.25;   % Sampling time
% Define all input values
VALL=[0.3; 0.28; 0.26; 0.27; 0.3; 0.28; 0.24;...
    0.21; 0.20; 0.20; 0.18; 0.19; 0.17; 0.18];
VALL=repmat(VALL,3,1);
% Obtain time instances
n=length(VALL);
time=[0:n]*monodParam.ts;

%% Solving for discrete-time system
YALL=zeros(3,n);           % For results
YALL(:,1)=[S0;X0;P0];      % Initial value

for k=1:n                  % Cycle over all time-steps
    yk=YALL(:,k);
    vk=VALL(k);
    % (i) ZOH with D constant at vk
    monodParam.D=vk;
    tSpan=[time(k), time(k+1)];
    % (ii) Solve for single time-step
    [tSol,YSol]=ode45(@(t,y) monodFun(t,y,monodParam),...
        tSpan, yk, opts);
    % Sampling to extract value at final time
    yk_plus_1=YSol(end,:);
    YALL(:,k+1)=yk_plus_1';
end
```

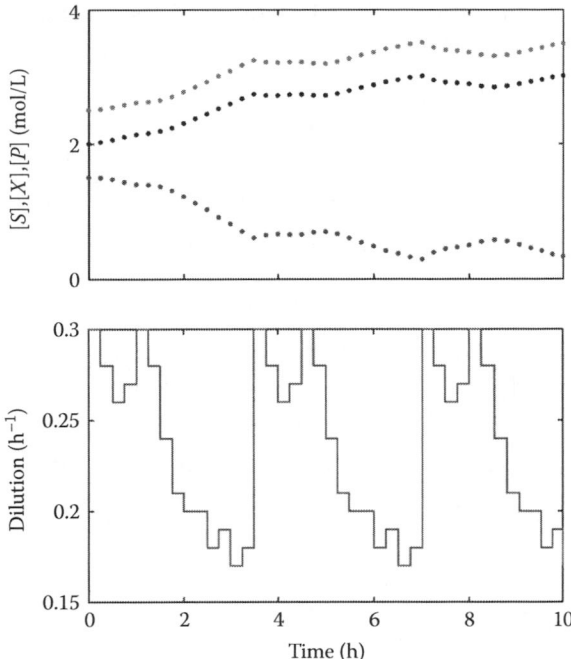

FIGURE 5.19   Simulation of chemostat with ZOH. Top plot: sampled output variables; bottom plot: manipulated input dilution rate.

```
%% Plotting the results
subplot(2,1,1);
plot(time,YALL);
ylabel('[S],[X],[P] (mol/L)');
subplot(2,1,2);
stairs(time(1:end-1),VALL);
ylabel('Dilution (hr^{-1})'); xlabel('Time (hr)');
```

Figure 5.19 shows the results of discrete-time simulation of the chemostat. The top panel shows the state variables, with symbols representing the actual measured values. Since this is a sampled-data system, there is no information between consecutive time-steps. The bottom panel represents the input. Here, a stairs plot is used for ZOH.

5.5 SIMULATING SYSTEM WITH BOUNDARY CONSTRAINTS

An adiabatic PFR was simulated in Section 4.5.1. Due to the exothermic heat of reaction, temperature at steady state increased from 450 K at the entrance to 575 K at the exit of the PFR. Since all the process parameters were known, it was possible to simulate this system and obtain the temperature and concentration distribution in the reactor. One of the more important sets of parameters in a PFR is the reaction kinetics (or broadly, the reaction mechanism). The process of identifying the kinetic mechanism and estimating the rate parameters is a crucial aspect of chemical engineering. Much of kinetic information is obtained from reactors where temperature and other conditions are carefully controlled.

However, in some cases, (for example, highly exothermic system) the temperature may vary within the system.

Analysis of reactors with strong temperature effects can indeed be done by solving mass and energy balances for the PFR simultaneously. However, at times, detailed analysis or parameter estimation in the presence of local temperature trends calls for an approach to decouple temperature and mass balance effects. In other cases, the boundary temperature may act as a constraint to the overall system.*

5.5.1 PFR with Temperature Profile Specified

In this case study, we consider the systems similar to the adiabatic PFR example from Section 4.5.1. Let us say that we conducted experiments on this reactor under controlled conditions, with temperature being measured at several axial locations along the reactor. The concentration profile in the reactor *with a given temperature profile* is to be simulated in this example. The steady state material balance in a PFR is given by

$$u \frac{dC_A}{dz} = -\frac{kC_A}{K_r + C_A} \tag{5.55}$$

$$k = k_0 \exp\left(-\frac{E}{RT}\right), \quad K_r = K_0 \exp\left(-\frac{\Delta E_r}{RT}\right) \tag{5.56}$$

The activation energy is $E = 60$ kJ/mol and $\Delta E_r = -10$ kJ/mol. The rate constants are specified at the temperature $T_1 = 450$ K as $k_1 = 2$ and $K_{r,1} = 1$. The inlet temperature was 450 K, concentration 1 mol/m³, and flowrate 0.4 m/min. The temperature profile was measured in the experiments at six different locations:

z (cm)	0.25	0.5	1.0	1.25	1.5	2.0
T (K)	454	460	480	520	595	605

Example 5.7 PFR with Specified Temperature Profile

Problem: Simulate the PFR to obtain a concentration profile along the length of the PFR for which the measured temperatures are specified above.

Solution: The temperatures specified in the problem are those measured in the PFR. We are not given other conditions (e.g., heat loss or heat gain if the PFR is in a furnace). Hence, we will solve the material balance Equation 5.55 with the given temperature profile superimposed. The PFR solution will require the temperature specified at various points within the reactor (and not just the six points above). Thus, we will need to use *interpolation* to obtain values within these spatial intervals.

---

* A simplified version of this was seen in the Graetz problem in Section 4.5.3, where the temperature at the external boundary (walls) was specified. An analogous problem of relevance in this case study is when the boundary temperature *profile* is specified. There are several examples of practical relevance where one needs to implement a boundary value profile as a constraint. This case study provides a way to solve problems of similar nature.

Interpolation is discussed in Appendix D. We will use the MATLAB `spline` command to interpolate within the data given above. This will be done in two parts: first, to fit the spline piecewise polynomial to data above and then to use this fit to obtain *T* at any location. The former is done as

```
pfrTProfile=spline(zMeas,TMeas);
```

The above command generates an interpolating polynomial structure, `pfrTProfile`. The value of *T* at any *z*, *T*(*z*), can be obtained using the `ppval` command:

```
T=ppval(pfrTProfile,z);
```

We will use this within the PFR simulations. The driver script is given below:

```
% Driver script for simulation of a PFR
% with specified temperature profile
%% Model Parameters
modelParam.k1=0.2;
modelParam.K1=1;
modelParam.T1=450;
modelParam.E=60*1000;
modelParam.DEr=-10*1000;
modelParam.Rgas=8.314;
% Reactor conditions
modelParam.u0=0.4;
L=2;
Cin=1;

% Measured profile
zMeas=[0,    0.25, 0.5, 1.0, 1.25, 1.5, 2.0];
TMeas=[450, 454,   460, 480, 520,  595, 605];
% Spline interpolating polynomial
modelParam.pfrTProfile=spline(zMeas,TMeas);

%% Simulating and Plotting
[Z,CA]=ode15s(@(z,C) pfrModelFun(z,C,modelParam), ...
     [0, L], Cin);
figure(1); plot(Z,CA);
xlabel('Length (m)'); ylabel('C_A (mol/m^3)');
```

The key difference from the earlier PFR example is that the temperature is specified. However, since the temperature is not available as an explicit function, we fit a `spline` interpolating polynomial to data. The PFR function, given below, uses this spline polynomial to simulate the reactor with specified boundary temperature:

```
function dC=pfrModelFun(z,Ca,modelParam)
% To compute dC/dz for a PFR with a single reaction
%    and reactor temperature specified
```

```
%% Model Parameters
u=modelParam.u0;
pfrTProfile=modelParam.pfrTProfile;
% Interpolate to obtain T(z)
T=ppval(pfrTProfile,z);
% Calculate reaction rate
r=rxnRate(T,Ca,modelParam);
% Model equation
dC=-1/u*r;
end
```

The following is a function for calculating reaction rate. This is same as the function used in transient PFR example from Chapter 4:

```
%% Get reaction rate
function r=rxnRate(T,C,par)
% Function to compute reaction rate
Rgas=par.Rgas;        % Gas constant
% Rate constants
k1=par.k1;
E=par.E;
T1=par.T1;
K1=par.K1;
DEr=par.DEr;
k =k1*exp(-E/Rgas*(1/T-1/T1));
Kr=K1*exp(-DEr/Rgas*(1/T-1/T1));

r=k*C/sqrt(1+Kr*C^2);
end
```

Figure 5.20 shows the concentration profile in the PFR. Since the temperature is rather low in the initial section of the PFR, only a modest amount of reaction takes place,

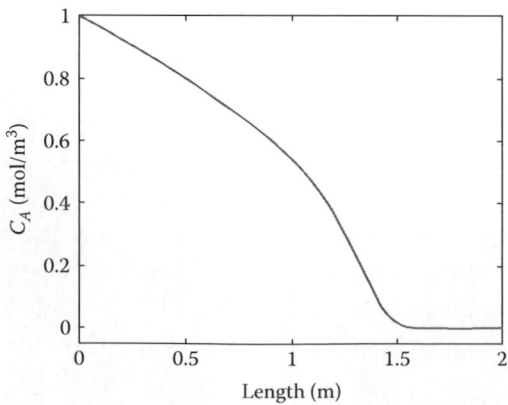

FIGURE 5.20   Concentration vs. length of the PFR, simulated with a given temperature profile.

and the conversion is about 43% at $z = 1$ m from the inlet. As we move further downstream, the temperature increases, which results in a rapid increase in conversion of A and complete conversion is observed in less than $z = 1.5$ m of the reactor.

5.6 WRAP-UP

This chapter built upon the case studies from previous chapters in Section I of the book. The previous chapters simulated cases where the system was represented by a set of ODEs (Chapter 3) or PDEs (Chapter 4) with specified initial and boundary conditions.

The first section of this chapter focused on simulating a *staged* operation. Like a distributed parameter system (DPS), the modeled variables of interest vary in time and location. However, unlike a traditional DPS, the model is not governed by PDEs; instead, the model consists of multiple interconnected stages, with the process variables' values specified only at these discrete stages.

The second section of this chapter focused on using linear algebra tools for *analysis and understanding* of transient systems. Specifically, we observed how rich dynamic behavior emerges for even a simple example of *linear* ODE in two dimensions. We limited ourselves to 2D since it is easy to visualize behavior in 2D pictorially. The concepts of linear system analysis are more general, and extendible to multiple dimensions as well. A thorough *nonlinear* analysis will have to wait until we cover additional topics in Section II of this book. The first two case studies in Chapter 9 will extend this linear analysis and cover nonlinear analysis and bifurcation. An interesting concept of nonnormal system analysis is also introduced in this section, as an extension of linear stability analysis. While most books on linear systems focus on long-term asymptotic response, this section brought out the importance of short-term transient growth.

Thereafter, we took a few examples to extend our analysis of transient systems. We considered a system where the energy balance leads to a parabolic PDE and material balance to an ODE-IVP. The overall system of equations was solved simultaneously. A combination of ODE and PDE occurs in practical examples but is usually not covered in numerical simulation textbooks.

Finally, the next two case studies demonstrated simulations of systems where (i) a process input or parameter varies with time and (ii) a boundary profile constraint is imposed on a system.

This completes our discussion of transient systems in Section I of this book. The next part will cover (nonlinear and linear) algebraic equations, ODE-BVPs (boundary value problems), elliptic PDEs, and differential algebraic equations (DAEs, which are a combination of ODEs and algebraic equations).

EXERCISES

Problem 5.1 Repeat the simulations for a binary distillation column from Example 5.1 for three stages each in rectification and stripping sections. Compare the results with a five-stage column for the same operating conditions.

Problem 5.2 Repeat the previous problem for a tougher distillation, with $\alpha = 1.75$. How many trays are required to get 95% purity (if it is possible) in the distillate? Let $D = 12.5$ mol/s and $V = 37.5$ mol/s.

Problem 5.3 A mass-spring-damper system consists of a mass suspended by a spring and a damper that damps the oscillations. The model is given by

$$mx'' + cx' + kx = 0$$

Show that the overall system can be written as

$$x'' + (2\zeta\omega)x' + \omega^2 x = 0$$

What is/are the conditions on ζ for the system to have oscillatory response?

Problem 5.4 Use the methodology of Section 5.2.3 for analysis of transient growth and phase portrait of the system, $\mathbf{y}' = A_2\mathbf{y}$, where

$$A_2 = \begin{bmatrix} 8.5 & -13.5 \\ 9 & -14 \end{bmatrix}$$

Problem 5.5 Consider a linear system given by

$$\mathbf{y}' = \begin{bmatrix} -1 & -\cot(\theta) \\ 0 & -2 \end{bmatrix} \mathbf{y}$$

The eigenvalues of this matrix are $\lambda = \{-1, -2\}$. Show that the eigenvectors of this matrix are $\mathbf{v}_1 = \begin{bmatrix} 1 & 0 \end{bmatrix}^T$, $\mathbf{v}_2 = \begin{bmatrix} \cos(\theta) & \sin(\theta) \end{bmatrix}^T$.

Verify transient growth in the system for different values of $\theta = \dfrac{\pi}{2}, \dfrac{\pi}{6}, \dfrac{\pi}{20}, \dfrac{\pi}{100}$. Using MATLAB simulations, show that the maximum transient growth is the highest when $\theta = \dfrac{\pi}{100}$.

Problem 5.6 Simulations for thermosetting polymer reaction in a 1D slab were discussed in Section 5.3. We ignored the heat of reaction. If the heat of reaction is included in the simulations, the source term in Equation 5.40 becomes

$$S(T, \mathbf{C}) = (-\Delta H_1)r_1 + (-\Delta H_2)r_2 \tag{5.57}$$

where $(-\Delta H_1) = 75$ kJ/mol and $(-\Delta H_1) = 100$ kJ/mol are exothermic heats of the two reactions. Repeat Example 5.3 by including the heat of reaction in the energy balance.

Problem 5.7 In this problem, we will simulate a condition similar to Example 5.5 but with variations in the substrate concentration. Consider that the chemostat is used for effluent treatment. The substrate concentration is measured every half hour and is found to be as follows:

```
t        0       0.5     1.0     1.5     2.0     2.5     3.0     3.5     4.0
Sf     5.72    3.67    5.23    4.96    4.15    4.58    4.97    5.59    5.18
t(cont.)       4.5     5.0     5.5     6.0     6.5     7.0     7.5     8.0
Sf(cont.) 4.13    5.32    5.16    4.77    5.16    4.70    4.74    5.54
```

Use `spline` to fit a smooth spline function to the above values, and hence use the spline interpolation to simulate the effect of the above variations in S_f on the performance of a chemostat. Choose $D=0.1$ and start with an initial condition of `Y0 = [0.1; 3.6; 3.2];`.

Problem 5.8 In Example 5.5, the dilution rate was ramped down in a series of steps. The problem was simulated with a continuously varying input. Instead, a discrete-time controller with a sampling time of 0.25 h is used. The trajectory with ZOH that is implemented is shown in the figure below:

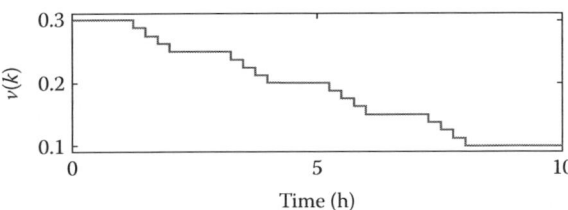

Repeat Example 5.5 with a discrete-time input signal reconstructed with zero-order hold. **Hint:** The first few terms of the discrete signal are

```
t 0      0.25    0.50    0.75    1.00    1.25      1.50      1.75
v 0.3    0.30    0.30    0.30    0.30    0.2875    0.275     0.2625
```

II

Linear and Nonlinear Equations and Bifurcation

CHAPTER 6

Nonlinear Algebraic Equations

6.1 GENERAL SETUP

Section I of this book focused on methods to solve ordinary differential equations (ODEs). This first chapter of Section II will deal with solving a set of algebraic equations of the type

$$\mathbf{g}(\mathbf{x}) = \mathbf{0} \tag{6.1}$$

where
   $\mathbf{x} \in \mathcal{R}^n$ is an $n \times 1$ solution vector
   $\mathbf{g}(\cdot)$ is a vector-valued function

I will retain the convention followed in Section I: All vectors (unless otherwise specified) will be column vectors. Thus, $\mathbf{x}$ is a $n \times 1$ column vector and $\mathbf{g}(\cdot)$ is an $n \times 1$ function vector.

Prior to generalizing the discussion to an n-variable problem, interpretation of equation solving for a single variable is first discussed. Solving algebraic equation is essentially finding the value(s) of variable x that satisfy Equation 6.1. The notation $x^\star$ will be used to denote the solution. Thus, $g(x)$ represents how the value of the function varies with x, whereas $g(x^\star)$ represents the value at a specific point $x = x^\star$. Since $x^\star$ is the solution, the function value at this point is, by definition, 0.

The geometric interpretation of solving a single nonlinear algebraic equation in one variable is shown in Figure 6.1. The function $g(x)$ varies with varying x and intersects the X-axis at two points, $x_1^\star$ and $x_2^\star$. The points are known as the roots of the function, since they satisfy Equation 6.1.

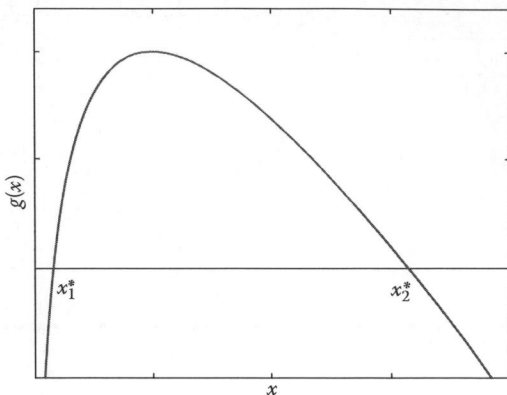

FIGURE 6.1 Schematic for a curve $g(x)$, which intersects the X-axis at two points, x_1^* and x_2^*.

6.1.1 A Motivating Example: Equation of State

The ideal gas law gives the relationship between volume occupied by a gas and its pressure and temperature:

$$Pv = RT$$

where

P is the pressure in Pa
v is the molar volume in mol/m³
R is the ideal gas constant
T is the temperature in K

The ideal gas law does not accurately capture the behavior of real gases. Equations of state (EOS) provide a better prediction of the volumetric properties of fluids. One such popular EOS is Redlich-Kwong (RK) EOS:

$$P = \frac{RT}{v-b} - \frac{a/\sqrt{T}}{v(v+b)} \tag{6.2}$$

The problem of finding the molar volume of a gas at a particular temperature and pressure using the ideal gas EOS is straightforward since it provides volume as an expression, $v = RT/P$. On the other hand, RK EOS is a cubic equation.

Both the EOS can be put in the standard form of Equation 6.1 as

$$\underbrace{Pv - RT}_{g_{\text{ideal}}(v)} = 0 \tag{6.3}$$

$$g(v) \equiv P(v-b) - RT + \frac{v-b}{v(v+b)} \frac{a}{\sqrt{T}} = 0 \tag{6.4}$$

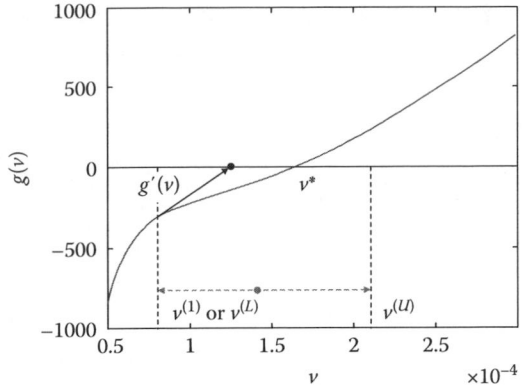

FIGURE 6.2   Plot of $g(v)$ vs. v for Redlich-Kwong EOS using Equation 6.4 for the chosen values of RK parameters at 100 bar and 340 K. The dashed arrow represents working of bisection rule while the solid arrow represents working of Newton-Raphson method.

The latter equation is obtained by multiplying the RK EOS by $(v-b)$ and rearranging. The value, $v^\star$, that satisfies the above equation is the molar volume of gas at the corresponding temperature and pressure.

In this chapter, the RK EOS will be used to compute the molar volume of a fluid at 340 K temperature and 100 bar pressure and hence demonstrate solution of nonlinear equations. The values of the two parameters in the RK EOS that I will use in this example (in SI units) are $a = 6.46$ and $b = 2.97 \times 10^{-5}$. Figure 6.2 shows plot of $g(v)$ vs. v for a range of values of v using Equation 6.4. The curve intersects the X-axis at $v^\star = 1.646 \times 10^{-4}$. This is the solution of the nonlinear equation and hence represented as $v^\star$.

The corresponding molar volume using ideal gas assumption is $v^\star_{ideal} = 2.827 \times 10^{-4}$. Solving the equation (6.3) signifies finding $v^\star_{ideal}$, the point at which the function $g_{ideal}(v)$ (which is a straight line) intersects the X-axis.

Numerical methods for solving a single nonlinear equation in one unknown are discussed in the next section using the above example. Thereafter, the methods are extended to a general n-variable case. Subsequently, MATLAB® functions fzero and fsolve used to solve algebraic equations are discussed, followed by a few exemplary case studies.

6.2 EQUATIONS IN SINGLE VARIABLE

Solution to nonlinear algebraic equations requires *iterative* numerical techniques to be used. The chosen numerical technique starts with an initial guess of the solution, and the chosen technique is used to move in the direction of the solution. This is shown in Figure 6.2 for bisection and Newton-Raphson method, which will be discussed presently. Two initial guesses are used in bisection method (denoted as $x^{(L)}$ and $x^{(U)}$), and a single initial guess is used in Newton-Raphson (denoted as $x^{(1)}$). The aim of both the methods is to "move" in the direction of the true solution $x^\star$. Note that typically this solution is not known *a priori*, and the intent of the numerical technique is to be able to find this solution efficiently (i.e., with low computational time).

The following notations are used in this chapter. Various guesses are represented as $x^{(i)}$, where i is the iteration index. Approximation error between subsequent iterations is $e^{(i)} = x^{(i)} - x^{(i-1)}$ and the true approximation error is $e_t^{(i)} = x^\star - x^{(i)}$. The corresponding relative errors will be represented as $\varepsilon^{(i)}$ and $\varepsilon_t^{(i)}$. The notation Δ will be used to represent difference, which will be explained based on the context in this chapter. The term $g^{(i)}$ will be used as a shorthand notation for $g(x^{(i)})$. By definition of the solution, $g^\star = g(x^{(\star)}) = 0$.

6.2.1 Bisection Method

Bisection method is an example of *bracketing method*, where two initial guesses are chosen. These are represented as $x^{(L)}$ and $x^{(U)}$. They are chosen such that they lie on either side of the desired solution. Thus, the function values for admissible initial guesses satisfy the relation

$$g\left(x^{(L)}\right)g\left(x^{(U)}\right) < 0 \tag{6.5}$$

The two guesses, $x^{(L)}$ and $x^{(U)}$, shown in Figure 6.2, lie on either side of the true solution, $x^\star$. The value of $g^{(L)}$ is negative and that of $g^{(U)}$ is positive, thus satisfying the condition (6.5). The new guess using bisection method is generated by simply bisecting the line segment joining $x^{(L)}$ and $x^{(U)}$. Thus

$$x^{(i+1)} = \frac{x^{(L)} + x^{(U)}}{2} \tag{6.6}$$

is the new iteration value. The condition (6.5) is verified for the new iteration value $x^{(i+1)}$, and either $x^{(L)}$ or $x^{(U)}$ is retained so that the current guesses still bracket the true solution. Thus, if $x^{(i+1)}$ lies on the same side as $x^{(L)}$, then $x^{(i+1)}$ replaces $x^{(L)}$ for the next iteration while $x^{(U)}$ remains unchanged; else, $x^{(L)}$ remains unchanged and $x^{(i+1)}$ replaces $x^{(U)}$. Numerically

$$\text{if } g\left(x^{(i+1)}\right)g\left(x^{(L)}\right) > 0, \quad x^{(L)} = x^{(i+1)}, \quad \text{else } x^{(U)} = x^{(i+1)} \tag{6.7}$$

Thereafter, the algorithm continues. At each iteration, Equation 6.6 is used to compute the next guess, whereas the condition in Equation 6.7 is used to determine whether this iterant replaces $x^{(L)}$ or $x^{(U)}$, based on the sign of $g^{(i+1)}$. This method is continued iteratively until the error $e^{(i+1)}$ falls below a predetermined tolerance threshold:

$$e^{(i+1)} \equiv \left| x^{(i+1)} - x^{(L)} \right| < e_{\text{tol}} \tag{6.8}$$

If the initial length of the line segment is $e^{(0)} = x^{(U)} - x^{(L)}$, then after the first iteration

$$e^{(1)} = \frac{e^{(0)}}{2}$$

Each subsequent iteration also halves the line segment. Thus, the bisection method has a *linear rate of convergence*. The box below discusses convergence order of an iterative numerical method.

DEFINITION: ORDER OF CONVERGENCE

An iterative numerical method generates new guess values of solution, $x^{(1)}, \ldots, x^{(i)}, \ldots$. If the consecutive errors between subsequent iterations satisfy

$$e^{(i+1)} = \kappa \left(e^{(i)} \right)^m$$

then m is known as order of convergence of the numerical method.
   If $m=1$, the method is said to have a *linear* rate of convergence.
   If $m=2$, the method is said to have a *quadratic* rate of convergence.
   For bisection method, $\kappa = 0.5$ and $m=1$.

Continuing for n iterations

$$e^{(n)} = \frac{e^{(0)}}{2^n} \tag{6.9}$$

If the error $e^{(n)}$ falls below the desired error tolerance threshold, e_{tol} in n iterations, then the value e_{tol} satisfies Equation 6.9. Thus, irrespective of the form of function, $g(x)$, the number of iterations required to meet the desired error threshold can be computed *a priori* as

$$n = \log_2 \frac{e^{(0)}}{e_{tol}} \tag{6.10}$$

The following example demonstrates the bisection method for finding the molar volume of a fluid represented by RK EOS.

Example 6.1 Redlich-Kwong Equation of State Using Bisection Method

Problem: Use the bisection rule to compute molar volume at 100 bar and 340 K for a fluid that obeys RK EOS (Equation 6.4), with parameters $a = 6.46$ and $b = 2.97 \times 10^{-5}$.

   Solution: Since this is the first example in Section II, I will use a single MATLAB script to solve the problem using the bisection method. Thereafter, I will modify the code to write in the recommended, modular coding.

   At the risk of sounding repetitive, this example is **not** an approach I recommend.

The core of the code involves computing the function $g(v)$ as per Equation 6.4 for $=340$, $P = 100 \times 10^5$, and the current guess value $v^{(i)}$. A new guess will be computed using Equation 6.6, $x^{(i)}$ will replace either $x^{(L)}$ or $x^{(U)}$ using condition (6.7), and the stopping condition will be verified using Equation 6.8.

```
% Molar volume using Redlich-Kwong EOS
a=6.46;
b=2.97e-5;
R=8.314;
% Operating conditions
P=1e7;   % Pressure (Pa)
T=340;   % Temperature (K)

%% Initial guesses and verification
vL=0.8e-4;
vU=2.2e-4;
gL=P*(vL-b) - R*T + (vL-b)/(vL*(vL+b))*a/sqrt(T);
gU=P*(vU-b) - R*T + (vU-b)/(vU*(vU+b))*a/sqrt(T);
if (gL*gU>0)
    error('Inconsistent initial guesses');
end
X=vU;    % Storing iterated values
G=gU;

%% Bisection method iterations
maxIter=50;
eTol=1e-8;
for i=1:maxIter
    v=(vL+vU)/2;
    g=P*(v-b) - R*T + (v-b)/(v*(v+b))*a/sqrt(T);

    % Verify and change bracket limits
    if (g*gL>0)
        gL=g; vL=v;
    else
        gU=g; vU=v;
    end

    % Store results and compute error
    G(i+1)=g;
    X(i+1)=v;
    err=vU-vL;
    if (err<eTol)
        break
    end
end
```

```
%% Display result
disp(['Molar volume, v = ', num2str(v)]);
```

The result using RK EOS is 1.65×10^{-4} m³/mol:

```
Molar volume, v = 0.0001646
```

The bisection method required `i=14` iterations to converge.

The two initial guesses were such that

$$e^{(0)} = v^{(U)} - v^{(L)} = 1.4 \times 10^{-4}$$

The desired error tolerance was $e_{tol} = 10^{-8}$. Thus, according to Equation 6.10

$$n = \log_2\left(1.4 \times 10^4\right) = 13.8 \tag{6.11}$$

Thus, the error falls below the tolerance threshold in 14 iterations, as seen in Example 6.1.

The bisection method is easy to use, and the number of iterations required for its convergence is known *a priori*. It is also robust because the two guesses always bracket the solution, and the new guess being midpoint of the current guesses can always be obtained (irrespective of the nature of $g(x)$). There are, however, certain cases where the bisection method fails. These are shown in Figure 6.3a: Although a root exists, there are no feasible starting points for the bisection method. Figure 6.3b shows that feasible starting points fall in a narrow range and are therefore difficult to obtain. Figure 6.3c shows where the feasible starting points bracket a singularity and not a solution.

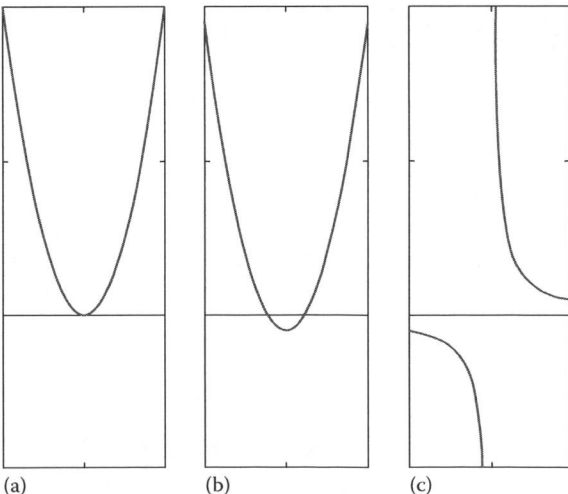

(a)   (b)   (c)

FIGURE 6.3   Three examples of functions where the bisection method "fails."

I will end this section with a modular code for solving Example 6.1. Hereafter, similar modular coding will be practiced.

Example 6.2 A Modular MATLAB® Code for Bisection Method

Problem: Redo Example 6.1 using a modular code.

Solution: The first part of writing the code is to separate out $g(v)$ as a separate function, as shown below:

```
function gVal = RK(v,par)
% Returns g(v) at a particular T and P
% All values are in SI units
% Parameters
R=par.R;
a=par.a; b=par.b;
% Operating conditions
T=par.T;
P=par.P;
% Function value
gVal=P*(v-b) - R*T + (v-b)/(v*(v+b))*a/sqrt(T);
end
```

Once the above function is written, it may be directly used for computing $g(v)$ within the main script. The main script is given below. The script has undergone only modest changes to make it modular: (i) $g(v)$ is computed through the above RK function and (ii) parameters are passed on to the function as a structure. Both these will make the code less prone to errors. The driver script is

```
% Molar volume using Redlich-Kwong EOS
% Model parameters for the EOS
param.a=6.46;
param.b=2.97e-5;
param.R=8.314;
% Operating conditions
param.P=1e7;  % Pressure (Pa)
param.T=340;  % Temperature (K)

%% Initial guesses and verification
vL=0.8e-4;
vU=2.2e-4;
gL=RK(vL,param);
gU=RK(vU,param);
if (gL*gU>0)
    error('Inconsistent initial guesses');
end
X=vU;    % Storing iterated values
G=gU;
```

```
%% Bisection method iterations
maxIter=50;
eTol=1e-8;
for i=1:maxIter
    v=(vL+vU)/2;
    g=RK(v,param);

    % Verify and change bracket limits
    if (g*gL>0)
        gL=g; vL=v;
    else
        gU=g; vU=v;
    end

    % Store results and compute error
    G(i+1)=g;
    X(i+1)=v;
    err=vU-vL;
    if (err<eTol)
        break
    end
end
disp(['Molar volume, v = ', num2str(v)]);
```

6.2.2 Secant and Related Methods

While the bisection method is easy to use and robust, it is often slow to converge. Hence, alternatives with faster convergence rates have been proposed. One such alternative is the secant method, where a straight line joins the two current guesses. If $x^{(i-1)}$ and $x^{(i)}$ are the two guesses, then the straight line joining the points $(x^{(i)}, g^{(i)})$ and $(x^{(i-1)}, g^{(i-1)})$ is

$$\frac{x - x^{(i)}}{y - g^{(i)}} = \frac{x^{(i)} - x^{(i-1)}}{g^{(i)} - g^{(i-1)}} \tag{6.12}$$

The point at which the line intersects the X-axis is given by $y = 0$. Substituting and rearranging, the following equation can be used to update the secant solution:

$$x^{(i+1)} = x^{(i)} - \frac{x^{(i)} - x^{(i-1)}}{g^{(i)} - g^{(i-1)}} \cdot g^{(i)} \tag{6.13}$$

The secant method starts with two initial guesses, $g^{(0)}$ and $g^{(1)}$, and uses the above expression to generate the next guess. The code for the secant method is given in the next example.

Example 6.3 Secant Method for Redlich-Kwong EOS

Problem: Solve Example 6.1 using the secant method.

Solution: The same function file, RK.m, will be used to generate $g(v)$. The driver script for the secant method is given below. Notice that the problem definition and preamble sections remain same as before. Only the core numerical part of the code changes.

As before, arrays X and G will be used to store the values of $v^{(i)}$ and $g(v^{(i)})$, respectively. Since MATLAB array indices start from 1, the two initial values are represented as X(1) and X(2). Therefore, the for loop starts from index i=2, instead of i=1 as before. The code for using the secant method is given below:

```
% Molar volume using Redlich-Kwong EOS
% Model parameters for the EOS
param.a=6.46;
param.b=2.97e-5;
param.R=8.314;
% Operating conditions
param.P=1e7;   % Pressure (Pa)
param.T=340;   % Temperature (K)

%% Initial guesses and verification
vL=0.8e-4;
vU=2.2e-4;
gL=RK(vL,param);
gU=RK(vU,param);
X=[vL, vU];    % Storing iterated values
G=[gL, gU];

%% Secant method iterations
maxIter=50;
eTol=1e-8;
for i=2:maxIter+1
    v=X(i) - (X(i)-X(i-1))/(G(i)-G(i-1))*G(i);
    g=RK(v,param);

    % Store results and compute error
    G(i+1)=g;
    X(i+1)=v;
    err=abs(X(i+1)-X(i));
    if (err<eTol)
        break
    end
end
disp(['Molar volume, v = ', num2str(v)]);
```

The result from the secant method is same as before. The secant method takes only five iterations (at value of i=6) to converge.

The locations of key differences between bisection and secant codes are underlined above. The main difference, of course, is how the next iteration value, $x^{(i+1)}$, is calculated. This is the first line in the `for` loop, where Equation 6.13 is used to compute the next iteration value and is stored in variable v. The value g(v) is calculated as before; however, the sign of $g(v)$ need not be compared with the sign of $g(v^{(L)})$.

The rate of convergence of the secant method is said to be super linear, since it converges faster than linear convergence. Specifically, it can be shown that for the secant method

$$e^{(i+1)} = \left(\frac{f''(\xi)}{2f'(\xi)} \right)^{\gamma-1} \left(e^{(i)} \right)^{\gamma}, \quad \gamma = \frac{\sqrt{5}+1}{2} = 1.618 \tag{6.14}$$

The exponent, γ, in the above equation is called the golden ratio.

6.2.2.1 Regula-Falsi: Method of False Position
The intermediate values obtained using the secant method in Example 6.3 were

```
1.5299e-4   1.6259e-4   1.6470e-4   1.6460e-4   1.6460e-4
```

The first two values, $v^{(2)}$ and $v^{(3)}$, in the list above both lie on the left of the solution 1.646×10^{-4}. This shows that the secant method is not a bracketing method since $g^{(2)} \times g^{(3)} > 0$. Method of false position, on the other hand, is a bracketing method. A new guess is generated in a similar way as Equation 6.13, though it uses $x^{(L)}$ and $x^{(U)}$ instead:

$$x^{(i+1)} = x^{(L)} - \frac{x^{(U)} - x^{(L)}}{g^{(U)} - g^{(L)}} \cdot g^{(L)} \tag{6.15}$$

and then using condition (6.7) to determine whether $x^{(L)}$ or $x^{(U)}$ gets replaced. Thus, calculation of the new approximation is similar to secant method, though it is a bracketing method where the approximations retained are the ones that have opposite signs like the bisection method.

6.2.2.2 Brent's Method
Equation 6.12 used in the derivation of the secant method involves fitting a straight line to two points, $(x^{(i)}, g^{(i)})$ and $(x^{(i-1)}, g^{(i-1)})$. A more accurate method can be derived by fitting a quadratic curve to three points, followed by the location where the curve intersects the X-axis. Using Lagrange interpolating polynomial (see Appendix D)

$$x = x^{(i-2)} \frac{\left(y - g^{(i-1)} \right)\left(y - g^{(i)} \right)}{\left(g^{(i-2)} - g^{(i-1)} \right)\left(g^{(i-2)} - g^{(i)} \right)} + x^{(i-1)} \frac{\left(y - g^{(i-2)} \right)\left(y - g^{(i)} \right)}{\left(g^{(i-1)} - g^{(i-2)} \right)\left(g^{(i-1)} - g^{(i)} \right)}$$
$$+ x^{(i)} \frac{\left(y - g^{(i-2)} \right)\left(y - g^{(i-1)} \right)}{\left(g^{(i)} - g^{(i-2)} \right)\left(g^{(i)} - g^{(i-1)} \right)} \tag{6.16}$$

The above quadratic curve intersects the X-axis when $y = 0$. Thus, the next approximation is

$$
\begin{aligned}
x^{(i+1)} = & \frac{x^{(i-2)} g^{(i-1)} g^{(i)}}{\left(g^{(i-2)} - g^{(i-1)}\right)\left(g^{(i-2)} - g^{(i)}\right)} + \frac{x^{(i-1)} g^{(i-2)} g^{(i)}}{\left(g^{(i-1)} - g^{(i-2)}\right)\left(g^{(i-1)} - g^{(i)}\right)} \\
& + \frac{x^{(i)} g^{(i-2)} g^{(i-1)}}{\left(g^{(i)} - g^{(i-2)}\right)\left(g^{(i)} - g^{(i-1)}\right)}
\end{aligned}
\tag{6.17}
$$

The above expression can be used directly, or may be simplified algebraically. If the function $g(x)$ is smooth in the region $[x^{(L)}, x^{(U)}]$, this quadratic method converges rapidly.

Brent's method is more popularly known as *inverse quadratic interpolation*, because a quadratic function of Equation 6.16 is employed for the next guess. It is called inverse quadratic because here x is interpolated as a function of $g^{(i)}$'s.

I end this section by mentioning that the four methods discussed herein, bisection, regula falsi, secant, and inverse quadratic interpolation, are methods that work with multiple initial guesses. Here, they are arranged in ascending order of their order of convergence. I will refer to these methods again when discussing MATLAB algorithm in Section 6.4.1. The next two methods discussed presently use a single initial guess.

6.2.3 Fixed Point Iteration

Fixed point iteration, also known as method of successive substitution, expresses the original nonlinear equation in the form

$$
x = G(x)
\tag{6.18}
$$

The above equation is used as an *expression* to generate subsequent approximate solutions to the nonlinear algebraic equation. In other words, starting with $x^{(0)}$, new guesses are generated using

$$
x^{(i+1)} = G\left(x^{(i)}\right)
\tag{6.19}
$$

One simple way to get (x) related to the original problem (6.1) is to write

$$
G(x) = x + \alpha g(x)
$$

though more often, it is better to rearrange the nonlinear equation appropriately. For example, the RK EOS (6.2) may be rewritten as

$$
v = b + \frac{RT}{P} - \frac{v - b}{v(v + b)} \frac{a}{P\sqrt{T}}
\tag{6.20}
$$

The code of fixed point iteration using the equation above is skipped. Starting with the initial guess of $v = 2.2 \times 10^{-4}$, it took 14 iterations for this method to converge. The linear rate of convergence is one disadvantage of fixed point iteration. A greater disadvantage is that it is not a robust method, since there are stringent requirements on $G(x)$ for stability of fixed point iteration.

These results can be derived by expanding the function $G(x)$ around the current guess $x^{(i)}$ using Taylor's series:

$$G(x) = G\left(x^{(i)}\right) + \left(x - x^{(i)}\right)G'\left(x^{(i)}\right) + \frac{\left(x - x^{(i)}\right)^2}{2}G''\left(x^{(i)}\right) + \cdots \qquad (6.21)$$

$$G(x) = x^{(i+1)} + \left(x - x^{(i)}\right)G'\left(x^{(i)}\right) + \frac{\left(x - x^{(i)}\right)^2}{2}G''\left(x^{(i)}\right) + \cdots$$

The above is obtained by using the expression (6.20) in the first term of Equation 6.21.

The solution $x^\star$ satisfies the condition $x^\star = G(x^\star)$. Substituting in the above

$$\underbrace{G\left(x^\star\right)}_{x^\star} = x^{(i+1)} + \left(x^\star - x^{(i)}\right)G'\left(x^{(i)}\right) + \frac{\left(x^\star - x^{(i)}\right)^2}{2}G''\left(x^{(i)}\right) + \cdots \qquad (6.22)$$

Applying the mean value theorem

$$x^\star - x^{(i+1)} = \left(x^\star - x^{(i)}\right)G'(\xi), \quad \xi \in \left[x^{(i)}, x^\star\right] \qquad (6.23)$$

Using the definition of true error

$$e_t^{(i+1)} = G'(\xi) \cdot e_t^{(i)} \qquad (6.24)$$

Thus, fixed point iteration has a linear rate of convergence, and the error converges if $|G'(\xi)| \leq 1$. If the value of $G'(\xi)$ is negative, the error will switch sign at every iteration. If $G'(\xi)$ is positive and less than 1, then the error will decrease monotonically.

The error analysis can be performed for Redlich-Kwong problem analyzing $G'(v^\star)$:

$$G'(v) = \frac{a}{P\sqrt{T}} \frac{(v-b)^2 - 2b^2}{v^2(v+b)^2} \qquad (6.25)$$

The computed value of $G'(v^\star) = 0.56$. Thus, it is estimated to take 15 iterations to converge, starting from $v^{(0)} = 2.2 \times 10^{-4}$.

One must exercise caution while using the condition for stability of fixed point iteration. If the derivative $G'(\xi)$ changes significantly in the vicinity of the solution, the estimates based on Equation 6.24 can be highly incorrect. Secondly, the stability for convergence of fixed point iteration in Equation 6.24 is a rather stringent condition. Although fixed point iteration has an advantage of simplicity, it finds limited real-world applications because the method is stable for only a restricted set of applications.

6.2.4 Newton-Raphson in Single Variable

Newton-Raphson or related methods are arguably the most widely used methods for solving nonlinear equations. Before proceeding to multivariable Newton-Raphson (next section), the single-variable version is discussed here. As seen in Figure 6.2, the slope of tangent to the curve at a point $(x^{(i)}, g(x^{(i)}))$ is $g'(x^{(i)})$. Thus, the equation for the tangent is

$$y - y^{(i)} = g'\left(x^{(i)}\right)\left(x - x^{(i)}\right) \tag{6.26}$$

The point where the above line intersects the X-axis is obtained by substituting $y = 0$. Newton-Raphson method uses this as the next iteration value for solving the nonlinear equation

$$x^{(i+1)} = x^{(i)} - \frac{g\left(x^{(i)}\right)}{g'\left(x^{(i)}\right)} \tag{6.27}$$

Note that I used the fact that the point $y^{(i)}$ on the curve is actually $g(x^{(i)})$. I will now use the Newton-Raphson method to solve the RK EOS problem.

Example 6.4 Newton Raphson

Problem: Use the Newton-Raphson method to obtain molar volume for the system in Example 6.1.

Solution: The function developed in Example 6.1 will be used to obtain $g(v)$. The derivative $g'(v)$ is

$$g'(v) = P + \frac{a}{\sqrt{T}}\left(\frac{v(v+b) - (v-b)(2v+b)}{v^2(v+b)^2}\right)$$

$$g'(v) = P + \frac{a}{\sqrt{T}}\left(\frac{2vb - v^2 + b^2}{v^2(v+b)^2}\right) \tag{6.28}$$

The function, RKdg.m, is written to obtain $g'(v)$ as below:

```
function dg = RKdg(v,par)
% Returns g'(v) at a particular T and P
% Parameters and operating conditions
a=par.a; b=par.b; R=par.R;
T=par.T; P=par.P;
% Function value
dg=P + a/sqrt(T)*(2*v*b-v^2+b^2)/v^2/(v+b)^2;
end
```

The main driver script for Newton-Raphson is similar to those of bisection and secant methods. The only change is that RKdg is used to calculate $g'(v)$ and $v^{(i+1)}$ is calculated from Equation 6.27. For the sake of brevity, only the Newton-Raphson loop is given below:

```
%% Newton-Raphson iterations
X=v;      % For storing v values
G=g;      % For storing g(v)
for i=1:maxIter
    dg=RKdg(v,param);
    v=v-g/dg;
    g=RK(v,param);
    % Store values & check convergence
    X(i+1)=v;
    G(i+1)=g;
    err=abs(X(i+1)-X(i));
    if (err<eTol)
        break
    end
end
```

The following are the values of $v^{(i)}$ generated, starting from $v^{(0)} = 2.2 \times 10^{-4}$:

```
1.7169e-4    1.6478e-4    1.6460e-4    1.6460e-4
```

In contrast to the bisection, secant, and fixed point iterations, Newton-Raphson converged in four iterations.

A formal error analysis of Newton-Raphson method, as seen before, involves using Taylor's series expansion of $g(x)$ around $x^{(i)}$:

$$g(x) = g\left(x^{(i)}\right) + \left(x - x^{(i)}\right)g'\left(x^{(i)}\right) + \frac{\left(x - x^{(i)}\right)^2}{2}g''\left(x^{(i)}\right) + \cdots \tag{6.29}$$

If $x = x^\star$, the solution of Equation 6.1, then

$$0 = g\left(x^{(i)}\right) + \left(x^\star - x^{(i)}\right) g'\left(x^{(i)}\right) + \frac{\left(x^\star - x^{(i)}\right)^2}{2} g''\left(x^{(i)}\right) + \cdots$$

$$\left(x^\star - x^{(i)}\right) g'\left(x^{(i)}\right) = g\left(x^{(i)}\right) + \frac{\left(x^\star - x^{(i)}\right)^2}{2} g''\left(x^{(i)}\right) + \cdots \tag{6.30}$$

$$\left(x^\star - x^{(i)}\right) = \frac{g\left(x^{(i)}\right)}{g'\left(x^{(i)}\right)} + \frac{\left(x^\star - x^{(i)}\right)^2}{2} \frac{g''\left(x^{(i)}\right)}{g'\left(x^{(i)}\right)} + \cdots$$

The above equation can be rearranged to give the Newton-Raphson method. The infinite series can be truncated to the first-order term (i.e., the first term on the right-hand side) and the remainder is the error. Thus

$$x^\star = \left[x^{(i)} - \frac{g\left(x^{(i)}\right)}{g'\left(x^{(i)}\right)}\right] + \left[-\frac{\left(x^\star - x^{(i)}\right)^2}{2} \frac{g'\left(x^{(i)}\right)}{g'\left(x^{(i)}\right)} + \cdots\right] \tag{6.31}$$

The term in the first bracket is the new iterated value $x^{(i+1)}$, whereas the term in the second bracket is the error, $e^{(i+1)}$:

$$x^{(i+1)} = x^{(i)} - \frac{g\left(x^{(i)}\right)}{g'\left(x^{(i)}\right)} \tag{6.27}$$

$$e^{(i+1)} = -\frac{\left(e^{(i)}\right)^2}{2} \frac{g''(\xi)}{g'(\xi)} \tag{6.32}$$

Note that the mean value theorem is used to convert the higher order terms of the infinite series, where $\xi \in [x^{(i)}, x^{(i+1)}]$. It is clear from the above expression that the Newton-Raphson method has a *quadratic rate of convergence*. Hence, in the vicinity of the solution, $x^\star$, Newton-Raphson converges faster than the other methods discussed in this section. Consequently, it emerged as one of the most popular methods for solving nonlinear algebraic equations.

6.2.5 Comparison of Numerical Methods

The five numerical methods considered here—bisection, secant, regula falsi, fixed point iteration, and Newton-Raphson—were analyzed for their error behavior. The error propagation equation is

$$e^{(i+1)} = \kappa \left(e^{(i)}\right)^m \tag{6.33}$$

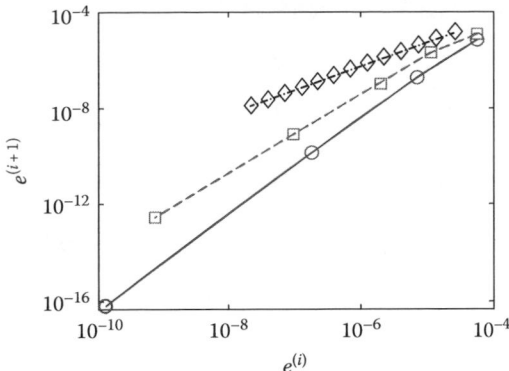

FIGURE 6.4 Plot of errors $e^{(i+1)}$ vs. $e^{(i)}$ for fixed point iteration (diamonds), secant (squares), and Newton-Raphson (circles). The Newton-Raphson method has significantly lower errors and shows a faster drop in error.

If the error in the $(i+1)^{\text{th}}$ iteration is plotted on X-axis against error in the (i)th iteration on a log-log plot, it will result in a straight line with a slope, m, equal to the order of convergence of the numerical method. Figure 6.4 shows this log-log plot for three methods. Bisection and regula falsi are skipped for ease of discussion. As is clear from the figure, all the three are straight lines, with slopes of 1, 1.6, and 2, respectively. The slope can either be computed from the figure or qualitatively confirmed. For example, $e^{(i)}$ and $e^{(i+1)}$ decrease by almost four orders of magnitude for the bisection method. For Newton-Raphson, on the other hand, $e^{(i+1)}$ decreases by 12 orders of magnitude when $e^{(i)}$ has decreased by six orders of magnitude (indicating a quadratic rate of convergence).

6.3 NEWTON-RAPHSON: EXTENSIONS AND MULTIVARIATE

The Newton-Raphson and related methods are investigated in more detail in this section, considering a general n-dimensional system of equations:

$$\mathbf{g}(\mathbf{x}) = 0, \quad \mathbf{x} \in \mathcal{R}^n, \ \mathbf{g} : \mathcal{R}^n \to \mathcal{R}^n \tag{6.1}$$

6.3.1 Multivariate Newton-Raphson

Derivation of a general Newton-Raphson method uses multivariate Taylor's series expansion. If the n elements of the n-dimensional function $\mathbf{g}(\mathbf{x})$ are $g_1(\mathbf{x}), \ldots, g_n(\mathbf{x})$, then

$$0 = g_1\left(\mathbf{x}^{(i)}\right) + \nabla g_1\left(\mathbf{x}^{(i)}\right)\left(\mathbf{x} - \mathbf{x}^{(i)}\right) + \cdots$$
$$\vdots \tag{6.34}$$
$$0 = g_n\left(\mathbf{x}^{(i)}\right) + \nabla g_n\left(\mathbf{x}^{(i)}\right)\left(\mathbf{x} - \mathbf{x}^{(i)}\right) + \cdots$$

where the gradient $\nabla g_k(\mathbf{x}^{(i)})$ is a $1 \times n$ row vector:

$$\nabla g_k\left(\mathbf{x}^{(i)}\right) = \left[\frac{\partial g_k}{\partial x_1} \quad \frac{\partial g_k}{\partial x_2} \quad \cdots \quad \frac{\partial g_k}{\partial x_n}\right] \tag{6.35}$$

These equations can be combined into the following matrix form:

$$0 = \mathbf{g}\left(\mathbf{x}^{(i)}\right) + \mathbf{J}\left(\mathbf{x}^{(i)}\right)\left(\mathbf{x} - \mathbf{x}^{(i)}\right) + \cdots \tag{6.36}$$

In the above, $\mathbf{J}(\mathbf{x}^{(i)})$ is called the Jacobian, which is an $n \times n$ matrix. Unlike the single variable case, the order of matrix multiplication is important. The Jacobian premultiplies the difference $(\mathbf{x} - \mathbf{x}^{(i)})$. It is easy to see that the above can be rearranged to obtain multivariate Newton-Raphson formula as below:

$$\mathbf{x}^{(i+1)} = \mathbf{x}^{(i)} - \mathbf{J}^{-1}\mathbf{g}\left(\mathbf{x}^{(i)}\right) \tag{6.37}$$

$$\mathbf{J} = \begin{bmatrix} \dfrac{\partial g_1}{\partial x_1} & \dfrac{\partial g_1}{\partial x_2} & \cdots & \dfrac{\partial g_1}{\partial x_n} \\[2ex] \dfrac{\partial g_2}{\partial x_1} & \dfrac{\partial g_2}{\partial x_2} & \cdots & \dfrac{\partial g_2}{\partial x_n} \\[2ex] \vdots & \ddots & \ddots & \vdots \\[2ex] \dfrac{\partial g_n}{\partial x_1} & \dfrac{\partial g_n}{\partial x_2} & \cdots & \dfrac{\partial g_n}{\partial x_n} \end{bmatrix}_{\mathbf{x}=\mathbf{x}^{(i)}} \tag{6.38}$$

In the rest of this chapter, I will drop the boldface. The examples considered hereafter will be multivariate. It bears repeating that all the vectors, unless stated otherwise, will be column vectors; subscripted number indicates corresponding element of the vector, and the iteration index is indicated by superscript in bracket.

The multivariate Newton-Raphson technique will be demonstrated using the chemostat example from Chapter 3. The model for the growth of microorganisms, substrate consumption, and product formation in a continuous stirred reactor is given by

$$0 = D\left(S_f - S\right) - r_g$$

$$0 = -DX + r_g Y_{xs}$$

$$0 = -DP + r_g Y_{ps} \tag{6.39}$$

$$r_g = \mu^{\max}\frac{SX}{K_s + S}$$

The model parameters and solution conditions may be used from the case study in Chapter 3. The solution is given in the next example.

Example 6.5 Steady State in a Chemostat (Using "Vanilla" Newton-Raphson)

Problem: Determine the steady state concentrations of substrate (S), biomass (X), and product (P) for the chemostat of Equation 6.39 using Newton-Raphson method.

   Solution: Two functions, funMonod and jacMonod, are written as below to compute $g(x)$ and $J(x)$, respectively. The former returns a 3×1 column vector, whereas the latter returns a 3×3 Jacobian matrix for current values of input parameter x. The Jacobian matrix is given by

$$
J = \begin{bmatrix}
-D - \dfrac{\partial r_g}{\partial S} & -\dfrac{\partial r_g}{\partial X} & 0 \\[2ex]
Y_{xs} \dfrac{\partial r_g}{\partial S} & -D + Y_{xs} \dfrac{\partial r_g}{\partial X} & 0 \\[2ex]
Y_{xs} \dfrac{\partial r_g}{\partial S} & Y_{ps} \dfrac{\partial r_g}{\partial X} & -D
\end{bmatrix}
$$

$$
\text{where} \quad \frac{\partial r_g}{\partial S} = \frac{\mu X K}{\left(K + S\right)^2}, \quad \frac{\partial r_g}{\partial X} = \frac{\mu S}{K + S}
$$

The function funMonod.m is similar to the one in Chapter 3:

```
function gVal=funMonod(C,monodParam)
mu=monodParam.mu;     K=monodParam.K;
Yxs=monodParam.Yxs;   Yps=monodParam.Yps;
D=monodParam.D;       Sf=monodParam.Sf;
%% Variables to solve for
S=C(1); X=C(2); P=C(3);
%% Model equations
rg= mu*S/(K+S)*X;
gVal(1,1)=D*(Sf-S) - rg;
gVal(2,1)=-D*X + Yxs*rg;
gVal(3,1)=-D*P + Yps*rg;
end
```

The only difference in the above function being that it is a function of only C, and not of time t, since a steady state solution is sought. The function jacMonod.m for calculation of numerical Jacobian is

```
function J=jacMonod(C,monodParam)
mu=monodParam.mu;     K=monodParam.K;
Yxs=monodParam.Yxs;   Yps=monodParam.Yps;
D=monodParam.D;       Sf=monodParam.Sf;
%% Variables to solve for
S=C(1); X=C(2); P=C(3);
```

```
%% Jacobian calculation
r_s=mu*X*K/(K+S)^2;      % drg/dS
r_x=mu*S/(K+S);          % drg/dX
J = [-D-r_s,    -r_x,       0;
      r_s*Yxs, -D+r_x*Yxs,  0;
      r_s*Yps,    r_x*Yps, -D];
end
```

The function $g(C)$ and Jacobian $J(C)$ can then be used in Newton-Raphson solver. The driver script, monodRun.m, is given below:

```
% Chemostat using Vanilla Newton-Raphson
% Reactor parameters
monodParam.mu=0.5;
monodParam.K=0.25;
monodParam.Sf=5;
monodParam.D=0.1;
monodParam.Yxs=0.75;
monodParam.Yps=0.65;
% Initial guess for Newton-Raphson
C0=[0.25; 4.0; 4.0];
g=monodFun(C0,monodParam);
X=C0; G=g;     % Storing results

% Newton Raphson Iterations
maxIter=25;
eTol=1e-8;
for i=1:maxIter
    J=monodJac(X(:,i),monodParam);
    X(:,i+1)=X(:,i)-inv(J)*g;
    err=abs(X(:,i+1)-X(:,i));
    if(max(err)<eTol)
        return
    end
    g=monodFun(X(:,i+1),monodParam);
    G(:,i+1)=g;
end
```

The value of the three state variables at steady state using the above code is

```
C = 0.0909    3.6818    3.1909
```

On the other hand, if the initial guess were changed to C0=[4; 1; 1], then

```
C = 5.0000    0.0000    0.0000
```

Note that the above signifies multiple coexisting steady states. This will be discussed in further detail in Case Studies (Section 6.5).

The above example demonstrates steady state multiplicity. When there are no microorganisms in the chemostat, no conversion of substrates into products occurs, and the outlet stream contains unconverted substrate (at concentration of s_f), without any biomass or product. On the other hand, for a range of values of dilution rate, D, there exists a steady state $\left(C = \begin{bmatrix} 0.091 & 3.68 & 3.19 \end{bmatrix}^T\right)$ with a significant conversion of the substrate to the product.

Numerically, Newton-Raphson suffers from a couple of problems. First, the intermediate solutions may violate bounds on variables. For example, with C0= [4; 1; 1], the various iterated values in the above Newton-Raphson example were

```
X =
     4.0000      5.0281      5.0000      5.0000      5.0000
     1.0000     -0.0211     -0.0000     -0.0000     -0.0000
     1.0000     -0.0183     -0.0000     -0.0000     -0.0000
```

Note that the values of X and P at the first iteration are negative. This does not affect convergence in Example 6.5. However, if the function $g(C)$ included square root or logarithm of $C(2)$ or $C(3)$, then Newton-Raphson would give complex values or fail. Another problem is that while Newton-Raphson converges rapidly in the vicinity of the solution, $x^\star$, its global convergence behavior is poor. That means starting from arbitrary initial guess, as would be the case when the solution $x^\star$ is unknown, it often does not converge. It also fails if the guess values or iterated values are close to local maxima or minima (where $g'(x)$ is zero or Jacobian is noninvertible). While this may not be a major issue for small problems (single to a few variables), it can be a significant problem for Newton-Raphson in multiple variables. Finally, the Jacobian may not be explicitly known. Some of these issues are discussed in the rest of this section.

6.3.2 Modified Secant Method

I will tackle the "simplest" issue first: when the Jacobian is not explicitly available. In such a scenario, one option is to compute numerical Jacobian (see Appendix B for numerical integration):

$$J_{k,l}^{(i)} = \frac{g_k\left(\{x\}_l^{(i)} + h\right) - g_k\left(\{x\}_l^{(i)}\right)}{h}, \quad h = \left|x_l^{(i)}\right| \times 10^{-8} \tag{6.40}$$

Here, the notation $\{x\}_l^{(i)}$ represents that only the lth element of the vector $\mathbf{x}$ is changed, and all other elements of $\mathbf{x}^{(i)}$ are kept at their original values. We have used the forward difference formula for numerical integration. While the central difference formula is more accurate, it involves an additional function calculation, $\mathbf{g}\left(\{x\}_l^{(i)} - h\right)$. The following example shows the use of Newton-Raphson with numerical Jacobian.

Example 6.6 Newton-Raphson with Numerical Jacobian

Problem: Solve Example 6.5 using Newton-Raphson with numerical Jacobian.

Solution: The driver script, monodRun.m, and the function, monodFun.m, remain similar to the previous solution. The only difference being the first line of the Newton-Raphson loop is replaced by numerical Jacobian calculation. Numerical Jacobian is calculated in the following code:

```
function J=monodJacNum(x,g,par)
% This function takes the current value x(i) and g(i)
% and returns the numerical Jacobian
for l=1:3     % Cycle over all x(l)
    xbracket=x;
    h=abs(xbracket(l))*1e-8;
    xbracket(l)=x(l)+h;
    g1=monodFun(xbracket,par);
    J(:,l)=(g1-g)/h;
end
```

The driver can now be executed with the numerical Jacobian replacing the previous case of analytical Jacobian. The code takes exactly the same number of iterations and converges to the same solution:

```
C = 0.0909    3.6818    3.1909
```

Since numerical Jacobian requires additional function calculation, multivariate secant method is another option. The reader can verify that this method can be executed as

$$\mathbf{x}^{(i+1)} = \mathbf{x}^{(i)} - \left[\mathbf{B}^{(i)}\right]^{-1}\mathbf{g}\left(\mathbf{x}^{(i)}\right)$$

$$B_{k,l}^{(i)} = \frac{g_k\left(x_l^{(i)}\right) - g_k\left(x_l^{(i-1)}\right)}{\left(x_l^{(i)} - x_l^{(i-1)}\right)} \tag{6.41}$$

Broyden's Method is a better method to calculate $\mathbf{B}^{(i)}$ instead:

$$\mathbf{B}^{(i)} = \mathbf{B}^{(i-1)} - \frac{\left(\left[\mathbf{g}^{(i)} - \mathbf{g}^{(i-1)}\right] - B^{(i-1)}\left[\delta^{(i)}\right]\right)\left[\delta^{(i)}\right]^T}{\left[\delta^{(i)}\right]^T\left[\delta^{(i)}\right]} \tag{6.42}$$

$$\text{where } \delta^{(i+1)} = \mathbf{x}^{(i+1)} - \mathbf{x}^{(i)} \tag{6.43}$$

Broyden's method is mentioned for the sake of completeness and will not be investigated further.

6.3.3 Line Search and Other Methods

The issue of poor global convergence is a major problem for Newton-Raphson method. A large portion of work in the literature is dedicated to this issue. Before discussing these methods, let us define the following:

$$s^{(i)} = \left[J\left(x^{(i)} \right) \right]^{-1} g\left(x^{(i)} \right) \tag{6.44}$$

Note that in "vanilla" Newton-Raphson method

$$x^{(i+1)} = x^{(i)} - s^{(i)} \tag{6.45}$$

Consider a scalar-valued function:

$$G\left(x \right) = \frac{1}{2} g^T g \tag{6.46}$$

Since the above function is a quadratic function of real-valued vector, its minimum value is 0. It is easy to see that the roots of $g(x)$ form the minima of $G(x)$. It can be shown that the vector $-s$ above provides the direction of steepest decrease in the value of $G(x)$, and thus the direction of steepest approach of $g(x)$ to a root. Instability in Newton-Raphson arises because the complete step may be too aggressive. In order to improve robustness, a partial step may be taken:

$$x^{(i+1)} = x^{(i)} - \omega s^{(i)} \tag{6.47}$$

There are various ways to obtain the *line search* parameter, ω. Several algorithms exist that try to obtain ω in an "optimal" way. The key idea is to "search along the direction" $-s$ such that the most optimal reduction in the value of $G(x^{(i+1)}(\omega))$ is obtained. Note that as per Equation 6.47, the next iteration value $x^{(i+1)}$ is written in terms of the parameter ω. This is mentioned for the sake of completeness; the discussion of these methods is beyond the scope of this text.

For the purpose of this section, I will discuss some heuristic ways to improve the performance of Newton-Raphson. One simple way is to start with the full Newton-Raphson step, that is, $\omega = 1$. If the $x^{(i)}$ is in bounds and $g(x^{(i)})$ has decreased, retain the step. Else, ω is halved. This can be progressively done until a reasonable step is obtained. This *back-tracing* process is repeated at each iteration.

Another option, which may work for certain specific problems (though it is not a general purpose algorithm) is to choose an appropriate ω that works for the problem under consideration. We often need to solve not a single problem but a family of problems involving the solution of nonlinear equations. For example, we may execute the simulation of Example 6.5 for various different values of dilution rates (D), feed substrate concentration (S_f), or rate constants (μ^{max}, K_s). One can set up a nominal case, and try the simulation with $\omega = 1$. If that does not yield a solution, choose $\omega = 0.5$ and keep it constant for all maxIter iterations. If that does not converge, repeat for $\omega = 0.25$, etc. It should be noted that using lower values of ω would necessitate a larger number of iterations; so maxIter should be a reasonably large number ($\sim 100 \times n$, where n is the number of equations to be solved). I have usually used $\omega > 0.1$, and very rarely have I had to use $\omega = 2^{-4}$ or 2^{-5}, for reasonably tough problems.

The next example demonstrates one simple way to use the line-search parameter ω to avoid negative values of concentrations. As we will see in Chapter 7, this idea of *under-relaxation* is used in other iterative numerical techniques as well. It is widely used in computational fluid dynamics (CFD) and finite element software to ensure stable performance of the solver.

Example 6.7 Newton-Raphson with Under-Relaxation

Problem: Repeat Example 6.5 with an appropriate value of ω to avoid negative values of concentration.

Solution: Consider the solution of Example 6.5, where the first iteration values starting from $[4 \quad 1 \quad 1]^T$ were

```
 5.0281
-0.0211
-0.0183
```

Note that the step taken in value of S was 1.0281, while the maximum allowable step is 1 (since S cannot exceed S_f). Based on this, we will redo Example 6.5 with $\omega = 0.9$. The following results are obtained:

```
4.0000    4.9253    4.9926    4.9993    4.9999 etc.
1.0000    0.0810    0.0080    0.0008    0.0001 etc.
1.0000    0.0835    0.0083    0.0008    0.0001 etc.
```

As can be seen above, negative values are avoided in the iterations. Moreover, it takes nine iterations for Newton-Raphson to converge.

PERSONAL NOTE:

I personally find the dichotomy between *line search* and *under-relaxation* quite interesting. I use the two terms replaceably, since the core idea is the same: to use the parameter ω to slow down or speed up an iterative numerical technique. The term "line search" is often used with the Newton-Raphson method (or optimization methods) when the aim is to find the value ω so that the one approaches the solution (or minima) in the "best" manner. The term "under-relaxation" is commonly used in Gauss-Siedel (Chapter 7) or in CFD, where the aim is to slow down the steps in order to ensure robust and stable algorithm.

If the above heuristic ideas do not solve the problem, more advanced techniques based on trust region or other methods (such as those used by MATLAB solvers) need to be used. Besides, there exist other methods, rather than the one shown in Example 6.7, that are designed for nonlinear equations in the presence of constraints. Either of these are not within the scope of this book.

6.4 MATLAB® SOLVERS

MATLAB solvers for nonlinear algebraic equations are provided in the Optimization Toolbox and are not available in the base MATLAB package. If optimization toolbox is not available, one may need to write one's own solution code based on algorithms discussed in the previous sections. This section discusses two MATLAB methods, `fzero` and `fsolve`, to solve single-variable and multivariable nonlinear equation problems, respectively.

6.4.1 Single Variable Solver: `fzero`

MATLAB function `fzero` is used to obtain a root of a single equation in one unknown. It uses a combination of Brent's inverse quadratic interpolation (Section 6.2.2) and the bisection method (Section 6.2.1). It typically requires two initial guesses that bracket the solution, but also works with a single initial guess as well. The following reproduces information `help fzero`, relevant to solving a nonlinear equation:

```
X = fzero(FUN,X0)
    Find a zero of the function FUN near X0, if X0 is scalar.
X = fzero(FUN,X0), where X0 is a vector of length 2,
    Assumes X0 is a finite interval where the sign of
    FUN(X0(1)) differs from the sign of FUN(X0(2)).
```

Calling `fzero` with an initial interval (using X0 as a vector of length 2) guarantees `fzero` will return a solution if it exists. The `fzero` algorithm first attempts an inverse quadratic step; if that results in $x^{(i+1)}$ that lies between $x^{(L)}$ and $x^{(U)}$, then the solution is accepted. If the value $x^{(i+1)}$ lies outside the interval, a bisection step is used instead. The solution $v^\star$ remains bracketed all the time. I recommend using `fzero` supplying an initial interval (rather than a scalar initial value), as seen in the next example.

Example 6.8 Using `fzero` for Redlich-Kwong EOS

Problem: Solve Example 6.1 using `fzero`.

Solution: The function file, `RK.m`, created to obtain $g(v)$, can be used with `fzero`. As before, the problem definition and preamble sections remain unchanged. For the sake of completeness, the full code is given below:

```
% Molar volume using Redlich-Kwong EOS
% Model parameters for the EOS
param.a=6.46;
param.b=2.97e-5;
param.R=8.314;
```

```
% Operating conditions
param.P=1e7;   % Pressure (Pa)
param.T=340;   % Temperature (K)

%% Solution using fzero
vL=0.8e-4;
vU=2.2e-4;
vSol=fzero(@(v) RK(v,param),[vL,vU]);
disp(['Molar volume, v = ', num2str(vSol)]);
```

Note that the underlined line was the only line that was added to solve the problem using `fzero`. The result using the above code is the same as those in the previous examples:

```
Molar volume, v = 0.0001646
```

The above call to `fzero` uses default optimization parameters. User-specified options can be supplied to `fzero` using

```
X = fzero(FUN,X0,OPTS)
```

where `OPTS` is a structure that can be created using the `OPTIMSET` command. Setting options will be covered in the next section, while discussing `fsolve`.

If more information regarding the solution is required, the following output arguments can be used with `fzero`:

```
[v,gVal,flag,info]=fzero(@(v) RK(v,param),[vL,vU]);
```

When used in the above example, the value returned for `flag = 1`, indicating that the solution was found successfully by `fzero`. The following information is obtained from the structure `info`: that `fzero` required seven iterations and it used a combination of bisection and Brent's inverse quadratic interpolation methods.

6.4.2 Multiple Variable Solver: `fsolve`

In Example 6.8, let us replace the underlined command with the following:

```
>> vSol=fsolve(@(v) RK(v,param),vU)
```

With the above expression, I am using another nonlinear equations solver, `fsolve`, available in MATLAB (optimization toolbox). `fsolve` solves the above problem and returns the value

```
vSol =
    1.6460e-04
```

This solution is same as the ones obtained using other techniques in this chapter.

The syntax for using `fsolve` is similar. Let us go over the similarities and differences in the usage of `fzero` and `fsolve`: (i) Like `fzero`, it requires users to supply function and an initial guess, and (ii) unlike `fzero` (and similar to Newton-Raphson), `fsolve` accepts only a single initial guess and not an interval. The second point is quite important because `fsolve` is a multivariate solver for nonlinear equations of the type

$$\mathbf{g}(\mathbf{x}) = 0, \quad \mathbf{x} \in \mathcal{R}^n, \quad \mathbf{g} : \mathcal{R}^n \to \mathcal{R}^n \tag{6.1}$$

The corresponding syntax for `fsolve` is

```
X = fsolve(FUN,X0)
```

where the size of vector `X0` is $n \times 1$, where n is the number of equations to be solved. Thus, as per the above equation, the size of `X0` and the size of function values returned by `FUN` should be the same.

Coming back to the Redlich-Kwong example, the solution of the nonlinear equation is

$$v^\star = 1.646 \times 10^{-4}$$

`fsolve` uses the following default tolerance values: `tolX=1e-6` and `tolFun=1e-6`. With this option, the solution is considered to be converged if

$$\left| x^{(i+1)} - x^{(i)} \right| < \text{tolX} \tag{6.48}$$

$$\left| g\left(x^{(i+1)}\right) - g\left(x^{(i)}\right) \right| < \text{tolFun} \tag{6.49}$$

Since the value of molar volume is expected to be of the order of 10^{-4} for this problem, tighter tolerance values—at least four orders of magnitude lower—are recommended. The tolerance values can be specified using the `optimset` command and then passing the options structure to `fsolve`:

```
opts=optimset('tolX',1e-10,'tolFun',1e-10);
vSol=fsolve(@(v) RK(v,param),vU,opts);
```

With this change, the solution is now obtained with a more stringent set of error tolerance values. I bring to your attention that in this particular example, higher tolerance values resulted in almost the same solution. However, this is not always the case, and the tolerance values should be carefully chosen based on the order of magnitude of the expected solution $x^\star$ and the order of magnitude of the function $g(x^{(0)})$.

Since `fsolve` is a multivariate solver, I will now use it for the chemostat example below.

Example 6.9 Chemostat Solution Using `fsolve`

Problem: Use `fsolve` to obtain steady state values of chemostat from Example 6.5.

   Solution: The function, `funMonod`, was written in Example 6.5 to compute **g(x)**. We will use this function with `fsolve` as shown in the code below:

```
% Driver script: monodRun_fsolve.m
% To demonstrate use of fsolve for Chemostat problem
%% Reactor parameters
monodParam.mu=0.5;
monodParam.K=0.25;
monodParam.Sf=5;
monodParam.D=0.1;
monodParam.Yxs=0.75;
monodParam.Yps=0.65;


%% Solving and displaying result
C0=[0.25; 4.0; 4.0];
X=fsolve(@(c) monodFun(c,monodParam),C0);
disp(X);
```

The result obtained using `fsolve` is the same as before:

```
    0.0909
    3.6818
    3.1909
```

Let us now analyze the above problem a bit further. First, the solutions are concentrations of substrate, biomass, and product. All of these are expected to be of the order $\sim \mathcal{O}(1)$. The initial guess was `C0=[0.25; 4.0; 4.0]`. The function, `monodFun`, evaluated at this initial condition gives

```
>> monodFun(C0,monodParam)
ans =
   -0.5250
    0.3500
    0.2500
```

The typical values of the function **g(x)** computed by `monodFun` are of the order $\mathcal{O}(0.1)$. Therefore, the default tolerance values of `tolX=1e-6` and `tolFun=1e-6` are sufficient and need not be changed for this example. Although the readers know this, it still is worth repeating that the initial guess and solution vectors `C0` and `X` (respectively) are 3×1 vectors; the function `monodFun` returns **g(x)** as a column vector of the exact same dimension.

6.5 CASE STUDIES AND EXAMPLES

6.5.1 Recap: Equation of State

The first example we considered was a demonstration of solving a single nonlinear equation in one unknown in finding the molar volume for a system described by RK EOS:

$$P = \frac{RT}{v-b} - \frac{a/\sqrt{T}}{v(v+b)} \tag{6.2}$$

The above equation was put in the standard form of nonlinear equations in one unknown as

$$\underbrace{P(v-b) - RT + \frac{v-b}{v(v+b)} \frac{a}{\sqrt{T}} = 0}_{g(v)} \tag{6.4}$$

The above nonlinear equation was solved for given values of T and P to obtain the corresponding molar volume, v.

Another popular EOS is Peng-Robinson EOS:

$$P = \frac{RT}{V-b} - \frac{a}{V(V+b) + b(V-b)} \tag{6.50}$$

An interested reader may use the above PR EOS and obtain the molar volume for given T and P. You may choose the parameter values of $a = 0.364$ and $b = 3 \times 10^{-5}$.

6.5.2 Two-Phase Vapor-Liquid Equilibrium

Calculation of vapor-liquid equilibrium involves calculating dew point and bubble point for a given system. Antoine's equation will be used for calculating the saturation conditions:

$$\ln\left(P_k^{sat}\right) = a - \frac{b}{T+c} \tag{6.51}$$

where
  P_k^{sat} is the saturation pressure in kPa
  T is the temperature in °C

Given a temperature, the above equation can be used to calculate the saturation vapor pressure, or vice versa.

Now consider a mixture of two components. If the mixture obeys Raoult's law, then the vapor pressure of the component in the gas phase is simply the saturation pressure from Equation 6.51 multiplied by its liquid-phase mole fraction:

$$p_k = x_k P_k^{sat} \tag{6.52}$$

This can be used for vapor-liquid equilibrium (VLE) calculations. At the boiling point of a pure fluid, the saturation pressure equals the total pressure.* In case of a mixture, we equivalently calculate bubble point and dew point temperatures for the mixture at a given pressure (or pressure at a given temperature) for various liquid- and gas-phase compositions, respectively.

6.5.2.1 Bubble Temperature Calculation

In order to find bubble point for a two-component mixture, the following iterative procedure is used. If T is the current guess of the bubble point, the saturation pressures of the two components are given by Antoine's equation (6.51). From Raoult's law, the net pressure is given by

$$P = x_1 P_1^{\text{sat}}(T) + (1 - x_1) P_2^{\text{sat}}(T) \tag{6.53}$$

The value of temperature that solves the above equation is the bubble point. Note that the function $g(T)$ is a single function in one variable, given by

$$\underbrace{1 - x_1 \frac{P_1^{\text{sat}}(T)}{P} + (1 - x_1) \frac{P_2^{\text{sat}}(T)}{P}}_{g_B(T)} = 0 \tag{6.54}$$

Additionally, to obtain the initial guesses, we will use the property that the bubble point of ideal mixture lies between the boiling points of the two individual species.

6.5.2.2 Dew Temperature Calculation

Given the vapor-phase composition and pressure, the partial pressure, $y_k P$, for each component is known. From Raoult's law

$$y_k P = x_k P_k^{\text{sat}}(T) \tag{6.55}$$

where T is the guess value of dew temperature. Rearranging,

$$x_1 = \frac{y_1 P}{P_1^{\text{sat}}(T)}, \quad 1 - x_1 = \frac{(1 - y_1) P}{P_1^{\text{sat}}(T)} \tag{6.56}$$

Adding the two, we get a single function in one variable:

$$\underbrace{1 - y_1 \frac{P}{P_1^{\text{sat}}(T)} + (1 - y_1) \frac{P}{P_2^{\text{sat}}(T)}}_{g_D(T)} = 0 \tag{6.57}$$

---

* For example, at its boiling point of 100°C, the saturation pressure of water is 1 atm.

6.5.2.3 Generating the T–x–y Diagram

The above two can be used to generate a *T–x–y* diagram for a binary mixture of acetonitrile and nitromethane. Antoine's coefficient for the two species is $a = 14.2724$, $b = 2945.47$, and $c = 224.0$ and $a = 14.2043$, $b = 2972.64$, and $c = 209.0$, respectively. The *T–x–y* diagram involves generating the bubble and dew point curves and plotting them on the same plot. This is demonstrated in the following example.

Example 6.10 *T–x–y* **Diagram for VLE Calculations**

Problem: Generate the *T–x–y* diagram for acetonitrile–nitromethane system for Antoine's vapor pressure parameters given above. Assume Raoult's law holds.

  Solution: We will use `fzero` to solve the two equations: Equation 6.54 to get the bubble point and Equation 6.57 to get the dew point. These are two different calculations, which will require separate function files (`bublFun.m` and `dewFun.m`, respectively). These functions are given below:

```
function gB=bublFun(T,par)
% Bubble point residual function (6.54) for fzero
pSat=antoine(T,par);  % Get pSat for both species
x=par.x;
P=par.P;
gB=1-x*pSat(1)/P-(1-x)*pSat(2)/P;
end
```

```
function gD=dewFun(T,par)
% Dew point residual function (6.57) for fzero
pSat=antoine(T,par);  % Get pSat for both species
y=par.y;
P=par.P;
gD=1-y*P/pSat(1)-(1-y)*P/pSat(2);
end
```

Both these functions require `antoine.m`, which takes in the temperature and calculates the saturation pressure of all the species using Equation 6.51. This function is given below:

```
function pSat=antoine(T,par)
a=par.a;
b=par.b;
c=par.c;
pSat=exp(a - b./(T+c));
end
```

Note that the underlined command is *vectorized* with respect to the species. Given Antoine's coefficients for *n* species, it will return the saturation pressures P_k^{sat} for all the species as a vector at the temperature *T*.

Both the bubble point and dew point calculations are performed in the same driver script VLEdriver.m. The calculations are performed in a loop, for a range of values of $x_1 \in [0, 1]$ for bubble point and $y_1 \in [0, 1]$ for dew point:

```
% To obtain the T-x-y curve
% for acetonitrile-nitromethane mixture
% Antoine constants
par.a = [14.2724,   14.2043];
par.b = [2945.47,   2972.64];
par.c = [224.0,     209.0];
% Operating points
par.P=101.3;  % 1 atm in kPa
n=20;
h=1/n;
for i=1:n+1
    par.x=(i-1)*h;
    par.y=(i-1)*h;
    Tbubl(i)=fzero(@(T) bublFun(T,par), [50 150]);
    Tdew(i) =fzero(@(T) dewFun(T,par), [50 150]);
end
% Plot the results
plot(0:h:1,[Tbubl; Tdew]);
```

Figure 6.5 shows the T–x–y curve for this system. The last point, corresponding to $x_1 = 1$, is the boiling point of acetonitrile, and the first point, corresponding to $x_1 = 0$, is the boiling point of nitromethane. The lower line is the bubble point curve, whereas the upper line is the dew point curve.

This method can be easily extended to any system in m-components. Standard thermodynamics texts use the example of ternary mixture of acetonitrile, nitromethane, and acetone

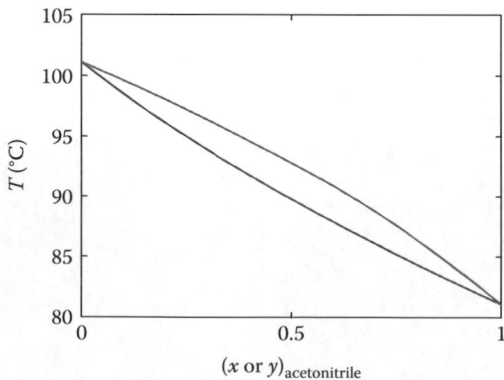

FIGURE 6.5 T–x–y diagram for acetonitrile-nitromethane mixture, with mole fraction of acetonitrile as the abscissa.

as a standard three-component example. Antoine's coefficients of acetone are $a = 14.5463$, $b = 2940.46$, and $c = 237.22$. This is left as an exercise.

6.5.3 Steady State Multiplicity in CSTR

An example of a nonisothermal continuous stirred tank reactor (CSTR) was discussed in Chapter 3. This same example will be further discussed in Chapter 9 in context of stability analysis and bifurcation in a formal way. Prior to that, obtaining steady state solutions for a range of operating parameters will be considered in this section. The system being modeled is shown in Figure 6.6. The model equations for the CSTR at *steady state* are given below (also see Chapter 3):

$$0 = F\left(C_{A0} - C_A\right) - \left[k_0 \exp\left(-\frac{E}{RT}\right)C_A\right]V$$

$$0 = F\rho c_p\left(T_0 - T\right) + \left(-\Delta H\right)\left[k_0 \exp\left(-\frac{E}{RT}\right)C_A\right]V - UA\left(T - T_j\right) \qquad (6.58)$$

$$0 = F_j\rho_j c_j\left(T_{j0} - T_j\right) + UA\left(T - T_j\right)$$

The parameters for this system are given in Table 6.1 (reproduced from Chapter 3). In Chapter 3, an ODE solver (ode45) was used to obtain the transient simulation results for the CSTR. After 50 s of operating time, the CSTR exit conditions were obtained as $C_A = 461.0$, $T = 647.6$, $T_j = 472.6$. Thereafter, the inlet temperature was reduced to 298 K, and the exit conditions obtained at the end of 200 s were $C_A = 3991$, $T = 298.8$, and $T_j = 298.4$.

Note the qualitative difference between the solutions obtained with $T_0 = 350$ K and $T_0 = 298$ K. In the former case, there is significant conversion of the reactant and a

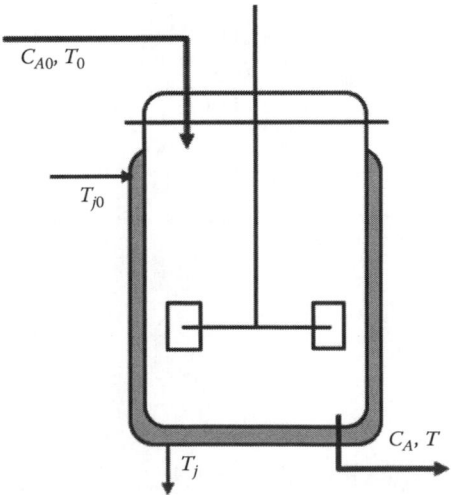

FIGURE 6.6   A schematic of jacketed CSTR modeled in this case study

TABLE 6.1   Model Parameters and Operating Conditions for Jacketed CSTR

CSTR Inlet	Cooling Jacket	Other Parameters
$F = 25$ L/s	$F_j = 5$ L/s	$C_{A0} = 4$ mol/L
$V = 250$ L	$V_j = 40$ L	$k_0 = 800$ s^{-1}
$\rho = 1000$ kg/m^3	$\rho_j = 800$ kg/m^3	$E/R = 4500$ K
$c_p = 2500$ J/kg·K	$c_p = 5000$ J/kg·K	$(-\Delta H) = 250$ kJ/mol
$T_0 = 350$ K	$T_{j0} = 25°$C	$UA = 20$ kW/K

corresponding increase in temperature. In the latter case, conversion is negligible, and the reactor exit temperature is very close to the inlet temperature itself. Thus, we observe a qualitative change in the nature of solution when an operating parameter, inlet temperature T_0 in this case, is varied. While the temporal evolution of modeled variables is interesting, it is also important to study various steady state solutions. Before we move on to study how the operating parameter affects steady state solution, I will discuss the use of nonlinear equation solver to obtain steady state values of the modeled variables in the next example.

Example 6.11 Steady State Solution of Nonisothermal CSTR

Problem: Obtain steady state solution for a CSTR described by Equation REF using fsolve.

Solution: The transient solution of CSTR was already discussed in Chapter 3. The development of the CSTR driver code and solver function follow similar lines as the examples before. The function file, `cstrFun.m`, to compute **g(x)** is given below.

Part-1: A first attempt at `cstrFunTry.m`:

```
function gVal=cstrFunTry(x,par)
%% Inlet and current conditions
C0=par.C0; T0=par.T0; Tj0=par.Tj0;
Ca=x(1);    T=x(2);    Tj=x(3);

%% Key Variables
rxnRate=par.k0*exp(-par.E/T)*Ca;
hXfer=par.UA*(T-Tj);
FRhoCp=par.F*par.rho*par.cp;
FjRhoC_j=par.Fj*par.rhoj*par.cj;

%% Model Equations
gVal=zeros(3,1);
gVal(1)=par.F*(C0-Ca)-rxnRate*par.V;
gVal(2)=FRhoCp*(T0-T)+par.DH*rxnRate*par.V-hXfer;
gVal(3)=FjRhoC_j*(Tj0-Tj)+hXfer;
```

Before discussing the CSTR driver script, let us check the values of $g(x^{(0)})$. Based on the solution obtained in the case study from Chapter 3, the value of initial guess is chosen as:

```
x0=[500; 600; 470];
```

The value of $g(x^{(0)})$ can be verified using

```
>> cstrFunTry(x0,modelPar)
ans =
    1.0e+06 *
      0.0000
     -4.3979
     -0.8400
```

The function file needs to be modified based on these observations, as discussed below.

The value of $g_1(x)$ is several orders of magnitude lower than the other two functions. As a result, the three equations are modified as follows:

$$0 = F\left(C_{A0} - C_A\right) - r_A V$$

$$0 = \left(T_0 - T\right) + \frac{\left(-\Delta H\right) r_A V - UA\left(T - T_j\right)}{F\rho c_p} \qquad (6.59)$$

$$0 = \left(T_{j0} - T_j\right) + \frac{UA\left(T - T_j\right)}{F_j \rho_j c_j}$$

CSTR function file is now modified. The modified file is shown below.

Example 6.11 (Continued) Steady State Solution of Nonisothermal CSTR

The CSTR function is now written as

```
function gVal=cstrFun(x,par)
%% Inlet and current conditions
C0=par.C0; T0=par.T0; Tj0=par.Tj0;
Ca=x(1);    T=x(2);     Tj=x(3);

%% Key Variables
rxnRate=par.k0*exp(-par.E/T)*Ca;
hXfer=par.UA*(T-Tj);
FRhoCp=par.F*par.rho*par.cp;
FjRhoC_j=par.Fj*par.rhoj*par.cj;

%% Model Equations
gVal=zeros(3,1);
gVal(1)=par.F*(C0-Ca)-rxnRate*par.V;
```

```
gVal(2)=(T0-T) + (par.DH*rxnRate*par.V-hXfer)/FRhoCp;
gVal(3)=(Tj0-Tj)+hXfer/FjRhoC_j;
```

Let us now check the values of $\mathbf{g(x)}$:

```
>> cstrFun(x0,modelPar)

ans =
    32.1916
   -70.3663
   -42.0000
```

Since the three elements of $\mathbf{g(x)}$ are of the same order of magnitude as the other, this function will be used in solving the CSTR model equations. The driver script, `cstrDriver.m`, is given below:

```
% Steady state solutions for jacketed CSTR
%    with exothermic first-order reaction
%% Define model parameters (SI units)
modelPar.F=25/1000; % CSTR parameters
modelPar.V=0.25;
modelPar.rho=1000;
modelPar.cp=2500;

modelPar.Fj=5/1000; % Cooling jacket
modelPar.Vj=40/1000;
modelPar.rhoj=800;
modelPar.cj=5000;

modelPar.k0=800;    % Rate constants
modelPar.E=4500;
modelPar.DH=250000;
modelPar.UA=20000;

%% Define inlet conditions
modelPar.C0=4000;
modelPar.T0=350;
modelPar.Tj0=298;
%% Solving using fsolve
x0=[500; 600; 470];
[X,gVal]=fsolve(@(x) cstrFun(x,modelPar), x0);
```

When the above script is executed, `fsolve` converges and value of X returned is:

```
X = [458.6 mol/m³;   648.1 K;   473.1 K]
```

The value of `gval` at the above conditions is of the order $\sim \mathcal{O}\left(10^{-10}\right)$, indicating that the solution is converged.

Furthermore, there isn't just one solution at this value of $T_0 = 350$ K. There are, in fact, three solutions, as demonstrated in the example below.

Example 6.11 (Continued) Steady State Solution of Nonisothermal CSTR (Multiple Steady States)

Problem: Find all the three solutions at $T_0 = 350$ K.

Solution: The other solution corresponds to the case of nearly no conversion of A. Thus, choose $\mathbf{x}^{(0)} = [C_0; T_0; T_{j0}]$:

```
>> x0=[4000; 350; 298];
>> [X,gVal]=fsolve(@(x) cstrFun(x,modelPar), x0);
```

This leads to the other solution, $\mathbf{x} = [3918.8$ mol/m^3; 349.8 K; 323.9 K$]$.

In fact, there exists a third solution as well, for which intermediate values are chosen as initial guess*:

```
>> x0 = [2000; 500; 400];
>> [X,gVal]=fsolve(@(x) cstrFun(x,modelPar), x0);
```

This leads to the third and final solution, $\mathbf{x} = [2450.2$ mol/m^3; 476.4 K; 387.2 K$]$.

In all the three cases, the value of gVal returned by fsolve was $\mathcal{O}(10^{-8})$, indicating that the function $\mathbf{g}(\mathbf{x}^\star)$ is converged successfully.

6.5.4 Recap: Chemostat

The chemostat example was covered in Section 6.4.2 (Example 6.9). It involved solving three equations in three unknowns simultaneously:

$$
\begin{aligned}
0 &= D(S_f - S) - r_g \\
0 &= -DX + r_g Y_{xs} \\
0 &= -DP + r_g Y_{ps} \\
r_g &= \mu^{max} \frac{SX}{K_s + S}
\end{aligned}
\tag{6.39}
$$

The code for solving this problem is given in Example 6.9.

Just like Section 6.5.3, multiple steady states can be shown for a chemostat as well. This is left as an exercise for the readers.

---

* At this stage, this is being presented as trial and error. The case study in Chapter 9 will provide specific details on how to obtain a reasonable initial guess close to the solution.

6.5.5 Integral Equations: Conversion from a PFR

Most of the problems in chemical engineering involve algebraic equations or differential equations. Another class of equations is known as integral equations. Ivar Fredholm extensively studied integral equations. The following is Fredholm integral equations of the first kind:

$$f(x) = \int_a^b K(x,s)\varphi(s)\,ds \tag{6.60}$$

Note that $f(x)$ is a function of x only, and not of s. A related equation is Volterra integral equation of the first kind:

$$f(x) = \int_a^x K(x,s)\varphi(s)\,ds \tag{6.61}$$

The main difference is that one of the limits of integration is also the variable x. The main crux of integral equations is that the left-hand side $f(x)$ is known. The function $K(x,s)$ is called kernel function and is also known.

Consider the design equation of a plug flow reactor (PFR) (see Chapter 3) with an inlet concentration of reactant as 1 mol/m³:

$$\frac{V}{Q} = -\int_1^{C_{out}} \frac{dC}{r(C)} \tag{6.62}$$

In Chapter 3, we integrated the above to find the volume of the PFR that achieves 75% conversion. In other words, Q and C_{out} were known, whereas V was unknown. This was therefore a straightforward *numerical integration* problem. We used an ODE solver in Chapter 3 when we asked the question, "find the conversion from a reactor of volume 0.125 m³." Let us now see an alternative way to pose the problem as *nonlinear integral equation* and solve the problem using techniques from this chapter.

If the volume is given, the right-hand side is known, whereas the upper limit of integration (C_{out}) is the unknown to be found. Thus, $x \equiv C_{out}$ in Equation 6.61. Rearranging the above equation

$$\underbrace{\frac{V}{Q} + \int_1^{C_{out}} \frac{dC}{r(C)}}_{g(C_{out})} = 0 \tag{6.63}$$

The aim of the nonlinear equation solver is to compute the C_{out} that is achieved in a PFR with residence time, $V/Q = 1.25$ s. We will do this for two cases: first-order reaction and a complex kinetic model from Chapter 3. In either case, the nonlinear equation (6.63) to be solved may be written as

$$g\left(C_{out}\right) \equiv \frac{V}{Q} + I\left(C_{out}\right) = 0 \tag{6.64}$$

In the first example, since the reaction is first-order, the integral $I(C_{out})$ can be solved analytically. The analytical and numerical ways of solving for first-order reaction will be considered in the next section for pedagogical reasons. Thereafter, complex kinetic expression that requires numerical solution is covered in Section 6.5.5.2. A more experienced reader may skip Section 6.5.5.1 and go directly to Section 6.5.5.2.

6.5.5.1 First-Order Kinetics

When the reaction is a first-order reaction ($r = kC$), the integral is given by

$$I\left(C_{out}\right) = \frac{1}{k}\left[\ln\left(C\right)\right]_{1}^{C_{out}} = \frac{1}{k}\ln\left(C_{out}\right) \tag{6.65}$$

When substituted in Equation 6.63, we get a nonlinear equation in one unknown (C_{out}), which can be solved using `fzero`, as shown in the next example.

Example 6.12 PFR Using Integral Equation Approach

Problem: Simulate the PFR using integral equation approach. The first-order reaction rate is given by $r = kC$, with $k = 0.5$ s^{-1}. The inlet concentration is 1 mol/m^3, PFR volume is 0.125 m^3, and flowrate is 0.1 m^3/s.

Solution (Part-1): We will solve this problem using `fzero`. From the residence time and rate constant, we know that the conversion lies somewhere between 1% and 99%. The parameters are specified in the driver script, `pfrIntegralSoln.m`, which uses `fzero`. The parameter definitions are similar to the ones we have seen in Chapter 3.

```
% Script for simulation of a PFR
% using integral equation approach
Acs=0.25;
L=0.5;
modelParam.k=0.5;
modelParam.Q=0.1;
modelParam.V=L*Acs;

%% Solve and obtain outlet concentration
C0= [0.01, 0.99];     % Initial guess for fzero
Csol=fzero(@(Cout) pfrIntFun(Cout,modelParam),C0);
```

```
X=(1-Csol)/1;        % Conversion
X=round(X*100,1);    % ... in percentage
disp(['Conversion = ',num2str(X),'%']);
```

We can compute the true value of concentration and error in numerical computation as

```
Ctrue = exp(-V*k/Q);
err=abs(Csol-Ctrue);
```

The above driver calls the function pfrIntFun.m, which calculates $g(C_{out})$ from Equation 6.64. This function is given below:

```
function gVal=pfrIntFun(Cout,par)
% To calculate g(Cout) for simulating a PFR
% using integral equation approach
% where, g = V/Q + I(Cout)
%         V/Q   is the residence time
%         I     is integral [dC/r] at given Cout

%% Calculating integral
intValue=log(Cout)/par.k;
%% Calculate g(Cout)
gVal=par.V/par.Q + intValue;
```

The conversion is found to be 46.5%. The error using the above approach is $\sim \mathcal{O}\left(10^{-16}\right)$.

One of the important features of numerical techniques is that the function ($g(C_{out})$ or $I(C_{out})$ in this example) does not need to be explicit; it can equivalently be implicitly calculated. The design problem was solved in Chapter 3, wherein the volume was obtained after calculating the integral using MATLAB function trapz.* Thus, the numerical solution that employs trapz can be used instead of the analytical solution. This would mean replacing the underlined sections of the function file pfrIntFun.m (in the code given above) with a numerical approach that uses trapz. This is shown in the example below.

Example 6.12 (Continued) PFR Using Integral Equation Approach

Solution (Part-2): The aim of the problem was to write a general code that computes $I(C_{out})$ by evaluating the integral *numerically* and use this to compute $g(C_{out})$. This is done by replacing the underlined lines of code with numerical integral. The driver

---

* See Appendix D for more information on the trapezoidal rule for integration. The MATLAB function trapz that implements the trapezoidal rule is also discussed in the appendix.

script, pfrIntegralSoln.m, remains unchanged. The actual pfrIntFun.m function that we will use is given below:

```
function gVal=pfrIntFun(Cout,par)
% To calculate g(Cout) for simulating a PFR
% using integral equation approach
% where, g = V/Q + I(Cout)
%          V/Q    is the residence time
%          I      is integral [dC/r] at given Cout
%% Calculating integral using trapz
n=50;              % number of intervals
h=(1-Cout)/n;      % step-size for integration
C=1:-h:Cout;
r=par.k * C;
intValue=trapz(C,1./r);
%% Calculate g(Cout)
gVal=par.V/par.Q + intValue;
```

Compare this modified code with the previous one to notice that the underlined section of the previous code (titled "Calculating integral") is replaced with the section titled "Calculating integral using trapz"). There is no other change. With this change, the error in computing C_{out} is expected to be higher, since it will be governed by the error in computing the integral. With $n = 50$ points used to compute the integral using trapezoidal rule, the following results were obtained:

```
Conversion = 46.5%
>> disp(err)
    9.5955e-06
```

It is also possible to check the effect of the number of intervals on the solution. When the number of intervals was increased by one order of magnitude to $n = 500$, the error in calculation of C_{out} decreased by two orders of magnitude. With $n = 500$

```
>> disp(err)
    9.5967e-08
```

This observation may be attributed to the fact that the global truncation error in trapezoidal rule* is $\mathcal{O}\left(h^2\right)$, where $h = (C_{in} - C_{out})/n$.

The primary take-home message from this example is that a numerical technique is agnostic to how $g(C_{out})$ was computed, either as a closed-form expression as per Equations 6.64 and 6.65, or whether integral required in $g(C_{out})$ is calculated numerically. Equipped with this, we proceed to solve the original problem of simulating a PFR with complex kinetics.

---

* See Appendix D for more information on the trapezoidal rule for integration. The MATLAB function trapz that implements the trapezoidal rule is also discussed in the appendix.

6.5.5.2 Complex Kinetics

We are now ready to solve the original problem, which was to find the variation in outlet concentration from a PFR as a function of length of the PFR, given the kinetic rate expression:

$$r = \frac{kC}{\sqrt{1+K_r C^2}} \tag{6.66}$$

$$\text{where } k = 2 \text{ s}^{-1}, \quad K_r = 1 \text{ mol}^{-2} \cdot \text{m}^6$$

The nonlinear equation to be solved, $(C_{out}) = 0$, is given by Equation 6.64, where the integral $I(C_{out})$ is computed numerically using `trapz`:

$$I(C_{out}) = \underbrace{\int_{1}^{C_{out}} \frac{dC}{r(C)}}_{\text{trapz}(C, 1./r)} \tag{6.67}$$

This is shown in the following example.

Example 6.13 PFR with LH Kinetics Using Integral Equation Approach

Problem: Modify Example 6.12 to solve the PFR with LH kinetics of Equation 6.66 to obtain outlet concentration as a function of PFR length.

Solution (Part-1): First, let us consider the problem of finding outlet concentration for 0.5 m length (volume = 0.125 m³). The driver script from Example 6.12 remains unchanged, except for providing the correct rate constants:

```
modelParam.k=2;
modelParam.Kr=1;
```

In the function file, the reaction rate term from first-order kinetics is modified for the rate expression (6.66), as was covered in Chapter 3. The modified code is given below, with the changes underlined:

```
function gVal=pfrIntFunLH(Cout,par)
% This function calculates g(Cout) for simulating
% a PFR using integral equation approach
% where, g = V/Q + I(Cout)
%          V/Q    is the residence time
%          I      is integral [dC/r] at given Cout
k=par.k;
Kr=par.Kr;
```

```
%% Calculating integral using trapz
n=500;              % number of intervals
h=(1-Cout)/n;       % step-size for integration
C=1:-h:Cout;
r=k*C ./ sqrt(1+Kr*C);
intValue=trapz(C,1./r);
%% Calculate g(Cout)
gVal=par.V/par.Q + intValue;
```

The solution obtained using the above code is same as that obtained in Chapter 3:

```
Conversion = 87.8%
```

Now that the above code gives the appropriate conversion from PFR, we now need to modify the driver script. We will solve the integral equation problem for 10 different lengths (L=0:0.05:5), store the solution in a vector (XOUT in the following script), and plot the conversion vs. length. The resulting script is given below.

Example 6.12 (Continued) PFR Using Integral Equation Approach

```
% Obtaining conversion vs. length of PFR
% using integral equation approach
Acs=0.25;
L=0:0.05:0.5;
modelParam.k=2;
modelParam.Kr=1;
modelParam.Q=0.1;

%% Solve for various reactor lengths
XOUT=zeros(size(L)); % For storing results
C0=[0.01, 0.99]; % Initial guess for fzero
for i=2:length(L)
    modelParam.V=L(i)*Acs;
    Csol=fzero(@(Cout) pfrIntFunLH(Cout,modelParam),C0);
    X=(1-Csol)/1;          % Conversion
    X=round(X*100,1);      % ... in percentage
    XOUT(i)=X;
end
plot(L,XOUT,'-bo');
xlabel('PFR length (m)'); ylabel('% conversion');
```

The results are shown in Figure 6.7. These results are the same as those obtained for the same problem using the ODE-solving approach in Chapter 3. This demonstrates the validity of the integral equation solving approach shown here.

I will end this section with some perspective about this problem. Simulating the behavior of reacting species in a PFR is more conveniently handled using ODE solution techniques.

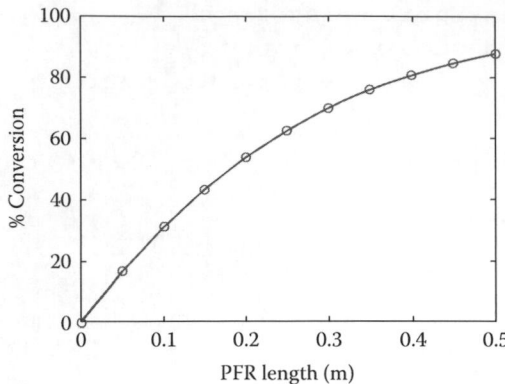

FIGURE 6.7   Conversion vs. length of PFR for LH kinetics using integral equation solving approach.

The intent of this case study was to demonstrate alternative routes to solving the same problem. However, more importantly, it was to introduce readers to the fact that nonlinear equation solving techniques are more general than what we would typically categorize as "root finding"; these methods can be used for problems that can be cast into the form $g(x) = 0$, where one needs to find x that satisfies these equations. The nature of $g(\cdot)$ need not be explicit; it could be implicitly specified or obtained as a solution using another numerical procedure. Finally, this case study also introduces the reader to integral equations. Although uncommon in chemical engineering, integral equations sometimes arise in some problems involving radiation.

6.6 EPILOGUE

Solving ODEs and nonlinear algebraic equations form two of the most important classes of examples of interest to chemical engineers. While Chapter 3 focused on ODEs, this chapter focused on solving nonlinear equations of the form

$$g(x) = 0 \tag{6.1}$$

I used the same approach in this chapter as in Chapter 3. An example of finding the molar volume of a real gas using EOS was used to motivate the problem. This is an example of a single-variable nonlinear equation solving problem.

Thereafter, several methods for solving algebraic equations were discussed in Section 6.2. Starting with single-variable problem, theoretical results for selected numerical methods were discussed. Rather than focusing on derivations, the aim of this chapter was to highlight the practical implications in solving problems of interest. Numerical methods for solving general algebraic equations of the type (6.1) fall into two categories: bracketing methods and open methods. In the case of the former, two initial guesses always "bracket" a solution, whereas this requirement is not applicable for open methods.

MATLAB provides several algorithms for solving nonlinear equations. The two more popular and versatile algorithms provided in MATLAB are `fzero` and `fsolve`. The former is a single-variable solver that preferably uses a bracketing method, whereas the latter is a multivariable general purpose solver. The single-variable bracketing methods were introduced in Section 6.2 to help readers make informed choices while using `fzero` algorithm. Newton-Raphson and its variants, which are the most popular nonlinear equation solving algorithms, were discussed in some detail. This highlights features and limitations of nonlinear equation solving algorithms and several variations used to address those limitations.

Solving coupled nonlinear equations for large-dimensional problems can sometimes be a tough problem to tackle. A few practical guidelines that I personally follow for MATLAB are summarized below.

(A) *Single-variable problem*: `fzero` is my preferred algorithm for single-variable problems. It is more robust when two permissible initial guesses are provided, although it works with a single initial guess as well.

The multivariable problems can be tougher, with `fsolve` being the primary algorithm geared toward solving nonlinear equations.

(B) *Choosing appropriate step tolerances*: Convergence metrics is an important point that needs to be stressed in this summary. The importance of using solver options was introduced in Section 6.4. Convergence is verified using tolerance on the step

$$\left| \mathbf{x}^{(i+1)} - \mathbf{x}^{(i)} \right| < \varepsilon \tag{6.48}$$

or the change in function values

$$\left| \mathbf{f}\left(\mathbf{x}^{(i+1)}\right) - \mathbf{f}\left(\mathbf{x}^{(i)}\right) \right| < \delta \tag{6.49}$$

The default values for both the tolerances in MATLAB are 10^{-6}. In the EOS example, since the molar volume was $\sim 1.6 \times 10^{-4}$, a more stringent stopping criterion for ε (`tolX` optimizer option in MATLAB) was used. Similar care is required with respect to the stopping criterion δ as well (`tolFun` optimizer option in MATLAB).

(C) *Choosing appropriate function tolerances*: Consider the following:

```
>> fsolve(@(x) (x-1)^3, 2)
Equation solved.
<...>
ans =
    1.0390
```

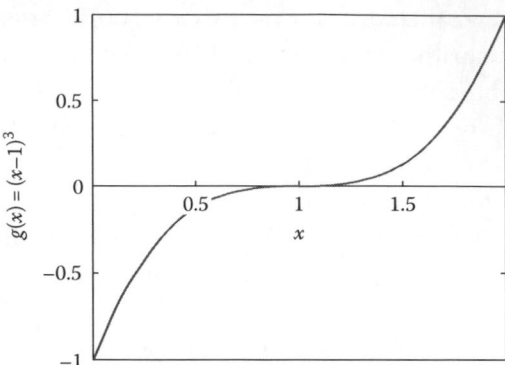

FIGURE 6.8 $g(x)$ vs. x for $g(x) = (x-1)^3$.

where a rather "simple" problem of solving $g(x) \equiv (x-1)^3 = 0$ gives an unexpected result. The solver has converged because the default function tolerance of 10^{-6} is met, even though there is ~3.9% error in the x^* value. The reason for this behavior can be traced to the nature of $g(x)$ plotted in Figure 6.8. Since the function value changes slowly near the solution, gradient-based methods such as Newton-Raphson or fsolve face issues close to the solution.

Two options* can be used to address the issue of this premature convergence: (i) specifying a more stringent function tolerance and/or (ii) trying to solve $\beta g(x) = 0$ instead. The optimization option

```
>> opt=optimset('tolFun',1e-10);
```

will set a more stringent error tolerance, whereas

```
>> fsolve(@(x) 1e6*(x-1)^3, 2, opt)
```

finally gives the expected result.

(D) *Verifying if converged solution is "reasonable"*: The next question that comes up is how do we know if the convergence is reasonable for a general function $\mathbf{g}(\mathbf{x})$? A reasonable rule of thumb is that we usually have some idea of the *expected* solution value, which is often provided as an initial guess, $\mathbf{x}^{guess}$. One can evaluate $\mathbf{g}(\mathbf{x}^{guess})$ and choose a value for $\delta \sim \mathcal{O}\left(10^{-5} \left| \mathbf{g}\left(\mathbf{x}^{guess}\right) \right|\right)$ or lower. While this provides a reasonable starting point, it is not guaranteed to work. For example, at the solution $x=1.039$, the value of $g(x) = 5.94 \times 10^{-5}$ seems reasonable. However, when the solver is executed with tolFun=1e-8, the solution changes to x=1.017; further reducing tolFun=1e-10, the solution changes further to x=1.0077. Although the value of

---

* Several users implicitly do not trust fsolve for its "inability" to solve such apparently simple problems. A problem that is easy to solve for humans may not be easy for numerical technique. Problems of the nature $g(x) = (x-1)^n$ or $g(x) = x^n - 1$ are *textbook* examples for exercising caution while using gradient-based solvers (Newton-Raphson, secant, or fsolve). This book is intended to equip readers to understand why the method appears to fail and how to make it work.

$g(x)$ computed was small, experimenting with different values of tolFun indicates whether solution returned by fsolve is reasonable.*

(E) *When solver does not converge*: Solver may not converge even after attempting to solve the problem with several different initial guesses. In such a case, the problem may be reformulated. For example, material balance equation in CSTR

$$0 = F\left(C_{A0} - C_A\right) - \left[k_0 \exp\left(-\frac{E}{RT}\right)C_A\right]V \tag{6.58}$$

may be rearranged as

$$0 = 1 - k_0 \tau \exp\left(-\frac{E}{RT}\right)\left(\frac{C_A}{C_{A0} - C_A}\right) \tag{6.68}$$

Alternatively, the model may be non-dimensionalized by defining $\xi = C_A/C_{ref}$ and $\theta = T/T_{ref}$. Converting variables into non-dimensional form and reformulating the balance equations is a popular method to attempt, if the original function shows difficulty converging.

(F) *Writing one's own code when all else fails*: When all else fails, writing one's own modified Newton-Raphson code (see Section 6.3.3) can be attempted. Alternatively, or additionally, analytically computed Jacobian may be provided to the solver. Analytical Jacobian can sometimes prove useful in convergence of system of equations that are otherwise difficult to converge. When I have chosen to write my own code, I have used an appropriate under-relaxation factor in the range $1/2$ to $1/16$.

This brings us to the end of this chapter, where we focused on solving a *generic* system of equations. Discretization of ODE-BVPs (boundary value problems) or partial differential equations (PDEs) result in a set of nonlinear equations with a specific structure. Methods that exploit the special structure and iterative methods that build upon the concept of under-relaxation will be the focus of Chapter 7.

EXERCISES

Problem 6.1 Peng-Robinson Equation of State

$$P = \frac{RT}{V-b} - \frac{a}{V(V+b) + b(V-b)} \tag{6.50}$$

is another EOS that overcomes some of the limitations of RK-EOS discussed in this chapter. It is especially accurate in predicting properties of

---

* Changing the tolFun will only affect the results when the function tolerance criterion causes fsolve to converge (i.e., change in the function value is less than the function tolerance). If instead fsolve convergence is due to other reasons, it is sufficient to verify if the function value at the solution, $g(x^*)$ is lower than a desired threshold.

nonpolar molecules and is able to predict liquid-phase properties accurately as well. Use the PR EOS to obtain the molar volume for given $T = 340$ K and $P = 100$ bar. You may choose the parameter values of $a = 0.364$ and $b = 3 \times 10^{-5}$.

Problem 6.2 Use fixed point iteration to solve the problem in Example 6.3 and compare with the results discussed in Section 6.2.3.

Problem 6.3 Wegstein's method is a numerical method used by several commercial process simulation packages such as Aspen Plus. Wegstein's method is an enhancement of the fixed point Iteration to solve problems of the form

$$x = G(x) \tag{6.18}$$

Starting with the initial condition, $x^{(0)}$, the next iteration value is obtained as $x^{(1)} = G(x^{(0)})$. Thereafter, subsequent iteration values are obtained as

$$x^{(i+1)} = qx^{(i)} + (1-q)G\left(x^{(i)}\right)$$
$$q = \frac{S}{S-1}, \quad S = \frac{G\left(x^{(i)}\right) - G\left(x^{(i-1)}\right)}{x^{(i)} - x^{(i-1)}} \tag{6.69}$$

Use Wegstein's method to solve the problem in Example 6.3.

Problem 6.4 Consider Equation 6.69 of Wegstein's method.

Easy: Show that when q is a constant, Equation 6.69 is fixed point iteration with relaxation.

Medium: When S is replaced by the slope, dG/dx, show that Wegstein's method in one variable is equivalent to Newton-Raphson for $g(x) = G(x) - x$.

Problem 6.5 The terminal settling velocity of a particle is given by the following force balance:

$$\frac{1}{2}\rho v^2 C_D = \frac{2}{3}d_p\left(\rho_p - \rho\right)g$$

where the drag coefficient is given by

$$C_D = \begin{cases} \dfrac{24}{Re} & \text{if } Re \leq 1 \\[2mm] \dfrac{24}{Re} + \dfrac{4}{Re^{0.33}} & \text{if } 1 < Re \leq 1000 \\[2mm] 0.44 & \text{if } 1000 < Re \end{cases}$$

Compute the terminal settling velocity of a spherical particle of diameter 2 mm falling through water ($\rho = 1000$ kg/m³, $\mu = 8 \times 10^{-4}$ kg/m·s). The density of the particle is $\rho_p = 2400$ kg/m³.

CHAPTER 7

Special Methods for Linear and Nonlinear Equations

7.1 GENERAL SETUP

Numerical methods for solving algebraic equations were covered in Chapter 6, wherein a generic problem of finding root(s) of the following set of equations was discussed:

$$\mathbf{g}(\mathbf{x}) = 0 \qquad (7.1)$$

where

$\mathbf{x} \in \mathcal{R}^n$ is an n-dimensional column vector

$\mathbf{g}: \mathcal{R}^n \to \mathcal{R}^n$ is an n-dimensional vector function

A specific case is when the above system of equations is linear. One could then write $g = Ax - b$, and the equations to be solved are a set of linear equations:

$$Ax = \mathbf{b} \qquad (7.2)$$

with

$A \in \mathcal{R}^{n \times n}$ is a matrix

$\mathbf{b} \in \mathcal{R}^n$ is a column vector

The standard methods for solving linear equations that are based on Gauss Elimination are introduced in Appendix C. A significant amount of time is devoted to Gauss Elimination and related methods in typical numerical techniques books. Since MATLAB® provides powerful tools for solving linear equations, we are simply going to use them in this book.

The methods for nonlinear equations in Chapter 6 and linear equations in Appendix C are general-purpose methods. These methods do not make any assumption about the structure of the function $\mathbf{g}(\mathbf{x})$ or the matrix A, respectively. However, in a large number of engineering problems of interest, $\mathbf{g(x)}$ and/or A have a particular *sparse* structure. These structures of the functions/matrix can be exploited to solve the problems efficiently. This chapter focuses on problems where $\mathbf{g(x)}$ and/or A have *banded* structures.

7.1.1 Ordinary Differential Equation–Boundary Value Problems

We will specifically look at examples of ordinary differential equation–boundary value problems (ODE-BVPs) in this chapter and solve them using a finite difference approach. A general second-order ODE-BVP is given by

$$\alpha \frac{d^2\phi}{dx^2} = f\left(x, \phi, \frac{d\phi}{dx}\right) \tag{7.3}$$

Two conditions are required for solving the above problem. If the two conditions are provided at the same initial location, the above can be converted into a set of two first-order ODE-IVP. However, there are several occasions when the conditions for the above are specified at two different boundaries. This leads to ODE-BVP.

The boundary conditions are of three types:

1. *Dirichlet boundary condition*, where the value of variable ϕ is specified at a location:

$$\phi(x_0) = a \tag{7.4}$$

2. *Neumann boundary condition*, where the derivative ϕ' is specified at a location:

$$\phi'(x_0) = b \tag{7.5}$$

3. *Mixed boundary condition*:

$$\phi'(x_0) = \alpha\phi(x_0) + \beta \tag{7.6}$$

Either of these boundary conditions is applied at each end of the computational domain.

7.1.2 Elliptic PDEs

Another set of problems that have qualitatively similar numerical features are elliptic PDEs, which were introduced in Chapter 4. Systems that are governed by elliptic PDEs are those where diffusive behavior is important, and therefore, conditions at all the boundaries affect the solution value at any point within the domain. Owing to these common physical features, a similar approach to solving and analyzing these systems can be used. A typical elliptic PDE in 2D is given by

$$\Gamma_x \frac{\partial^2\phi}{\partial x^2} + \Gamma_y \frac{\partial^2\phi}{\partial y^2} = f\left(x, y, \phi, \frac{\partial\phi}{\partial y}, \frac{\partial\phi}{\partial y}\right) \tag{7.7}$$

The above PDE is solved subject to two boundary conditions in each direction. As in the case of ODE-BVP, these may be Dirichlet (Equation 7.4), Neumann (Equation 7.5), or mixed (Equation 7.6) boundary conditions.

If convection dominates in one of the directions (e.g., $\Gamma_x \approx 0$), the problem reduces to parabolic PDE; if diffusion is negligible in both directions, we get first-order hyperbolic PDEs. Methods for solving hyperbolic and parabolic PDEs were discussed in Chapter 4.

7.1.3 Outlook of This Chapter

Before starting with the numerical techniques, I would like to draw the attention of the readers to one difference between handling of sparse and banded systems in this book. In standard numerical techniques textbook, the tridiagonal methods (Section 7.2) are discussed in the Linear Equations section immediately after Gauss Elimination, almost exclusively in the context of linear equations. Iterative methods for solving linear equations (Section 7.3), such as Gauss-Siedel, are traditionally discussed independently of iterative nonlinear equations methods. Since this book takes a practical outlook toward solving engineering problems, banded methods and iterative methods are clubbed together in this "Special Methods" chapter.

The next section introduces the reader to problems involving such banded structures, followed by so-called direct methods to solve the problems. Direct methods are related to Gauss Elimination (see Appendix C), but where the banded structure of matrix A is exploited to solve the linear problem (7.2) more efficiently. Thereafter, Section 7.3 discusses iterative methods for solving such problems, where both linear as well as nonlinear equations of the form (7.1) will be discussed. Finally, Section 7.4 expands the linear banded diagonal methods to specific examples of nonlinear equations. Finally, case study examples are discussed in Section 7.5. Advanced readers interested in learning MATLAB-based problem solving for chemical engineers may directly skip to this section for examples.

7.2 TRIDIAGONAL AND BANDED SYSTEMS

7.2.1 What Is a Banded System?

Consider the linear equation (7.2). The matrix A is said to be sparse if only a small fraction of its elements $A_{i,j}$ are nonzero, whereas the rest of them are zero. Furthermore, it is said to be banded if these nonzero elements occur "close" to the diagonal, that is, $A_{i,i\pm\ell}$ are nonzero. In this case, ℓ is the bandwidth of the system.* Let us take a closer look at what this means for ith equation:

$$A_{i,i-\ell}x_{i-\ell} + \cdots + A_{i,i}x_i + \cdots + A_{i,i+\ell}x_{i+\ell} = b_i \tag{7.8}$$

Thus, the elements in the ith equation depend only on the previous and subsequent ℓ elements. Extending this idea to nonlinear equations, the ith function $g_i(\mathbf{x})$ is such that it depends only on $(x_{i-\ell}, \ldots, x_{i+\ell})$, and it does not depend on $(x_1, \ldots, x_{i-\ell-1})$ or $(x_{i+\ell+1}, \ldots, x_n)$. Alternatively, the Jacobian matrix, at any $\mathbf{x} \in R^n$, has the sparse structure with bandwidth ℓ.

---

* The number of subdiagonal and superdiagonal elements need not be equal. I have taken the two to be equal for ease of discussion. The band could span $i - \ell_{\text{sub}}$ to $i + \ell_{\text{super}}$ for ith row.

7.2.1.1 Tridiagonal Matrix

A tridiagonal matrix is a matrix with $\ell = 1$, that is, a matrix where only the diagonal element, one subdiagonal element, and one superdiagonal element are nonzero. An example of a tridiagonal system was seen in Chapter 4, when we implemented method of lines, which used the finite difference formula on the spatial derivative $\partial/\partial z$ and $\partial^2/\partial z^2$.

Sometimes, it becomes important to view the structure of a banded/sparse matrix while solving or debugging a problem. MATLAB provides a command, spy(A), that can be used to visually inspect the structure of a matrix. This generates a plot with n rows and columns, with the nonzero elements of A indicated with markers on the plot. I will use this command to investigate some of the common banded structures we are likely to encounter in chemical engineering problems.

Figure 7.1 shows some common sparse structures. Tridiagonal and banded structures were discussed earlier. For the specific band-diagonal example in the figure, $\ell = 3$. The third common example is that of block-diagonal structure. This is quite similar to the band-diagonal structure but with some additional elements zero as well. A closer look at Figure 7.1 (bottom left) reveals that the overall structure contains a single block that is repeated four times. The reason it is called block-diagonal is because one can imagine a 4×4 block, each block containing a matrix. Only the diagonal blocks are nonzero, whereas the others are a matrix of zeros. These repeating diagonal blocks have full or sparse structures. Finally, a multibanded structure is self-explanatory in that it contains additional nonzero bands, as seen in Figure 7.1.

7.2.2 Thomas Algorithm a.k.a TDMA

I will follow the same pattern in this chapter as well: Rather than a general description of the Thomas algorithm, or the tridiagonal matrix algorithm (TDMA), we will first

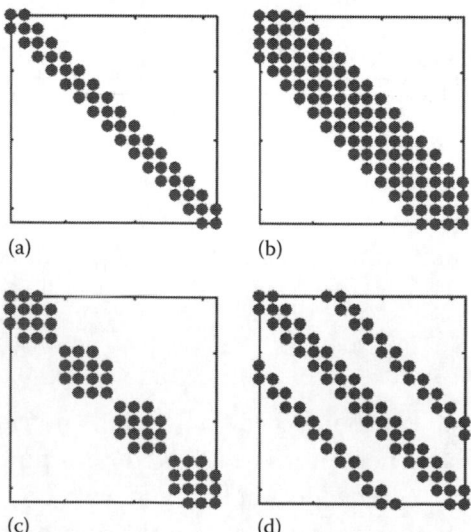

(a)   (b)

(c)   (d)

FIGURE 7.1 Some of the common sparse structures, examined by MATLAB® spy command. (a) Tridiagonal, (b) banded, (c) block-diagonal, and (d) multiband.

discuss an example and use it to set up the problem. Thereafter, the TDMA will be discussed in the context of this problem.

7.2.2.1 Heat Conduction Problem

Heat conduction in a 1D structure is a commonly used example in various disciplines of engineering and physics. Metallic fin or mesh structures are routinely used for improved heat dissipation. You would have seen such structures in car radiators, old refrigerators,* CPU fans with heat sinks in a computer, etc. For the sake of simplicity, let us model a single rod in this assembly. The idealized model consists of a single metal, which has its one end connected to the hot equipment and the other end exposed to atmosphere such that it dissipates heat to the surroundings. This is schematically shown in Figure 7.2.

The heat transfer with heat loss in this rod is modeled as

$$k\frac{d^2T}{dz^2} = h_\infty a_v (T - T_\infty) \tag{7.9}$$

$$T(0) = T_0, \quad T(L) = T_\infty \tag{7.10}$$

where
  T is the temperature at any location in °C
  T_∞ is the temperature of the ambient
  k is the thermal conductivity of the rod
  h_∞ is the heat loss coefficient
  a_v is the external area of the rod per unit volume

For a cylindrical rod, $a_v = 2/r$. The above problem is an ODE-BVP, since the conditions are specified at two separate boundary points, $z = 0$ and $z = L$. Before moving to the numerical solution, let us discuss the true solution that can be obtained analytically since the above is a linear ODE.

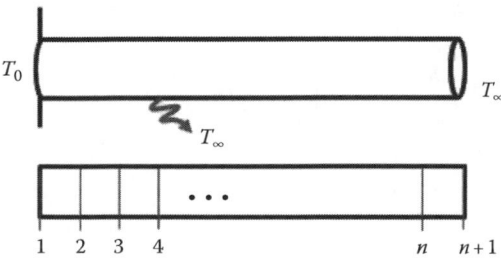

FIGURE 7.2   Heat conduction in a metal rod with heat loss to the surroundings and the domain with finite difference discretization.

---

* The modern refrigerators use evaporative cooling, the model for which would be somewhat more complex than what we
   need in this chapter.

7.2.2.1.1 *Analytical Solution* The above problem can be solved analytically by defining a deviation variable, $\theta = (T - T_\infty)$, which results in the following ODE-BVP:

$$\frac{d^2\theta}{dz^2} - \underbrace{\frac{h_\infty a_v}{k}}_{m^2}\theta = 0$$

$$\theta(0) = \theta_0, \quad \theta(L) = 0$$

(7.11)

The above is a linear homogeneous second-order ODE, which is a standard problem in an undergraduate calculus course. The general solution to the above ODE is*

$$\theta = C_1 \sinh(mz) + C_2 \cosh(mz)$$

(7.12)

The solution to the ODE-BVP of Equation 7.11 is given by

$$\theta = \theta_0 \frac{\sinh(m(L-z))}{\sinh(mL)}$$

(7.13)

which can be expressed in terms of the original variables:

$$T = T_\infty + (T_0 - T_\infty)\frac{\sinh(m(L-z))}{\sinh(mL)}$$

(7.14)

The numerical solutions can be compared against the above analytical solution of the ODE-BVP. In order to solve the original problem (7.9) numerically, the domain is discretized into n-intervals using the finite difference approach. Figure 7.2 shows the discretized domain. Since the array indices in MATLAB start with 1, the vertices of the discretized domain are numbered from $i = 1$ to $i = (n+1)$. The interval size is given by $\Delta = L/n$, and application of the central difference formula (see Appendix B) results in the following equations:

$$k\frac{T_{i+1} - 2T_i + T_{i-1}}{\Delta^2} = h_\infty a_v(T_i - T_\infty), \quad i = 2 \text{ to } n$$

(7.15)

The two boundary conditions are applied at $i = 1$ and $i = (n+1)$. Defining parameter $\beta = h_\infty a_v \Delta^2 / k$, the resulting $(n+1)$ linear equations are given by

$$T_1 = T_0$$
$$T_{i+1} - 2T_i + T_{i-1} = \beta(T_i - T_\infty), \quad i = 2 \text{ to } n$$
$$T_{n+1} = T_\infty$$

(7.16)

---

* Please refer to any standard engineering math text for the derivation of the solution for this and related problems.

7.2.2.1.2 *Tridiagonal Matrix Equation* The above set of equations can be written in the standard linear equations form by defining the solution vector $\mathbf{x} = [T_1\ T_2\ \cdots\ T_n\ T_{n+1}]^T$:

$$
\underbrace{\begin{bmatrix}
1 & 0 & 0 & \cdots & \cdots & 0 & 0 \\
1 & -(2+\beta) & 1 & \cdots & \cdots & 0 & 0 \\
0 & 1 & -(2+\beta) & \cdots & \cdots & 0 & 0 \\
\vdots & \vdots & \ddots & \ddots & \ddots & \vdots & \vdots \\
0 & 0 & 0 & \cdots & 1 & -(2+\beta) & 1 \\
0 & 0 & 0 & \cdots & 0 & 0 & 1
\end{bmatrix}}_{A}
\underbrace{\begin{bmatrix}
T_1 \\ T_2 \\ T_3 \\ \vdots \\ T_n \\ T_{n+1}
\end{bmatrix}}_{\mathbf{x}}
=
\underbrace{\begin{bmatrix}
T_0 \\ -\beta T_\infty \\ -\beta T_\infty \\ \vdots \\ -\beta T_\infty \\ T_\infty
\end{bmatrix}}_{\mathbf{b}}
\tag{7.17}
$$

The above problem can be solved by specifying the matrix A and vector $\mathbf{b}$ and solving the resulting equations in a standard manner: $X = A \backslash b$. This is demonstrated in the next example.

Example 7.1 Conduction with Known Boundary Temperatures: Standard Solution

Problem: Heat conduction in a steel rod is given by Equation 7.9. The rod, 2.5 m long, is maintained at 100°C and 30°C at the two ends. The thermal conductivity of steel is 15 W/m·K, heat loss coefficient is 2 W/m²·K, area per unit volume is 10 m⁻¹, and the ambient temperature is 30°C. Convert the above ODE-BVP into a set of linear equations and solve them using MATLAB linear equation solver.

Solution: The first step is to define the matrix A and vector b, as given in Equation 7.17, and hence solve the original ODE-BVP. The MATLAB script is given below:

```
% Heat conduction in a rod with
% heat loss to the surroundings
%% Model Parameters
hinf=2.0;    abyV=10;
ks=15;       L=2.5;
Ta=30;       % Ambient T
T_0=100;     % T(z=0)
T_L=Ta;      % T(z=L)
%% Discretization & Solution
n=40;
delta=L/n;
beta=hinf*abyV/ks * delta^2;
Z=[0:delta:L]';
m=sqrt(hinf*abyV/ks);
Ttrue=Ta+(T_0-Ta)*sinh(m*(L-Z))/sinh(m*L);

%% Numerical solution and comparison
A=zeros(n+1,n+1);
b=zeros(n+1,1);
```

```
A(1,1)=1;           b(1)=T_0;
for i=2:n
    A(i,i-1)=1;
    A(i,i)=-(2+beta);
    A(i,i+1)=1;
    b(i)=-Ta*beta;
end
A(n+1,n+1)=1;     b(n+1)=T_L;
X=A\b;
plot(Z,Ttrue,'-k',Z,X,'-.m');
err=max(abs(Ttrue-X)./Ttrue);
```

The results are plotted in Figure 7.3. The solid (black) line is the true solution, whereas the two lines are the solutions for $n = 5$ and $n = 40$ nodes, respectively. The latter solution closely matches the true solution. The solution with $n = 5$ nodes is also of reasonable accuracy.

The relative errors in temperature values were also calculated. Starting with $n = 5$, the number of nodes was doubled until reasonable solution was obtained. The errors were

```
n       err
5       0.0060
10      0.0015
20      3.8759e-04
40      9.7041e-05
```

The solution with $n = 20$ and $n = 40$ nodes are both reasonably accurate. It may also be of interest to the reader to note that each time the step-size is halved (i.e., the number of nodes is doubled), the error decreases by a factor of 4. This is because the $\mathcal{O}(h^2)$ accurate central difference formula governs the overall errors in this problem.

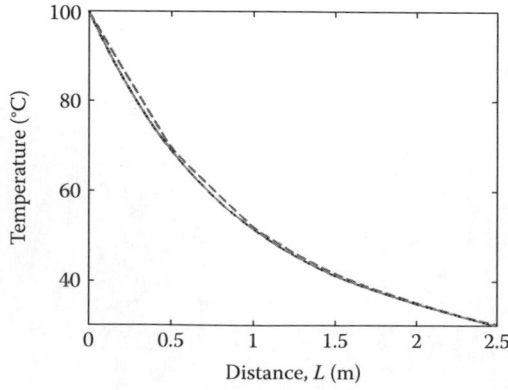

FIGURE 7.3 Temperature profile in a rod with heat conduction and heat losses. Dirichlet boundary conditions are applied, that is, the temperatures at the two ends of the rod are specified.

The standard method of solving a general linear equation was used in the previous example. This method works, and for a small-dimensional system as the one in Example 7.1, the computational requirements are manageable. However, for larger-dimensional systems and in cases where the problem needs to be solved multiple times, solving the full problem is very inefficient. There are several methods that exploit the banded structure of matrix A to solve the linear equation more efficiently. The TDMA, which is applicable to tridiagonal systems, will be discussed in the following text.

7.2.2.2 Thomas Algorithm

Thomas algorithm, or TDMA, can be used to efficiently solve a linear system of equations of the type (7.2), where A has a tridiagonal structure. A tridiagonal structure is when only three diagonals of matrix A are nonzero:

$$
\begin{bmatrix}
d_1 & u_1 & 0 & \cdots & 0 & 0 \\
l_2 & d_2 & u_2 & \cdots & 0 & 0 \\
0 & l_3 & d_3 & \cdots & 0 & 0 \\
\vdots & \vdots & \ddots & \ddots & \vdots & \vdots \\
0 & 0 & \cdots & \cdots & d_{n-1} & u_{n-1} \\
0 & 0 & \cdots & \cdots & l_n & d_n
\end{bmatrix}
\mathbf{x} =
\begin{bmatrix}
b_1 \\
b_2 \\
b_3 \\
\vdots \\
b_{n-1} \\
b_n
\end{bmatrix}
\tag{7.18}
$$

Let the three diagonals be represented as vectors $\mathbf{l}, \mathbf{d}, \mathbf{u}$. Thus, any equation may be written as

$$
l_i x_{i-1} + d_i x_i + u_i x_{i+1} = b_i
\tag{7.19}
$$

As in the case of Gauss Elimination (see Appendix C), the TDMA aims to convert matrix A into a lower triangular matrix using row operations, starting with the first row:

$$
d_1 x_1 + u_1 x_2 = b_1
\tag{7.20}
$$

In the first step, this is the pivot row and d_1 is the pivot element. The following two-row operations are done: (i) divide the entire row with the pivot element, and (ii) use the pivot row to make subdiagonal element in the first column 0. These operations yield

$$
x_1 + \underbrace{u_1/d_1}_{\bar{u}_1} x_2 = \underbrace{b_1/d_1}_{\bar{b}_1}
\tag{7.21}
$$

The first row operation requires only two computations on the first row: $\bar{u}_1 = u_1/d_1$ and $\bar{b}_1 = b_1/d_1$, and the diagonal element is then set to unity, $\bar{d}_1 = 1$. The only nonzero element in the pivot column is the element l_2, while all other elements are already zero. Thus, the row operation with the row R_1 as the pivot row is $\bar{R}_2 = R_2 - l_2 R_1$. The only computations required in this row are $\bar{d}_2 = d_2 - l_2 \bar{u}_1$ and $\bar{b}_2 = b_2 - l_2 \bar{b}_1$. Thus, at the end of the first sequence of operations, we have the first two rows as

$$
\left[\begin{array}{ccccccc|c}
1 & \bar{u}_1 & 0 & 0 & \cdots & 0 & \bar{b}_1 \\
0 & \bar{d}_2 & u_2 & 0 & \cdots & 0 & \bar{b}_2 \\
\vdots & \vdots & \vdots & \vdots & \vdots & \vdots & \vdots
\end{array}\right]
\tag{7.22}
$$

and the remaining are unchanged.

In the second step, row R_2 becomes the pivot row, with $\bar{d}_2$ as the pivot element:

$$
\bar{d}_2 x_2 + \bar{u}_2 x_2 = \bar{b}_2
\tag{7.23}
$$

This modified row has the exact same form as Equation 7.20. Hence, the exact same procedures are to be repeated. Generalizing, the operations involved in the ith step are

$$
R_i \leftarrow \frac{R_i}{d_i}, \quad R_{i+1} \leftarrow R_{i+1} - l_{i+1} R_i
\tag{7.24}
$$

I have dropped the over-bars and replaced the equality with an assignment operator "$\leftarrow$" to indicate that the right-hand side replaces the values on the left-hand side. Recall that the above two operations have two calculations each:

$$
u_i \leftarrow u_i/d_i, \quad b_i \leftarrow b_i/d_i, \quad d_i \leftarrow 1
\tag{7.25}
$$

$$
d_{i+1} \leftarrow d_{i+1} - l_{i+1} u_i, \quad b_{i+1} \leftarrow b_{i+1} - l_{i+1} b_i, \quad l_{i+1} \leftarrow 0
\tag{7.26}
$$

The above calculations are repeated for all pivot rows, that is, $i = 1$ to $(n-1)$.

The number of mathematical operations required for TDMA is as follows: Equation 7.25 requires two division operations (the third one is simply an assignment since the result $d_i \leftarrow 1$ is known a priori), whereas Equation 7.26 requires two multiplication and two subtraction operations (again $l_{i+1} \leftarrow 0$ is known a priori). Thus, the total number of operations in the forward elimination steps is $4(n-1)$ multiplication/division and $2(n-1)$ addition/subtraction operations. This is significantly lower than $\mathcal{O}(n^3)$ operations required in Gauss Elimination.

At the end of elimination operations, the equations are converted to the following form:

$$
\left[\begin{array}{cccccc}
1 & \bar{u}_1 & 0 & \cdots & 0 & 0 \\
0 & 1 & \bar{u}_2 & \cdots & 0 & 0 \\
0 & 0 & 1 & \cdots & 0 & 0 \\
\vdots & \vdots & \ddots & \ddots & \vdots & \vdots \\
0 & 0 & \cdots & \cdots & 1 & \bar{u}_{n-1} \\
0 & 0 & \cdots & \cdots & 0 & d_n
\end{array}\right] \mathbf{x} =
\left[\begin{array}{c}
\bar{b}_1 \\
\bar{b}_2 \\
\bar{b}_3 \\
\vdots \\
\bar{b}_{n-1} \\
\bar{b}_n
\end{array}\right]
\tag{7.27}
$$

Now that the equations are in lower triangular form, the solutions can be obtained using back-substitution. Starting for the last row, the back-substitution method (again see Appendix C and compare with full back-substitution that follows Gauss Elimination) first provides the solution x_n:

$$x_n = \bar{b}_n / \bar{d}_n \tag{7.28}$$

The next equation is

$$x_{n-1} + \bar{u}_{n-1} x_n = \bar{b}_{n-1} \tag{7.29}$$

Also notice that all the rows, starting from the second-last row, have the same structure. Thus, the following back-substitution steps are used to obtain the solution:

$$x_i = \bar{b}_i - \bar{u}_i x_{i+1}, \quad i = (n-1) \text{ to } 1 \tag{7.30}$$

Thus, the overall back-substitution steps require n multiplication/division steps and $(n-1)$ addition/subtraction steps. The next example shows implementation of TDMA.

Example 7.2 Heat Conduction Using TDMA

Problem: Solve the problem in Example 7.1 using TDMA.

Solution: The driver script in Example 7.1 needs to be suitably modified to extract the vectors $\mathbf{l}, \mathbf{d}, \mathbf{u}$. This is left as an exercise to the reader. Thereafter, the following TDMA function, thomasSolve.m, is used to obtain the solution:

```
function x=thomasSolve(l,d,u,b)
% Implementation of Thomas algorithm
% BACKGROUND
% ----------
% Solves the equation of the form
%         Ax = p
% Where A is a tri-diagonal matrix:
%   [d(1) u(1)                            ]
%   [l(2) d(2) u(2)                       ]
%   [     l(3) d(3) u(3)                  ]
%   [          ...  ...  ...              ]
%   [                ...    ...      ...  ]
%   [                  l(n-1) d(n-1) u(n-1)]
%   [                         l(n)    d(n)]
% The function takes in three vectors for A:
%       l  (n*1) sub-diagonal vector
%       d  (n*1) diagonal element vector
%       u  (n*1) super-diagonal vector
```

```
% and (n*1) vector b and returns solution x
% ================================================

% Check vector lengths
n=length(d);
if (length(l)~=n)
    error('Vectors should be of same length');
end
if (length(u)~=n)
    error('Vectors should be of same length');
end
if (length(b)~=n)
    error('Vectors should be of same length');
end

% Forward elimination
for i=1:n-1
    % Divide row by diagonal
    u(i)=u(i)/d(i);
    b(i)=b(i)/d(i);
    d(i)=1;
    % Get zeros in pivot column
    d(i+1)=d(i+1)-u(i)*l(i+1);
    b(i+1)=b(i+1)-b(i)*l(i+1);
    l(i+1)=0;
end

% Back-substitution
x=zeros(n,1);
x(n)=b(n)/d(n);
for i=n-1:-1:1
    x(i)=b(i)-u(i)*x(i+1);
end
```

Solving the above equations results in the exact same solution as that obtained in Example 7.1, albeit with lower computational effort.

A key element of Gauss Elimination is pivoting. Pivoting is a procedure where a pivot row is exchanged with a row below it such that the pivot element becomes the largest element (in terms of its absolute value) in that column. In contrast, pivoting is not used in TDMA since it will break the nice tridiagonal structure of the problem. This works well in practice because often the diagonal elements are dominant elements for a large number of problems of practical interest. For example, in Equation 7.17, not only is the diagonal element $d_i = -(2+\beta)$ the largest element, the matrix A is, in fact, diagonally dominant. Even when the diagonal element is not largest, TDMA *usually* gives good performance without pivoting.

7.2.3 ODE-BVP with Flux Specified at Boundary

The problem introduced in Example 7.1 was an ODE-BVP with Dirichlet boundary conditions since the temperature was specified at both ends. If the rod were insulated at one end (or if heat flux was known), we will have an example of Neumann boundary condition at that end, whereas balancing heat loss at the boundary with conductive heat transfer gives rise to mixed boundary condition. For example, in Example 7.1, if the end of the rod $(z = L)$ is insulated:

$$\left. \frac{dT}{dz} \right|_{z=L} = 0 \tag{7.31}$$

whereas if it loses heat to the surroundings, the following mixed boundary condition is obtained:

$$-k \left. \frac{dT}{dz} \right|_{z=L} = h_\infty \left(T_L - T_a \right) \tag{7.32}$$

For implementing the boundary conditions at the boundary, two approaches can be used: (i) three-point backward difference formula, which is used to discretize Equations 7.31 and 7.32, or (ii) ghost-point approach. We will use the latter in this example. Recall that the discretized equation for the $(N + 1)$th node is

$$T_{n+2} - \left(2 + \beta \right) T_{n+1} + T_n = -\beta T_a \tag{7.33}$$

As discussed in Chapter 4, this introduces a "ghost-point" T_{n+2} outside the solution domain. The central difference formula applied to the boundary condition will be used to rewrite Equation 7.33. In case of an insulated boundary, Equation 7.31 reduces to $T_{n+2} = T_n$. On the other hand, substituting the central difference formula for Equation 7.32 yields

$$T_{n+2} - T_n = \frac{h_\infty \left(2\Delta \right)}{-k} \left(T_{n+1} - T_a \right) \tag{7.34}$$

$$T_{n+2} = T_n - \frac{h_\infty \left(2\Delta \right)}{k} \left(T_{n+1} - T_a \right) \tag{7.35}$$

Substituting the above in Equation 7.33 yields the following:

$$2T_n - \left(2 + \gamma \right) T_{n+1} = -\gamma T_a$$
$$\text{where } \gamma = \frac{h_\infty}{k} \left(2\Delta + \Delta^2 a_v \right) \tag{7.36}$$

The equation above replaces the last equation in Equation 7.16. The resulting system of linear equations is

$$
\underbrace{\begin{bmatrix}
1 & 0 & 0 & \cdots & \cdots & 0 & 0 \\
1 & -(2+\beta) & 1 & \cdots & \cdots & 0 & 0 \\
0 & 1 & -(2+\beta) & \cdots & \cdots & 0 & 0 \\
\vdots & \vdots & \ddots & \ddots & \ddots & \vdots & \vdots \\
0 & 0 & 0 & \cdots & 1 & -(2+\beta) & 1 \\
0 & 0 & 0 & \cdots & 0 & 2 & -(2+\gamma)
\end{bmatrix}}_{A}
\underbrace{\begin{bmatrix}
T_1 \\ T_2 \\ T_3 \\ \vdots \\ T_n \\ T_{n+1}
\end{bmatrix}}_{x}
=
\underbrace{\begin{bmatrix}
T_0 \\ -\beta T_\infty \\ -\beta T_\infty \\ \vdots \\ -\beta T_\infty \\ -\gamma T_\infty
\end{bmatrix}}_{b}
\qquad (7.37)
$$

Note that only the last row of matrix A has changed compared to Equation 7.17. Let us solve this problem in the next example.

Example 7.3 Heat Conduction with Mixed Boundary Condition

Problem: Solve the problem of heat conduction in a rod from Example 7.1, where one end of the rod is held at 100°C and the other end is governed by the mixed boundary condition given in Equation 7.32.

Solve: The problem to be solved is shown in Equation 7.37. The code for this using MATLAB's inbuilt linear equation solver is given below*:

```
% Heat conduction in a rod with heat losses
% using mixed boundary condition at one end
% Model Parameters
hinf=2.0;
abyV=10;
ks=15;
L=2.5;
Ta=30;          % Ambient T
T_0=100;        % T(z=0)
% Discretization
n=20;
delta=L/n;
beta=hinf*abyV/ks*delta^2;
gamma=hinf/ks*(abyV*delta^2+2*delta);
Z=[0:delta:L]';

%% Solving using matrix inverse
A=zeros(n+1,n+1);
```

---

* A more appropriate way to solve this problem is to use the TDMA code from Example 7.2. However, the reader is urged to solve this using TDMA as an exercise (and compare with results in Example 7.3). The TDMA code from Example 7.2 takes in the three diagonal elements of the matrix A and the vector b as four vectors, l, d, u, b.

```
b=zeros(n+1,1);
A(1,1)=1;          b(1)=T_0;
for i=2:n
    A(i,i-1)=1;
    A(i,i)=-(2+beta);
    A(i,i+1)=1;
    b(i)=-Ta*beta;
end
A(n+1,n)   =2;
A(n+1,n+1)=-(2+gamma);
b(n+1)=-Ta*gamma;
T=A\b;
plot(Z,T); hold on
xlabel('distance, L (m)'); ylabel('temperature (^oC)')
%% Solving using TDMA
T1=newThomasSolve(A,b);
% <newThomasSolve.m is left as student exercise>
plot(Z,T1,'ro');
```

Figure 7.4 shows the variation in temperature profile for an axial rod with mixed boundary condition. The end temperature is slightly higher than 30°C that was assumed in the previous example.

Finally, Figure 7.5 shows the structure of matrix A, investigated using the MATLAB spy command. As is clear from the figure, the matrix A indeed has a tridiagonal structure.

The solution is repeated for different values of thermal conductivity, $k = 5, 15, 50$ W/m · K. Figure 7.6 shows the effect of thermal conductivity on the temperature profile in the rod. The temperature in the rod becomes more uniform as the thermal conductivity is increased from 5 to 50 W/m · K. Higher thermal conductivity also causes greater heat transfer through the rod, which is then lost to the surroundings.

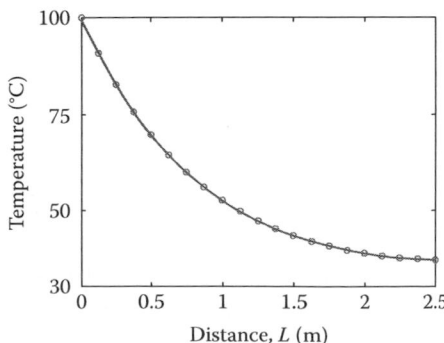

FIGURE 7.4 Temperature profile in the rod with heat transfer to the surroundings. The line shows solution using the general linear equation solved (A\b in MATLAB®), whereas the circles show the solution using the TDMA solver from Example 7.2.

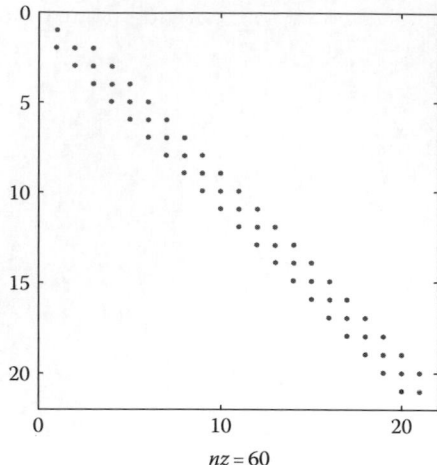

FIGURE 7.5   The structure of matrix A observed through MATLAB® spy(A) command.

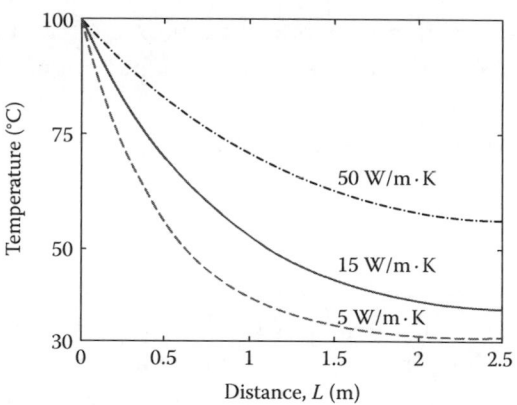

FIGURE 7.6   Temperature profiles for different values of thermal conductivity.

7.2.4 Extension to Banded Systems

An example of banded system will be considered in case studies. Let us now consider a band-diagonal system of equations. If the storage space for storing matrix A is not an issue, it will be easier to solve a band-diagonal system as a specific implementation of Gauss Elimination. For ease of discussion, matrix A with equal sub- and superdiagonal elements is considered. The method can be easily extended to a system with a different number of sub- and superdiagonal elements. This method is a modification of Gauss Elimination, discussed in Appendix C.

In the kth sequence of steps, $A_{k,k}$ is the pivot element. The first step is to divide the first row with $A_{1,1}$. Thereafter, row-1 is used to make zeros in subdiagonal elements of the pivot column. Unlike Gauss Elimination, there are only ℓ nonzero elements on the pivot row, and only ℓ rows have nonzero elements in the pivot column. Thus, the first row operation is executed as follows:

$$A_{k,k+i} \leftarrow \frac{A_{k,k+i}}{A_{k,k}}, \quad i = 1 \text{ to } \ell \tag{7.38}$$

$$b_k \leftarrow \frac{b_k}{A_{k,k}}, \quad A_{k,k} \leftarrow 1 \tag{7.39}$$

Note that Equation 7.38 is implemented for only ℓ elements and not for all elements since the remaining elements of the row are already zero. Since only ℓ subdiagonal elements are nonzero, the row operations for eliminating subdiagonal elements in the pivot column need to be done for only ℓ operations:

$$R_{k+i} \leftarrow R_{k+i} - A_{k+i,k} R_i, \quad i = 1 : \ell \tag{7.40}$$

At this stage, it should be noted that the only nonzero elements in the kth row are elements $A_{k,k}$ to $A_{k,k+\ell}$. Thus, the row operations need to be performed for only these elements and not all. Thus, the expansion of Equation 7.40 in terms of individual elemental operations is

$$\text{For } i = 1 : \ell \quad \bar{\alpha}_{k+i,k} = A_{k+i,k} \tag{7.41}$$

$$A_{(k+i),(k:k+\ell)} \leftarrow A_{(k+i),(k:k+\ell)} - \bar{\alpha}_{k+i,k} A_{k,(k:k+\ell)} \tag{7.42}$$

$$b_{k+i} \leftarrow b_{k+i} - \bar{\alpha}_{k+i,k} b_k \tag{7.43}$$

Note that Equation 7.41 is different from standard Gauss Elimination (Appendix C) because we had already used Equation 7.39 to ensure $A_{k,k} \leftarrow 1$.

Writing a code for the above operations is left as an exercise for the reader. An example of band-diagonal system will be taken in Case Studies section.

Block-diagonal system: The block-diagonal systems can be treated as band-diagonal. For the example in Figure 7.1, the bandwidth is $\ell = 2$ for superdiagonals and $\ell = 1$ for subdiagonals. The procedure described above can be used for a block-diagonal system as well.

7.2.5 Elliptic PDEs in Two Dimensions

Heat conduction in a 2D geometry is a standard problem that leads to the following elliptic PDE:

$$0 = k\nabla^2 T + S(T) \tag{7.44}$$

which, when written in a rectangular coordinate system in 2D, becomes

$$0 = k\left(\frac{\partial^2 T}{\partial x^2} + \frac{\partial^2 T}{\partial y^2}\right) + S(T) \tag{7.45}$$

The model needs to be solved subject to boundary conditions. Since it is a second-order PDE in both dimensions, boundary conditions at four boundaries are required to solve the problem. These may be the Dirichlet boundary condition (Equation 7.4) if temperature is specified, the Neumann boundary condition (Equation 7.5) if flux is specified or the boundary is insulated, or the mixed boundary condition (Equation 7.6) if heat loss is specified at the boundary.

The term $S(T)$ is the source term; it is included to model any source or sink of heat within the domain. Unlike the 1D model of Equation 7.9, heat loss typically appears as a boundary condition and not a source term. Elliptic PDEs are also encountered in systems with flow:

$$\sigma \nabla \phi = \Gamma \nabla^2 \phi + S(\phi)$$

where

σ is the coefficient of convection

Γ is the diffusive coefficient

As long as the Péclet number, $Pc = \sigma L/\Gamma$ (where L is the characteristic length), is not very large, diffusion may not be neglected and the PDE is elliptic in nature. Besides heat transfer problems, elliptic PDEs are encountered in a large number of problems: diffusion in multiple dimensions, diffusion and reaction, momentum balance (Navier-Stokes), and so on.

Due to qualitative similarity in the behavior of elliptic PDEs, we take a similar approach as the in previous section to solve them. The domain is discretized in both x and y directions. The temperature at any location (i, j) is related to its neighbors through the central difference formula, which yields the following discretized equation:

$$0 = k\frac{T_{i+1,j} - 2T_{i,j} + T_{i-1,j}}{\Delta x^2} + k\frac{T_{i,j+1} - 2T_{i,j} + T_{i,j-1}}{\Delta y^2} + S(T_{i,j}) \tag{7.46}$$

where $\Delta x = L/n_x$ and $\Delta y = H/n_y$, where L and H are length and height of the domain, respectively. Including the boundary nodes, this will lead to $(n_x + 1)(n_y + 1)$ equations in as many unknowns. The rest of the solution procedure follows on similar lines as discussed in the previous sections.

7.3 ITERATIVE METHODS

I now switch my attention to a completely different class of numerical methods: iterative methods. As seen in Chapter 6, iterative methods start with an initial guess of the solution and use a sequence of equations to iteratively improve the solution. Fixed point iteration,

Newton-Raphson, secant method, etc., from Chapter 6 are examples of iterative methods. We now focus our attention back to iterative methods, for both linear and nonlinear systems.

7.3.1 Gauss-Siedel Method

Gauss-Siedel method (and related Jacobi method) is the linear equations equivalent of fixed point iteration. In case of fixed point iteration, the nonlinear equation was modified appropriately, and this was then used as an update equation to iteratively improve the current solution guess:

$$\mathbf{x}^{(i+1)} = \mathbf{h}\left(\mathbf{x}^{(i)}\right)$$

The Gauss-Siedel works in a similar manner for linear systems. Consider the kth equation:

$$A_{k,k}x_k = b_k - \left(A_{k,1}x_1 + \cdots + A_{k,k-1}x_{k-1} + A_{k,k+1}x_{k+1} + \cdots + A_{k,n}x_n\right) \qquad (7.47)$$

Similar to the fixed point iteration, the above equation is used as an iterative update equation if the overall system of equations is linear. Based on how the update equation is used, there are two possibilities: (i) the previous values $\mathbf{x}^{(l)}$ is used for all the update equations in the *Jacobi* method, or (ii) the most recent values are used for the update equations in the *Gauss-Siedel* method.

First consider the Jacobi iteration. When Equation 7.45 is used to update x_k, we have the previous iteration values $x_{k+1}^{(i)}, \ldots, x_n^{(i)}$ because the updated values have not yet been calculated. However, for the preceding elements, both the updated values, $x_1^{(i+1)}, \cdots, x_{k-1}^{(i+1)}$, or the previous iteration values, $x_1^{(i)}, \cdots, x_{k-1}^{(i)}$, are available. Jacobi iteration uses the previous iteration values for all elements of $\mathbf{x}$, thus resulting in the following update equations:

$$x_k^{(i+1)} = \frac{b_k - \sum_{\substack{j=1, \\ j \neq k}}^{n} A_{k,j}x_j^{(i)}}{A_{k,k}} \qquad (7.48)$$

Gauss-Siedel, on the other hand, uses the most updated values. Thus, $x_1^{(i+1)}$ is computed using $x_2^{(i)}, \ldots, x_n^{(i)}$. However, computation of $x_2^{(i+1)}$ uses $x_1^{(i+1)}$ since that is the most updated value of iterated solution vector. Likewise, $x_k^{(i+1)}$ is computed using $x_1^{(i+1)}, \ldots, x_{k-1}^{(i+1)}$ (most recently update values) as well as $x_k^{(i)}, \ldots, x_n^{(i)}$ (which are not yet updated). Mathematically, the updated equations for Gauss-Siedel are

$$x_k^{(i+1)} = \frac{b_k - \sum_{j=1}^{k-1} A_{k,j}x_j^{(i+1)} - \sum_{j=k+1}^{n} A_{k,j}x_j^{(i)}}{A_{k,k}} \qquad (7.49)$$

In Chapter 6, I derived the error propagation equation for fixed point iteration in a single variable. This can be extended to multivariate fixed point iteration as

$$e_k^{(i+1)} = \frac{\partial h_k}{\partial x_1} e_1^{(i)} + \cdots + \frac{\partial h_k}{\partial x_n} e_n^{(i)} \tag{7.50}$$

Extending the arguments from the single-variable case of Chapter 6, the criterion for convergence of fixed point iteration is

$$\sum_i \left| \frac{\partial h_k}{\partial x_i} \right| < 1 \tag{7.51}$$

This leads to the following diagonal dominance criterion for the Gauss-Siedel method:

$$\sum_{i \neq k} \left| \frac{A_{k,i}}{A_{k,k}} \right| < 1$$
$$|A_{k,k}| > \sum_{i \neq k} |A_{k,i}| \tag{7.52}$$

The following example, which can be worked out with hand calculations, demonstrates this criterion for stability of the Gauss-Siedel method.

Example 7.4 Stability and Diagonal Dominance of Gauss-Siedel

Problem: Solve the following problem using Gauss-Siedel method, with $\mathbf{x} = [0 \quad 0]^T$ as the initial guess:

$$x_1 + 2x_2 = 1$$

$$x_1 - x_2 = 4$$

Solution-1: The above system of equations is not diagonally dominant. The Gauss-Siedel iterations for this system of equations are

$$x_1^{(i+1)} = 1 - 2x_2^{(i)}$$

$$x_2^{(i+1)} = x_1^{(i+1)} - 4$$

The first five iterations of Gauss-Siedel method give the following results:

```
0    1    7   -5    19   -29
0   -3    3   -9    15   -33
```

As can be concluded from the above results, the iterations are diverging. Hence, the equations can first be rearranged to make them diagonally dominant, followed by using the Gauss-Siedel method.

Solution-2: The two equations are switched to make the system diagonally dominant. With this change, the Gauss Siedel iterations are given by

$$x_1^{(i+1)} = 4 + x_2^{(i)}$$

$$x_2 = \frac{1}{2}\left(1 - x_1^{(i+1)}\right)$$

The first five iterations for this diagonally dominant system of equations is given by

```
0    4.0000    2.5000    3.2500    2.8750    3.0625
0   -1.5000   -0.7500   -1.1250   -0.9375   -1.0312
```

Clearly, the above is converging to the solution, $x = \begin{bmatrix} 3 & -1 \end{bmatrix}^T$. This example shows the improved convergence behavior of the Gauss-Siedel method when the same system of equations is rearranged in a diagonally dominant form.

Let us turn to the implications of the above observation to solving a sparse system of equations. The tridiagonal system of Section 7.2.2 already satisfies the diagonal dominance condition. This is true for a large number of systems of practical interest.

Since only one subdiagonal and superdiagonal elements are nonzero, the Gauss-Siedel iteration for the kth element becomes

$$x_k^{(i+1)} = \frac{b_k - l_k x_{k-1}^{(i+1)} - u_k x_{k+1}^{(i)}}{d_k} \tag{7.53}$$

The extension to the banded system, with a bandwidth of ℓ, is also straightforward:

$$x_1^{(i+1)} = \frac{b_1 - \sum_{j=2}^{\ell+1} A_{1,j} x_j^{(i)}}{A_{1,1}}$$

$$x_2^{(i+1)} = \frac{b_2 - A_{2,1} x_1^{(i+1)} - \sum_{j=3}^{\ell+1} A_{1,j} x_j^{(i)}}{A_{2,2}} \tag{7.54}$$

$$\vdots$$

$$x_k^{(i+1)} = \frac{b_k - \sum_{j=k-\ell}^{k-1} A_{k,j} x_j^{(i+1)} - \sum_{j=k+1}^{k+\ell} A_{k,j} x_j^{(i)}}{A_{k,k}}$$

$$\vdots$$

The next example demonstrates the use of Gauss-Siedel method for a heat conduction problem.

Example 7.5 Gauss-Siedel Method for Tridiagonal System

Problem: Use the Gauss-Siedel method to solve Example 7.2.

  Solution: Given below is the Gauss-Siedel code for the tridiagonal system:

```
% Heat conduction in a rod using Gauss-Siedel method
%% Model Parameters
hinf=2.0;
abyV=10;
ks=15;
L=2.5;
Ta=30;        % Ambient T
T_0=100;      % T(z=0)
T_L=Ta;       % T(z=L)
%% Discretization
n=40;
delta=L/n;
Z=[0:delta:L]';
beta=hinf*abyV/ks * delta^2;
%% Numerical solution: Gauss-Siedel
A=zeros(n+1,n+1);
b=zeros(n+1,1);
A(1,1)=1;          b(1)=T_0;
for i=2:n
    A(i,i-1)=1;
    A(i,i)=-(2+beta);
    A(i,i+1)=1;
    b(i)=-Ta*beta;
end
A(n+1,n+1)=1;    b(n+1)=T_L;
T=T_L*ones(n+1,1);
aTol=1e-3;
maxIter=5000;
for iter=1:maxIter
    Told=T;
    T(1)=(b(1)-A(1,2)*T(2))/A(1,1);
    for k=2:n
        T(k)=(b(k)-A(k,k+1)*T(k+1)-A(k,k-1)*T(k-1))/A(k,k);
    end
    T(n+1)=(b(n+1)-A(n+1,n)*T(n))/A(n+1,n+1);
    err=max(abs(T-Told));
    if (err<aTol)
        break
    end
end
plot(Z,T);
```

The above method took 486 iterations to converge to the desired tolerance. The solution obtained using this approach is close to the temperature values obtained using the TDMA-based approach of the previous example.

The matrix A in this example was diagonally dominant, and the criterion in Equation 7.50 is satisfied. If the system is not diagonally dominant, it is possible to "slow down" the iterations using an under-relaxation factor. Conversely, an over-relaxation factor may be used to "speed up" the convergence. These relaxation methods are introduced next.

7.3.2 Iterative Method with Under-Relaxation

In Chapter 6, the line-search parameter was introduced in Newton-Raphson to improve its convergence properties. Relaxation methods, discussed here, have the same idea. Instead of taking a full step toward the solution, we take a partial step. If Δx is the full step taken in the direction of the solution, then the next iteration value is obtained as

$$x^{(i+1)} = x^{(i)} + \omega \Delta x \qquad (7.55)$$

The update equation (7.46) in Jacobi iteration can be used to define Δx_k:

$$\Delta x_k^{(i+1)} = \frac{b_k - \sum_{\substack{j=1 \\ j \neq k}}^{n} A_{k,j} x_j^{(i)}}{A_{k,k}} - x_k^{(i)} \qquad (7.56)$$

Thus, Jacobi iteration with relaxation is written as

$$x_k^{(i+1)} = (1-\omega) x_k^{(i)} + \omega \frac{b_k - \sum_{\substack{j=1 \\ j \neq k}}^{n} A_{k,j} x_j^{(i)}}{A_{k,k}} \qquad (7.57)$$

Here, ω is known as a relaxation parameter. The value of parameter ω is kept between 0 and 2. When the value of ω is less than unity, the method is called the Jacobi under-relaxation method. In contrast, the value of ω is chosen between $1 < \omega < 2$ to improve the speed of convergence for a slowly converging system.

Likewise, the Gauss-Siedel method with relaxation is given by

$$x_k^{(i+1)} = (1-\omega) x_k^{(i)} + \omega \left(\frac{b_k - \sum_{j=1}^{k-1} A_{k,j} x_j^{(i+1)} - \sum_{j=k+1}^{n} A_{k,j} x_j^{(i)}}{A_{k,k}} \right) \qquad (7.58)$$

The term *successive over-relaxation* is used for the Gauss-Siedel method with the relaxation factor, $1 < \omega < 2$.

7.4 NONLINEAR BANDED SYSTEMS

A second-order 1D ODE discussed so far is of the form

$$\alpha \frac{d^2 \phi}{dx^2} = f\left(x, \phi, \frac{d\phi}{dx}\right) \tag{7.59}$$

Here α is a coefficient, which itself can be a function of ϕ and/or x.

Typical chemical process simulation problems involve diffusive/conductive as well as convective transport terms. Such systems are governed by a specific type of ODE-BVP:

$$\alpha \frac{d^2 \phi}{dx^2} - u \frac{d\phi}{dx} - f\left(x, \phi\right) = 0 \tag{7.60}$$

The above ODE is nonlinear if either α or u (or both) depend on ϕ or if $f(x, \phi)$ is nonlinear.

7.4.1 Nonlinear ODE-BVP Example

Consider the heat conduction problem of Section 7.2.2. Since the balance is for a solid rod, $u = 0$. The term $f(x, \phi)$, called the *source term*

$$f\left(x, T\right) = h_\infty a_v \left(T - T_\infty\right) \tag{7.61}$$

is linear. For a narrow range of temperature (303 K $\leq T \leq$ 373 K), the thermal conductivity was considered constant. This led to a linear ODE-BVP with linear boundary conditions, as seen in Equation 7.11. Discretizing using finite difference approximation resulted in a tridiagonal set of linear equations, as described in the chapter so far.

At times, the thermal conductivity may not be a constant but expressed as a power law or a polynomial function of temperature:

$$k = k_0 + k_1 T + k_2 T^2 + k_3 T^3, \quad \text{or} \quad k = aT^\eta \tag{7.62}$$

Compared to problems involving temperature-dependent coefficients, the most common form of nonlinearity in process simulation examples arises when the source term $f(x, \phi)$ is nonlinear. For example, if one end of the rod was at a very high temperature, then radiation becomes important. Radiation flux is given by the following fourth-order term:

$$\text{Flux} = \varepsilon\sigma\left(T^4 - T_\infty^4\right) \tag{7.63}$$

where
  ε is the emissivity factor
  $\sigma = 5.67 \times 10^{-8}$ W/m$^2 \cdot$ K^4 is Stefan's constant

In fact, nonlinear source terms are common in the majority of chemical engineering examples, due to nonlinear heat loss as radiation, a nonlinear reaction or heat release term, bilinear (or nonlinear) interaction/adsorption term in two-phase systems, etc.

When the nonlinear differential equation(s) described here are discretized, we obtain a set of nonlinear algebraic equations. These equations have a tridiagonal (sparse) structure, which can be exploited for efficient solving of the nonlinear equations. Modifications to the TDMA, Gauss-Siedel, or Jacobi method and `fsolve` options are discussed in this section. The nonlinearities are of two types: (i) variable diffusive/convective coefficients and (ii) source term nonlinearity. The latter will be discussed at length in this section. The methods discussed are rather general and are often applicable when coefficients are varying as well. However, an example of variable coefficient will be left as an exercise for the reader.

7.4.1.1 Heat Conduction with Radiative Heat Loss

Let us focus back on the heat conduction problem with radiation. The ODE-BVP model for this system is given by

$$k\frac{d^2T}{dz^2} = \varepsilon\sigma a_v\left(T^4 - T_\infty^4\right) \tag{7.64}$$

$$T(0) = T_0 \tag{7.65}$$

$$T(L) = T_\infty \tag{7.66}$$

Discretizing the above using finite difference approximation, as before

$$T_{k-1} - 2T_k + T_{k+1} = \underbrace{\frac{\varepsilon\sigma\Delta^2 a_v}{k}\left(T_k^4 - T_\infty^4\right)}_{\beta(T_k)}, \quad k = 2 \text{ to } n \tag{7.67}$$

Unlike the previous example, $\beta(T)$ is a nonlinear function of T. The discretized boundary conditions are

$$\begin{aligned} T_1 &= T_0 \\ T_{n+1} &= T_\infty \end{aligned} \tag{7.68}$$

Thus, we can form the following tridiagonal system of nonlinear equations:

$$\underbrace{\begin{bmatrix} 1 & 0 & 0 & \cdots & \cdots & 0 & 0 \\ 1 & -2 & 1 & \cdots & \cdots & 0 & 0 \\ 0 & 1 & -2 & \cdots & \cdots & 0 & 0 \\ \vdots & \vdots & \ddots & \ddots & \ddots & \vdots & \vdots \\ 0 & 0 & 0 & \cdots & 1 & -2 & 1 \\ 0 & 0 & 0 & \cdots & 0 & 0 & 1 \end{bmatrix}}_{A} \underbrace{\begin{bmatrix} T_1 \\ T_2 \\ T_3 \\ \vdots \\ T_n \\ T_{n+1} \end{bmatrix}}_{\mathbf{x}} = \underbrace{\begin{bmatrix} T_0 \\ \beta(T_2) \\ \beta(T_3) \\ \vdots \\ \beta(T_n) \\ T_\infty \end{bmatrix}}_{\mathbf{b}(\mathbf{x})} \tag{7.69}$$

In the next couple of sections, we will discuss numerical techniques to solve this problem.

7.4.2 Modified Successive Linearization–Based Approach

Since the above is a set of nonlinear equations, an iterative method needs to be used to solve it. Recall that in fixed point iteration, we rewrite the original Equation 7.1 as

$$\mathbf{x}^{(i+1)} = \mathbf{G}\left(\mathbf{x}^{(i)}\right) \tag{7.70}$$

The above is used as an update equation. In the same manner, it is easy to rewrite Equation 7.66 as

$$\mathbf{x}^{(i+1)} = A^{-1}\mathbf{b}\left(\mathbf{x}^{(i)}\right) \tag{7.71}$$

Instead of inverting the matrix A, a TDMA solver can be used to solve the problem (7.66) with a fixed value of $\mathbf{b}^{(i)}$ to obtain $\mathbf{x}^{(i+1)}$, substitute this value to obtain $\mathbf{b}^{(i+1)} = \mathbf{b}(\mathbf{x}^{(i+1)})$ and iterate until convergence, that is, $\|\mathbf{x}^{(i+1)} - \mathbf{x}^{(i)}\| < E_{tol}$.

It is also possible to combine the above approach with under-relaxation. The above TDMA-based successive substitution is a rather naïve way to solve a nonlinear system of equations, and it may not work for examples with highly nonlinear source term. In such cases, TDMA with under-relaxation may be used, where the updated equation becomes

$$\mathbf{x}^{(i+1)} = \left(1 - \omega\right)\mathbf{x}^{(i)} + \omega A^{-1}\mathbf{b}\left(\mathbf{x}^{(i)}\right) \tag{7.72}$$

In the above, the same relaxation factor ω was used for all variables x_k. However, it is possible to choose different values of ω for different elements of $\mathbf{x}$:

$$\mathbf{x}^{(i+1)} = \left(I - \Omega\right)\mathbf{x}^{(i)} + \Omega A^{-1}\mathbf{b}\left(\mathbf{x}^{(i)}\right)$$
$$\Omega = \text{diag}\left(\begin{bmatrix} \omega_1 & \omega_2 & \cdots & \omega_{n+1} \end{bmatrix}\right) \tag{7.73}$$

Unlike the standard fixed point iteration, the above methods do not have a general requirement that guarantees convergence. So, this is a rather heuristic combination of linear equation solving with method of successive substitution. The next example demonstrates the use of this approach to heat conduction with radiative heat loss.

Example 7.6 Heat Conduction with Radiative Heat Loss

Problem: Solve the problem of heat conduction in a metal rod that loses heat to surroundings by radiation. The model is described by Equation 7.61 and the two ends of the rod are kept at 600 K and 300 K, respectively. Solve this for two values of emissivity, $\varepsilon = 0.1$ and $\varepsilon = 0.9$.

Solution: First we will solve this using successive substitution employing a TDMA-based solution approach. The matrix A is constructed according to Equation 7.66, and vector $\mathbf{b}(\mathbf{x}^{(i)})$ is computed at the current guess of temperature. The Thomas solver explained earlier is used to obtain the next guess, and the process is repeated iteratively.

```
% Heat conduction in a rod with radiation heat loss
%% Model Parameters
epsi=0.1;
sigma=5.67e-8;
abyV=10;
ks=15;
L=2.5;
Ta=300;      % Ambient T
T_0=600;     % T(z=0)
T_L=Ta;      % T(z=L)
%% Discretization
n=50;
delta=L/n;
Z=[0:delta:L]';
betaFactor=epsi*sigma*delta^2*abyV/ks;


%% Iterative solution with TDMA
iv=ones(n,1);
A=-2*eye(n+1)+diag(iv,1)+diag(iv,-1);
A(1,:)=[1,zeros(1,n)];
A(n+1,:)=[zeros(1,n),1];
T=T_L*ones(n+1,1);
maxIter=5000;
tol=1e-3;
for iter=1:maxIter
    b(1,1)=T_0;
    b(2:n) = betaFactor*(T(2:n).^4-Ta^4);
    b(n+1,1)=T_L;
    Tnew=newThomasSolve(A,b);
    TSTORE(:,iter)=Tnew;
    err=max(abs(Tnew-T));
    T=Tnew;
    if (err<tol)
        break
    end
end
plot(Z,T-273); hold on
xlabel('distance, L (m)'); ylabel('temperature (^oC)')
```

The above code took 30 iterations to converge to a solution, starting with initial guess of $T = T_\infty$, for $\varepsilon = 0.1$. While the temperature could be expressed either in °C or K in the convective heat loss case, we have to use K in this example. The temperature profile is plotted as solid line in Figure 7.7. The temperature profiles with heat loss via

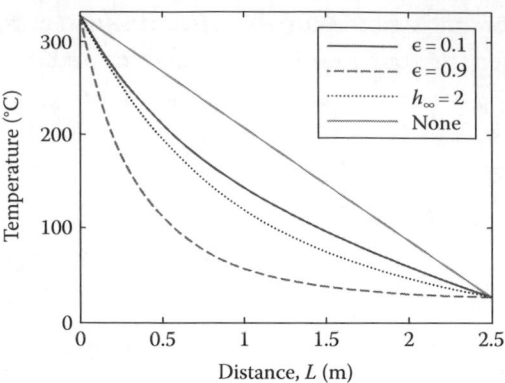

FIGURE 7.7   Heat conduction in a metal rod with heat loss to the surroundings through radiation for two different emissivity values ($\varepsilon = 0.1$: solid; $\varepsilon = 0.9$: dashed). The temperature profile when heat loss is only by convection is plotted as a dotted line for comparison. Also plotted is the case with no heat loss (straight line).

convection (solved using code from Example 7.2) and with no heat loss are also plotted as dotted line and solid straight line for comparison.

The above code does not converge for the more relevant emissivity value of $\varepsilon = 0.9$. The temperature diverges and we get NaN as the resulting temperature profile. This is because the fixed point iterations do not converge owing to the nonlinear radiation term. Consequently, under-relaxation-based approach, shown in Equation 7.69, was used with the under-relaxation factor $\omega = 0.25$.

Modification of the code above for implementing this under-relaxation is left to the reader (see Problem 7.3).

The above approach combined linear equation solving with successive linearization in a heuristic algorithm. Alternatively, a regular fixed point iteration may be used, treating each equation in Equation 7.66 as a regular nonlinear equation. Thus, the update expression

$$x_k^{(i+1)} = G_k\left(\mathbf{x}^{(i)}\right) \tag{7.74}$$

can be written in a manner similar to the Gauss-Siedel iteration[*]:

$$x_k^{(i+1)} = \frac{\beta\left(x_k^{(i)}\right) - A_{k,k-1}x_{k-1}^{(i+1)} - A_{k,k+1}x_{k+1}^{(i)}}{A_{k,k}} \tag{7.75}$$

or successive under-relaxation approach:

$$x_k^{(i+1)} = \left(1 - \omega_k\right)x_k^{(i)} + \omega_k\frac{\beta\left(x_k^{(i)}\right) - A_{k,k-1}x_{k-1}^{(i+1)} - A_{k,k+1}x_{k+1}^{(i)}}{A_{k,k}} \tag{7.76}$$

---

[*] In the same manner, Jacobi iteration or Jacobi under-relaxation may also be used.

In either case, the iterations are repeated until convergence, that is, $\|\mathbf{x}^{(i+1)} - \mathbf{x}^{(i)}\| < E_{tol}$. This is the fixed point iteration approach (either with or without under-relaxation), which has been modified for the tridiagonal problem represented in Equation 7.66.

Previously, the diagonal dominance of matrix A guaranteed convergence for a linear tridiagonal system. However, due to the nonlinear source term, the Gauss-Siedel iterations for Equation 7.66 do not have convergence guarantees. If the Gauss-Siedel method diverges, the successive under-relaxation approach may be used to improve the convergence behavior of the system. The nonlinear heat conduction with a radiative heat loss problem is solved using Gauss-Siedel (with under-relaxation) in the next example.

Example 7.7 Heat Conduction with Radiative Heat Loss Using Gauss-Siedel

Problem: Solve the problem from Example 7.6 using Gauss-Siedel method.

Solution: The code for Gauss-Siedel for this problem is given below:

```
% Heat conduction in a rod with radiation heat loss
%% Model Parameters
epsi=0.9;
sigma=5.67e-8;
abyV=10;
ks=15;
L=2.5;
Ta=300;      % Ambient T
T_0=600;     % T(z=0)
T_L=Ta;      % T(z=L)
%% Discretization
n=50;
delta=L/n;
Z=[0:delta:L]';
betaFactor=epsi*sigma*delta^2*abyV/ks;

%% Solving using GS-Under-Relaxation
iv=ones(n,1);
A=-2*eye(n+1)+diag(iv,1)+diag(iv,-1);
A(1,:)=[1,zeros(1,n)];
A(n+1,:)=[zeros(1,n),1];
T=T_L*ones(n+1,1);    % Initial guess
maxIter=50000;
tol=1e-3;
omega=1;
for iter=1:maxIter
    b(1,1)=T_0;
    b(2:n)=betaFactor*(T(2:n).^4-Ta^4);
    b(n+1,1)=T_L;
    Told=T;    % For verifying convergence
    for k=1:n+1
```

```
        if (k==1)
            dT=(b(1)-A(1,2)*T(2))/A(1,1);
        elseif (k==n+1)
            dT=(b(n+1)-A(n+1,n)*T(n))/A(n+1,n+1);
        else
            dT=( b(k)-A(k,k+1)*T(k+1)-...
                A(k,k-1)*T(k-1) ) / A(k,k);
        end
        T(k)=(1-omega)*T(k) + omega*dT;
    end
    TSTORE(:,iter)=T;
    err=max(abs(T-Told));
    if (err<tol)
        break
    end
end
```

The above code gives the exact same results as those shown in Figure 7.7. It took 394 Gauss-Siedel iterations for convergence when $\varepsilon = 0.9$ and 1059 iterations when $\varepsilon = 0.1$. These reports are for $\omega = 1$ (i.e., without under-relaxation).

When the above code was modified for Jacobi iteration, the iterations did not converge for $\varepsilon = 0.9$, whereas it took as many as 8089 iterations for convergence for $\varepsilon = 0.1$. In order to improve convergence behavior, Jacobi under-relaxation was used. With the under-relaxation factor of $\omega = 0.9$, the method converged in 796 iterations for $\varepsilon = 0.9$. The modification of the above code for Jacobi iterations is left as an exercise to the reader.

7.4.3 Gauss-Siedel with Linearization of Source Term

The convergence of fixed point iteration for tridiagonal systems can sometimes be slow, or the method may not be robust. Adjusting the relaxation parameter, as described in the previous section, is one option to improve convergence behavior. Another option to do so is by linearizing the source term. Before going to the tridiagonal example, let us consider a single-variable case:

$$x = G(x) \tag{7.77}$$

The function can be linearized around $x^{(i)}$ using Taylor's series:

$$
\begin{aligned}
G(x) &\approx G\left(x^{(i)}\right) + G'\left(x^{(i)}\right)\left[x - x^{(i)}\right] \\
&= \underbrace{G\left(x^{(i)}\right) + x^{(i)}G'\left(x^{(i)}\right)}_{G_0} + \underbrace{G'\left(x^{(i)}\right)}_{G_1}x
\end{aligned} \tag{7.78}
$$

Substituting in Equation 7.74 yields the following update equation:

$$x^{(i+1)} = \frac{G_0}{1 - G_1}$$

$$\text{where } G_0 = G\left(x^{(i)}\right) - x^{(i)}G'\left(x^{(i)}\right) \quad \text{and} \quad G'\left(x^{(i)}\right) \tag{7.79}$$

Using equivalence between this and the Newton-Raphson method from the previous chapter, it can be shown that the linearization-based method has the quadratic rate of convergence for problems in single variable. This is left as an exercise to the reader.

The same concept can be expanded to a multivariable case. The equation with a nonlinear source term (such as that in Equation 7.64) can be written in the general form of Equation 7.19 as

$$l_k x_{k-1} + d_k x_k + u_k x_{k+1} = \beta(x_k) \tag{7.80}$$

Using a single-variable Taylor's series expansion for linearizing the nonlinear source term around x_k yields

$$\beta(x_k) \approx \beta\left(x_k^{(i)}\right) + \beta'\left(x_k^{(i)}\right)\left[x_k - x_k^{(i)}\right]$$

$$= \underbrace{\beta\left(x_k^{(i)}\right) + x_k^{(i)}\beta'\left(x_k^{(i)}\right)}_{\beta_{0,k}} + \underbrace{\beta'\left(x_k^{(i)}\right)}_{\beta_{1,k}} \cdot x_k \tag{7.81}$$

Note that since single-variable Taylor's series expansion is used, that is

$$\beta'\left(x_k^{(i)}\right) = \left.\frac{\partial \beta}{\partial x_k}\right|_{\mathbf{x} = \mathbf{x}^{(i)}} \tag{7.82}$$

Substituting this in Equation 7.77, the tridiagonal equation is written as

$$l_k x_{k-1} + \left(d_k - \beta_{1,k}\right)x_k + u_k x_{k+1} = \beta_{0,k}$$

$$\text{where } \beta_{0,k} = \beta\left(x_k^{(i)}\right) - x_k^{(i)}\beta'\left(x_k^{(i)}\right), \quad \beta_{1,k} = \beta'\left(x_k^{(i)}\right) \tag{7.83}$$

Thus, source term linearization often results in improved convergence properties by ensuring the tridiagonal system is diagonally dominant. For small- to medium-size problems, this approach may not provide significant benefits over the regular Gauss-Siedel approach. However, this approach is quite popular in large-scale commercial solvers, as it tends to make the iterative solver more robust without slowing down the rate of convergence.

7.4.4 Using `fsolve` with Sparse Systems

This brings me to the method of choice for a nonlinear banded system of equations: to use `fsolve`. While the `fsolve` solver can be used as in Chapter 6, a full-order solver will be used. However, it is possible to exploit the sparse/banded structure of the Jacobian in `fsolve` using the option `'jacobPattern'` and specifying the pattern of the Jacobian. The syntax for this is

```
>> opt=optimoptions('jacobPattern',J);
```

The options are appropriately passed to `fsolve`, as discussed earlier.

We will extensively make use of `fsolve` in the case studies discussed in the next section.

7.5 EXAMPLES

7.5.1 Heat Conduction with Convective or Radiative Losses

The problem of heat conduction in a rod with convective or radiative heat losses was discussed earlier in this book. Recall that the problem solved in Example 7.1 was that of heat conduction in a rod, modeled as the following ODE-BVP:

$$k\frac{d^2T}{dz^2} = h_\infty a_v \left(T - T_\infty\right) \tag{7.9}$$

$$T(0) = T_0, \quad T(L) = T_\infty \tag{7.10}$$

where

T is the temperature at any location in °C

T_∞ is the temperature of the ambient

k is the thermal conductivity of the rod

h_∞ is the heat loss coefficient

a_v is the *external* area of the rod per unit volume

This problem was modified in Section 7.2.3 by replacing the Dirichlet boundary condition with either a Neumann or a mixed boundary condition. If the far end of the rod is insulated, a Neumann boundary condition is used:

$$\left.\frac{dT}{dz}\right|_{z=L} = 0 \tag{7.31}$$

On the other hand, if it loses heat to the surroundings, a mixed boundary condition is used instead:

$$-k\left.\frac{dT}{dz}\right|_{z=L} = h_\infty \left(T_L - T_a\right) \tag{7.32}$$

Example 7.3 solved the problem (Equation 7.9) with a mixed boundary condition at one end of the rod.

Both these problems were linear; Section 7.4 extended this to a nonlinear system. Specifically, the problem of heat conduction with radiative heat losses was solved in Example 7.6. The ODE-BVP model for this system, along with the relevant Dirichlet boundary conditions, is given by

$$k\frac{d^2T}{dz^2} = \varepsilon\sigma a_v\left(T^4 - T_\infty^4\right) \tag{7.64}$$

$$T(0) = T_0 \tag{7.65}$$

$$T(L) = T_\infty \tag{7.66}$$

The interested reader is referred to these sections for details on these three simulation case studies on the 1D heat conduction problem.

7.5.2 Diffusion and Reaction in a Catalyst Pellet

A standard problem in reaction engineering is diffusion and reaction in a solid catalyst pellet. A reaction taking place in a spherical catalyst pellet of radius R is modeled as

$$0 = \frac{\mathcal{D}_e}{r^2}\frac{d}{dr}\left(r^2\frac{dC_A}{dr}\right) - r_A \tag{7.84}$$

where

C_A is the concentration of the reactant in the pellet
$\mathcal{D}_e$ is the effective diffusivity
r_A is the rate of reaction

The above equation can be written as

$$\mathcal{D}_e\frac{d^2C_A}{dr^2} + \frac{2\mathcal{D}_e}{r}\frac{dC_A}{dr} - r_A = 0 \tag{7.85}$$

We will consider two different cases for solving and analyzing the system: (i) first-order reaction and (ii) Langmuir-Hinshelwood (LH) rate kinetics.

7.5.2.1 Linear System and Thiele Modulus

The above equation can be non-dimensionalized and written as

$$\frac{d^2\psi}{d\xi^2} + \frac{2}{\xi}\frac{d\psi}{d\xi} - \phi^2\psi = 0 \tag{7.86}$$

where

$\phi^2 = kR^2/\mathcal{D}_e$ is known as the Thiele modulus
$\psi = C_A/C_{A0}$ is the dimensionless concentration
$\xi = r/R$ is the dimensionless radius

The boundary conditions are no-flux condition at the center $\left(\left[C_A'\right]_{r=0} = 0\right)$ and the concentration at the outer surface of the pellet is $[C_A]_R = C_{A0}$. The interested reader may refer to any reaction engineering text (see Fogler, 2008) to verify that the concentration at any point in the pellet is given by

$$\psi = \frac{1}{\xi} \frac{\sin h(\phi\xi)}{\sin h(\phi)} \tag{7.87}$$

We will write a MATLAB code to simulate mass transfer and reaction in the spherical pellet using TDMA and compare the results with the above solution. As Equations 7.85 and 7.86 are equivalent, either of them can be simulated to obtain the same results. We will simulate the dimensionless model in this example, and the original model (Equation 7.84) will be simulated in the next section.

Example 7.8 Reaction in Catalyst Pellet: Effect of Thiele Modulus

Problem: Solve Equation 7.86 for diffusion and reaction in a spherical pellet of radius $R = 2$ mm. Effective diffusivity of the species in the pellet is $\mathcal{D}_e = 10^{-5}$ cm²/s, and reaction rate constant is $k = 6.25$ s⁻¹. Also verify the effect of the Thiele modulus by changing k.

Solution: The model is discretized as follows:

$$\frac{\psi_{i+1} - 2\psi_i + \psi_{i-1}}{h^2} + \frac{\psi_{i+1} - \psi_{i-1}}{\xi_i h} - \phi^2 k = 0 \tag{7.88}$$

where $h = 1/n$, since we are working with dimensionless quantities, and $\xi_i = (i-1)/n$. Collecting appropriate terms together, the tridiagonal equation for the internal nodes is

$$\underbrace{\left(\frac{1}{h^2} - \frac{1}{\xi_i h}\right)}_{\beta_1}\psi_{i-1} + \underbrace{\left(-\frac{2}{h^2} - \phi^2\right)}_{\beta_2}\psi_i + \underbrace{\left(\frac{1}{h^2} + \frac{1}{\xi_i h}\right)}_{\beta_3}\psi_{i+1} = 0 \tag{7.89}$$

At the first node, the ghost-point boundary condition yields $\psi_{i+1} = \psi_0$. Thus, β_1 is not relevant and $\beta_3 = 2/h^2$.

The following code solves the ODE-BVP:

```
% Simulation of porous catalyst pellet
% as a function of Thiele Modulus
%% Model parameters
diffCoeff=0.25;
R=0.2;
```

```
k=6.25;
C_A0=5e-5;
phi=sqrt(k/diffCoeff)*R;
%% Discretization and Precomputing
n=20;
h=1/n;       % Dimensionless
xi=0:h:1;    % Dimensionless

%% Coefficients for various terms
beta1=1/h^2 - 1./(xi*h);
beta1(1)=0; beta1(end)=0;       % Boundaries
beta2=-2/h^2 - phi^2;
beta2=beta2*ones(1,n+1);           % At center
beta2(end)=1;                   % Outer surface
beta3=1/h^2 + 1./(xi*h);
beta3(1)=2/h^2; beta3(end)=0; % Boundaries
%% Solving and displaying results
b=[zeros(n,1); 1];
C_A=thomasSolve(beta1,beta2,beta3,b);
psi=sinh(phi*xi)/sinh(phi)./xi;   % Analytical
plot(xi,C_A,'-b', xi,psi,'r.');
```

The results are shown in Figure 7.8. Solid lines represent the numerical solution, and dots represent the analytical result as per Equation 7.87. The effect of varying the Thiele modulus is seen in the figure. Since the Thiele modulus is the ratio of reaction rate to diffusion rate, large values of the modulus result in diffusion limitation. Conversely, when $\phi \lesssim 1$, diffusion is fast and there is no diffusion limitation in the system.

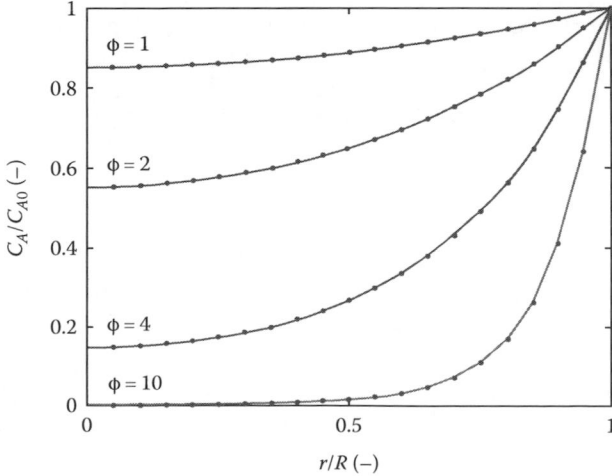

FIGURE 7.8 Simulation of reaction and diffusion in a spherical pellet for various values of the Thiele modulus. Lines show numerical results, and symbols show the analytical solution.

7.5.2.2 Langmuir-Hinshelwood Kinetics in a Pellet

The catalytic reaction is often given by Langmuir-Hinshelwood rate kinetics. As before, the reaction rate will be given by

$$r = \frac{kC_A}{\sqrt{1 + K_r C_A^2}} \tag{3.92}$$

This problem will be solved using the dimensional Equation 7.85 in the CGS system. The concentration at the surface of the pellet is $C_{A0} = 5 \times 10^{-5}$ mol/cm³, the reaction rate constant is 100 s⁻¹, and the constant $K_r = 10^9$ cm⁶/mol². We will solve this using $\texttt{fsolve}$. The equations after discretization are given by

$$\frac{D_e}{h^2}\left(C_{i+1} - 2C_i + C_{i-1}\right) + \frac{D_e}{r_i h}\left(C_{i+1} - C_{i-1}\right) - r\left(C_i\right) = 0, \quad \text{for } i = 2 \text{ to } n$$

$$\frac{D_e}{h^2}\left(2C_{i+1} - 2C_i\right) - r\left(C_i\right) = 0, \quad \text{for } i = 1$$

$$C_{n+1} - C_{A0} = 0 \tag{7.90}$$

An important aspect while writing codes is to benchmark them with known solutions. Since the ODE-BVP with LH kinetics is complex, it is difficult to benchmark them. So, we will take an indirect approach here. We will write the code for the original problem and execute it with $K_r = 0$. With this change, the LH kinetics will be equivalent to the first-order reaction that we saw in Example 7.7. After comparing these results with those from the previous numerical code, we will analyze the effect of various parameters. The next example demonstrates this approach.

Example 7.9 Diffusion and Reaction with Langmuir-Hinshelwood Kinetics

Problem: Solve the problem for the spherical pellet of Equation 7.85, with reaction rate given by the Langmuir-Hinshelwood model of Equation 3.92.

Solution: First, we will write the MATLAB function to compute the residuals as per Equation 7.90. This function file $\texttt{thieleLHfun.m}$ is given below:

```
function gval=thieleLHfun(Y,par)
%% Getting parameters
n=par.n;
h=par.h;
k=par.k;
Kr=par.Kr;
Da=par.diffCoeff;
```

```
%% Building the model
gval=zeros(n+1,1);
for i=1:n
    C=Y(i);
    % Reaction term
    rRxn=k*C/sqrt(1+Kr*C^2);
    % Diffusion terms
    if (i==1)
        diffTerm=Da/h^2*(2*Y(i+1)-2*Y(i));
    else
        diffTerm=Da/h^2*(Y(i+1)-2*Y(i)+Y(i-1)) ...
            + Da/((i)*h*h)*(Y(i+1)-Y(i-1));
    end
    gval(i)=diffTerm-rRxn;
end
gval(n+1)=Y(n+1)-par.C0;
```

The *driver script* will be written in a similar manner as before, using fsolve to obtain the solution as the concentration profile.

```
% Simulation of porous catalyst pellet
% with Langmuir-Hinshelwood kinetic model
%% Model parameters
R=0.2;
n=100;
h=R/n;
catParam.R=R;
catParam.n=n;
catParam.h=h;

catParam.diffCoeff=0.25;
catParam.k=100;
catParam.Kr=1e10;
catParam.C0=5e-5;

%% Initializing and Solving
Yguess=catParam.C0*ones(n+1,1);
opt=optimset('tolx',1e-12,'tolfun',1e-12);
Ysol=fsolve(@(Y) thieleLH1(Y,catParam), Yguess);
%% Plotting the results
xi=0:1/n:1;
psi=Ysol/catParam.C0;
plot(xi,psi,'-.r');
xlabel('r/R (-)'); ylabel('C_A/C_{A0} (-)');
```

Figure 7.9 shows the results of benchmarking the nonlinear code, by changing the underlined line of code to vary K_r. The results with $K_r = 0$ are compared with the first-order reaction case (with $\phi = 4$). The numerical results from the nonlinear solver match the analytical solution very well.

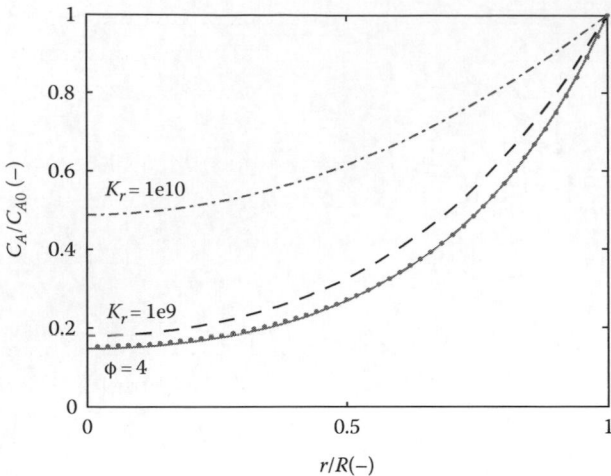

FIGURE 7.9   Benchmarking of code with LH kinetics with $K_r = 0$ (solid line) by comparing it with the analytical solution for $\phi = 4$ (symbols). The effect of varying K_r is also shown.

Figure 7.9 also presents the effect of varying K_r on the concentration profile in the pellet. Note that the baseline case considered in Chapter 3 was $K_r = 1$ m^6/mol^2, which equals $K_r = 10^{12}$ cm^6/mol^2. At this value of K_r, the diffusion is very fast, and nearly the entire pellet concentration is the same as the surface concentration. This is because of the small pellet size. For larger pellets (or smaller pores, which reduces $\mathcal{D}_e$), mass transfer effects gain importance. However, with highly porous catalyst particles of size ~1 mm, internal diffusion may usually be assumed to be fast.

Finally, Figure 7.10 shows the effect of concentration at the pellet surface on concentration profiles inside the pellet. Four cases are considered, with C_{A0} calculated using the ideal gas law at 300 K and partial pressure, p_A, in the gas phase being 1, 2, 5, and 10 bar.

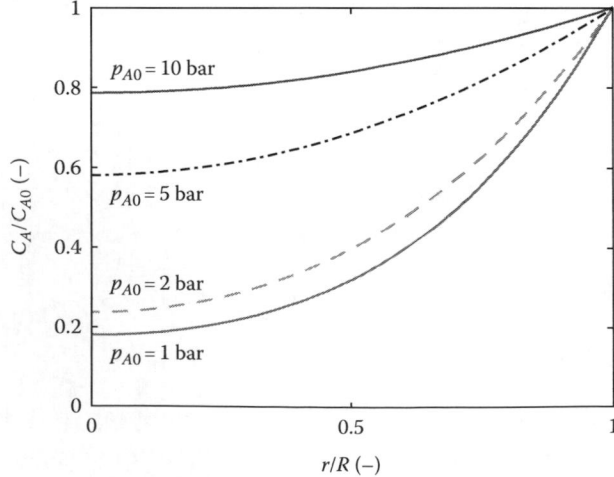

FIGURE 7.10   Effect of varying the concentration at the catalyst surface.

As the concentration increases, the inhibition term (denominator in the kinetic rate expression) increases and the pellet increasingly becomes reaction-limited.

The above example brings us to the end of this section. Irrespective of the coordinate system, the key approach in solving ODE-BVPs, elliptic PDEs, and other similar problems remains the same: using appropriate numerical differentiation to convert the differential equations into a set of algebraic equations and solving these equations either directly or by exploiting their sparse structure.

7.6 EPILOGUE

This chapter dealt with numerical techniques for solving linear or nonlinear equations with sparse structure. This implies that the matrix A in the linear equation

$$A\mathbf{x} = \mathbf{b}$$

or the Jacobian, in case of the nonlinear function, $J = \nabla \mathbf{g}(\mathbf{x})$ has a banded structure, as described in Figure 7.1. A banded system of equations most commonly arises in process simulation due to the discretization of BVPs.

In the case of linear system of equations, we focused on two "special" methods to efficiently solve the linear equations: (i) the TDMA and (ii) the Gauss-Siedel method. From a numerical techniques perspective, we discussed how TDMA (or a banded solver) is more efficient than a general-purpose Gauss Elimination–based solver. Gauss-Siedel is introduced as one of the key iterative methods for linear systems.

From a nonlinear systems perspective, a method of successive substitution was introduced, which iteratively employs a TDMA algorithm to solve a tridiagonal set of model equations. This method was heuristically motivated and is straightforward to use; however, it may not provide a desirable convergence behavior when solving several nonlinear problems.

The Gauss-Siedel method, on the other hand, was introduced for both linear and nonlinear equations. This iterative method is a general-purpose method, but it is also easy to exploit banded structure of matrix A to improve the computational requirement on Gauss-Siedel. This method benefits from the fact that the band-diagonal matrix obtained after finite difference discretization is often diagonally dominant. Convergence properties can further be improved using the under-relaxation approach or source term linearization.

Finally, we introduced the option `'jacobPattern'` in MATLAB's `fsolve` solver.

Based on the discussion in this chapter and experience, I make the following recommendations: (i) TDMA or banded solver for linear systems, (ii) `fsolve` as a method of choice for nonlinear banded systems, and (iii) Gauss-Siedel with appropriate under-relaxation if the problem does not give the desired results with `fsolve`.

EXERCISES

Problem 7.1 Consider a nonlinear equation $x = G(x)$, which is to be linearized and solved using fixed point iteration. The linearized equation is given by

$$x = G_0^{(i)} + G_1^{(i)} x$$

$$G_0^{(i)} = G\left(x^{(i)}\right) - x^{(i)} G'\left(x^{(i)}\right), \quad G_1^{(i)} = G''\left(x^{(i)}\right)$$

Show that solving the fixed point iteration of a linearized model at each iteration

$$x^{(i+1)} = \frac{G_0^{(i)}}{1 - G_1^{(i)}}$$

is equivalent to using the Newton-Raphson method for the original problem.

Problem 7.2 The code for TDMA solver `thomasSolve.m` was given in Example 7.2. This code takes vectors `l,u,d,b` and solves the tridiagonal linear equations problem and returns the solution `x`. In the other examples, we created matrix `A` and vector `b`. Convert the `thomasSolve` code into a new function `newThomasSolve`, which would take tridiagonal matrix `A` and vector `b` as inputs and return `x`.

Problem 7.3 Modify the code from Example 7.6 to implement the TDMA-based successive substitution method with under-relaxation. Solve for $\omega = 0.25$ and $\omega = 0.1$.

Problem 7.4 Modify the Gauss-Siedel code from Example 7.7 for Jacobi iterations with and without under-relaxation. Hence, solve the heat conduction problem.

CHAPTER 8

Implicit Methods

Differential and Differential Algebraic Systems

8.1 GENERAL SETUP

The underlying theme of this chapter is *implicit* methods for solving differential equations. Examples of systems covered in this chapter include ordinary differential equations (ODEs) that are stiff in nature, partial differential equations (PDEs) that require implicit time stepping, and systems governed by differential algebraic equations (DAEs, such as differential equations with algebraic constraints). However, DAEs will remain a *major focus* of this chapter.

This is an advanced chapter that builds upon the analyses of ODEs in Chapter 3, PDEs in Chapter 4, linear system dynamics in Chapter 5, and algebraic equations in Chapter 6.

The final case study of Chapter 3 presented an example where ode45 is unable to converge for default ODE solver options. This is an example of a stiff system of ODE-IVP. The Crank-Nicolson method was introduced in Chapter 4 for solving hyperbolic PDEs, but we deferred it until this chapter. Finally, DAEs, which consist of both differential and algebraic equations, are a class of problems of interest to chemical engineers. Traditionally, these three topics are covered separately in textbooks. Since a unifying theme in this broad spectrum of topics is the use of *implicit* methods, I present a unified coverage in this chapter.

The remainder of this section will provide more details on each of the three topics. Since ODE-IVP and PDEs were covered at length in Chapters 3 and 4, respectively, most of this chapter will be devoted to DAEs.

8.1.1 Stiff System of Equation

Stiffness is a property of a single ODE or a system of ODEs where fast and slow dynamics appear simultaneously. Stiff ODEs are difficult to solve using explicit Runge-Kutta (RK) methods. For example, the simulation of a chemostat in Section 3.5.4 showed that when the value of saturation constant K_s was changed from 0.25 to 0.005, ode45 gave unstable response for the latter value. I mentioned in the example that this unstable behavior of ode45 was because the set of ODEs becomes *stiff*.

This issue and its implication on numerical methods are analyzed presently.

A linear approximation of this nonlinear chemostat model around its steady state was obtained in Chapter 5. The linearized model is given by

$$\frac{d}{dt}\overline{\mathbf{y}} = J\overline{\mathbf{y}} \tag{8.1}$$

where $\overline{\mathbf{y}}$ represents deviation variables, that is, $\overline{\mathbf{y}} = \begin{bmatrix} S - S_{ss} & X - X_{ss} \end{bmatrix}^T$, where S_{ss} and X_{ss} are the steady state exit concentrations of the substrate and biomass, respectively. For the stiff ODE case (i.e., $K_s = 0.005$), the steady state concentrations are $S_{ss} = 0.002$ mol/L and $X_{ss} = 3.749$ mol/L. Under these conditions, the Jacobian is

$$J = \begin{bmatrix} -201.7 & -0.133 \\ 151.2 & 0 \end{bmatrix} \tag{8.2}$$

The eigenvalues of the Jacobian are −201.6 and −0.1, and the ratio of the two eigenvalues is ~2000. The system is stable because the two eigenvalues are negative.

Recall that the transient solution of a system $y' = \lambda y$ is $y(t) = y_0 e^{\lambda t}$. We discussed in Chapter 5 that, for a linear ODE, each eigenvalue governs the dynamic response of the system along the corresponding eigenvector. The larger the eigenvalue, the faster is the response. Thus, the eigenvector corresponding to $\lambda_1 = -201.6$ is the fast eigenmode. The time required to attain steady state is ~50 min, since it is determined by the slow eigenvalue, $\lambda_2 = -0.1$. Compared to the overall process dynamics, the dynamics along the first eigenvector is almost instantaneous. The phase-plane plot for this system is shown in Figure 8.1. Note that the dynamics along the direction $v_1 \equiv [-0.8; 0.6]$ stabilize rapidly, and the system then moves along the slow eigenvector $v_2 \equiv [0.0007; 1]$.

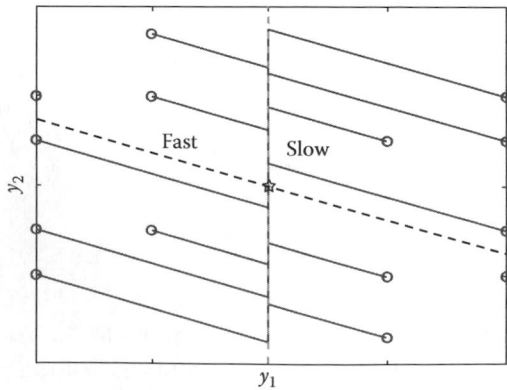

FIGURE 8.1 Phase-plane plot for a moderately stiff system obtained by linearizing the chemostat bioreactor model at its steady state value.

The upper limit on the step-size for an explicit ODE-IVP method is determined by the largest eigenvalue. For example, the largest step-size for Euler's explicit method is given by $h < 2/\lambda_1$. Thus, the step-size, h, needs to be less than 0.01, for Euler's explicit method to be stable. However, the time to reach steady state is ~50 min, since it is governed by the slower λ_2. Therefore, at least 5000 steps will be required by an explicit solver to reach the final solution. In fact, for the above example, ode45 required ~12,000 steps.

An implicit solver, ode15s, on the other hand, required only 93 steps. I will use Figure 8.1 to qualitatively explain the "mechanism" behind this observation. In the first 0.05 s, the system responds along the first eigenvector (dashed line marked as "fast"), and the solver reaches the slow eigenvector (dashed line marked as "slow"). The solver takes small steps during this phase for accuracy. Now the response along eigenvector v_1 has settled, and the system now moves along the slow eigenvector v_2. It is now possible to take larger steps, and still maintain the desired accuracy. This is exactly what ode15s does. In contrast, ode45 takes a very large number of steps because there is a stringent upper limit on step-size for stability.

The analysis of the original nonlinear chemostat from Section 3.5.4 follows the same arguments as above, but is complicated by the nonlinear reaction term. After the evolution of the system along the fast eigenvector, the solver attempts to take larger steps. Unlike the linear case, ode45 does not account for the rapidly changing nonlinear reaction term and fails to converge. Thus, the presence of fast and slow dynamics simultaneously made the implicit ode15s the preferred solver for stiff ODEs.

8.1.1.1 Stiff ODE in Single Variable

Let us consider the system

$$y(t) = e^{-1000t} + e^{-t}$$

which also has fast and slow dynamics, simultaneously. Differentiating

$$y' = -1000e^{-1000t} + (-1000 + 999)e^{-t}$$

gives us the ODE

$$\frac{dy}{dt} = -1000y + 999e^{-t}, \quad y(0) = 2 \tag{8.3}$$

Solving the above ODE using ode45 takes 12000 steps, whereas ode15s takes only 67 steps. The cause for this is the same as the multivariable case: $\lambda = -1000$ dictates the step-size while e^{-t} determines the time required for the dynamics to settle to a steady state.

The original chemostat problem of Section 3.5.4 showed how a stiff system of ODEs can make an explicit solver unstable. Several examples of practical interest may be significantly more stiff than chemostat. I used this section to define stiffness and to demonstrate it from a practical perspective. As in Chapter 3, we will devote Section 8.2 to implicit methods for solving ODEs.

8.1.2 Implicit Methods for Distributed Parameter Systems

The Crank-Nicolson method is an implicit method for solving hyperbolic and parabolic PDEs. A generic hyperbolic PDE discussed in Chapter 4 is given by

$$\frac{\partial \phi}{\partial t} + u \frac{\partial \phi}{\partial z} = \mathcal{S}(\phi) \tag{8.4}$$

Finite difference approximation is used in both spatial and temporal directions. The spatial domain is split in n intervals and the interval size is $\Delta_z = L/n$; a step-size of Δ_t is used in time, and $\phi_{p,i}$ represents the solution at time p and location i. The solution is known at all locations at time p, and this information is used to obtain the solution at next time, $p + 1$.

The Crank-Nicolson method was derived in Chapter 4. A central difference formula is used for the spatial derivative and an implicit formula similar to the trapezoidal rule is used in time. This results in the following implicit formula for the *Crank-Nicolson method*:

$$\frac{\phi_{p+1,i} - \phi_{p,i}}{\Delta_t} + \frac{u}{2}\left(\frac{\phi_{p,i+1} - \phi_{p,i-1}}{2\Delta_z} + \frac{\phi_{p+1,i+1} - \phi_{p+1,i-1}}{2\Delta_z}\right) - \frac{1}{2}\left(\mathcal{S}_{p,i} + \mathcal{S}_{p+1,i}\right) = 0 \tag{8.5}$$

The Crank-Nicolson method will be demonstrated in Section 8.3.

8.1.3 Differential Algebraic Equations

This brings us to the third set of problems: differential algebraic equations or DAEs, which form a major focus of this chapter. Consider a system whose behavior is defined by n-dependent variables, $\mathbf{Y} \in \mathcal{R}^n$. Some dependent variables are governed by differential equations:

$$\frac{d}{dt}\mathbf{y} = \mathbf{f}(t,\mathbf{y},\mathbf{x}) \tag{8.6}$$

where $\mathbf{y} \in \mathcal{R}^{n_d}$ and $\mathbf{Y}^T = \begin{bmatrix} \mathbf{y}^T & \mathbf{x}^T \end{bmatrix}$. The system is underdefined and needs additional $(n - n_d)$ equations. In the case of DAEs, these additional equations are algebraic:

$$0 = \mathbf{g}(\mathbf{y},\mathbf{x}) \tag{8.7}$$

The variables, $\mathbf{y}$, that appear in differential form in Equation 8.6 are differential variables, whereas the remaining variables, $\mathbf{x}$, are algebraic.

DAEs of the form of Equations 8.6 and 8.7 form one class of DAEs, known as *semiexplicit DAEs*. A general form for representing DAEs is

$$F(t,\mathbf{Y},\mathbf{Y}') = 0 \tag{8.8}$$

While several numerical methods are available to solve the above general form of DAE, we will restrict ourselves to semiexplicit DAEs in this chapter.

Typically, fully implicit methods are used to solve DAEs. Some of the popular, so-called *linear multi-step methods* will be discussed in Section 8.2. Analysis and classification of DAEs and their implementation in MATLAB will be discussed in Section 8.4.

8.2 MULTISTEP METHODS FOR DIFFERENTIAL EQUATIONS*

Due to their importance in modeling physical phenomena, differential equations have received a lot of attention. All ODE-solving methods work thusly: Given the values of dependent variables y_k at any time k and the functional form of the differential equation, how does one efficiently compute the value y_{k+1} at the next time instance?

Various classes of techniques have been developed for solving ODEs. Explicit methods are easier to use and provide good accuracy. They are however unsuitable for highly stiff problems, problems with algebraic constraints, and in problems involving convection without attenuation. The RK class of methods, discussed in Chapter 3, form an important class of *explicit* ODE-solving methods. The distinguishing feature of RK methods is that the slope of the curve $y(t)$ is computed as a weighted average of function values computed at various points between k and $(k + 1)$. It involves *projecting* the solution within this time interval and using that information to improve the accuracy of the technique.

When reaching the solution $y(k)$ at time k, we have computed an entire history of solutions, $y_0, y_1, \ldots, y_k$. RK methods, however, do not use this historical information in computing the next solution. Multistep methods, on the other hand, use the past information to compute the solution at the next time-step. The update equation for solving an ODE using a general *linear multistep* method is written as

$$y_{k+1} = \left[a_1 y_k + a_2 y_{k-1} + \cdots + a_n y_{k \pm n+1} \right] + b_0 h f \left(t_{k+1}, y_{k+1} \right)$$
$$+ h \left[b_1 f \left(t_k, y_k \right) + \cdots + b_m f \left(t_{k-m+1}, y_{k-m+1} \right) \right] \tag{8.9}$$

The above equation leads to an explicit method if $b_0 = 0$; if b_0 is nonzero, then the method is implicit. The most popular multistep methods are the following:

- *Adams-Moulton* (AM) methods are implicit methods, with $a_j = 0$.

- *Adams-Bashforth* (AB) methods are explicit methods, with $a_j = 0$ and $b_0 = 0$.

- *Backward difference formula* (BDF) methods are implicit methods, with $b_j = 0$ ($j \geq 1$).

While there are other solvers, we will primarily focus on these three classes of methods. I will end this discussion by mentioning one more key property of multistep methods.

---

* This section is added for completeness. It may be skipped without any loss in continuity.

Non-self-starting property: The term "multistep" implies that the information at past n points is used for computing the next point, $\mathbf{y}_{k+1}$. Let us say that we are at the initial time t_0 with $\mathbf{y}_0$ as the initial value. In order to compute the next solution, $\mathbf{y}_1$, we need the previous data points $\mathbf{y}_0$, $\mathbf{y}_{-1}$, and so on. Since the values $\mathbf{y}_{-1}$ (and earlier) are not available, it is not feasible to directly compute $\mathbf{y}_1$. Thus, the higher-order multistep methods are non-self-starting; they require a procedure to compute past values and start the ODE solution.

I will now briefly summarize the three classes of methods. The focus of this chapter is not on numerical techniques per se; rather, we are interested in using differential equation solvers in MATLAB® for solving problems of practical interest. A brief discussion of these methods is provided to equip readers to make an educated choice of solvers in one's work.

8.2.1 Implicit Adams-Moulton Methods

The core idea behind Adams' family of methods (both Adams-Moulton as well as Adams-Bashforth) is that the function $f(t,y)$ is approximated as an interpolated polynomial $\bar{f}(t,y(t))$. Thus

$$\int_{k}^{k+1} dy = \int_{kh}^{(k+1)h} \bar{f}(\tau,y(\tau))d\tau \tag{8.10}$$

The left-hand side evaluates simply as $(y_{k+1}-y_k)$. The polynomial on the right-hand side varies depending on the number of past terms included in interpolation, giving rise to different Adams-Moulton methods of varying order.

The first-order method, AM-1, is simply the Euler's implicit method since it uses a single point, $\bar{f} = f_{k+1}$, for interpolation, resulting in

$$y_{k+1} = y_k + hf_{k+1} \tag{8.11}$$

where a shorthand notation

$$f_i \triangleq f(t_i,y_i) \tag{8.12}$$

is used to represent the data point.

The second-order method, AM-2, uses a linear interpolation between f_{k+1} and f_k:

$$y_{k+1} = y_k + \int_{kh}^{(k+1)h} \left(f_k + \frac{f_{k+1} - f_k}{h}(\tau - kh) \right) d\tau \tag{8.13}$$

Solving this, we get

$$y_{k+1} = y_k + \left[f_k h \right] + \left(\frac{f_{k+1} - f_k}{h} \right) \left(\left[\frac{\tau^2}{2} \right]_{kh}^{(k+1)h} - kh \left[\tau \right]_{kh}^{(k+1)h} \right) \tag{8.14}$$

The last term in the brackets expands as

$$h^2 \left(\frac{(k+1)^2 - k^2}{2} - k \right)$$

Since this evaluates to $h^2/2$, the final equation for AM-2 method is given by

$$y_{k+1} = y_k + \frac{h}{2} \left(f(t_k, y_k) + f(t_{k+1}, y_{k+1}) \right) \tag{8.15}$$

The second-order Adams-Moulton method is also known as the trapezoidal rule since it *looks like* the trapezoidal rule from numerical integration (see Appendix E).

8.2.2 Higher-Order Adams-Moulton Method

A formal derivation of Adams-Moulton methods, including error analysis, can be more conveniently from polynomial interpolation formulae. Let us consider the derivation of AM-2 method again, using Newton's backward difference formula. Refer to Appendix D, where we derived Newton's forward difference formula; the backward difference formula works in a similar manner. Nondimensional time, α, is a real valued parameter defined as

$$\alpha = \frac{t - t_{k+1}}{h} \tag{8.16}$$

Thus, $t_{k+1} \to (\alpha = 0)$, $t_k \to (\alpha = -1)$, $t_{k-1} \to (\alpha = -2)$, and so on. In the case of AM-2 method, a second-order polynomial is used. Using Newton's backward difference formula

$$\bar{f}_B(t) = f_{k+1} + \alpha \nabla f_{k+1} + \cdots + \frac{\alpha(\alpha+1)\cdots(\alpha+n-1)}{n!} \nabla^n f_{k+1} + R_B \tag{8.17}$$

where

$$R_B = \frac{\alpha(\alpha+1)\cdots(\alpha+n)}{(n+1)!} h^{n+1} f^{n+1}(\xi) \tag{8.18}$$

AM-2 method is derived with $n = 1$, that is, by retaining the first two terms in Equation 8.17. Thus, the equation for AM-2 is given by substituting this in Equation 8.10. Since $d\tau = h\, d\alpha$

$$y_{k+1} = y_k + h \int_{-1}^{0} \left[f_{k+1} + \alpha \nabla f_{k+1} + \frac{\alpha(\alpha+1)}{2} h^2 f''(\xi) \right] d\alpha \qquad (8.19)$$

Using the definition of ∇f_{k+1}, we get

$$y_{k+1} = y_k + h \left[f_{k+1} \left[\alpha \right]_{-1}^{0} + \left(f_{k+1} - f_k \right) \left[\frac{\alpha^2}{2} \right]_{-1}^{0} + \frac{h^2}{2} f''(\xi) \left[\frac{\alpha^3}{3} + \frac{\alpha^2}{2} \right]_{-1}^{0} \right]$$

$$= y_k + h \left[f_{k+1} + \left(f_{k+1} - f_k \right) \left(-\frac{1}{2} \right) + \frac{h^2}{2} f''(\xi) \left(-\frac{1}{6} \right) \right] \qquad (8.20)$$

Thus, the AM-2 method with the local truncation error is given by the following equation:

$$y_{k+1} = \underbrace{y_k + \frac{h}{2} \left(f\left(t_k, y_k\right) + f\left(t_{k+1}, y_{k+1}\right) \right)}_{\text{AM-2}} + \underbrace{\left(-\frac{h^3}{12} f''(\xi) \right)}_{\mathcal{O}(h^3)} \qquad (8.21)$$

Thus, AM-2 has a local truncation error as shown above. As we observed in Chapter 3, the global truncation error (GTE) in the ODE solver drops by one order, making AM-2 method $\mathcal{O}(h^2)$ accurate with respect to the GTE.

Higher-order Adams-Moulton formulae are derived by progressively including an additional term in the backward difference formula of Equation 8.17. The first few Adams-Moulton formulae are given in Table 8.1. Note that the AM-n method is nth order accurate since the GTE is $\sim\mathcal{O}(h^n)$.

8.2.3 Explicit Adams-Bashforth Method

Explicit Adams-Bashforth methods follow a similar principle as Adams-Moulton methods, except that b_0 in Equation 8.9 is 0, making them explicit ODE-solving methods. Thus, a

TABLE 8.1   Coefficients for Adams-Moulton Methods

Method	b_0	b_1	b_2	b_3	b_4	GTE
AM-2	1/2	1/2				$\mathcal{O}(h^2)$
AM-3	5/12	8/12	−1/12			$\mathcal{O}(h^3)$
AM-4	9/24	19/24	−5/24	1/24		$\mathcal{O}(h^4)$
AM-5	251/720	646/720	−264/720	106/720	−19/720	$\mathcal{O}(h^5)$

Only methods up to AM-5 are shown. Interested readers are referred to numerical ODE texts for coefficients of higher-order methods.

Newton's backward difference polynomial is fitted to the past n points. Readers can easily notice that the first-order Adams-Bashforth method (AB-1) is simply the explicit Euler's method. As in Adams-Moulton methods, the following real variable will be defined for deriving Adams-Bashforth formulae:

$$\alpha = \frac{t - t_k}{h} \tag{8.22}$$

A backward interpolating polynomial is fitted to the data $f_k, f_{k-1}, \ldots, f_{k-n+1}$ and substituted in Equation 8.10 to obtain

$$y_{k+1} = y_k + \int_0^1 \overline{f}_{B,k}(\alpha) h \, d\alpha \tag{8.23}$$

Clearly, since the polynomial is fitted to the past data points

$$\overline{f}_B(t) = f_k + \alpha \nabla f_k + \cdots + \frac{\alpha(\alpha+1)\cdots(\alpha+n-1)}{n!} \nabla^n f_k + R_B \tag{8.24}$$

The *mathematical* contrast between AB-n and AM-n methods will be evident on comparing Equation 8.16 with (8.22) and Equation 8.17 with (8.24). The *practical* contrast indeed is that they are explicit and implicit methods, respectively. Notice also that integrating from k to $k + 1$ implies integrating from $\alpha = 0$ to 1, in AB-n method. Substituting and integrating, we get

$$y_{k+1} = y_k + h \left[f_k [\alpha]_0^1 + (f_k - f_{k-1}) \left[\frac{\alpha^2}{2} \right]_0^1 + \frac{h^2}{2} f''(\xi) \left[\frac{\alpha^2}{2} + \frac{\alpha^3}{3} \right]_0^1 \right] \tag{8.25}$$

Thus, the AB-2 method is given by the following formula:

$$y_{k+1} = \underbrace{y_k + \frac{h}{2}\left(3f(t_k, y_k) - f(t_{k-1}, y_{k-1})\right)}_{\text{AB–2}} + \underbrace{\left(\frac{5h^3}{12} f''(\xi)\right)}_{\mathcal{O}(h^3)} \tag{8.26}$$

Since the right-hand side does not depend on $(k + 1)$, AB-2 is an explicit method. AB-2 is not a self-starting method because y_1 depends on y_{-1}, and hence the method can only start from computing y_2 onward.

   Higher-order AB-n methods include a larger number of terms in the formula as well as yield higher accuracy. These are summarized in Table 8.2.

TABLE 8.2   Coefficients for Adams-Bashforth Method from Orders 2–5

Method	b_0	b_1	b_2	b_3	b_4	b_5	GTE
AB-2	0	3/2	−1/2				$\mathcal{O}(h^2)$
AB-3	0	23/12	−16/12	5/12			$\mathcal{O}(h^3)$
AB-4	0	55/24	−59/24	37/24	−9/24		$\mathcal{O}(h^4)$
AB-5	0	1901/720	−2774/720	2616/720	−1274/720	251/720	$\mathcal{O}(h^5)$

Interested readers are referred to numerical ODE texts for coefficients of higher-order methods. Since AB-n methods are explicit, the coefficient b_0 is zero.

8.2.4 Backward Difference Formula

The backward difference formula (BDF) methods work in a different way than Adams' families of ODE solvers. In Adams methods, a polynomial was fitted to past function values, f_i. Instead, in BDF, a polynomial is fitted to the past solution values y_i and yet-to-be-determined value y_{k+1}. Thus, an nth-order polynomial is fitted to the past n points* to obtain an interpolating polynomial, $\bar{y}_B$. The BDF is derived by differentiating this polynomial and substituting it in the ODE.

For example, the first-order BDF uses the data points y_{k+1}, y_k to obtain the polynomial as

$$\bar{y}_B = y_{k+1} + \frac{y_{k+1} - y_k}{t_{k+1} - t_k}(t - t_{k+1}) \tag{8.27}$$

Clearly, $\bar{y}_B'$ is just the slope of the above line and the denominator is simply h. Thus, we obtain the first-order BDF as simply the Euler's implicit formula:

$$\frac{y_{k+1} - y_k}{h} = f(t_{k+1}, y_{k+1})$$

$$\text{i.e., } y_{k+1} = y_k + h f(t_{k+1}, y_{k+1}) \tag{8.28}$$

How does one derive higher-order BDF? They are derived in quite a similar manner as the Adams-Moulton method, except that (i) $\bar{y}_B(t)$ is fitted instead of $\bar{f}_B(t)$, and (ii) the derivative $\bar{y}_B'$ is substituted in the original ODE instead of integrating $\bar{f}_B$. The derivation of error for first-order BDF is left as an exercise.

Second-order Newton's backward difference formula is given by

$$\bar{y}_B(t) = y_{k+1} + \alpha \nabla y_{k+1} + \frac{\alpha(\alpha+1)}{2} \nabla^2 y_{k+1} + \frac{\alpha(\alpha+1)(\alpha+2)}{6} h^3 y'''(\xi) \tag{8.29}$$

---

* Recall that nth-order polynomial requires $(n + 1)$ points: y_{k+1} and the previous n points.

Since

$$\frac{dy}{dt} = \frac{1}{h}\frac{dy}{d\alpha}$$

$$\bar{y}'_B = \frac{1}{h}\left(\nabla y_{k+1} + \frac{2\alpha+1}{2}\nabla^2 y_{k+1} + \frac{h^3}{6}\left(3\alpha^2 + 6\alpha + 2\right)y'''(\xi)\right) \tag{8.30}$$

Note that the derivative is computed at time $(k+1)$, that is, $\alpha = 0$. Substituting in the ODE yields

$$\left(y_{k+1} - y_k\right) + \frac{1}{2}\left(y_{k+1} - 2y_k + y_{(k-1)}\right) + \frac{h^3}{3}y'''(\xi) = hf\left(t_{k+1}, y_{k+1}\right) \tag{8.31}$$

Rearranging, we get the BDF-2:

$$y_{k+1} = \underbrace{\frac{4y_k - y_{k-1}}{3} + \frac{2h}{3}f\left(t_{k+1}, y_{k+1}\right)}_{\text{BDF-2}} + \underbrace{\frac{2h^3}{9}y'''(\xi)}_{\mathcal{O}\left(h^3\right)} \tag{8.32}$$

It is often customary to write the BDF as

$$\underbrace{y_{k+1} - \frac{4y_k - y_{k-1}}{3} = \frac{2h}{3}f\left(t_{k+1}, y_{k+1}\right)}_{\text{BDF-2}} + \underbrace{\frac{2h^3}{9}y'''(\xi)}_{\mathcal{O}\left(h^3\right)} \tag{8.33}$$

Table 8.3 shows various BDFs of various orders.

One of the most popular numerical methods for solving stiff ODEs and DAEs is Gear's method, which uses BDF of varying orders, from first to fifth order. BDF methods are popular for solving highly stiff problems because they have good stability (up to fifth-order), reasonably high accuracy, and capability to solve differential equations with algebraic constraints. In case of DAE problems, the differential equations retain the entire formula from Table 8.3, whereas the left-hand side is zero for algebraic equations.

TABLE 8.3   Formula for BDF Methods with Various Accuracy

Method	b_0	a_1	a_2	a_3	a_4	a_5	a_6	GTE
BDF-2	2/3	4/3	–1/3					$\mathcal{O}\left(h^2\right)$
BDF-3	6/11	18/11	–9/11	2/11				$\mathcal{O}\left(h^3\right)$
BDF-4	12/25	48/25	–36/25	16/25	–3/25			$\mathcal{O}\left(h^4\right)$
BDF-5	60/137	300/137	–300/137	200/137	–75/137	12/137		$\mathcal{O}\left(h^5\right)$
BDF-6	60/147	360/147	–450/147	400/147	–225/147	72/147	–10/147	$\mathcal{O}\left(h^6\right)$

Most commercial solvers used BDF methods of orders 1–5. Methods beyond BDF-6 are not used since they are not zero stable.

NUMERICAL DIFFERENCE FORMULA (NDF)–BASED SOLVERS

NDF is a modification of the BDF method for solving ODEs and DAEs. Let us take a closer look at BDF. The qth-order term of the polynomial expansion of Equation 8.29 is given by

$$P_q = \frac{\nabla^q y_{k+1}}{q!} \alpha (\alpha + 1) \cdots (\alpha - q + 1) \tag{8.34}$$

and $\bar{y}_B = \sum_{q=0}^{n} P_q$. Recall that in deriving the BDF, we differentiated $\bar{y}_B$ since we required

$$\left[\frac{d}{dt} \bar{y}_B \right]_{(k+1)}$$

in the derivation of BDF. Moreover, at $(k + 1)$, $\alpha = 0$. Differentiating Equation 8.34, we get

$$\frac{d}{d\alpha} P_q = \frac{\nabla^q y_{k+1}}{q!} \left[(\alpha + 1)(\alpha + 2) \cdots (\alpha - q + 1) + \left(\text{terms with } \alpha \right) \right] \tag{8.35}$$

Since we compute the above at $\alpha = 0$, all the terms in α disappear. The remaining term in the bracket at $\alpha = 0$ is $(q - 1)!$. Thus

$$\left[\frac{d}{dt} P_q \right]_{(k+1)} = \frac{1}{h} \left(\frac{\nabla^q y_{k+1}}{q} \right) \tag{8.36}$$

Thus, the BDF of nth-order may also be written as

$$\sum_{q=1}^{n} \frac{\nabla^q y_{k+1}}{q} = hf\left(t_{k+1}, y_{k+1} \right) \tag{8.37}$$

NDF method is a modification of the above:

$$\sum_{q=1}^{n} \frac{\nabla^q y_{k+1}}{q} = hf\left(t_{k+1}, y_{k+1} \right) + \kappa \gamma \left(y_{k+1} - y_{k+1}^{(0)} \right) \tag{8.38}$$

where
   κ is a constant
   $\gamma = [1 + (1/2) + \cdots + (1/n)]$ and
   $y_{k+1}^{(0)}$ is an initial guess of the solution

The final term on the right-hand side is added to improve the convergence behavior of the numerical method.
   MATLAB's `ode15s` solver uses a combination of BDF and NDF.

8.2.5 Stability and MATLAB® Solvers

8.2.5.1 Explicit Adams-Bashforth Methods

The stability of explicit Euler's method was discussed in Chapter 3. For an ODE, $y' = -ay$, the stability requirement for Euler's explicit method was $0 \leq ah < 2$. The analysis of linear ODEs was extended to multivariable systems in Chapter 5. Using the same analogy, the stability condition for explicit Euler's method for multivariable system is

$$-2 < \lambda h \leq 0 \tag{8.39}$$

where λ denotes eigenvalues of the multivariable system. The largest eigenvalue determines the limit on step-size, h. Eigenvalues can be complex numbers. Thus, if we plot λh on a complex plane, the stability region for Euler's explicit method is a circle in the left-half plane, with the circle intersecting the real axis at -2.

A narrow stability region limits the applicability of the ODE solver, as observed for Euler's explicit method. Unfortunately, the stability region for Adams-Bashforth methods shrinks with the increasing order of the method. Thus, higher-order Adams-Bashforth methods are not very stable. While their higher accuracy can be exploited for nonstiff systems, they are unsuitable for stiff ODEs and DAEs. They can be used as an alternative to RK methods when a high degree of accuracy is required.

8.2.5.2 Implicit Euler and Trapezoidal Methods

Euler's implicit method is highly stable. It is said to be A-stable, which implies that it is stable for all values of λh in the left-half plane. In other words, ODE solution using Euler's implicit method is stable for any $h > 0$ if the original ODE itself was stable. However, it is not very useful in practice since it is $\sim\mathcal{O}(h^1)$ accurate. The trapezoidal method, with its stability region spanning the left-half plane, is also A-stable and hence preferred over Euler's implicit method.

8.2.5.3 Implicit Adams-Moulton Methods of Higher Order

Trapezoidal method is AM-2 method. While it has a wide range of stability, it is only $\mathcal{O}(h^2)$ accurate. However, as the order of Adams-Moulton methods increases, the stability region decreases. The stability region of AM-n method, though, is significantly broader than the stability region of AB-n methods. Thus, higher-order AM-n methods present in a trade-off between accuracy and stability of the method. The use of variable-order AM solvers is, therefore, popular.

8.2.5.4 BDF/NDF Methods

Both BDF-1 (implicit Euler's method) and BDF-2 are A-stable. Like AM-n method, the stability region of BDF-n also shrinks as the order n increases. Stability regions for BDF-3 and BDF-4 are also very large, but shrink for BDF-5 and BDF-6. Interestingly, BDF methods beyond the sixth order are not zero-stable.* Consequently, BDF beyond BDF-6 cannot be used. In fact, most commercial solvers use BDF methods of order BDF-1 to BDF-5.

---

* Zero stability implies that a numerical technique is stable as $h \to 0$. This is a very important property as a very small step-size may be needed for higher accuracy. Hence, a method that is not zero stable is not useful in practice.

8.2.5.5 MATLAB® Nonstiff Solvers

The default and most popular ODE-IVP solver in MATLAB, ode45, uses an adaptive step-size fourth-order RK method, with GTE $\sim\mathcal{O}\left(h^4\right)$. As discussed in Chapter 3, ode45 uses a combination of RK-4 and RK-5 methods for step-size adaptation.

ode23 is the other RK solver. It uses a combination of second- and third-order Bogacki-Shampine formulae. It has a GTE $\sim\mathcal{O}\left(h^2\right)$.

The popularity of ode45 is due to its high accuracy. Adaptive step-sizing is seamless in RK methods, making ode45 very efficient and accurate for nonstiff problems.

MATLAB provides ode113, a variable order solver that uses Adams-Bashforth-Moulton (ABM) methods of orders 1 to 13. Variable order implies that the order of the ABM solver is varied based on stability and accuracy. High-order RK methods are not popular because the effort in computing $f(t, y)$ at multiple locations increases rapidly for higher-order methods. However, multistep methods rely on already computed past data. Thus, there is only a modest increase in computational requirement in higher-order ABM methods. If very high accuracy is needed and the system is nonstiff, ode113 may be used.

In my experience, the primary choice of *nonstiff* ODE-IVP solver is ode45.

8.2.5.6 MATLAB® Stiff Solvers

MATLAB also provides a few stiff solvers. The most popular and versatile of them is ode15s, which uses a combination of orders 1–5 NDF or BDF solvers. ode15s is a variable order solver, which means that the order of the solver is varied from first to fifth order based on the stability and accuracy of the solution. It attempts to use constant step-sizes, only changing the step-size if the best method does not provide adequate accuracy. It is also capable of solving moderately stiff index-1 DAE problems as well. BDF and NDF methods are indeed very popular for solving stiff ODEs and DAE problems.

There are some stiff problems that ode15s is unable to solve. In some examples, ode23s may be used. It is a second-order Rosenbrock solver. I did not cover Rosenbrock solvers in our discussions. I often get asked whether ode23s is better than ode15s. It is a different category of solvers, so one cannot conclude whether it is "better" or "worse" than ode15s.

However, since ode15s is a higher-order solver and BDF/NDF methods are robust, it is a preferred choice over all other stiff solvers. One may use ode23s if ode15s does not work.

Finally, we come to two solvers that use the trapezoidal rule: ode23t (trapezoidal rule) and ode23tb (trapezoidal rule followed by BDF-2). Since the trapezoidal rule is A-stable, these methods may be used for highly stiff problems if both ode15s and ode23s fail. The disadvantage of these two methods is that they are only $\mathcal{O}\left(h^2\right)$ accurate. It should be noted that A-stable does not mean *unconditionally* stable; these methods may still fail to converge due to strong nonlinearity, physical constraints, and error propagation at very small values.

In my experience, the primary choice of *nonstiff* ODE-IVP solver is ode15s.

8.3 IMPLICIT SOLUTIONS FOR DIFFERENTIAL EQUATIONS

This section will demonstrate the application of the AM-2 (trapezoidal) method to solve stiff ODE problems (Section 8.3.1) and also demonstrate the Crank-Nicolson method for hyperbolic PDEs (Section 8.3.2). Recall that I had mentioned in Chapter 4 that Crank-Nicolson method for hyperbolic and parabolic PDEs is equivalent to using the trapezoidal method in time. I have therefore included these two examples in this chapter.

8.3.1 Trapezoidal Method for Stiff ODE

The AM-2 trapezoidal method is given by

$$y_{k+1} = \underbrace{y_k + \frac{h}{2}\left(f\left(t_k, y_k\right) + f\left(t_{k+1}, y_{k+1}\right)\right)}_{\text{AM-2}} + \underbrace{\left(-\frac{h^3}{12} f''(\xi)\right)}_{\mathcal{O}\left(h^3\right)} \tag{8.21}$$

At time k, the solution y_k is known and y_{k+1} is the unknown quantity to solve for. Thus, this is a nonlinear equation, the solution of which is the value y_{k+1}. Rewriting the above equation in the form used in Chapter 6 gives

$$\underbrace{y_{k+1} - \left[y_k + \frac{h}{2}\left(f\left(t_k, y_k\right) + f\left(t_{k+1}, y_{k+1}\right)\right)\right]}_{g(y_{k+1})} = 0 \tag{8.40}$$

The stiff ODE (8.3) in single variable will be solved using AM-2 method in the next example. The function $f(t, y)$ in the above equation is simply the ODE function in $y' = f(t, y)$ and h is the step-size, exactly as was seen in the explicit methods in Chapter 3.

Example 8.1 Implicit Trapezoidal Method for Stiff ODE in Single Variable

Problem: Solve the ODE-IVP in Equation 8.3 using the AM-2 trapezoidal method. Compare the results with `ode45`.

Solution: The function $f(t, y) = -1000y + 999e^{-t}$ will be defined using MATLAB anonymous function definition:

```
f=@(t,y) -1000*y+999*exp(-t);
```

This ODE function is used to obtain $g(y_{k+1})$ in Equation 8.40. If the current value y_k is known at time t_k, then

```
fk=f(tk,yk);
```

will give the value $f(t_k, y_k)$, which is a known quantity. This quantity can be precomputed and used in $g(y_{k+1})$ defined in Equation 8.40. If we denote y_{k+1} as variable yNew, then function $g(y_{k+1}) \equiv g$ (yNew) is given by the following function:

```
g=@(yNew) yNew-yk - h/2*(fk+f(tNew,yNew));
```

Note that `tk`, `yk`, and `fk` are known since they are the past values; while `tNew`, the time point at which yNew is calculated is also known (`tNew=tk+h`). Thus, the unknown is yNew. We will use `fsolve` to find the root of Equation 8.40.

The function `f(t,y)` remains the same throughout the simulation, whereas the function `g(yNew)` changes at each time point: The function is updated with most recent values of `tk`, `yk`, and `fk`. The complete code for this problem is given below:

```
% Solving single ODE using
% Adam-Moulton Second Order (Trapezoidal)
y0=2;
t0=0; tN=1;
f=@(t,y) -1000*y+999*exp(-t);

%% Solving using ode45 for comparison
[tSol,ySol]=ode45(f,[t0 tN],y0);
plot(tSol,ySol); hold on
xlabel('t'); ylabel('y(t)');

%% Setting up AM-2 Trapezoidal method
h=1e-3;
n=(tN-t0)/h;
tALL=0:h:tN;
yALL=zeros(1,n+1); yALL(1)=y0;

%% Solving using AM-2 Trapezoidal method
for k=1:n
    % Current time-point
    yk=yALL(k);
    tk=tALL(k);
    fk=f(tk,yk);
    % Next time-point
    tNew=tALL(k+1);
    % Defining g(yNew)
    g=@(yNew) yNew-yk - h/2*(fk+f(tNew,yNew));
    % Solve and store results
    yNew=fsolve(g,yk);
    yALL(k+1)=yNew;
end
plot(tALL,yALL,'--');
```

Figure 8.2 shows the results from the AM-2 trapezoidal method at two different step-sizes. Since the fast dynamics stabilize within $t = 0.003$, a small step-size is required for accuracy. When `h=0.01` was used as the step-size, oscillatory behavior was observed initially. Thus, `h=0.001` was required for stability. With this step-size, AM-2 required 1000 iterations. Despite being an adaptive step-size solver, `ode45` required 1229 iterations.

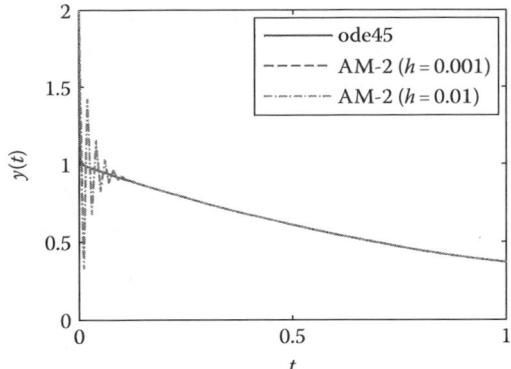

FIGURE 8.2   Comparison of AM-2 trapezoidal method (with two different step-sizes) and ode45 solution for stiff ODE (8.3) in single variable.

8.3.1.1 Adaptive Step-Sizing

Adaptive step-size methods exist for trapezoidal and other Adams-Moulton methods. These can be used to give better accuracy. The faster dynamics, in this example, have a time constant of 0.001 (since $\lambda = 1000$). An adaptive step-size ode45 was unable to take large steps. We now investigate whether it is possible to increase step-size of AM-2 method and achieve stable performance. Adaptive step-size AM-2 methods are beyond the scope of this text. But I will give a flavor of adaptive step-sizing in using this example.

An "adaptive step-sizing" will be implemented in an ad hoc manner. We know that a small step-size is needed for accuracy until the fast dynamics settle down, after which we are at liberty to increase the step-size* for implicit AM-2 method. We will use this knowledge for the next simulation, which shall be performed in three steps:

- Run the code in Example 8.1 with h=0.0005 for 20 iterations to yield solutions of the ODE until tN=0.01.

- Now that the solutions at $t=0$ and $t=0.01$ are available, use a larger step-size of h=0.01 for the next 10 iterations.

- Now with the solutions at $t=0$ and $t=0.1$, use a still larger step-size of h=0.1 for the remaining 10 iterations until we reach tN.

The results from this procedure are not shown since they exactly overlap with the solution from AM-2 with a step-size of h=0.001. No oscillatory behavior is obtained, even in the third stage. With this procedure, an excellent solution can be obtained within 40 iterations.

---

* If we were using an explicit method, such as explicit Euler or RK-2, stability requirement will still limit us from using a large step-size. For example, RK-2 method will require $h \leq 2\lambda$, that is, (h < 0.002) for stability. Implicit method such as trapezoidal method allows a larger h.

MATLAB's `ode15s`, `ode23s`, `ode23t`, and `ode23tb` are all adaptive step-size solvers. Adaptive step-size implicit solvers can be used to obtain accurate solutions with good computational efficiency.

8.3.1.2 Multivariable Example

We are now equipped to solve the bioreactor example from Section 3.5.4 using the AM-2 trapezoidal rule. The case with smaller value of saturation coefficient, $Ks = 0.001$, was stiff and resulted in an unstable solution with `ode45`. The same example (with same values of coefficients and initial condition) will now be simulated using the AM-2 trapezoidal rule.

The function `funMonod.m` remains the same as in the previous example. The anonymous function `f` can be defined as

```
f=@(t,y) monodFun(t,y,monodParam);
```

Since MATLAB naturally works with vectors, the definition of the function "g" remains unchanged. Since the nonlinear functions to be solved are represented by $g(y_{k+1})=0$, one simply needs to ensure that $g(y)$ takes a 3×1 vector argument yNew and returns a 3×1 vector. Thus, `g(yNew)` is defined as

```
g=@(yNew) yNew-yk - h/2*(fk+f(tNew,yNew));
```

This is *exactly* the same as before. The only difference is that yNew, yk, and `fk` are all 3×1 vectors. The actual code for this example is left as an exercise.

Figure 8.3 compares the simulation results using the AM-2 trapezoidal method and `ode15s`. The two results are very close to each other. Since the trapezoidal method is implicit, it captures the sharp change in process trends observed ~19 h. A step-size of `h=0.05` was used for the trapezoidal method; a lower value of *h* gives unstable results.

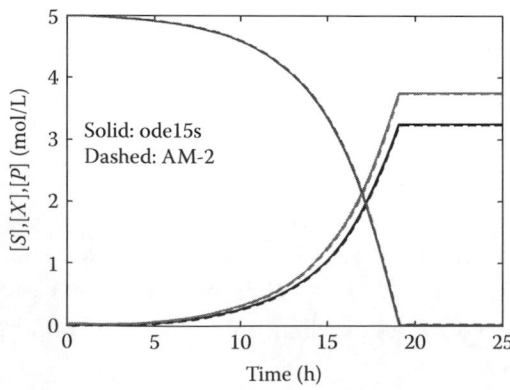

FIGURE 8.3   Bioreactor with Monod kinetics simulated using AM-2 trapezoidal method (dashed lines). The solution using ode15s is plotted as solid lines for comparison.

8.3.2 Crank-Nicolson Method for Hyperbolic PDEs

Various numerical techniques to solve hyperbolic PDEs of the form

$$\frac{\partial \phi}{\partial t} + u \frac{\partial \phi}{\partial z} = \mathcal{S}(\phi) \tag{8.4}$$

were discussed in Chapter 4. The most versatile method is to convert the above PDE into a set of ODEs in time by discretizing the spatial derivative. With the spatial domain split in n intervals, with interval size $\Delta_z = L/n$, the ODE for the solution variable at the ith location is given by

$$\frac{d}{dt}\phi_i = -C_i(\phi) + \mathcal{S}(\phi_i) \tag{8.41}$$

The convection term, $u\phi_z$, is represented with finite difference approximation. The above expression for $C_i(\cdot)$ is written in a very general form. In reality, $C_i(\cdot)$ does not depend on an entire vector ϕ but only on a few terms $\phi_{i-\ell}, \ldots, \phi_i, \ldots, \backslash phi_{i+\ell}$ (as we had seen in the previous chapter). For example, if central difference formula is used, then $\ell = 1$ and

$$C_i(\phi) = u \frac{\phi_{i+1} - \phi_{i-1}}{2\Delta_z} \tag{8.42}$$

Let the right-hand side of Equation 8.41 be represented as $f_i(\phi)$. Thus, the set of ODEs obtained after discretization in space is given by

$$\frac{d}{dt}\phi = \begin{bmatrix} f_1(\phi) \\ f_2(\phi) \\ \vdots \\ f_n(\phi) \end{bmatrix} \tag{8.43}$$

One of the options for solving the above set of ODEs is the AM-2 trapezoidal method. If Δ_t is the time-step size and ϕ_p represents the solution vector at time p, then

$$\phi_{p+1} = \phi_p + \frac{\Delta_t}{2}\left(\mathbf{f}(\phi_{p+1}) + \mathbf{f}(\phi_p)\right) \tag{8.44}$$

The ith equation above is given by

$$\phi_{p+1,i} = \phi_{p,i} + \frac{\Delta_t}{2}\left(f_i(\phi_{p+1}) + f_i(\phi_p)\right) \tag{8.45}$$

Substituting the central difference formula for $C_i(\cdot)$ from Equation 8.42 in the above equation yields us the Crank-Nicolson equation:

$$\phi_{p+1,i} = \phi_{p,i} + \frac{\Delta_t}{2}\left(\left[S_{p+1,i} - u\frac{\phi_{p+1,i+1} - \phi_{p+1,i-1}}{2\Delta_z}\right] + \left[S_{p,i} - u\frac{\phi_{p,i+1} - \phi_{p,i-1}}{2\Delta_z}\right]\right) \qquad (8.46)$$

This can be more conveniently arranged in the form of Equation 8.5:

$$\frac{\phi_{p+1,i} - \phi_{p,i}}{\Delta_t} + \frac{u}{2}\left(\frac{\phi_{p,i+1} - \phi_{p,i-1}}{2\Delta_z} + \frac{\phi_{p+1,i+1} - \phi_{p+1,i-1}}{2\Delta_z}\right) - \frac{1}{2}\left(S_{p,i} + S_{p+1,i}\right) = 0 \qquad (8.5)$$

Thus, the Crank-Nicolson method may be visualized as an implicit trapezoidal method implementation of a PDE with central discretization in space. This derivation also allows the users to extend the Crank-Nicolson method to higher-order formulae. To summarize, at this stage, I have done two things: introduced the Crank-Nicolson method as second-order in time and second-order in space implicit method, and reinterpreted it as an implicit in time implementation of the method of lines approach. The final order of business is to solve the example of a transient plug flow reactor (PFR) from Chapter 4, which yields a hyperbolic PDE. Recall that the PFR model that was solved in Chapter 4 is

$$\frac{\partial C_A}{\partial t} + u\frac{\partial C_A}{\partial z} = -r$$
$$C_A(t=0) = C_{A0}, \quad C_A(t,z=0) = C_{A,\text{in}} \qquad (8.47)$$

A first-order reaction, with $r = kC_A$ and $k = 0.2$ min^{-1}, takes place in the reactor.

Example 8.2 Solving a PFR Problem Using the Crank-Nicolson Method

Problem: Solve the PFR problem from Equation 8.47. The inlet velocity is $u = 0.1$ m/min, length of the PFR is 0.5 m, and inlet concentration is 1 mol/L. The initial concentration is the same as 1 mol/L.

Solution: Let us define individual terms in the overall equation. The convection term, given by Equation 8.42, is

$$C_{p,i} = \begin{cases} (u/2\Delta_z)(\phi_{p,i+1} - \phi_0) & i = 1 \\ (u/\Delta_z)(\phi_{p,i} - \phi_{p,i-1}) & i = n \\ (u/2\Delta_z)(\phi_{p,i+1} - \phi_{p,i-1}) \end{cases} \qquad (8.48)$$

whereas the reaction term is $r_{p,i} = k\phi_{p,i}$. A function file can be written to return the reaction and convection terms for all locations, given the vector ϕ_p at any time p.

(i) *The function to calculate* $C_{p,i}$ *and* $r_{p,i}$ (`pfrDiscreteFun.m`) *is given below:*

```
function [reac,convec] = pfrDiscreteFun(Phi,par)
% This function calculates two key terms of PFR
% model: reaction and convection terms
%% Get the parameters
deltaZ=par.deltaZ;
n=par.n;
k=par.k;
C0=par.Cin;
u=par.u0;

%% Computing individual values
reac=zeros(n,1);
convec=zeros(n,1);
% First node
reac(1)=k*Phi(1);
convec(1)=u/(2*deltaZ) * (C(2)-C0);
% Other internal nodes
for i=2:n-1
    reac(i)=k*Phi(i);
    convec(i)=u/(2*deltaZ) * (C(i+1)-C(i-1));
end
% End-node
reac(n)=k*Phi(n);
convec(n)=u/deltaZ * (C(n)-C(n-1));
```

Since the Crank-Nicolson method uses a single previous value $\boldsymbol{\phi}_p$ to compute $\boldsymbol{\phi}_{p+1}$, the above functions will be calculated at the new (guess) point and the previous solution. Using the definitions of convection and reaction terms in Equation 8.48, the Crank-Nicolson equation may be written as

$$\underbrace{\left(\phi_{p+1,i}-\phi_{p,i}\right)+\frac{\Delta_t}{2}\left(\left[C_{p,i}+r_{p,i}\right]+\left[C_{p+1,i}+r_{p+1,i}\right]\right)=0}_{g\left(\phi_{p+1}\right)} \tag{8.49}$$

Since $\mathbf{r}_p$ and $\mathcal{C}_p$ are computed at known (previous) values of $\boldsymbol{\phi}_p$, they can be calculated once and passed to the function $\mathbf{g}(\boldsymbol{\phi})$ as constants.

Let us use `phi` to represent the previous solution, $\boldsymbol{\phi}_p$, and $\mathbf{Y}$ to represent the function argument in Equation 8.49, that is, $\mathbf{g}(\mathbf{Y})$. The MATLAB function that calculates $\mathbf{g}(\mathbf{Y})$ consists of the following calculation:

$$\mathbf{g}(\mathbf{Y})=\left(\mathbf{Y}-\phi\right)+\frac{\Delta_t}{2}\left(\left[\mathbf{f}^{old}\right]+\left[\mathbf{f}^{new}\right]\right)$$

$$\mathbf{f}^{old}=\left[\mathcal{C}_p+\mathbf{r}_p\right], \quad \mathbf{f}^{new}=\left[\mathcal{C}_{p+1}+\mathbf{r}_{p+1}\right]$$

(ii) *The function to calculate* $\mathbf{g(Y)}$ *is given below:*

```
function gVal = pfrCrankNicolFun(Y)
    %% Function to calculate g(Y)
    [reacNew,convecNew]=pfrDiscreteFun(Y,par);
    fNew=reacNew+convecNew;
    gVal = (Y-Phi) + deltaT/2*(fNew+fOld);
end
```

The above function, pfrCrankNicolFun, will be supplied to fsolve solver to obtain the root, $\boldsymbol{\phi}_{p+1}$. The variable fOld (underlined above) is the value $\mathbf{f}^{old}$ above, which is calculated at $\phi_{p,i}$. Since it remains constant within the fsolve iterations, it is calculated outside the pfrCrankNicolFun function.

(iii) *Function for time stepping in Crank-Nicolson*: pfrCN_Stepper is a function that will take (t_p, y_p) as input arguments and return the next iteration value (t_{p+1}, y_{p+1}). It will invoke fsolve, with $\mathbf{g(Y)}$ provided by pfrCrankNicol-Fun, as given below:

```
function [tNew,PhiNew]= pfrCN_Stepper (t,Phi,par)
% Function file for transient PFR problem
% to be solved using Crank-Nicolson Method
%% Perform time-stepping
deltaT=par.deltaT;
tNew=t+deltaT;

% Get function values at previous time
[reacOld,convecOld]=pfrDiscreteFun(Phi,par);
fOld=reacOld+convecOld;
% Use old value as the initial guess
Yguess=Phi;
% Call fsolve for one-step of Crank Nicolson
opt=optimoptions(@fsolve);
PhiNew=fsolve(@(Y) pfrCrankNicolFun(Y),Yguess,opt);

    function gVal = pfrCrankNicolFun(Y)
        %% Function to calculate g(Y)
        [reacNew,convecNew]=pfrDiscreteFun(Y,par);
        fNew=reacNew+convecNew;
        gVal = (Y-Phi) + deltaT/2*(fNew+fOld);
    end
end
```

The above function is self-explanatory, where the steps shown in various equations are implemented sequentially. Since pfrCrankNicolFun is an embedded function, the variables in pfrDiscreteFun are accessible to it; thus, allowing us to avoid transferring the values fOld, Phi, and par to this function.

(iv) *Overall driver function* for solving the PFR problem follows a similar structure as before. The function, `pfrCN_Stepper`, is called at each time-step to compute ϕ_{p+1} in a loop, starting from the initial condition:

```matlab
% Simulating transient PFR using Crank-Nicolson
%% Model Parameters
modelParam.k=0.2;
modelParam.u0=0.1;
C0=1;
modelParam.Cin=C0;
modelParam.L=0.5;
%% Initializing the PFR
n=25;      h=modelParam.L/n;
phi=ones(n,1)*C0;
t=0;       tN=10;
dt=0.1;
modelParam.n=n;
modelParam.deltaZ=h;
modelParam.deltaT=dt;
% For storing results
Z=[0:h:h*n];
T=zeros(tN/dt+1,1);
PHI=zeros(n+1,tN/dt+1);
PHI(:,1)=[C0;phi];

% Jacobian pattern for efficient fsolve
Jvec=ones(n-1,1);
Jpat=diag(Jvec,-1)+eye(n)+diag(Jvec,1);
modelParam.Jpat=Jpat;
%% Solving the PFR using Crank-Nicolson
for p=1:tN/dt
    [tNew,phiNew]= pfrCN_Stepper(t,phi,modelParam);
    T(p+1)=tNew;
    PHI(:,p+1)=[C0;phiNew];
    t=tNew;
    phi=phiNew;
end

%% Plotting results
% Steady state plot
CModel=C0*exp(-modelParam.k/modelParam.u0*Z);
C_ss=PHI(:,end);
figure(1)
plot(Z,C_ss); hold on
plot(Z,CModel,'.');
xlabel('Length (m)'); ylabel('C_A (mol/L)');
```

```
% Exit concentration vs. time
figure(2)
plot(T,PHI(end,:)); hold on
xlabel('time (min)'); ylabel('C_A (mol/m^3)');
```

The underlined part of the code above will be discussed later.

The steady state results for $\Delta_t = 0.1$ (solid line) are compared with the analytical solution (symbols) in Figure 8.4. The steady state results are reproduced with sufficient accuracy. The discretization in time was verified by running the code at another step-size of $\Delta_t = 0.2$, which is double the original step-size. The steady state results are not shown because the two curves lie exactly on top of one another. Figure 8.5 shows the comparison for transients in the outlet concentration (dashed line represents $\Delta_t = 0.2$). The results indicate that the chosen step-sizes give adequate accuracy for this system.

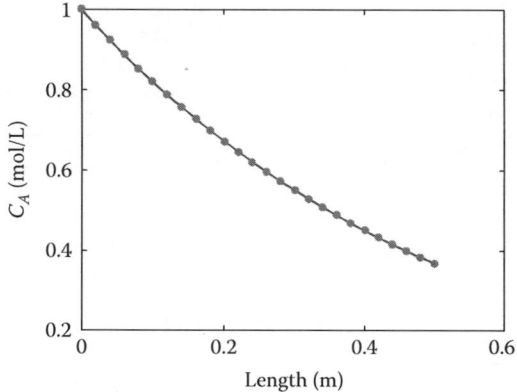

FIGURE 8.4   Steady state results using the Crank-Nicolson method, and comparison with analytical solution (as symbols; see Chapter 4 for derivation).

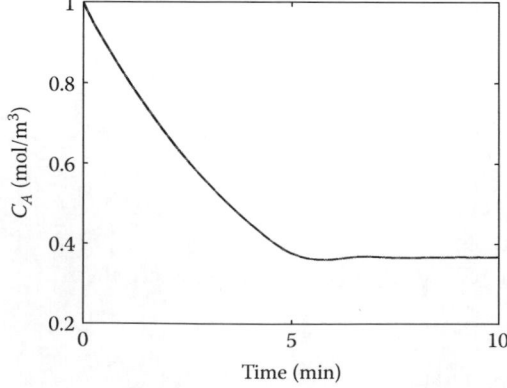

FIGURE 8.5   Transient variation in outlet concentration for two different step-sizes: $\Delta_t = 0.1$ (solid line) and $\Delta_t = 0.2$ (dashed line).

8.3.2.1 Exploiting Sparse Structure for Efficient Simulation

The 14th line in `pfrCN_Stepper` code above (see item (iii) in Example 8.2) was

```
opt=optimoptions(@fsolve);
```

This structure `opt` can be used to provide various options for `fsolve`. Since Equation 8.5 resulting from the Crank-Nicolson method is tridiagonal, one of the methods discussed in the previous chapter can be used to improve the computational efficiency by exploiting the sparse nature of Equation 8.5. This can be done by providing "JacobPattern" option for fsolve. The above underlined code is modified as

```
opt=optimoptions(@fsolve,'jacobPattern',par.Jpat);
```

The variable `Jpat` was defined in the driver script (see underlined part) as

```
Jpat=diag(Jvec,-1)+eye(n)+diag(Jvec,1);
```

Recall from Chapter 7 that the above command indicates to the solver that the problem has a tridiagonal structure. The sparsity pattern can be verified using `spy(Jpat)`.

The original code was executed again with $\Delta_t = 0.01$ (to highlight the computational gains clearly). Without Jacobian pattern, it took 33.7 s for the 1000 steps to reach $t_N = 10$. With the Jacobian pattern included, the computation time reduced* to 29.8 s. The computation time can be reduced even further by suppressing the display at the end of each iteration using

```
opt=optimoptions(opt,'display','none');
```

A combination of specifying Jacobian pattern and suppressing display resulted in a three-fold reduction of computation time to 11.2 s.

This concludes our discussion on the Crank-Nicolson method. A key outcome from the stiff system and Crank-Nicolson examples in Section 8.3 was to demonstrate the use of implicit method for solving ODEs. We now turn our attention to using implicit solvers for DAEs in the remainder of this chapter.

8.4 DIFFERENTIAL ALGEBRAIC EQUATIONS

DAEs were briefly introduced in Section 8.1.3 as system models that have both an algebraic and a differential equation representation:

$$\mathbf{F}\left(t,\mathbf{Y},\frac{d\mathbf{Y}}{dt}\right)=0 \tag{8.8}$$

---

* `fsolve` uses various different algorithms; only some of these can exploit Jacobian pattern. The codes were run on Macbook Pro with 2.6 GHz Intel i5 processor, running MATLAB R2016a.

where $\mathbf{Y} \in \mathcal{R}^n$. Fully implicit methods are used to solve the generic DAE of the form above. Rather than discussing a general form of DAEs as above, this chapter focuses on the most important class of DAEs for chemical engineers—semiexplicit DAEs:

$$\frac{d}{dt}\mathbf{y} = \mathbf{f}(t,\mathbf{y},\mathbf{x})$$

(8.6)

$$\mathbf{0} = \mathbf{g}(t,\mathbf{y},\mathbf{x})$$

(8.7)

As described before, $\mathbf{y} \in \mathcal{R}^{n_d}$ are differential variables since they are governed by ODEs, whereas the remaining variables, $\mathbf{x} \in \mathcal{R}^{n-n_d}$, are algebraic. When $n_d = 0$, we have a purely algebraic system of equations, whereas when $n_d = n$, we have ODEs. Since these were covered in Chapters 6 and 3, respectively, we focus solely on DAEs.

8.4.1 An Introductory Example

Consider a continuous stirred tank reactor (CSTR) example that was solved in Chapter 3:

$$\frac{d}{dt}C_A = \frac{F}{V}\left(C_{A,\text{in}} - C_A\right) - r\left(C_A\right)$$

$$C_A(0) = C_{A0}$$

(8.50)

Here, C_A is a differential variable. Since $n = n_d = 1$, the above is an example of ODE-IVP. Let us consider that the inlet concentration is not constant but varies according to some function:

$$C_{A,\text{in}} = \xi(t)$$

(8.51)

There are two ways to solve this problem.

8.4.1.1 Direct Substitution

It is possible to substitute Equation 8.51 in Equation 8.50:

$$\frac{d}{dt}C_A = \frac{F}{V}\left(\xi(t) - C_A\right) - r\left(C_A\right)$$

(8.52)

leading to ODE-IVP. This was an approach we took in Chapter 5. For example, we considered a case study of chemostat in Section 5.4 where inlet concentration varied with time, whereas a boundary constraint in the form of PFR temperature profile was imposed in Section 5.5. This approach may be used if algebraic variables are explicit functions of time and/or differential variables, or if they can be converted to explicit functions with some algebraic manipulations.

8.4.1.2 Formulating and Solving a DAE

Rewriting Equation 8.51 as

$$0 = \underbrace{C_{A,\text{in}} - \xi(t)}_{g(t,C_A,C_{A,\text{in}})} \tag{8.53}$$

The above equation along with Equation 8.50 defines DAE, with C_A as the differential variable and $C_{A,\text{in}}$ as the algebraic variable.

8.4.1.2.1 Case Where Algebraic Variable Is Specified Implicitly   Instead of being specified as an explicit function, the differential variable, $C_{A,\text{in}}$, may be specified implicitly in the form, $g(t, C_A, C_{A,\text{in}}) = 0$. There are several instances when such a case arises. During the start-up of a reactor, the inlet condition may need to be varied based on a certain desired profile, borne out of some design optimization procedure. In highly exothermic systems or in bioreactors to avoid toxicity, $C_{A,\text{in}}$ may be specified implicitly. For example

$$\left[F\left(C_{A,\text{in}} - C_A\right) \right]^{\eta} + \alpha C_{A,\text{in}} = 0 \tag{8.54}$$

for an arbitrary $\eta \neq 1$ is a case where the DAE comprising Equations 8.50 and 8.54 needs to be solved using a combined approach.

8.4.1.2.2 Case Where Algebraic Variable Is Specified Indirectly   There are several problems where the algebraic constraints do not contain the algebraic variables, **x**. For example, the inlet composition may be varied so that the exit concentration meets a specific profile. The DAE consists of differential Equation 8.50 and algebraic Equation 8.55 below:

$$\frac{d}{dt} C_A = \frac{F}{V} \left(C_{A,\text{in}} - C_A \right) - r\left(C_A\right) \tag{8.50}$$

$$0 = C_A - \xi(t) \tag{8.55}$$

Since C_A appears in the differential equation, it is the differential variable; the algebraic variable is $C_{A,\text{in}}$ even though it does not appear in Equation 8.55. Thus, the algebraic variable gets specified indirectly due to the algebraic constraint in Equation 8.55.

Such cases are routinely seen in design optimization problems and (off-line) control formulations. For example, during reactor start-up, the concentration or temperature of CSTR may need to follow a certain trajectory; inlet concentration $C_{A,\text{in}}$ is determined by solving the DAE (8.50) with (8.55). Another example is that of a surge tank or blender kept upstream of a critical reactor or equipment. The aim of this tank or blender is to dampen the variations and ensure the downstream reactor or equipment is provided the process stream at desired conditions. This is achieved by varying the algebraic variable $C_{A,\text{in}}$ to achieve the constraint in Equation 8.55.

8.4.2 Index of a DAE and More Examples

Three cases were discussed in the above example: Equation 8.50 was the differential equation and the possible algebraic constraints were

1. $C_{A,in}$ specified explicitly as in Equation 8.51

2. $C_{A,in}$ specified implicitly in an algebraic equation (8.53) or (8.54)

3. $C_{A,in}$ obtained indirectly by solving Equation 8.50 subject to constraint in (8.55)

The first is an example of ODE if direct substitution approach is followed, the second is index-1 DAE, and the third is a higher-index DAE (specifically, index-2).

DEFINITION: INDEX OF A DAE

The *index of a DAE* is defined as the number of times the algebraic equations need to be differentiated in order to convert the DAE into ODEs.

There are several different definitions of index: differentiation index, perturbation index, and tractability index. The above is convenient to use and relevant to solving DAEs numerically.

The first case is where the inlet concentration is specified, as given in Equation 8.53. Differentiating the equation yields us

$$\frac{d}{dt} C_{A,in} - \xi'(t) = 0 \tag{8.56}$$

Since a single differentiation was required to convert the DAE into ODE, this is an example of index-1 DAE.

The second case is where the outlet concentration is specified. Differentiating Equation 8.55

$$\frac{d}{dt} C_A = \xi'(t) \tag{8.57}$$

The variable $C_{A,in}$ is still an algebraic variable. Substituting Equation 8.50, we get

$$\frac{F}{V}\left(C_{A,in} - C_A\right) - r\left(C_A\right) = \xi'(t) \tag{8.58}$$

Differentiating this a second time

$$\frac{d}{dt} C_{A,in} - C_A' - \frac{V}{F} \frac{\partial r}{\partial C_A} C_A' = \xi''(t) \tag{8.59}$$

$$\frac{d}{dt} C_{A,\text{in}} = \xi' \left(1 + \frac{V}{F} \frac{\partial r}{\partial C_A} \right) + \xi''(t) \tag{8.60}$$

Differentiating the algebraic equation twice converts the original DAE into a set of two ODEs—Equations 8.50 and 8.60. Therefore, this is an example of index-2 DAE.

8.4.2.1 Example 2: Pendulum in Cartesian Coordinate System

We are all familiar with the pendulum model. It is a rather simple model in polar coordinate system since there is a change in only one direction, θ. The force balance in Cartesian coordinates results in a DAE as shown below:

$$x'' = -T \sin \theta, \quad y'' = g - T \cos \theta \tag{8.61}$$

Here, $\sin \theta = x/L$ and $\cos \theta = y/L$. Let u represent the velocity in x-direction and v represent the velocity in y-direction. Tension in the connecting string/rod is also unknown and is to be found out. However, T is indirectly specified because of the constraint that the string is rigid, and hence the pendulum traverses a circular path with constant radius. Thus, the algebraic equation for pendulum is given by Equation 8.63 below and the differential equations (8.61) are written as a set of four ODEs in (8.62) below:

$$
\begin{aligned}
x' &= u \\
u' &= -\frac{Tx}{L} \\
y' &= v \\
y' &= g - Ty/L
\end{aligned}
\tag{8.62}
$$

$$0 = x^2 + y^2 - L \tag{8.63}$$

The overall solution vector is $\mathbf{Y} = \begin{bmatrix} x & u & y & v & T \end{bmatrix}^T$, tension T is the algebraic variable, whereas the remaining four are differential variables. Clearly, this is an example of higher-index DAE. Computing the index of the DAE is left as an exercise to the reader.

8.4.2.2 Example 3: Heterogeneous Catalytic Reactor

Previously, we simulated plug flow and packed bed reactors, where concentration was assumed constant throughout any axial slice of the reactor. Heterogeneous reactor models are models for catalytic reactors that explicitly account for molecular diffusion from the bulk gas to the catalyst surface. The concentration at the catalyst surface is different from the bulk concentration, and the rate of diffusion is given by the product of mass transfer coefficient and concentration driving force:

$$J_{\text{diff}} = k_g \left(C_A - C_{As} \right) \tag{8.64}$$

The balance for bulk gas gives rise to the differential equation

$$u\frac{dC_A}{dz} = -k_g \underline{a}(C_A - C_{As})$$

(8.65)

whereas balance at the surface gives the algebraic equation

$$0 = k_g \underline{a}(C_A - C_{As}) - r(C_{As})\underline{a}$$

(8.66)

With this introduction, we now focus on methods to solve semiexplicit DAEs in MATLAB. We further assume that the algebraic equation **g** is such that its Jacobian

$$\frac{\partial \mathbf{g}}{\partial \mathbf{x}}$$

is nonsingular. Most common DAE problems in chemical engineering are either in this form or can be converted easily to this form.

8.4.3 Solution Methodology: Overview

The above example of heterogeneous catalytic reactor model will be used to demonstrate various options for solving ODEs. Initially, while discussing solution methods, first-order kinetics will be assumed:

$$u\frac{dC_A}{dz} = -k_g \underline{a}(C_A - C_{As})$$

(8.65)

$$0 = k_g(C_A - C_{As}) - kC_{As}$$

(8.66)

The above problem can be easily solved because Equation 8.66 is linear in the algebraic variable C_{As}. The first step in solving the DAE by hand is to obtain C_{As} for *any* given value of concentration C_A by solving Equation 8.66 analytically to yield

$$C_{As} = \frac{k_g C_A}{k_g + k}$$

(8.67)

Thus, the value of C_{As} can be known for any value of C_A. Knowing the value of C_{As}, Equation 8.65 becomes

$$\frac{d}{dz}C_A = -\frac{1}{u}\left(\frac{kk_g}{k+k_g}\underline{a}\right)C_A$$

(8.68)

The analytical solution of the DAE is therefore

$$C_A(t) = C_{A0}e^{-\alpha z}, \quad \alpha = \frac{1}{u}\left(\frac{kk_g}{k+k_g}a\right)$$

$$C_{As}(t) = \frac{k_g}{k_g+k}C_{a0}e^{-\alpha z}$$

(8.69)

This solution will be used in subsequent examples to compare the solution obtained using numerical approaches.

Since the reaction term, $r(C_{As})$, can be complex for most practical problems, numerical techniques need to be general enough to solve in cases where Equation 8.66 cannot be rearranged to give an explicit expression for C_{As}. Two approaches for solving the DAE will be discussed presently: nested approach and combined approach.

8.4.3.1 Solving Algebraic Equation within ODE

Owing to the discussion in Chapter 3, we are familiar with methods to solve ODE of the type (8.65), with the only challenge in case of DAE being that C_{As} is not known. Since the above is an index-1 DAE, the algebraic equation can be solved to obtain the value of C_{As} for any given value of C_A. This was indeed the approach followed in analytical calculation of C_{As} in Equation 8.67. This hand-computed expression in (8.67) will be replaced by a numerically computed solution. Indeed, fsolve is an appropriate way to do so (though any nonlinear equation solver may be used). Since the aim is to compute the value of $x = C_{As}$ for a given value of $y = C_A$, the function $g(y,x)$ in algebraic equation (8.7) is given by the MATLAB function call:

```
@(Cas) funName(Ca,Cas,modelParam)
```

The above anonymous function call is passed to fsolve, which then computes C_{As} for the specified value of C_A, given the parameters in modelParam structure.

The versatility of numerical techniques is that the function to calculate the algebraic variable need not be explicit $x = g(y)$; it can equally easily be presented as a numerical solution to $g(y,x) = 0$. The following example demonstrates this *nested approach* of solving DAE.

Example 8.3 Heterogeneous PFR with Nested Approach

Problem: Solve the 1-D heterogeneous reactor model of Equations 8.65 and 8.66. The reactor length is $L = 1$, the surface area to volume ratio is $\underline{a} = 100$, and the inlet velocity is $u = 1$. A first-order reaction A → B takes place with $r = 0.02\,C_A$. The mass transfer coefficient is $k_g = 0.01$. Obtain the axial profiles of concentration C_A and C_{As} in the reactor.

Solution: The following function calculates the catalytic reaction rate:

```
%% Function to calculate reaction rate
function r=catalRxnRate(C,par)
    r=par.k*C;
end
```

Calculation of C_{As} by solving nonlinear equation: The nonlinear equation (8.66) will be solved using MATLAB `fsolve` solver to compute C_{As} for any value of C_A. This can be compared with the analytically calculated C_{As} as per Equation 8.67. The function to compute the algebraic function $g(\cdot)$ is given below:

```
%% Function for the algebraic model
function gVal=surfaceBal(CaBulk,Cas,par)
rxnTerm=catalRxnRate(Cas,par);
gVal=par.kg*(CaBulk-Cas) - rxnTerm;
end
```

In my experience, the underlined line of the code above is the source of a majority of errors. The reaction rate is calculated at C_{As}; a common error is when the rate is calculated at C_A instead. Another potential source of error is the order of input arguments.[*] Please be careful when setting up the problem!

Let us now verify how this code works. Defining the parameters at the command line

```
>> par.k=0.01;
>> par.kg=0.02;
>> Ca0=1;     CaTry=Ca0;
```

we solve to obtain $C_{As,0}$ for a given value of $C_{A,0}$:

```
>> Cas0=fsolve(@(Cas) surfaceBal(Ca0,Cas,par), CaTry)
Cas0 =
     0.6667
```

The analytically calculated C_{As} as per Equation 8.67 is 2/3, which is what we obtained. In the `fsolve` call above, the underlined entity (`Cas`) is a *variable* that `fsolve` solves for, and the result (`Cas0`) is the solution found for any given *constant* value of `Ca0`. To summarize what `fsolve` does: (i) `Ca0` is treated as a constant within the `fsolve` call, (ii) the value of `Cas` is varied using the root-finding algorithm, (iii) the result `Cas0` is the solution that satisfies $g(y, x) = 0$ for that particular value `Ca0`, and (iv) `CaTry` is the initial guess for the solver. If C_{A0} is changed to another value, C_A, `fsolve` will find the corresponding value of C_{As} as the root of the algebraic equation (8.66).

Function for differential equation with nested algebraic solver: The differential equation (8.65) will be provided to the solver. Let us say that C_{As} was calculated analytically as per Equation 8.67. The ODE function would simply involve the following two lines:

```
Cas=kg/(k+kg)  * Ca;
dC=-(kg*abyv/u)*(Ca-Cas);
```

---

[*] This is another example why I personally recommend using the powerful *anonymous function* approach of MATLAB. In the older versions of MATLAB, the first variable in the input arguments was the variable solved for by `fsolve`. The anonymous function allows us to write a MATLAB function in a manner similar to how we write analytically.

Since the code developed needs to be general for any other forms of reaction rate $r(C)$, the underlined line will be replaced by the `fsolve` solution described on the previous page. The final function file used for the differential equation is given below:

```
function dC=pfrNestedFun(t,Ca,par)
% Differential function for heterogeneous
% reactor using nested approach
k=par.k;      kg=par.kg;
u=par.u0;     abyv=par.abyv;

CaTry=Ca;     % Guess value for fsolve
Cas=fsolve(@(Cas) surfaceBal(Ca,Cas,par), CaTry);
dC=-(kg*abyv/u)*(Ca-Cas);
end
```

Driver script and results: This completes the main part of our code. The driver script to run this code and the results are given below:

```
% Driver script for 1D heterogeneous model
% for a reactor with first-order reaction
%% Model parameters
L=1;
Ca0=1;
u=1;          modelParam.u0=u;
k=0.01;       modelParam.k=k;
kg=0.02;      modelParam.kg=kg;
abyv=200;     modelParam.abyv=abyv;

%% Solving the DAE: Nested approach
[zSol,CaSol]=ode45(@(z,Ca)...
    pfrNestedFun(z,Ca,modelParam),[0 L],Ca0);
plot(zSol,CaSol); hold on;
```

The computation of the C_A values analytically is not shown in the code above.

Figure 8.6 shows the concentration profile calculated numerically (solid line) and is compared with the analytically calculated value of C_A as per Equation 8.69.

The nested approach is easy to follow and is a convenient learning tool. However, it is computationally intensive since the nonlinear equation is solved at each integration step.

8.4.3.2 Combined Approach

The combined approach uses an appropriate implicit ODE-solving method to convert the differential equations (8.6) into algebraic equations (as was done in Sections 8.2.1, 8.2.3, and 8.3). Since the AM-2 trapezoidal method was used in Section 8.3.1, I will use it to

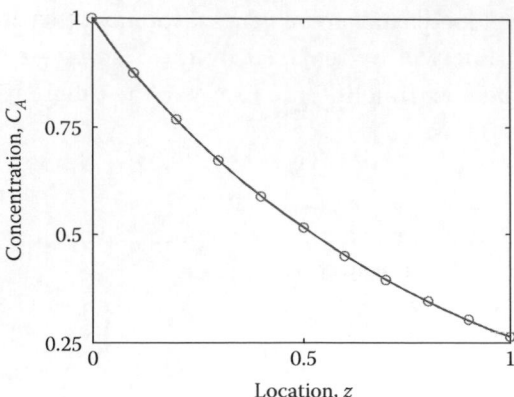

FIGURE 8.6 Solution of DAE for heterogeneous reactor using nested approach (solid line), and comparison with the analytical solution (symbols).

illustrate the combined approach. The same arguments work for other implicit methods for ODE-IVP also. Applying the trapezoidal method to Equation 8.6 yields

$$\mathbf{y}_{k+1} = \mathbf{y}_k + \frac{h}{2}\Big[\mathbf{f}_k + \mathbf{f}\big((k+1)h,\mathbf{y}_{k+1},\mathbf{x}_{k+1}\big)\Big] \tag{8.70}$$

where h is the step-size and $\mathbf{f}_k = \mathbf{f}(hh,\mathbf{y}_k,\mathbf{x}_k)$ is a known value. Equation 8.7 is written as

$$0 = \mathbf{g}\big((k+1)h,\mathbf{y}_{k+1},\mathbf{x}_{k+1}\big) \tag{8.71}$$

The two sets of equations, written together as

$$\mathbf{F}(\mathbf{Y}) \triangleq \begin{bmatrix} \mathbf{y}_{k+1} - \mathbf{y}_k - \dfrac{h}{2}\Big[\mathbf{f}_k + \mathbf{f}\big((k+1)h,\mathbf{y}_{k+1},\mathbf{x}_{k+1}\big)\Big] \\ \mathbf{g}\big((k+1)h,\mathbf{y}_{k+1},\mathbf{x}_{k+1}\big) \end{bmatrix} = 0 \tag{8.72}$$

can be solved to obtain the solution $\mathbf{Y}_{k+1}$ at each time. Viewed thusly, both ODE and non-linear algebraic equations may be considered as a subset of DAEs.

Example 8.4 DAE Using Implicit AM-2 Trapezoidal Rule

Problem: Solve the problem in Example 8.3 using the implicit AM-2 trapezoidal rule.

Solution: The function file developed in Example 8.3 will be used in this problem. As per Equation 8.72, two functions are needed: the algebraic function $g(\cdot)$ and the differential function $f(\cdot)$. The algebraic function `surfaceBal` and catalytic reaction rate calculation `catalRxnRate` will be used from the previous example.

Function to compute f(.) for the differential function: As per Equation 8.65, the differential function is given by

$$f(C_A) = -\frac{k_g a}{u}(C_A - C_{As})$$

which is implemented in the function below:

```
%% Function for differential model
function fVal=gasBalance(Ca,Cas,par)
kg=par.kg;      abyv=par.abyv;
u=par.u0;
fVal=-(kg*abyv/u)*(Ca-Cas);
end
```

AM-2 Solver for forward stepping: The solver that uses AM-2 trapezoidal rule will be built on the same lines as Example 8.1. The first element of the vector **F(Y)** will use the differential model above within an AM-2 discretization, whereas the second element is the algebraic equation `surfaceBal`. Specifically, the two elements of **F(Y)** will be

```
FVAL(1,1) = Ca-CaOld - par.h/2*(fNew+fOld);
FVAL(2,1) = surfaceBal(Ca,Cas,par);
```

where `fNew` is $\mathbf{f}((k+1)h, \mathbf{y}_{k+1}, \mathbf{x}_{k+1})$ and `fOld` is $\mathbf{f}_k$ in Equation 8.70. As in Example 8.1, an embedded function that calculates FVAL is passed to `fsolve` solver to obtain the solution at the next location, $\mathbf{Y}_{k+1}$. The function is given below:

```
function [zNew,YNew]=pfrDAE_AM2(z,YOld,par)
% Function for single step for solving DAE PFR
% problem using AM-2 trapezoidal method
CaOld=YOld(1);
CasOld=YOld(2);
fOld=gasBalance(CaOld,CasOld,par);
% Compute next values using AM-2
zNew=z+par.h;
YNew=fsolve(@(Y) pfrAM2Fun(Y), YOld);

    %% Function for AM-2 as per Eq. (8.72)
    function FVAL=pfrAM2Fun(Y)
        Ca=Y(1);    Cas=Y(2);
        fNew=gasBalance(Ca,Cas,par);
        FVAL(1,1) = Ca-CaOld-par.h/2*(fNew+fOld);
        FVAL(2,1) = surfaceBal(Ca,Cas,par);
    end
end
```

The driver script and results: Only the part of the driver script that solves the DAE using AM-2 method is shown below. The script is self-explanatory.

```
%% Solving the DAE: Using AM-2 method
h=0.01;        modelParam.h=h;
n=L/h;         modelParam.n=n;
% Variables for storing results
Z=0:h:L;
Y=zeros(2,n); Y=[Y0, Y];
for k=1:n
    [zNew,yNew]=pfrDAE_AM2(Z(k),Y(:,k),modelParam);
    Z(k+1)=zNew;
    Y(:,k+1)=yNew;
end
plot(Z,Y(1,:),'-.');
```

The results obtained in this example were the same as that in Example 8.4 (Figure 8.6). Hence, they are skipped for brevity.

The above example showed the feasibility of using implicit method for solving a DAE. While this approach indeed works very well, it is also possible to use MATLAB implicit solvers, such as ode15s, to solve semiexplicit DAEs. This will be discussed in Section 8.4.4.

8.4.4 Solving Semiexplicit DAEs Using ode15s in MATLAB®

MATLAB solver ode15s uses fully implicit NDF and is therefore able to solve the DAE. Specifically, ode15s has an option to provide the so-called mass matrix, M, which is an $n \times n$ matrix. If the mass matrix is provided, it solves the ODE:

$$M\frac{d\mathbf{Y}}{dt} = \mathbf{F}(t, \mathbf{Y}) \tag{8.73}$$

When M is an identity matrix (which is a default option in ode15s, when M is not specified), the above is an ODE-IVP. For a DAE consisting of n_d differential equations and $n_a = n - n_d$ algebraic equations, the mass matrix is

$$M = \begin{bmatrix} I_{n_d} & 0_{n_d \times n_a} \\ 0_{n_a \times n_d} & 0_{n_a \times n_a} \end{bmatrix} \tag{8.74}$$

where I_{nd} is an $n_d \times n_d$ identity matrix and $0_{m \times n}$ is a zero matrix whose size is specified in the subscript. The mass matrix is specified using the option 'Mass'.

Example 8.5 Using `ode15s` to Solve DAE

Problem: Solve the problem in Example 8.3 using `ode15s`.

Solution: The DAE comprising Equations 8.65 and 8.66 may be written in the standard form required for ode15s as

$$
\underbrace{\begin{bmatrix} 1 & 0 \\ 0 & 0 \end{bmatrix}}_{M} \frac{d}{dt} \begin{bmatrix} C_A \\ C_{As} \end{bmatrix} = \underbrace{\begin{bmatrix} -\dfrac{k_g a}{u}\left(C_A - C_{As}\right) \\ k_g \left(C_A - C_{As}\right) - k C_{As} \end{bmatrix}}_{\mathbf{f}(\mathbf{Y})} \tag{8.75}
$$

The above equation is treated in a similar manner as the ODEs, the difference being that an appropriate mass matrix is specified, and we have consistent initial conditions. Since we have already created functions for gas and surface balance earlier, we will simply use them. The function file for the DAE is given below:

```
function dY=pfrDAEFun(t,Y,par)
Ca=Y(1);
Cas=Y(2);
dY(1,1)=gasBalance(Ca,Cas,par); % Differential
dY(2,1)=surfaceBal(Ca,Cas,par); % Algebraic
end
```

Driver script: The mass matrix is specified in the driver script as follows:

```
MassMat=[1 0; 0 0];
opt=odeset('Mass',MassMat);
```

whereas the consistent initial condition for the algebraic variable is obtained by solving the algebraic equation for inlet concentration, C_{A0}:

```
Cas0=kg/(k+kg)*Ca0;
```

Thus, the overall driver script is given below (for the sake of completeness, the entire script is given here):

```
% Driver script for 1D heterogeneous model
% for a reactor with first-order reaction
%% Model parameters
L=1;
Ca0=1;
u=1;          modelParam.u0=u;
k=0.01;       modelParam.k=k;
```

```
kg=0.02;        modelParam.kg=kg;
abyv=200;       modelParam.abyv=abyv;

%% Solving the DAE: ode15s approach
MassMat=[1 0; 0 0];
opt=odeset('Mass',MassMat);
Cas0=kg/(k+kg)*Ca0;     % Consistent initial condition
Y0=[Ca0;Cas0];
[zSol,YSol]=ode15s(@(z,Ca) pfrDAEFun(z,Ca,modelParam),...
    [0 L], Y0, opt);

%% Analyzing and plotting
plot(zSol,YSol); hold on
xlabel('Location, z'); ylabel('Concentration, C_A');

alpha=k*kg/(k+kg) * (abyv/u);
ZModel=[0:0.1:1]';
Ca_Model=Ca0*exp(-alpha*ZModel);
Cas_Model=kg/(k+kg)*Ca_Model;
plot(ZModel,[Ca_Model, Cas_Model],'o');
```

The results are shown in Figure 8.7. The solid line represents concentration C_A and the dashed line represents surface concentration C_{As}. The analytical solution is provided as symbols for comparison. The numerical results are very close to the analytical solution. Since $k_g \approx k$, the system is governed by both mass transfer and surface reaction.

This completes our discussion on DAEs and numerical methods to solve the DAEs. Specific case studies will be shown in the next section to demonstrate the principles described in this chapter.

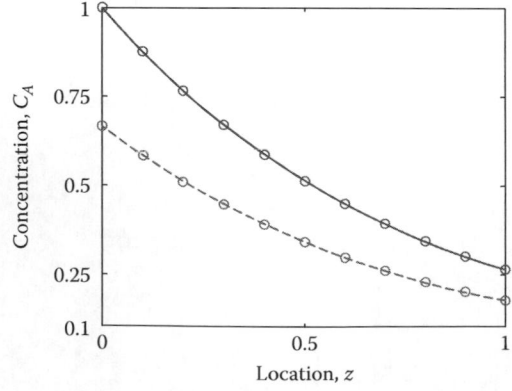

FIGURE 8.7   Solution of the 1D heterogeneous reactor model using ode15s (lines), with analytical solution (symbols) presented for comparison.

8.5 CASE STUDIES AND EXAMPLES

8.5.1 Heterogeneous Catalytic Reactor: Single Complex Reaction

The 1D heterogeneous reactor model discussed in Example 8.5 considered a first-order reaction to benchmark the various numerical solutions with analytical results. The model will be extended to the case where the reaction rate is given by a complex kinetic rate expression:

$$r(C_A) = \frac{kC_A}{\sqrt{1 + K_r C_A^2}} \tag{8.76}$$

The 1D heterogeneous reactor model is given by the following equations:

$$u\frac{dC_A}{dz} = -k_g a (C_A - C_{As}) \tag{8.65}$$

$$0 = k_g (C_A - C_{As}) - r(C_{As}) \tag{8.66}$$

The results of the 1D heterogeneous model will be compared with the PFR model from Section 3.5.1. The reaction rate above is in terms of $mol/(m^2_{cat} \cdot s)$, which needs to be converted to $mol/(m^3 \cdot s)$ as required in the PFR model. Thus, the PFR model is given by

$$u\frac{dC_A}{dz} = -a\, r(C_A) \tag{8.77}$$

The main difference between Equations 8.65 and 8.77 is that the reaction rate is computed at the surface concentration C_{As}, in the former, and at bulk concentration C_A, in the latter.

Example 8.6 Heterogeneous Reactor with Single Complex Reaction

Problem: Solve the 1D heterogeneous reactor model above for the reaction given by (8.76). The parameter values are $k = 0.005\ m^3/(m^2_{cat} \cdot s)$, $K_r = 1\ m^6/mol^2$, $a = 400\ m^2_{cat}/m^3$, and $k_g = 0.01\ m/s$.

    Solution: Comparing with the first-order reaction, the only difference in the problem statement is that a Langmuir-Hinshelwood-Hougen-Watson (LHHW) model type kinetic rate expression (8.76) is used in this example. Due to the modular approach emphasized in this book, the only change from Example 8.5 is in the function that calculates the rate of reaction:

```
function r=catalRxnRate(C,par)
k=par.k;    Kr=par.Kr;
r=k*C/sqrt(1+Kr*C^2);
end
```

Additionally, the parameter, Kr, must be specified in the driver script. No other change is required. Note that the DAE was initialized in Example 8.5 by specifying

$$C_{As,0} = \frac{k_g}{k+k_g} C_{A,0} \tag{8.78}$$

In this problem, ode15s is able to solve the DAE even when the problem is initialized as $Y_0 = [C_{A,0}; \quad C_{As,0}]$. Although this C_{As} does not provide consistent initial condition, the ode15s solver computes it internally before solving the DAE.

Figure 8.8 shows the results of heterogeneous reactor model. The solid and dashed lines represent bulk concentration C_A and surface concentration C_{As}, respectively. The results from a PFR simulation (see Section 3.5.1 for details) are shown as symbols for comparison. Due to mass transfer effects, the conversion from the heterogeneous reactor model is less than that from a PFR, where complete mixing (uniformity) in the radial direction is assumed.

The mass transfer affects the net rate of reaction, as evident from Figure 8.8. The effect of mass transfer coefficient, k_g, on reactant concentration along the length of the reactor is shown in Figure 8.9. The bulk concentration C_A is shown as solid line and the surface concentration C_{As} as dashed line for two different values of k_g. As the mass transfer coefficient increases, the bulk and surface concentrations get closer to each other.

Furthermore, conversion of A also increases with an increase in the value of k_g. This is evident in Figure 8.10, where bulk concentration is plotted for various values of k_g. The value of k_g increases in the direction of the arrow. As the value of k_g is increased, conversion from the heterogeneous reactor model progressively increases (i.e., C_A decreases), with the results approaching that of the PFR. Indeed, when the Damköhler number $Da \sim k/k_g \ll 1$, the bulk concentration from the heterogeneous model closely follows that from the PFR.

The concepts developed in this example will be further used in Section 9.4 to solve the problem of simulating methanol synthesis in a tubular reactor.

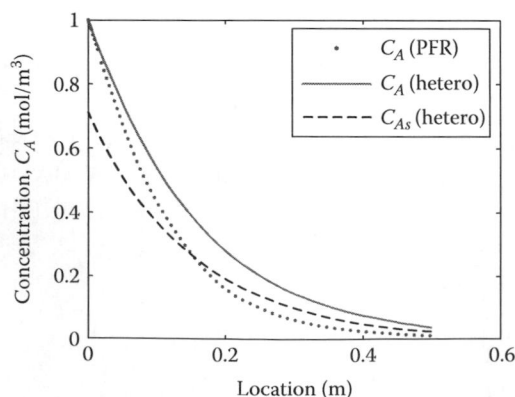

FIGURE 8.8 Simulation of heterogeneous reactor (lines) and comparison with PFR (symbols).

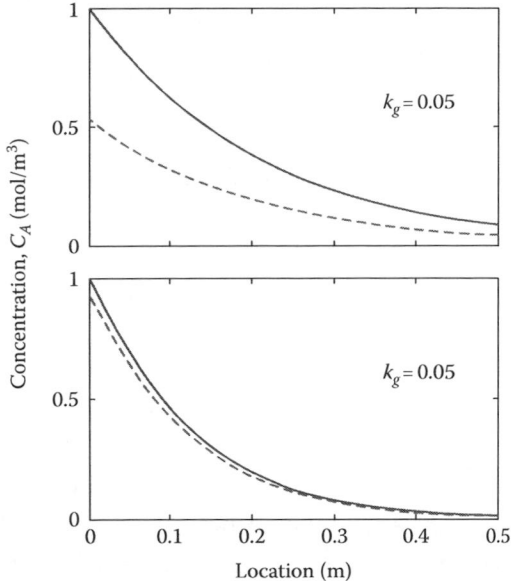

FIGURE 8.9   Bulk and surface concentrations (C_A and C_{As}) for two different values of k_g.

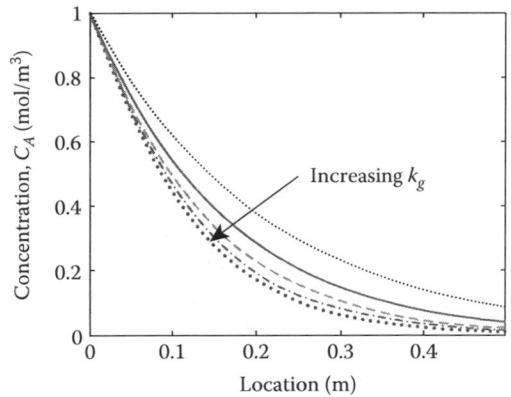

FIGURE 8.10   Effect of varying k_g on bulk concentration, C_A. The four lines in the direction of the arrow represent $k_g = 0.005$, 0.01, 0.02, and 0.5 m/s, respectively.

8.5.2 Flash Separation/Batch Distillation

A binary distillation column was simulated in Section 5.5.1, and a $T–x–y$ curve for a mixture of nitrobenzene and acetonitrile was generated in Section 6.5.2. Simulation of a binary batch distillation will be considered in this case study. In the continuous distillation column example, we assumed the relative volatility to be constant. This may not always be the case, especially in the case of batch distillation. In this problem, we will simulate binary batch distillation of a mixture of benzene (B) and ethylbenzene (EB). Instead of assuming a constant relative volatility, the vapor-liquid equilibrium information will be obtained through Antoine's coefficients (similar to the $T–x–y$ problem). First, the overall model equations will be derived.

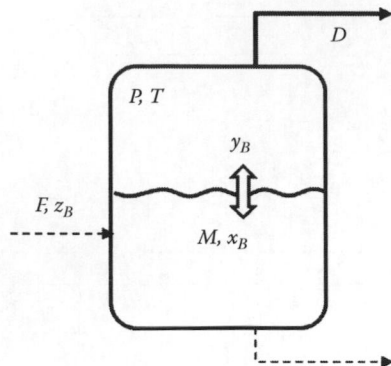

FIGURE 8.11 Schematic of flash separation, which will be used to develop binary distillation model.

Due to familiarity with a single stage of distillation (from Chapter 5), let us consider the overall process as shown in Figure 8.11. If M is the molar holdup in the distillation tank, and D and B are the molar flowrates, then the overall balance is given by

$$\frac{d}{dt}M = F - D - L \tag{8.79}$$

whereas the balance on B is given by

$$\frac{d}{dt}Mx_B = Fz_B - Dy_B - Lx_B \tag{8.80}$$

We have assumed that both the liquid and vapor phases are well mixed so that the outlet conditions are the same as that existing in the flash/distillation tank. The mole fractions in gas and liquid phases are in equilibrium. By Raoult's law

$$p_B = x_B P_B^{sat}, \quad p_{EB} = \left(1 - x_B\right) P_{EB}^{sat} \tag{8.81}$$

where P_i^{sat} represents the saturation pressure of species i. The latter equation is for binary mixture, as $x_{EB} = 1 - x_B$. The saturation pressure is related to the temperature by Antoine's equation:

$$\log_{10}\left(P_i^{sat}\right) = A_i - \frac{B_i}{C_i + T} \tag{8.82}$$

The mixture in the tank will be at its bubble point during the entire operation. For a given liquid composition, x_B, the temperature can be calculated as shown in Section 6.5.2. The vapor is in equilibrium at that temperature and satisfies the following condition:

$$P = p_B + p_{EB} \tag{8.83}$$

$$P - x_B P_B^{sat} - \left(1 - x_B\right) P_{EB}^{sat} = 0 \tag{8.84}$$

$$\underbrace{\phantom{P - x_B P_B^{sat} - \left(1 - x_B\right) P_{EB}^{sat}}}_{g(T)}$$

and the vapor fraction of B is

$$y_B = \frac{x_B P_B^{sat}}{P} \tag{8.85}$$

The differential equations (8.79) and (8.80), the algebraic equation (8.84), and the expression (8.85) together form the model for any of these processes: continuous flash separation, single-stage continuous distillation, and batch/semibatch binary distillation.

Batch distillation model can be further simplified because there is no feed or liquid outlet. Thus

$$\frac{dM}{dt} = -D \tag{8.86}$$

whereas the balance on B is given by

$$M \frac{d}{dt} x_B + x_B \frac{dM}{dt} = -D y_B$$
$$\frac{dx_B}{dt} = \frac{D}{M} \left(x_B - y_B\right) \tag{8.87}$$

The solution variables for the binary batch distillation are defined as

$$Y = \begin{bmatrix} M \\ x_B \\ T \\ y_B \end{bmatrix}$$

with the first two being the differential variables and the last two algebraic variables. Note that Equation 8.85 is simply an expression for calculating y_B. It is therefore not necessary to include y_B as an algebraic variable.

Antoine's coefficients for B and EB are

$$A_B = 4.7, \quad B_B = 1660, \quad C_B = -1.5$$

$$A_{EB} = 4.1, \quad B_{EB} = 1419, \quad C_{EB} = -60.5$$

Example 8.7 Binary Batch Distillation

Problem: Solve the binary distillation example for B-EB mixture. The distillation tank is charged with 100 mol of equimolar mixture initially. After the mixture reaches the bubble point, heating is maintained such that the distillate is obtained at a constant rate of $D = 0.5$ mol/min. Compute how distillation proceeds in time.

Solution: We start with the core of the MATLAB function used for generating the model equations as per Equations 8.84 through 8.87:

```
dPhi(1)=-D;
dPhi(2)=D/M*(xB-yB);
dPhi(3)=P - xB*PsatB-(1-xB)*PsatEB;
dPhi(4)=yB - xB*PsatB/P;
```

where D is the rate at which the distillate is drawn from the system. Antoine's equation is used to compute PsatB and PsatEB, the saturation pressures of B and EB, respectively, at the operating temperature, T. The operating temperature is equal to the bubble point of the liquid in the distillation tank, whereas the operating pressure, P, is taken as the atmospheric pressure.

The function file, batchDistFun.m, for this problem is given below:

```
function dPhi=batchDistFun(t,phi,par)
% Model for binary batch distillation
% of Benzene-Ethyl Benzene mixture
%% Model Parameters
P=par.P;          % Operating pressure (bar)
D=par.D;          % Distillate rate
% Antoine's coefficients
Ab=par.Ab;        Bb=par.Bb;        Cb=par.Cb;
Aeb=par.Aeb;      Beb=par.Beb;      Ceb=par.Ceb;
% Model variables
M=phi(1);
xB=phi(2);
T=phi(3);
yB=phi(4);

%% Model equations
PsatB=10^(Ab-Bb/(T+Cb));
PsatEB=10^(Aeb-Beb/(T+Ceb));      % Saturation P
dPhi=zeros(4,1);
dPhi(1)=-D;
dPhi(2)=D/M*(xB-yB);
dPhi(3)=P - xB*PsatB-(1-xB)*PsatEB;
dPhi(4)=yB - xB*PsatB/P;
```

The above function requires pressure, distillate rate, and Antoine's coefficients as the model parameters. These will be defined in the driver script and passed to the function above. The driver script for batch distillation is given below:

```
% Binary batch distillation of benzene--ethylbenzene
%% Model parameters
distParam.D=0.5;     % Distillate rate
distParam.P=1;       % 1 bar pressure
% Antoine's constants: Benzene
distParam.Ab =4.7;    distParam.Bb =1660;
distParam.Cb =-1.5;
% Antoine's constants: EB
distParam.Aeb=4.1;    distParam.Beb=1419;
distParam.Ceb=-60.5;

%% Initializing / Solving the DAE
M0=100;    xB0=0.5;
T0=378;    % Guess T0 as avg. of B and EB boiling points
yB0=0.7;
Phi0=[M0;xB0;T0;yB0];
% Define mass matrix
massMat=diag([1 1 0 0]);
opt=odeset('Mass',massMat);
% Solve the DAE
[tSol,PhiSol]=ode15s(@(t,y) ...
    batchDistFun(t,y,distParam),[0:5:100],Phi0,opt);
%% Getting output
% Liquid left behind in distillation tank
molesB_tank=PhiSol(:,1) .* PhiSol(:,2);
molesEB_tank=PhiSol(:,1) .* (1-PhiSol(:,2));
figure(1)
plot(tSol,molesB_tank,tSol,molesEB_tank);
xlabel('Time (min)'); ylabel('Moles left in tank');
% Temperature
figure(2);
plot(tSol,PhiSol(:,3));
xlabel('Time (min)'); ylabel('Temperature (K)');
% Distillate flow rate
distFlow_B=distParam.D*PhiSol(:,4);
distFlow_EB=distParam.D*(1-PhiSol(:,4));
figure(3);
plot(tSol,distFlow_B, tSol,distFlow_EB);
xlabel('Time (min)'); ylabel('Distillate flow rate');
```

As time progresses, the vapors in the tank are drawn out at the rate of D mol/min. The vapor phase is in equilibrium with the liquid in the tank. The vector Phi0 contains the guess for initial conditions. Since the boiling point of B is 80°C and that of EB is 136°C, we chose 105°C = 378 K as the initial guess for the temperature. The `ode15s` solver internally computed the correct initial conditions as

```
>> disp(PhiSol(1,:))
  100.0000    0.5000   371.8551     0.8256
```

The first two elements are $M = 100$ and $x_B = 0.5$, which are the initial conditions provided. The value of temperature that satisfies the algebraic Equation 8.84 is $T = 371.86$ K. The corresponding value of $y_B = 0.826$. Thus, the equimolar mixture of B and EB has its bubble point at 372 K, and the vapor in equilibrium with the liquid contains 82.6% B.

When this vapor is drawn out, the distillate contains more amount of B and the liquid left in the distillation tank is richer in EB. Thus, the bubble point temperature increases, as shown in the top panel of Figure 8.12. Since both B and EB are drawn out from the distillation tank, the amount of B and EB falls. The amount of B in the tank equals Mx_B, whereas that of EB is $M(1 - x_B)$. After the 100 min distillation, 12.85 mol of B and 37.15 mol of EB are remaining in the tank.

Since there is more B in the distillate than EB, the amount of B in the tank falls faster than that of EB. Equivalently, the flowrate of B in the distillate stream is higher than that of EB. D is the net distillate flowrate and the vapors exiting the distillation tank are in equilibrium with the liquid in the tank, the flowrates of B and EB in the distillate are Dy_B and $D(1 - y_B)$, respectively. Figure 8.13 shows the flowrates of B and EB in the distillate.

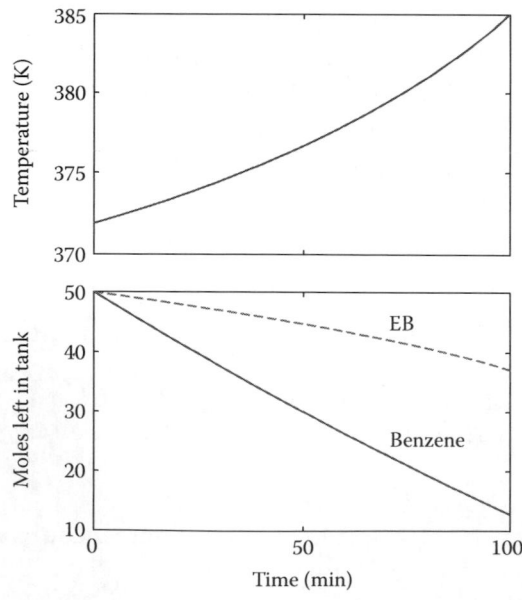

FIGURE 8.12 Temporal variations in equilibrium temperature and the amount of B and EB left behind in the distillation tank.

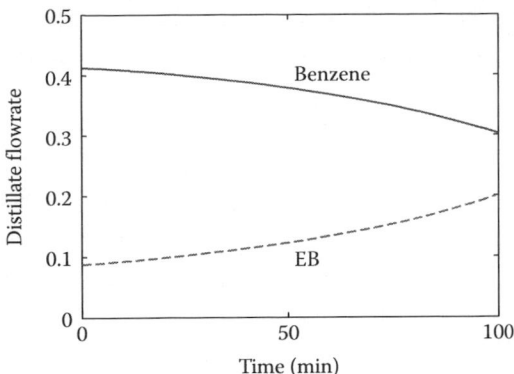

FIGURE 8.13   Flowrate of B and EB in distillate as a function of distillation time.

The total number of moles of B left in the distillation tank equals $Mx_B = 12.85$ mol. Since the batch distillation process started with 50 mol each of B and EB, the rest 37.15 mol of B are collected as distillate over the 100 min operation.

The flowrates of B and EB in distillate are plotted in Figure 8.13. The total number of moles collected in the distillate over the 100 min operation is also equal to the area under the respective curves. Using numerical integration (trapezoidal rule with `trapz` command), the amount of B in the distillate is computed as 37.13 mol.

Distillate flowrate was used as a design parameter. Distillate flowrate is increased to 1 mol/min by changing $D = 1.0$, which means that 50 mol of the mixture will be collected in 50 min. Executing the above code for 50 min, one can observe that the same amount of B (i.e., 37.15 moles) is distilled as before. Likewise, the temperature at the end of 50 min is $T = 385$ K; this is the same temperature reached at the end of the batch distillation process in Example 8.7.

Finally, I would like to draw attention to the initial conditions. The initial conditions are specified for the differential variables, M and x_B. Since the algebraic variables, T and y_B, are not free variables, but depend implicitly on M and x_B, they cannot be chosen arbitrarily. In general, it is a good idea to provide *consistent* initial conditions to the DAE solver `ode15s`. This involves solving the algebraic Equation 8.84 to obtain T and using Equation 8.85 to obtain y_B. Since this problem was not very stiff, it worked even when the initial conditions were arbitrary.

With this, we come to the end of the case studies for this chapter. Next section summarizes the findings and provides some tips based on my experiences.

8.6 EPILOGUE

The focus of this chapter was solving various problems that require an implicit time-stepping solver to determine how the solution evolves in time.

The trifecta of Chapters 3 and 4 and the current one covers problems involving differential equations that evolve in time. Chapters 3 and 4 focused on explicit time-stepping approaches to solve differential equations: ODE initial value problems in Chapter 3, whereas hyperbolic and parabolic PDEs in Chapter 4. That left us with three types of problems in this chapter: stiff ODEs, hyperbolic/parabolic PDEs, and DAEs that have both differential and integral components. The glue connecting these seemingly disparate examples was that all of them require implicit-in-time solution procedure, which forms the central theme of this chapter.

The ODE-BVPs (boundary value problems) and elliptic PDEs form the final set of differential equation problems. Discretization of the domain led to conversion of these differential equations to linear or nonlinear equations. Methods to solve these equations, which have special sparse matrix structure, were analyzed in Chapter 7.

I end this chapter with a brief from my personal experience. I have found one of the two methods as most reliable in MATLAB: using `ode15s` as described in several examples in this chapter or writing my own time-stepping solver based on the Crank-Nicolson method.

While `ode15s` has worked for me in a majority of examples,* there are typical stumbling blocks when one attempts to solve a new problem. Assuming that the code is free of errors and typos, the first issue is "poor" problem formulation. Sometimes, reformulating the problem in dimensionless or normalized quantities makes it numerically more tractable, for example, converting from partial pressures to mole fractions. Alternatively, a change of coordinate system or a linear transformation may make the problem more tractable. It is also possible to use a physical constraint to replace one of the ODEs. Problem 8.9 shows one such example, where one may use an algebraic constraint in lieu of a differential equation.

The second issue may be poorly specified initial conditions (in case of DAEs). It is therefore a good strategy to solve the algebraic equations

$$\mathbf{g}\left(\mathbf{y}_0, \mathbf{x}_0\right) = 0$$

to obtain consistent initial conditions. Providing consistent initial conditions to a DAE solver is always a recommended practice. It is sometimes beneficial to tweak the error tolerances to get `ode15s` to solve the problem.

At certain times, the PDEs or partial DAEs may still be unsolvable using `ode15s` to solve the discretized ODEs. The Crank-Nicolson method implemented with my own code is my alternative. Having said that, the `ode15s`–based approach has often given good performance in my work.

This brings to a close this chapter on DAEs.

EXERCISES

Problem 8.1   Derive the error estimate for Euler's implicit method using the procedure for the Adams-Moulton method (Section 8.2.1) and BDF methods (Section 8.2.3). This can be done by treating Euler's implicit method as AM-1/BDF-1.

---

* As explained in Chapters 3 and 4, and earlier in this chapter, `ode45` is the preferred ODE-IVP solver. Thus, `ode15s` is a "go-to solver" if I suspect a problem to be stiff or if the nature of the problem (PDE with highly nonlinear source term or DAE) calls for an implicit solver.

Problem 8.2 Solve the nonlinear chemostat problem from Section 8.3.1 using the second-order Adams-Moulton method (also known as trapezoidal method). You may modify the code from Example 8.1 or write a fresh code. The chemostat model is

$$
\begin{aligned}
S' &= D(S_f - S) - r_g \\
X' &= -DX + r_g Y_{xs} \\
P' &= -DP + r_g Y_{ps}
\end{aligned}
\tag{8.88}
$$

where $r_g = \left(\mu^m \dfrac{S}{K_S + S}\right) X$, $\mu^m = 0.5$, $K_S = 0.005$, $Y_{xs} = 0.75$, and $Y_{ps} = 0.65$. The feed concentration of the substrate is $S_f = 5$, and the dilution rate is $D = 0.1$ h^{-1}. The initial conditions at $t = 0$ are $S(0) = 5$, $X(0) = 0.02$, and $P(0) = 0$.

Problem 8.3 Modify the Crank-Nicolson solution of Example 8.2 for packed bed reactor with axial dispersion from Section 4.4. The nondimensional model for the packed bed reactor is given by

$$
\frac{\partial \phi}{\partial \tau} + \frac{\partial \phi}{\partial \zeta} = \frac{1}{Pe} \frac{\partial^2 \phi}{\partial \zeta^2} - Da \cdot \phi
$$

where
$\phi = C/C_0$ is dimensionless concentration
$\tau = tu/L$ is dimensionless time
$\zeta = z/L$ is dimensionless length
Pe is Péclet number
Da is Damköhler number

The initial condition is $[\phi]_{\tau=0} = 1$ and boundary conditions are $\phi(t, \zeta = 0) = 1$ and $\left[\dfrac{\partial \phi}{\partial \zeta}\right]_{t, \zeta=1} = 0$. The parameter values are Pe = 1 and Da = 2 and residence time is 10. Solve this problem using the Crank-Nicolson method and compare with the solved example from Chapter 4.

Problem 8.4 What is the index of DAE for the model of pendulum in Cartesian coordinates, discussed in Section 8.4.2 and Equations 8.62 and 8.63?

Problem 8.5 The following DAE represents an idealized separation process, where C is the concentration of the solute being separated. The simplest representation that retains the DAE nature of the complex dynamics is given by

$$
\frac{dC}{dt} = -0.1\xi^{0.7}
$$

$$\xi^2 - 3.45(C - \xi) = 0$$

with the initial condition $C(0) = 1$. Solve the above problem in two ways:

(i) Rewrite the second equation to get $\xi = \xi(C)$ and substitute in the ODE.

(ii) Solve the DAE using `ode15s` and compare the results.

Problem 8.6 Repeat Example 8.7 for a transient flash separation process. Assume the tank is initially at bubble point of equimolar benzene-ethylbenzene mixture. An equimolar mixture enters the flash as feed at 10 mol/min. The amount of vapor drawn is $D = 5$ mol/min and liquid drawn is $B = 5$ mol/min. Compute the transient behavior of the binary flash column.

Problem 8.7 Repeat the above problem for various values of the distillate $D = \{1, 3, 5, 7, 9\}$ and with $B = F - D$.

Problem 8.8 Formulate and solve transient flash separation problem for a ternary mixture of benzene-toluene-ethylbenzene. Antoine's coefficients for toluene are $A_T = 4.1$, $B_T = 1344$, and $C_T = -53.8$.

Problem 8.9 The following reaction takes place in a batch reactor:

$$A \rightarrow B \rightleftharpoons C$$

where $k_1 = 0.01$, $k_2 = 10^6$, and $k_3 = 10^4$. The reactor initially has $C_{A0} = 1$ mol/L of species A, and no B or C. The mass balance equations lead to three ODEs in three unknowns, C_A, C_B, and C_C.

1. Solve the set of three ODEs simultaneously.

2. The three species also satisfy the condition: $C_A + C_B + C_C = C_{A0}$. Replace the ODE for C_B with the above algebraic constraint and solve the resulting DAE.

CHAPTER 9

Section Wrap-Up

Nonlinear Analysis

This chapter will provide a wrap-up of concepts discussed in Section II of the book. The concepts for solving algebraic and differential equations, covered in the preceding chapters, will be used for simulation and analysis of nonlinear systems. The first three sections of this chapter will focus on methods to analyze the dynamic behavior of nonlinear systems, and the subsequent sections will focus on specific simulation examples of relevance to process engineers.

The first three sections will use specific examples of nonlinear dynamical systems to introduce the readers to nonlinear analysis techniques. Nonlinear systems display very rich dynamic behavior. Strogatz's book *Nonlinear Dynamics and Chaos* is an excellent introductory text on this topic. The linear stability analysis (Section 5.2) and the examples in the next three sections cover some basic and important concepts of nonlinear analysis. Specifically, this involves studying the stability, dynamics, and bifurcation behavior of nonlinear systems. The examples of nonisothermal continuous stirred tank reactor (CSTR) and chemostat (from Chapter 3) that show multiple steady state behavior will be discussed at length to introduce the turning point and transcritical bifurcations, respectively. The third section will study systems with cyclic dynamics (limit cycle).

Finally, two simulation case studies will culminate this part of the book. A tubular/fixed bed reactor has been discussed at several points throughout this text. An example of *gasphase* reactor with multiple species and multiple reactions will be discussed in Section 9.4. Both steady state and dynamical models, in the presence of volume change due to reaction, will be contrasted. The final section will solve the problem stated in the first chapter: to compute trajectory of a cricket ball.

9.1 NONLINEAR ANALYSIS OF CHEMOSTAT: "TRANSCRITICAL" BIFURCATION

The nonlinear model of chemostat and its numerical solution using ode45 was discussed in Section 3.5.4, and the linear stability around its steady states was analyzed in Section 5.2. A nonlinear analysis of the chemostat and an investigation of bifurcation behavior are considered in this section.

9.1.1 Steady State Multiplicity and Stability

The model for chemostat was written as

$$\mathbf{y}' = \mathbf{f}(\mathbf{y}) \tag{9.1}$$

where

$$\begin{bmatrix} f_1 \\ f_2 \end{bmatrix} = \begin{bmatrix} D(S_f - S) - \dfrac{\mu S X}{K + S} \\ -DX + \dfrac{\mu S X}{K + S} Y_{xs} \end{bmatrix} \tag{9.2}$$

At steady state, since $\mathbf{y}' = \mathbf{0}$, the steady state solutions can be found by solving the nonlinear equations $\mathbf{f}(\mathbf{y}) = \mathbf{0}$. The standard method in undergraduate textbooks involves using the second equation to get the two steady state values. Recall from Chapter 5 that these two values are given by

$$S = \frac{DK}{\mu Y_{xs} - D}, \quad X = 0 \tag{9.3}$$

Here, we will follow a slightly different procedure that will allow us to qualitatively perform the so-called *bifurcation analysis* later.

Adding $(Y_{xs}f_1 + f_2)$ at steady state gives us

$$0 = D(S_f - S)Y_{xs} - DX \tag{9.4}$$

Thus, the biomass concentration can be expressed in terms of substrate concentration at steady state as

$$X_{ss} = (S_f - S)Y_{xs} \tag{9.5}$$

Substituting this in the first equation gives

$$f_{1,ss} = (S_f - S)\left(D - \frac{\mu S}{K + S}Y_{xs}\right) \tag{9.6}$$

At steady state, since $f_1 = 0$, the two solutions are given by

$$\mathbf{y}_{ss} = \begin{bmatrix} \dfrac{DK}{\mu Y_{xs} - D} \\ (S_f - S)Y_{xs} \end{bmatrix}, \quad \mathbf{y}_{ss} = \begin{bmatrix} S_f \\ 0 \end{bmatrix} \tag{9.7}$$

The solutions in the above equation are the same as those obtained earlier in Equation 9.3. Numerically, the two solutions exist for all conditions. However, negative values of either substrate or biomass concentration are not physical. Thus, we will get nonphysical values for steady state if $0 \leq S \leq S_f$ is not satisfied.

The linear stability analysis was performed in Chapter 3. The Jacobian was calculated as

$$J = \begin{bmatrix} -D - \dfrac{\mu X K}{(K+S)^2} & -\dfrac{\mu S}{K+S} \\ \dfrac{\mu X K}{(K+S)^2} Y_{xs} & -D + \dfrac{\mu S}{K+S} Y_{xs} \end{bmatrix} \tag{9.8}$$

The nominal value of the dilution rate was chosen as $D = 0.1$. The two steady states under these conditions are

$$\mathbf{y}_{ss} = \begin{bmatrix} 0.091 \\ 3.68 \end{bmatrix}, \quad \mathbf{y}_{ss} = \begin{bmatrix} 5 \\ 0 \end{bmatrix} \tag{9.9}$$

whereas the eigenvalues of the Jacobian corresponding to the two steady states are

$$\Lambda = \{-3.96, \ -0.1\}, \quad \Lambda = \{-0.1, \ 0.257\} \tag{9.10}$$

The linear stability analysis indicates that the first steady state is a *stable node*, whereas the second is an *unstable (saddle) node*.

The linear stability analysis above provides a *local dynamical behavior* of the system in the vicinity of the steady state. We will use these results to present analysis of the dynamic behavior of the overall system, followed by the effect of varying the parameter (dilution rate, in this example) on *qualitative* behavior of system dynamics.

9.1.2 Phase-Plane Analysis

A phase-plane plot is a plot that *visually* depicts dynamics of the nonlinear system on a 2D plane, with the two state variables as the two axes. Starting at arbitrary initial points within the space, the ordinary differential equations (ODE) will be solved to obtain variation in the state variables with time. A trajectory in the 2D phase plane then tracks the dynamic evolution of the system starting from that particular initial condition. A phase-plane plot is a collection of such trajectories, which thus maps out the system dynamics in a visual manner.

The substrate concentration can take values between 0 and $S_f(=5)$, whereas the biomass concentration takes values between 0 and 3.75. We will choose 25 points on a 5×5 grid in the S–X plane as initial conditions, solve the ODE, and plot the trajectory on the phase plane.

The phase-plane plot is shown in Figure 9.1. The circles represent starting points (initial conditions), and the stars represent the two steady states. Based on the linear stability analysis, the first steady state is linearly stable, implying that all trajectories starting from the vicinity of this steady state converge toward it. The second steady state ("trivial" steady state because $X = 0$

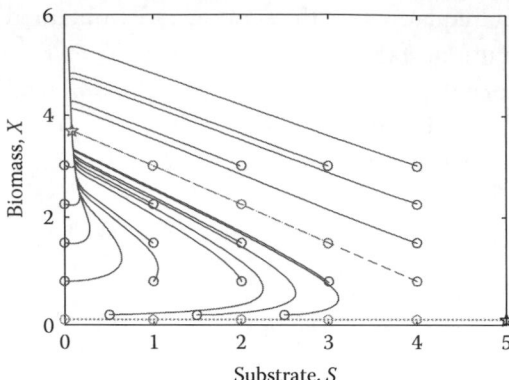

FIGURE 9.1 Phase-plane plot of nonlinear chemostat. Circles represent initial points, and the two stars represent the two steady states.

and no conversion of substrate) is a saddle node. Recall that the S-axis was the stable subspace at the saddle node. Thus, if the initial condition lies on this axis, the nonlinear chemostat will go to the trivial steady state; all other initial conditions in the S–X space will converge to the first steady state. As can be seen from Figure 9.1, the nonlinear system shows these dynamic patterns as well. Starting at $X(0) = 0$, all the trajectories converge to the trivial steady state. These trajectories are shown by dotted lines. This axis remains the stable subspace for the nonlinear system also. All other initial conditions converge to the first steady state. In fact, even the initial conditions starting at a very small value of initial $X(0) = 0.01$ reach the first steady state. It is also interesting to see that trajectories starting on the eigenvector $v_1 = [-0.8 \quad 0.6]^T$ stay on the eigenvector itself, while converging to the first steady state.

In summary, we have (i) obtained multiple steady states in chemostat, (ii) analyzed the linear stability around each steady state, and (iii) used phase portrait to analyze the dynamics of the nonlinear system. The next task is to determine how the steady states and stability change with change in the operating parameters of the system.

9.1.3 Bifurcation with Variation in Dilution Rate

Bifurcation may be defined as a sudden change in behavior of a system with a small change in the system parameter(s). When the parameter is varied smoothly, the system dynamics also show a smooth change. However, at a certain point, there is an abrupt change in the *qualitative* behavior of the system.

In the chemostat example, the dilution rate may be considered a key parameter. Bifurcation analysis involves monitoring qualitative behavior with variations in the parameter to identify topological changes in the system dynamics. The two steady states in a chemostat are characterized by the following solutions:

$$\mathbf{y}_{ss} = \begin{bmatrix} \dfrac{DK}{\mu Y_{xs} - D} \\ \left(S_f - S\right)Y_{xs} \end{bmatrix}, \quad \mathbf{y}_{ss} = \begin{bmatrix} S_f \\ 0 \end{bmatrix} \tag{9.7}$$

The first steady state is physically relevant only when the two concentrations are positive. For example, when the dilution rate is increased to $D > \mu Y_{xs}$, substrate concentration becomes negative. This value is not physically relevant. Although the system attains a physically irrelevant value, the numerical solution still exists.

Nonetheless, the system shows an interesting behavior numerically as well. Let ε be a small positive number. According to Equation 9.7, when $D = \mu Y_{xs} - \varepsilon$, S is a large positive number. However, when the dilution rate is increased slightly to $D = \mu Y_{xs} + \varepsilon$, S becomes a large negative number. Thus, there is a discontinuity in the solution at $D = \mu Y_{xs}$. Thus, this point represents a *bifurcation point* because there is a change in the qualitative behavior of the system.

DEFINITION

A bifurcation point is defined as the value of the bifurcation parameter where a qualitative change in the behavior of the system is observed.

The discontinuity in the root is not the only change that happens in the system. Consider the Jacobian at the second (trivial) steady state:

$$J = \begin{bmatrix} -D & -\dfrac{\mu S_f}{K + S_f} \\ 0 & -D + \dfrac{\mu S_f}{K + S_f} Y_{xs} \end{bmatrix} \tag{9.11}$$

This is just Equation 9.8, with $S_{ss} = S_f$ and $X_{ss} = 0$. Since $D > \mu Y_{xs}$, the second diagonal term is also negative. Thus, both eigenvalues are negative, and the trivial steady state becomes stable. This is shown in the next example.

Example 9.1 Bifurcation Analysis

Bifurcation analysis of the chemostat will be performed by varying the dilution rate from the original value $D = 0.1$ by 0.05 at each step. The two solutions and respective eigenvalues will be monitored. The following code monitors the solution and eigenvalues. The code is run multiple times, each time changing the dilution rate:

```
mu=0.5;       K=0.25;
Yxs=0.75;     Sf=5;
D=0.1;
% First solution and Jacobian
S=D*K/(mu*Yxs-D);       X=(Sf-S)*Yxs;
a1=mu*S/(K+S);
b1=mu*X*K/(K+S)^2;
J=[-D-b1,    -a1;
    b1*Yxs,  -D+a1*Yxs];
```

```
sol1=[S;X];
l1=eig(J);
% Second solution
X=0;      S=Sf;
a1=mu*S/(K+S);
b1=mu*X*K/(K+S)^2;
J=[-D-b1,   -a1;
     b1*Yxs,  -D+a1*Yxs];
sol2=[S;X];
l2=eig(J);
```

The results from varying the dilution rate are presented in the table below:

D	0.1	0.2	0.3	0.35	0.4	0.45	0.5
SS-1	0.0909	0.286	1	3.5	−4	−1.5	−1
	3.682	3.536	3	1.125	6.75	4.875	4.339
Λ	−3.96	−1.54	−0.3	−0.35			
	−0.1	−0.2	−0.24	−0.01			
SS-2	5	5	5	5	5	5	5
	0	0	0	0	0	0	0
Λ	−0.1	−0.2	−0.3	−0.35	−0.4	−0.45	−0.5
	0.257	0.157	0.057	0.007	−0.043	−0.093	−0.143

The bifurcation point is somewhere between $D = 0.35$ and 0.4 (we know it is at 0.375). The first steady state becomes unphysical. However, more importantly, the second steady state exchanges stability; it goes from being unstable to stable when the dilution rate is increased beyond the bifurcation point.

9.1.4 Transcritical Bifurcation

The bifurcation observed in a chemostat is known as *transcritical bifurcation*, and the point $D=0.375$ where the trivial solution becomes stable is the *bifurcation point* for the system (as we will see in this section, the dynamic behavior of the chemostat is *even more* complex and interesting!). The salient features of transcritical bifurcation are that both the solutions continue to exist numerically on either side of the bifurcation point; however, there is an exchange of stability in one or both types of solutions. Recall that the two types of solutions for a chemostat are the regular solution (with generation of biomass) and the trivial solution (no conversion or generation of biomass). At low values of dilution rate, the former is a stable solution whereas the latter is unstable. At higher values of dilution rate, however, the trivial solution becomes stable.

A graphical method will be used to further demonstrate transcritical bifurcation. We will do so in a single dimension since multiple dimensions are difficult to visualize. The treatment here follows on the same lines as that in the nonlinear dynamics book by Strogatz; however, I have extended his graphical to two dimensions and to a practical system of interest to chemical engineers. At steady state

$$X_{ss} = (S_f - S)Y_{xs} \tag{9.12}$$

and

$$\bar{f_1} = \left(S_f - S\right)\left(D - \frac{\mu S}{K + S} Y_{xs}\right)$$ (9.13)

Let us say that the system was moved slightly from steady state point by introducing a very small change from the steady state concentrations. We wish to analyze how the system responds to this change. The form of the equation

$$S' = \left(S_f - S\right)\left(D - \frac{\mu S}{K + S} Y_{xs}\right)$$ (9.14)

broadly gives *qualitative* insights into the dynamic response of the system in the vicinity of the steady state. The points of intersection of this curve $\bar{f_1}(S)$ with the S-axis gives the steady state solutions of the system. When $\bar{f_1}(S) > 0$, S' is positive and substrate concentration S will increase with time. Conversely, S will decrease with time when $\bar{f_1}(S) < 0$.

Figure 9.2 shows the plot of $\bar{f_1}(S)$ vs. S for four different values of D. The top panel represents dilution rates below $\mu Y_{xs} = 0.375$, whereas the bottom panels represent dilution rates exceeding this value. The arrows indicate the direction in which S would move for small deviations from the steady state. If S' is positive, the value of S increases. At $D = 0.3$, the two steady states are $S = 0.09$ and $S = 5$. When the value of S is decreased slightly (or conversely, X is increased slightly) from $S = 0.09$, the dynamics cause S to increase and reach steady state. Likewise, increasing S slightly will again move the system back to the steady state. This makes $S = 0.09$ a stable steady state. When the dilution rate is increased to 0.3, the

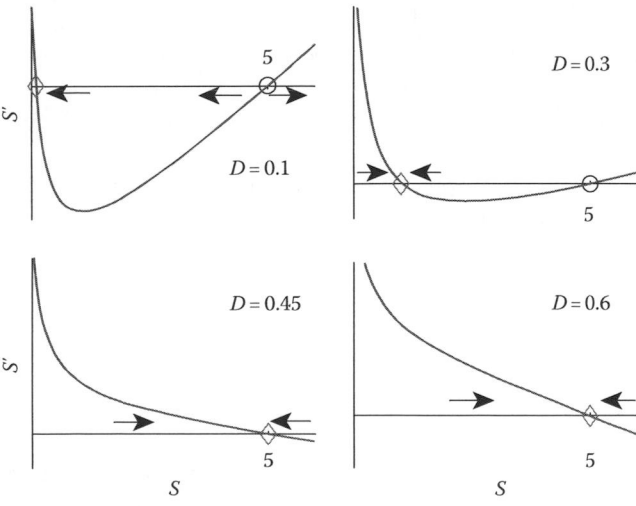

FIGURE 9.2 Plot of $\bar{f_1}(S)$ vs. S for four different values of dilution rate, D. Arrows indicate the direction in which S moves with time. Stable and unstable steady states are marked by a diamond and a circle, respectively.

first steady state moves toward the second steady state. The steady state value of S keeps increasing with D. At around $D = 0.357$, the second steady state almost reaches $S = 5$; when D is increased further, the solution becomes nonphysical. From a numerical viewpoint, the trivial steady state becomes stable at this bifurcation point.

THE RICHER WORLD OF BIFURCATION

This box may be skipped without loss in continuity of discussion.

In the previous discussion involving steady states and *linear* stability analysis of the chemostat, we identified $D = 0.375$ as a bifurcation point. Physically, the first steady state becomes meaningless because the first root gives $S < 0$. However, in the discussion above, between $0.3571 < D < 0.375$, the first root $S > S_f$, which results in $X < 0$ becoming a nonphysical solution. So, to summarize our discussion so far

(i) There exist two steady states for $D < 0.3571$. The first steady state is stable and the trivial steady state is unstable.

(ii) Between $0.3571 < D < 0.375$, the trivial steady state is the only physically relevant steady state. Numerically, however, both the steady states continue to exist, though the first one is nonphysical because $S > S_f$ and $X < 0$.

(iii) For $0.375 > D$, again the first steady state is nonphysical because $S < 0$.

This brings me to a nuance, which is skipped in the undergraduate material. We often make the assumption that the saturation constant $K \ll S_f$. In this case, washout is observed very close to $D > \mu Y_{xs}$. However, for larger values of K (say 1.0), the washout will happen reasonably below the $D = \mu Y_{xs}$ condition.

The question then remains, which of the two points, $D = 0.3571, 0.375$, is the bifurcation point. In order to answer this question, let us look at the plot of $\bar{f_1}(S)$ at an intermediate value of $D = 0.36$ shown in Figure 9.3.

When the dilution rate is increased from 0.3 to 0.36, the two steady states collide at the bifurcation point $D = 0.3571$ and exchange stability and continue. This is a classic example of *transcritical bifurcation*. The first steady state is stable and the trivial steady state unstable (saddle node) below the bifurcation point. At this bifurcation, the two steady states exchange stability. The first steady state becomes a saddle node, whereas the trivial steady state becomes stable.

Another interesting phenomenon occurs as D is increased further and approaches 0.375. As described before, between $0.3571 < D < 0.375$, S is positive and X is negative. When the

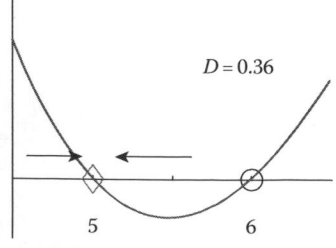

FIGURE 9.3 Plot of $\bar{f_1}(S)$ at an intermediate value of $D = 0.36$. The X-axis is zoomed in to clearly show the points of intersection of the curve.

value of D is increased beyond 0.375, the steady state S jumps to a large negative value, X becomes positive, and the second solution becomes stable.

Let us analyze and verify the results using linear stability analysis in the table below:

D	0.3	0.36	0.4
SS-1	1	6.0	−4
	3	−0.75	6.75
Λ	−0.3	−0.36	−0.4
	−0.24	**0.0024**	**−0.06**
SS-2	5	5	5
	0	0	0
Λ	−0.3	−0.36	−0.4
	0.057	−0.0029	−0.043

Thus, the first steady state goes from stable to saddle node and back to stable when the dilution rate is increased across the two bifurcation points, $D=0.3571$ and $D=0.375$.

Figure 9.4 is the *bifurcation plot* for the system. It represents various steady states of the system and how they vary with the dilution rate. This plot is thus a locus of all steady state points as the dilution rate is varied. Solid lines represent stable nodes, and dashed lines represent unstable (saddle) nodes. The locus of the first steady state is plotted as thick lines. Before the bifurcation point, this steady state is stable. After the bifurcation point, this steady state is only numerical; it does not have a physical meaning. It is shown in Figure 9.4 for the sake of completeness. The thin lines represent the trivial steady state. This is unstable until the bifurcation point. At the *transcritical* bifurcation, the two steady states exchange stability and the trivial steady state becomes stable beyond $D>0.3571$.

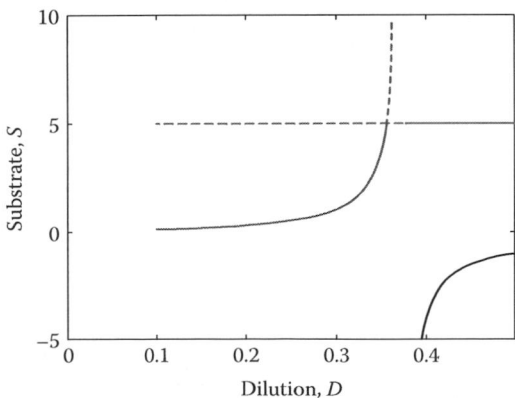

FIGURE 9.4   Bifurcation diagram for the chemostat. Thick lines represent the first steady state and thin lines represent the trivial steady state. Solid lines imply stable and dashed lines unstable steady state.

This is just a brief introduction to the wide world of nonlinear dynamics. The next example, nonisothermal CSTR, displays another type of bifurcation behavior, called turning-point bifurcation.

9.2 NONISOTHERMAL CSTR: "TURNING-POINT" BIFURCATION

A nonisothermal stirred tank reactor with a cooling jacket was modeled in Section 3.5.2. The reactant concentration, reactor temperature, and temperature of the cooling fluid were three state variables, that is, $\mathbf{y} = [C_A \; T \; T_j]^T$. The model equations are given by

$$\frac{dC_A}{dt} = \frac{F}{V}\left(C_{A,\text{in}} - C_A\right) - r_A \tag{9.15}$$

$$V\rho c_p \frac{dT}{dt} = F\rho c_p \left(T_{\text{in}} - T\right) + \left(-\Delta H\right) r_A V - UA\left(T - T_j\right) \tag{9.16}$$

$$V_j \rho_j c_j \frac{dT_j}{dt} = F_j \rho_j c_j \left(T_{j,\text{in}} - T_j\right) + UA\left(T - T_j\right) \tag{9.17}$$

The existence of two qualitatively different steady states was briefly presented in Chapter 3. In this section, a nonlinear analysis of this system will be presented. The standard multiple steady state behavior observed in this system is categorized as *turning-point bifurcation*. The qualitative features of this bifurcation will be analyzed, and the results contrasted with the transcritical bifurcation from the previous section. Our analysis will follow a familiar pattern from the previous section.

9.2.1 Steady States: Graphical Approach

A graphical method, similar to the one used in the previous section, will be used to find the steady state solutions. The steady state solutions are the roots of the equation $\mathbf{f}(\mathbf{y}) = \mathbf{0}$. Like before, the two state variables C_A and T_j will be expressed in terms of T:

$$T_j = \frac{\left(F_j \rho_j c_j\right) T_{j,\text{in}} + \left(UA\right) T}{F_j \rho_j c_j + UA} \tag{9.18}$$

The reaction is first order so that $r = k_0 \exp\left(-\dfrac{E}{RT}\right) C_A$. The first equation, $f_1(\mathbf{y}) = 0$, is used to express the concentration as a function of temperature. Substituting the value of T_j and C_A in the second equation, we get

$$\bar{f}_2(T) = \left(-\Delta H\right) \frac{k\tau}{1 + k\tau}\left(FC_{A,\text{in}}\right) + F\rho c_p \left(T_{\text{in}} - T\right) - \frac{UA}{1 + \beta}\left(T - T_{j,\text{in}}\right) = 0 \tag{9.19}$$

$$\text{where } \tau = \frac{V}{F}, \quad \beta = \frac{UA}{F_j \rho_j c_j} \tag{9.20}$$

The derivation of the above equation is left as an exercise. Depending on the values of the various parameters, the above equation can have one, two, or three solutions. These will be investigated shortly.

MULTIPLE STEADY STATES IN CSTR: FROM UNDERGRADUATE REACTION ENGINEERING TEXT

Multiple steady state behavior is also considered in an undergraduate reaction engineering course. The conversion is defined as $X = (C_{A,\text{in}} - C_A)/C_{A,\text{in}}$. With this definition, the mass balance equation (9.15) becomes

$$0 = \frac{X_A}{\tau} - k_0 \exp\left(-\frac{E}{RT}\right)(1 - X_A) \tag{9.21}$$

whereas the energy balance equation (9.16) becomes

$$0 = F\rho c_p (T_{\text{in}} - T) + (-\Delta H) F C_{A,\text{in}} X_A - UA \frac{(T - T_{j,\text{in}})}{1 + \beta} \tag{9.22}$$

The two equations can be rearranged to express X_A as a function of T. In his book on reaction engineering, Fogler denotes these two as $X_{A,\text{MB}}$ and $X_{A,\text{EB}}$ to represent mass and energy balance equations, respectively. Thus

$$X_{A,\text{MB}} = \frac{k_0 \exp\left(-\dfrac{E}{RT}\right)\tau}{1 + k_0 \exp\left(-\dfrac{E}{RT}\right)\tau} \tag{9.23}$$

$$X_{A,\text{EB}} = \frac{UA\dfrac{(T - T_{j,\text{in}})}{1 + \beta} + F\rho c_p (T - T_{\text{in}})}{F C_{A,\text{in}} (-\Delta H)} \tag{9.24}$$

When X_A is plotted against T, the former is an S-shaped curve, whereas the latter is a straight line (see Figure 9.5). It is clear that the roots of the equations are the points of intersection of

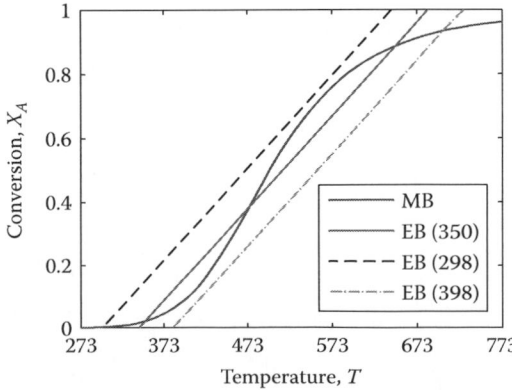

FIGURE 9.5 Conversion vs. temperature from material balance ("MB" from Equation 9.23) and energy balance ("EB" from Equation 9.24). The inlet temperature, T_{in} is shown in brackets.

TABLE 9.1   Model Parameters and Operating Conditions for a Jacketed CSTR

CSTR Inlet	Cooling Jacket	Other Parameters
$F = 25$ L/s	$F_j = 5$ L/s	$C_{A,in} = 4$ mol/L
$V = 250$ L	$V_j = 40$ L	$k_0 = 800$ s^{-1}
$\rho = 1000$ kg/m^3	$\rho_j = 800$ kg/m^3	$E/R = 4500$ K
$c_p = 2500$ J/kg·K	$c_p = 5000$ J/kg·K	$(-\Delta H) = 250$ kJ/mol
$T_{in} = 350$ K	$T_{j,in} = 25°$C	$UA = 20$ kW/K

the two curves. The parameters used in Chapter 3 are given in Table 9.1. The solid line and solid S-shaped curve in Figure 9.5 represent the above two equations for the nominal values of parameters shown in the table. The two curves intersect at three different points, which are the three steady states of the system. When the inlet temperature is decreased to 298 K, the material balance curve (denoted as MB) remains the same, while the energy balance curve shifts to the left (dashed line, "EB (298)"). Now, there is only a single point of intersection of the two curves (and hence a single solution). Likewise, when the inlet temperature is increased to 398 K (dash-dot line, "EB (398)"), again the MB and EB curves intersect at a single point.

This analysis from standard textbooks is nothing but an analysis of multiple steady states with the CSTR inlet temperature as a bifurcation parameter.

The discussion in the box is a popular approach to introduce multiple steady state behavior in higher dimensions taken in standard undergraduate texts in reaction engineering.

We will use another approach, similar to the one in the previous section, by plotting $\bar{f}_2(T)$ vs. T. The points of intersection of this curve with the T-axis give the steady state solutions of the original nonlinear system of equations. The curve $\bar{f}_2(T)$ vs. T is plotted (see Table 9.1 for nominal values of parameters) in Figure 9.6. The curve intersects T-axis at three points, $\{350, 478, 648$ K$\}$, which represents the three steady states. The corresponding values of C_A, T_j can be calculated from Equation 9.18.

Based on the *approximate* graphical method, the three steady state values are

$$\mathbf{y}_{ss} = \begin{bmatrix} 3918 \\ 350 \\ 324 \end{bmatrix}, \quad \mathbf{y}_{ss} = \begin{bmatrix} 2420 \\ 478 \\ 388 \end{bmatrix}, \quad \mathbf{y}_{ss} = \begin{bmatrix} 459 \\ 648 \\ 473 \end{bmatrix} \tag{9.25}$$

The actual steady state values can be computed using fsolve with these as initial guesses.

9.2.2 Stability Analysis at Steady States

As before, two different approaches to analyzing the steady states will be discussed. First is the qualitative graphical approach. Consider the $\bar{f}_2(T)$ vs. T curve in Figure 9.6. When the

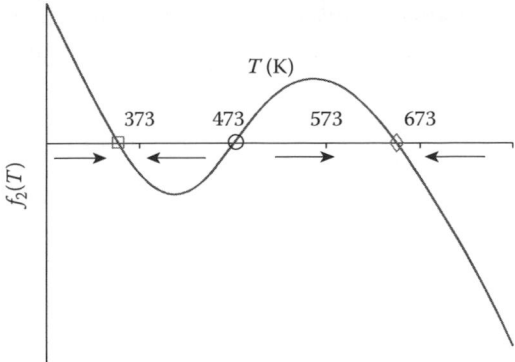

FIGURE 9.6 Function $\bar{f}_2(T)$ from Equation 9.19 plotted against temperature. Steady state solutions are indicated by roots of $\bar{f}_2(T)$. The arrows represent the direction of change in T for small variations from steady state value.

CSTR state variables are moved slightly from their steady state values, the approximate dynamics of the system is qualitatively governed by

$$T' = \frac{\bar{f}_2(T)}{V\rho c_p}$$

This is because assuming Equation 9.15 to be at steady state yielded $C_A(T)$ and assuming Equation 9.17 at steady state yielded $T_j(T)$. The function $\bar{f}_2(T)$ was obtained by substituting these expressions in Equation 9.16. Thus, if the value of $\bar{f}_2(T)$ is positive in the vicinity of a steady state, the temperature is likely to increase, whereas if it is negative, then the temperature will decrease. As can be seen in Figure 9.6, $\bar{f}_2(T)$ is positive for $T < 350$ and negative between $350 < T < 478$. So, if the temperature is nudged away from the low-conversion steady state in either direction, it will return back to the steady state. This *qualitatively* indicates that the first steady state is stable. Similarly, $\bar{f}_2(T)$ is positive slightly below 648 K and is negative slightly above 648 K. Again, this high-conversion steady state is stable because nudging the system from this steady state in either direction will cause the system to return to the steady state. Finally, the steady state at 478 K is unstable. This is because $\bar{f}_2(T)$ is negative between $350 < T < 478$ and positive between $478 < T < 648$, indicating that nudging the system away from the steady state will make the system diverge away from the steady state.

This is the qualitative analysis of system stability. Since the 3D space is projected on a single dimension, this analysis is only approximate. This needs to be corroborated with stability analysis at each steady state.

The function `cstrFun.m`, discussed in Chapter 3, can be used with the `fsolve` solver to find the actual steady state values. The approximate values in Equation 9.25 will be used as initial guesses for `fsolve` solver. The reader can verify that the steady state solutions are indeed very close to the values shown in Equation 9.25. Once these values are known, the next task is computing the Jacobian at each of the three steady states. One may either compute the Jacobian analytically or numerically. In case of the latter, the function cstrFun.m itself can be used to compute the numerical Jacobian as

$$J(i,j) = \frac{f_i(\mathbf{y}_{ss} + \delta_j) - f_i(\mathbf{y}_{ss} + \delta_j)}{2\varepsilon} \qquad (9.26)$$

where δ_j is a 3×1 vector such that its jth element is ε and the other elements are zero. For example

$$\delta_2 = \begin{bmatrix} 0 \\ \varepsilon \\ 0 \end{bmatrix}$$

The actual steady state was computed using fsolve, followed by computing the Jacobian numerically for each of the three steady states. The results are given below:

$$\text{Low } J = \begin{bmatrix} -0.102 & -0.299 & 0 \\ 0.0002 & -0.102 & 0.032 \\ 0 & 0.125 & -0.25 \end{bmatrix}, \quad \lambda = \{-0.099, -0.081, -0.273\} \qquad (9.27)$$

$$\text{Mid } J = \begin{bmatrix} -0.163 & -3.072 & 0 \\ 0.006 & 0.175 & 0.032 \\ 0 & 0.125 & -0.25 \end{bmatrix}, \quad \lambda = \{\mathbf{0.117}, -0.099, -0.256\} \qquad (9.28)$$

$$\text{High } J = \begin{bmatrix} -0.872 & -3.794 & 0 \\ 0.077 & 0.247 & 0.032 \\ 0 & 0.125 & -0.25 \end{bmatrix}, \quad \lambda = \{-0.108, -0.402, -0.365\} \qquad (9.29)$$

The qualitative predictions based on Figure 9.6 are confirmed by the linear stability analysis above. The low- and high-conversion steady states are stable, whereas the middle steady state is unstable (in fact, a saddle node).

9.2.3 Phase-Plane Analysis

Phase-plane analysis is very useful to get a global picture of the system behavior. The concentration values are in the range $0 < C_A < 4000$, whereas temperature values are in the range $300 < T < 700$. Choosing several points in the state space as initial conditions, simulations are performed. The dynamical trajectories are then plotted on a C_A–T plane. Circles represent initial starting points, and stars represent the three steady states. The phase portrait for the nonlinear CSTR is shown in Figure 9.7.

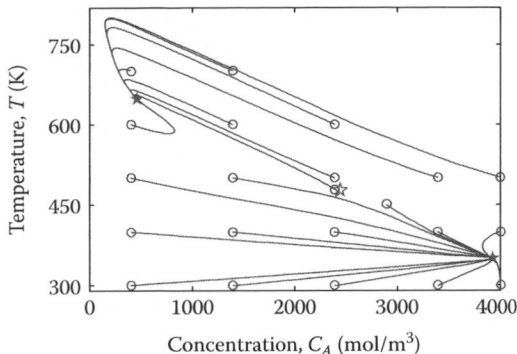

FIGURE 9.7   Phase portrait of a nonlinear CSTR with T_{in} = 350 K. The filled stars represent stable steady state, open star represents unstable steady state, and circles represent initial conditions.

9.2.4 Turning-Point Bifurcation

Like in the bioreactor case, the bifurcation analysis of the nonisothermal CSTR with respect to the inlet temperature, T_{in}, as the bifurcation parameter will be discussed in this section. Inlet conditions, which can be varied easily in practice, become the natural choice as bifurcation parameters. The curve $\bar{f}_2(T)$ intersects the T-axis at three points, indicating that three different steady states coexist at the value of parameter T_{in} = 350 K. Since the inlet temperature appears in Equation 9.19 as

$$F\rho c_p \left(T_{in} - T\right)$$

increasing this value will simply move the curve $\bar{f}_2(T)$ upward. As the temperature T_{in} is increased, the curve keeps moving upward. As seen in Figure 9.8, when the inlet temperature is increased to about 380 K, the lower and middle steady states merge. At this *bifurcation point*, there are only two steady states. The high-conversion steady state is qualitatively the same as before.* When the inlet temperature is further increased, the high-conversion steady state remains the only steady state. Thus, the characteristic of turning-point bifurcation is that two steady states come to merge at the bifurcation point and eventually disappear when the bifurcation parameter is varied further. The turning-point bifurcation occurs between the low- and mid-conversion steady states; the high-conversion steady state does not undergo bifurcation when T_{in} is increased from 350 K.

A similar turning-point bifurcation behavior is seen when the inlet temperature T_{in} is decreased from 350 K. Specifically, when this parameter reaches ~313 K, the high-conversion and mid-steady states merge; further reducing the inlet temperature eliminates these two steady states and only the low-conversion steady state remains. Thus, this is a second turning point, often called *extinction* point, where two steady states (mid and high) disappear, leaving only the low-conversion steady state.

---

* The high steady state remains stable as before. The characteristic of turning-point bifurcation in one dimension is that one stable node (low steady state) and one unstable node (middle steady state) merge to give a saddle node, according to Figure 9.7. For this reason, the turning-point bifurcation is also known as *saddle node* bifurcation. However, I think this is a misnomer because in multiple dimensions, the middle steady state was not unstable; it was already a saddle node!

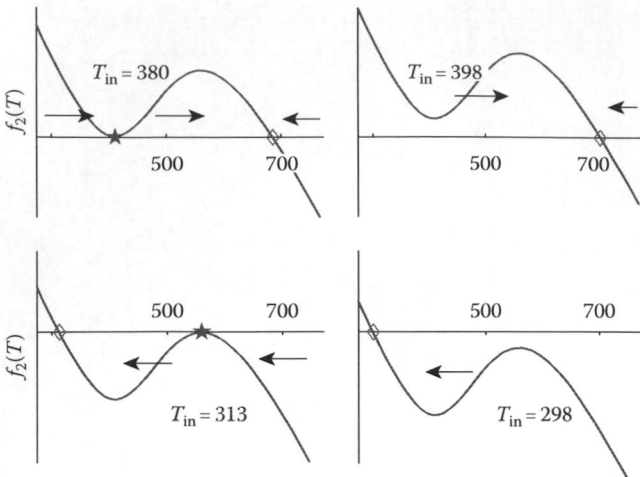

FIGURE 9.8 Effect of varying T_{in} on the multiple steady state behavior of CSTR. The top panels show increase in T_{in}, and the bottom panels show decrease in T_{in} from the nominal value of 350 K. At the bifurcation point, two steady states merge (represented as star in left panels) and then disappear.

We are now equipped to make a bifurcation diagram, which is shown in Figure 9.9. The actual numerics behind constructing a complete bifurcation diagram is beyond the scope of this text. Nonetheless, we will use the tools studied so far to make this plot. Between the two bifurcation points, $313 \lesssim T_{in} \lesssim 380$, all the three steady states exist simultaneously; below the lower bifurcation point, only the low steady state exists, whereas above the higher bifurcation point, only the high steady state exists. Starting with $T_{in} = 298$, T_{in} is increased progressively, each time using `fsolve` to obtain the low steady state. The previous `fsolve` solution may be used as an initial guess. This continues until $T_{in} \approx 380$, beyond which the solution jumps to the higher branch (denoted by the upward arrow in Figure 9.9). Increasing T_{in} further, the solution continues marching along the high-conversion branch.

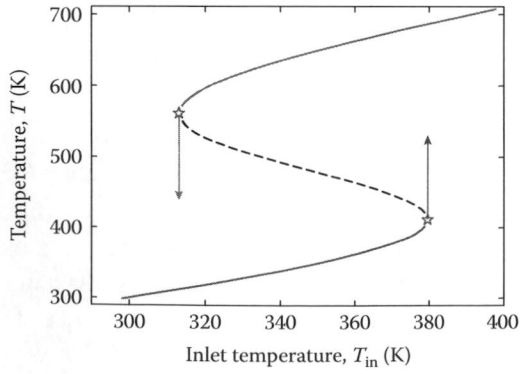

FIGURE 9.9 Bifurcation plot for a nonlinear CSTR with T_{in} as the bifurcation parameter. The two turning points are marked as stars.

After reaching 398 K, the temperature is decreased again. The procedure of incrementally decreasing T_{in}, each time using fsolve to obtain the steady state value, may be repeated. At this stage, a hysteresis is observed; the system does not jump to the low-conversion steady state at 380 K. Instead, it continues along the high-conversion branch until the other turning point is reached at $T_{in} = 313$ K.

The high- and low-conversion branches correspond to stable steady states and are therefore shown as solid lines. The bifurcation points are marked as stars in Figure 9.9. The middle steady state branch is denoted by dashed lines since the middle steady state is unstable.

Obtaining the middle steady state branch can be a little tricky. At the first bifurcation point of 380 K, the steady state solution jumps from $T = 402$ K to $T = 685$ K. Thus, at $T_{in} = 379$, fsolve can be used to obtain the middle steady state by assuming the initial guess of the temperature to be $T^{(0)} = 450$ K. Indeed, the solution at the middle branch is obtained as $T = 420.8$ K. Thereafter, the inlet temperature T_{in} may be progressively decreased, each time using fsolve to chart out the middle steady state branch.

This procedure to chart out the bifurcation plot is known as *parametric continuation*. Recall that we were already introduced to *natural parameter continuation* in Chapter 3. There, we increased the inlet temperature T_{in} in the same step-wise manner but used an ODE solver instead. That procedure will allow us to trace the two stable branches; the algebraic solver with the procedure elaborated in this section is required to trace the unstable branch.

This completes our introduction to steady state multiplicity. There are two other types of local bifurcation—subcritical and supercritical pitchfork bifurcation. We will skip them in this book. The next interesting example is that of systems exhibiting oscillations.

9.3 LIMIT CYCLE OSCILLATIONS

9.3.1 Oscillations in Linear Systems

The mass-spring-damper system, introduced in Chapter 3, as an example of second-order ODE-IVP is modeled as

$$mx'' + cx' + kx = 0 \qquad (9.30)$$

where

x'' is the acceleration

$v = x'$ is the velocity

The ODE models the motion of a body with mass m attached to a spring. The mass is displaced initially to $x = x_0$ and released. The second-order ODE is converted to the following set of first-order ODE-IVP:

$$\mathbf{y}' = \begin{bmatrix} v \\ -\dfrac{c}{m}v - \dfrac{k}{m}x \end{bmatrix}, \quad \mathbf{y}(0) = \begin{bmatrix} x_0 \\ 0 \end{bmatrix} \qquad (9.31)$$

This can be written in standard linear ODE form, $\mathbf{y}' = A\mathbf{y}$, where

$$A = \begin{bmatrix} 0 & 1 \\ -\bar{k} & -\bar{c} \end{bmatrix}$$

The eigenvalues of the above matrix can be computed to be

$$\lambda = \frac{-\bar{c} \pm \sqrt{\bar{c}^2 - 4\bar{k}}}{2}$$

Consider the case where the spring constant $\bar{k} = 1$ and the damping coefficient is varied. When the damping coefficient is large, that is, $\bar{c} > 2\bar{k}$, the system is overdamped. The origin is a center and trajectories are attracted to the center. When the damping coefficient is decreased, the response becomes oscillatory. For example, with $\bar{c} = 0.5$, the eigenvalues are $\lambda = -0.25 \pm 0.968i$. The system is a stable spiral. As the damping coefficient value is decreased, system response becomes increasingly underdamped and oscillations increase. This happens until the damping coefficient becomes zero, in which case, the system oscillates without any damping. We had analyzed such linear dynamical systems in Section 5.2. The time response and phase portrait for the system are shown in Figure 9.10.

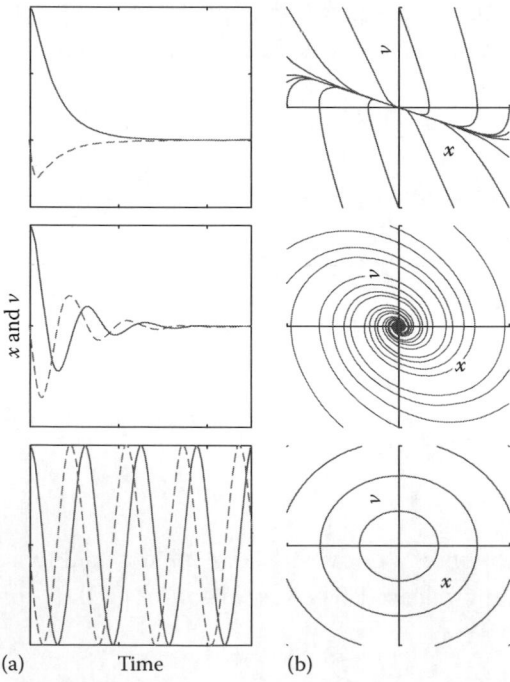

FIGURE 9.10 Response of mass-spring-damper system for various values of damping coefficient. (a) Show location (solid line) and velocity varying with time and (b) phase-plane plots. The top, middle, and bottom rows indicate $\bar{c} = 3$, $\bar{c} = 0.5$, and $\bar{c} = 0$, respectively.

The closed circles are a characteristic response of linear systems with purely imaginary eigenvalues. The oscillating system charts out multiple concentric circles. Thus, with an initial displacement of 1, the system oscillates between $-1 \leq x \leq 1$ without any attenuation, and the amplitude remains equal to 1 even after a long time. Likewise, if the initial displacement were 0.5, the system would oscillate between $-0.5 \leq x \leq 0.5$, and so on.

9.3.2 Limit Cycles: van der Pol Oscillator

The documentation for MATLAB® ODE solvers describes a van der Pol oscillator. It is a standard example of nonlinear oscillating systems. Unlike the linear system described above, van der Pol oscillator has a more interesting behavior, known as the *limit cycle*. Before describing the limit cycle, let us consider the simulation of the van der Pol system. The model for the system is given by

$$x'' + \mu\left(x^2 - 1\right)x' + x = 0 \qquad (9.32)$$

A Dutch electrical engineer, Balthasar van der Pol, proposed this model and showed that the system exhibited stable oscillatory behavior. He called these dynamics "relaxation oscillations," as will be described after the following example. In this example, the transient and phase-plane plots of a van der Pol oscillator will be obtained, followed by a discussion of their dynamics.

Example 9.2 van der Pol Oscillator

Problem: Simulate the van der Pol oscillator of Equation 9.32.

Solution: The ODE-IVP is written as the following set of two ODEs:

$$\begin{aligned} x' &= v \\ v' &= -\mu\left(x^2 - 1\right)v - x \end{aligned} \qquad (9.33)$$

This becomes a 2D ODE-IVP problem, with the function defined as

```
function dy=vanderPolFun(t,y,mu)
     dy(1,1)=y(2);
     dy(2,1)=-mu*(y(1)^2-1)*y(2)-y(1);
end
```

The resulting problem is solved using ode45 for two different values of μ, 1 and 10. The initial condition is chosen as $[x \quad v]^T = [0.1 \quad 0]^T$. Figure 9.11 shows the transient response and phase-plane plot of the van der Pol oscillator for two different values of μ, that is, $\mu = 1$ and $\mu = 10$. Periodic oscillations are observed for both the values of parameter μ.

Note the difference between these oscillations and those in Figure 9.10: The van der Pol oscillator has a single stable closed trajectory. This is shown in the top right panel, where the response to three different initial conditions is plotted in the phase plane

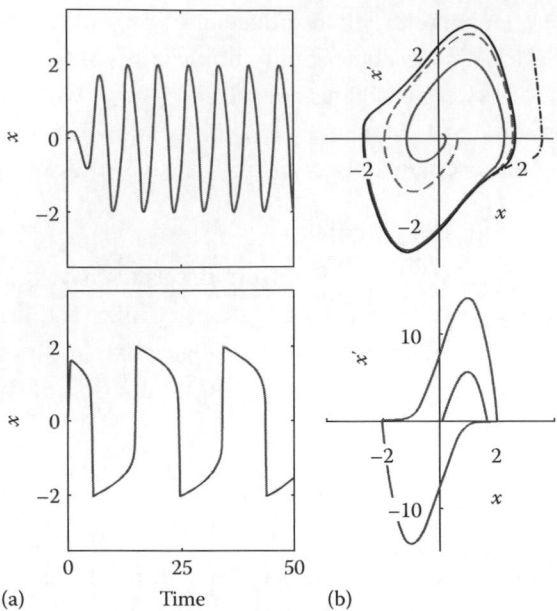

FIGURE 9.11   Transient response of x (a) and phase-plane plots (b) for van der Pol oscillator. The top row represents $\mu = 1$ and bottom row represents $\mu = 10$.

(as solid, dashed, and dash-dot lines). Irrespective of the initial condition, the system settles at the same closed trajectory. This is known as *limit cycle* behavior.

The top row represents $\mu = 1$. Notice how the trajectory is nearly uniform. However, when the parameter value is increased to $\mu = 10$, the transient trajectory gets a peculiar qualitative behavior. There are long stretches of slowly changing response followed by a short period of rapid change. This observation is discussed in this section.

9.3.2.1 Relaxation vs. Harmonic Oscillations

van der Pol coined the term "relaxation oscillations" to describe the behavior of this oscillator. This may be contrasted with harmonic oscillations of the mass-spring-damper system. When $\mu = 0$, the model becomes equivalent to a (undamped) mass-spring system. When μ is positive, damping is introduced in the system. Unlike the mass-spring-damper system, the damping is nonlinear. This has significant consequence on the system response near the origin. When the van der Pol equation (9.33) is linearized at the origin, we get

$$J = \begin{bmatrix} 0 & 1 \\ -1 & \mu \end{bmatrix}$$

The characteristic equation of the Jacobian

$$\lambda^2 - \mu\lambda + 1 = 0 \tag{9.34}$$

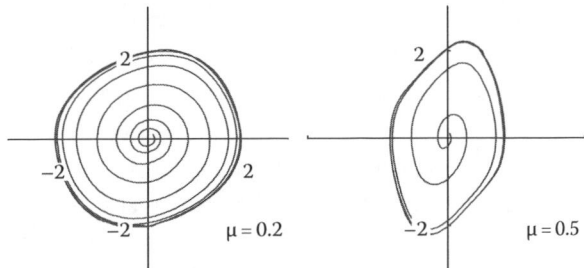

FIGURE 9.12   Phase portrait of a van der Pol oscillator for small values of $\mu = 0.2$ and 0.5.

indicates that the origin is an unstable node. Starting from a point close to the origin, the trajectory will diverge. However, the nonlinear damping coefficient, $c \equiv \mu(x^2 - 1)$ grows as x increases. Thus, the state variables are not allowed to grow in an unbounded manner. This is the physical interpretation of the closed cyclic trajectories seen in Figure 9.11.

Figure 9.12 shows the phase portrait when the damping coefficient is small. When $\mu = 0.2$, the damping term is small. The eigenvalues from Equation 9.34 are $0.1 \pm 0.995i$. The origin is now an unstable spiral. As seen in the left-hand panel of Figure 9.12, the trajectory spirals out from the origin. As the value of x increases beyond 1, the damping, $c \equiv \mu(x^2 - 1)$ increases. The trajectory spirals multiple times around the origin and settles at the limit cycle. When μ is increased to 0.5 (and then to 1), the system still grows slowly away from the origin and settles at the limit cycle. When the coefficient is increased further, damping is increased and the shape of the limit cycle becomes highly skewed (Figure 9.11, bottom-right panel).

Finally, let's discuss what the term "relaxation oscillation" implies. Consider the case when $\mu = 10$, indicating that the damping is quite high. Around time $t = 15$, as seen in the bottom-left panel of Figure 9.11, $x \approx 2$. The damping term is large, making the system response highly overdamped (see previous example) and hence very sluggish. As x approaches 1, the damping term reduces and the speed of response increases. When x falls below 1, the damping $c \equiv \mu(x^2 - 1)$ becomes negative and the system tries to become unstable. Thus, there is a brief period when the value of x falls rapidly. Once it goes below -1, the damping term becomes positive, the van der Pol system becomes stable, and the response again becomes overdamped when $x < -2$. Thus, the system is characterized by a very long "relaxation time" (when the response is overdamped), followed by a short "impulsive period" where the state variable changes rapidly. The relaxation time governs the period of oscillations for the system.

The van der Pol system shows *limit cycle* behavior, where a *unique* stable closed trajectory exists. This manifests as constant-amplitude oscillations, irrespective of the initial condition.

9.3.3 Oscillating Chemical Reactions

Boris Belusov was a Russian chemist who is credited for discovering oscillatory chemical reactions, which then led to the development of nonlinear analysis of chemical systems. While working on citric acid cycle, Belusov discovered an oscillating chemical reaction. His discovery was ahead of its time and was met with skepticism within the scientific community. This work was picked up by Anatol Zhabotinsky, who further investigated the reaction. Their work spawned a lot of activity in similar systems. Their oscillating reaction between citric acid, bromate, and cerium as catalyst is now popular as Belusov-Zhabotinsky (B-Z) reaction.

Since then, several other oscillating chemical reactions have been discovered. The Briggs-Rauscher reaction is a popular reaction for demonstrating this phenomenon. It is a reaction between malonic acid, potassium iodate, and hydrogen peroxide in presence of Mn^{+2} catalyst. Bromate solution from the original B-Z system is replaced with an iodide solution. With starch added as indicator, the demo reaction is visually striking because it oscillates between dark-blue color in the presence of iodide ion and amber color when it gets converted to iodine.*

The basic mechanism that reproduces oscillations observed in the B-Z reaction is

$$A \xrightarrow{k_1} X$$
$$2X + Y \xrightarrow{k_2} 3X \tag{9.35}$$
$$B + X \xrightarrow{k_3} Y + C$$
$$X \xrightarrow{k_4} D$$

This mechanism is called the Brusselator scheme. The second reaction in the above mechanism is an *autocatalytic* reaction. The overall reaction scheme remains oscillatory so long as A and B are in significant excess. When A and B are in excess, we can assume C_A, C_B to be constant. Since C and D are products, they do not affect the material balance of X and Y. The ODEs for the two species X and Y are

$$C_X' = k_1 C_{A0} + k_2 C_X^2 C_Y - (k_3 C_{B0} + k_4) C_X$$
$$C_Y' = k_3 C_{B0} C_X - k_2 C_X^2 C_Y \tag{9.36}$$

Dividing throughout by $k_1 C_{A0}$ and defining $x = C_X / k_1 C_{A0}$ and $y = C_Y / k_1 C_{A0}$, the two ODEs can be written as

$$x' = 1 + ax^2 y - (\beta + k_4) x$$
$$y' = \beta x - ax^2 y \tag{9.37}$$

where

$$\alpha = k_2 (k_1 C_{A0})^2, \quad \beta = k_3 C_{B0}$$

The following example shows the code for the Brusselator system.

---

* A nice demonstration of this and other oscillating reactions is available on YouTube. For example, see https://www.youtube.com/watch?v=IggngxY3riU [Last accessed: January 10, 2017].

Example 9.3 Simulations of the Brusselator System

Problem: Perform dynamic simulations of the Brusselator system, given by Equation 9.37 for $\alpha = 1, \beta = 1, k_4 = 1$.

Solution: The function file for the Brusselator system is straightforward and given below:

```
function dY=brusselatorFun(t,Y,par)
% Function file for Brusselator system
a=par.alpha; b=par.beta;
k4=par.k4;
x=Y(1); y=Y(2);
%% Model equations
dx=1+a*x^2*y-(b+k4)*x;
dy=b*x-a*x^2*y;
dY=[dx;dy];
```

The above function can be passed on to an `ode45` solver to obtain the transient dynamics. The transients of the system obtained for the initial conditions $x_0 = 1, y_0 = 0.5$ are shown in the top panel (Figure 9.13). The bottom panel shows the phase portrait, which was obtained starting from various initial conditions in the phase plane. The thick dashed line in the phase-plane plot corresponds to the initial conditions $[1 \quad 0.5]^T$ of the top panel. The point, $(1,1)$ is a stable steady state point. It is a spiral since the trajectories show some oscillations before settling down to the steady state.

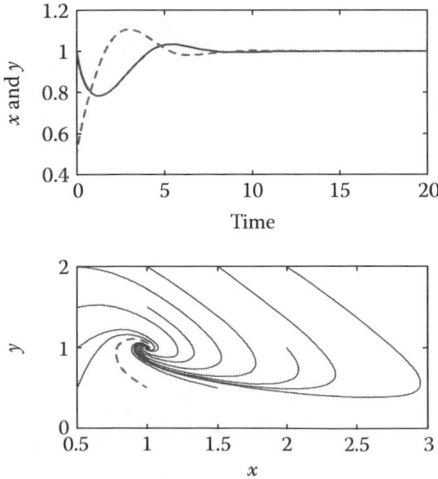

FIGURE 9.13   Transients of the Brusselator reactor and the phase portrait for $\alpha = 1, \beta = 1, k_4 = 1$. In the top panel, the initial conditions are $x_0 = 1, y_0 = 0.5$.

Let us analyze the Brusselator system. The steady state for the system can be obtained by equating the two transient equations to zero and adding them up. For constant values of $\alpha = 1, k_4 = 1$, the steady state is

$$x_{SS} = 1, \quad y_{SS} = \beta \qquad (9.38)$$

Linearizing the Brusselator equation at this steady state, we get the Jacobian as

$$J = \begin{bmatrix} \beta - 1 & 1 \\ -\beta & -1 \end{bmatrix} \qquad (9.39)$$

The characteristic equation of the above matrix is

$$\lambda^2 + \lambda(2 - \beta) + 1 = 0 \qquad (9.40)$$

When $\beta = 1$, as in Example 9.3, the steady state $[1 \quad 1]^T$ is a stable attractor. The eigenvalues of the Jacobian are $\lambda = -0.5 \pm 0.866i$, which confirms our observation that the steady state is a stable spiral.

As in Section 9.1, we will analyze the effect of varying the inlet concentration C_{B0} on the dynamics of the system. Changing C_{B0} affects the parameter β. When the bulk concentration of B is increased threefold, the value of the parameter β is increased to 3. Under these conditions, the system shows interesting dynamics. The steady state can be calculated from Equation 9.38 as $\mathbf{Y}_{SS} = [1 \quad 3]^T$. From the characteristic equation (9.40), one can compute the eigenvalues at the steady state to be $\lambda = 0.5 \pm 0.866i$. This indicates the steady state is an unstable spiral. Although this steady state is unstable, it is circumscribed by a limit cycle, as shown in Figure 9.14. Any trajectory starting close to the steady state diverges

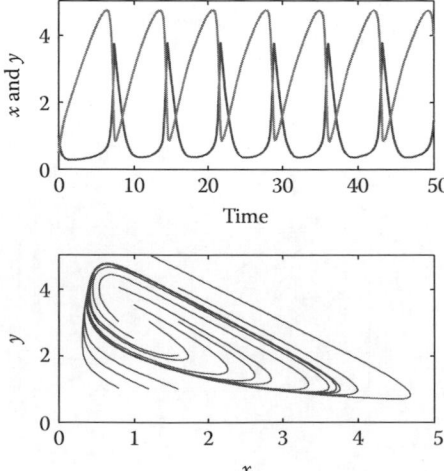

FIGURE 9.14 Transient response of the Brusselator system and phase portrait for $\beta = 3$.

outward. However, as we have seen in the van der Pol oscillator, the trajectory does not grow unbounded. Instead, it settles as a single closed trajectory, the limit cycle.

Thus, from $\beta = 1$ to 3, the system goes from a stable spiral to an unstable spiral, with the unstable spiral being enclosed by a stable limit cycle. At $\beta = 2$, the eigenvalues of the Jacobian are $\lambda = \pm i$. When $\beta < 2$, the real part of the eigenvalue is negative and the steady state is a stable spiral. When $\beta > 2$, the steady state becomes an unstable spiral. Thus, when the eigenvalues of the Jacobian are plotted in a complex plane, they shift from the left-half plane (stable) to the right-half plane (unstable), with the eigenvalues crossing over from stable to unstable at the bifurcation point $\beta = 2$. Such a crossing-over of eigenvalues at the imaginary axis (instead of origin) is indicative of the presence of limit cycle oscillations. Specifically, the above qualitative behavior of a stable spiral transitioning to an unstable spiral combined with the appearance of a stable limit cycle is an example of *supercritical Hopf bifurcation*. When C_{B0} is increased across the Hopf bifurcation point, the stable system suddenly transitions into limit cycle oscillations.

This ends our brief introduction to oscillations and limit cycles.

9.4 SIMULATION OF METHANOL SYNTHESIS IN TUBULAR REACTOR

Methanol synthesis is carried out in a catalytic reactor at high pressure and moderate temperature. The synthesis gas is feedstock for this process. The primary reaction taking place in the presence of a supported copper/zinc catalyst is

$$CO + 2H_2 \rightarrow CH_3OH$$

Since there is carbon dioxide also present in the syn gas, the water-gas shift reaction

$$CO_2 + H_2 \rightarrow CO + H_2O$$

and direct reduction of CO_2

$$CO_2 + 3H_2 \rightarrow CH_3OH + H_2O$$

also take place on the catalyst. The three reactions above are not linearly independent. The third reaction is the sum of the first two reactions. Various kinetic models for different catalysts, operating conditions, and reactor types have been developed in the past. In this case study, the kinetic model of van den Bussche and Froment (1996) will be used.

Their model considers the methanol synthesis process comprising direct CO_2 reduction and water-gas shift reactions. The kinetics of the two reactions are given by

$$r_1 = \frac{k_1 p_{CO_2} p_{H_2}}{G^3} \left(1 - \frac{1}{K_{eq,1}} \prod_i p_i^{v_{i1}} \right) \tag{9.41}$$

$$r_2 = \frac{k_2 p_{CO_2}}{G} \left(1 - \frac{1}{K_{eq,2}} \prod_i p_i^{v_{i2}} \right) \tag{9.42}$$

where p_i are partial pressures in bar, the reaction rate is in mol/kg$_{cat}$ · s, and the rate constants are given by

$$k_1 = 1.07 \exp\left(\frac{36,696}{RT}\right), \quad k_2 = 1.22 \times 10^{10} \exp\left(-\frac{94,765}{RT}\right) \tag{9.43}$$

The equilibrium constants are obtained as

$$\log_{10}\left(K_{eq,1}\right) = \frac{3066}{T} - 10.592, \quad \log_{10}\left(K_{eq,2}\right) = 2.029 - \frac{2073}{T} \tag{9.44}$$

The inhibition term in the above rate expressions is given by

$$G = 1 + K_a \frac{p_{H_2O}}{p_{H_2}} + K_b \sqrt{p_{H_2}} + K_e p_{H_2O}$$

$$K_a = 3453.4, \quad K_b = 0.499 \exp\left(\frac{17,197}{RT}\right), \tag{9.45}$$

$$K_e = 6.62 \times 10^{-11} \exp\left(\frac{124,119}{RT}\right)$$

In this section, we will consider two cases: a steady state and transient simulations of an isothermal plug flow reactor (PFR). In the case of the former, pressure drop through the packed bed is considered, whereas pressure drop will be neglected in the transient simulations. We will adopt a modular approach, so that the codes developed in the first part can be reused in the transient simulations.

9.4.1 Steady State PFR with Pressure Drop

The model equations for the steady state PFR will be discussed in Section 9.4.1.3. Before that, the calculation of surface reaction rate and processing of input parameters will be discussed in Sections 9.4.1.1 and 9.4.1.2, respectively.

9.4.1.1 Reaction Kinetics

The first step is to write a function that calculates the rates of the two reactions. The reaction rates depend on the temperature of the catalyst and partial pressures of various species. These two form the input arguments for the function. The output argument could either be rate of reactions, r_j (as per Equations 9.41 and 9.42), or rates of formation/consumption, $R_k = \sum_j \nu_{kj} r_j$. We choose the latter in this example. The following function computes the rates as per the model of vanden Bussche and Froment (1996).

The following function, which is common to both cases, calculates reaction rates:

```
function rConv=surfRxn(T,pPress,par)
% To calculate rate for methanol synthesis
% Ref: vanden Bussche and Froment (1996) J. Catal.
```

```
R=par.R;
nu=par.stoich;
pPress=pPress/1e5;    % Pressure in bar

% Rate constants
k1=1.07*exp(36696/(R*T));
k2=1.22e10*exp(-94765/R/T);
% Inhibition term
K1=3453.4;
K2=0.499 * exp(17197/R/T);
K5=6.62e-11 * exp(124119/R/T);
G=1 + K1*pPress(5)/pPress(2) + ...
    K2*sqrt(pPress(2)) + K5*pPress(5);
% Forward rate
rate    =zeros(2,1);
rate(1)=k1/G^3 *pPress(2)*pPress(3);
rate(2)=k2/G    *pPress(3);

% Equilibrium
Keq(1)=10^(3066/T-10.592);
Keq(2)=10^(-2073/T+2.029);
for j=1:2
    eqTerm=prod(pPress.^nu(:,j));
    rate(j)=rate(j)*(1-eqTerm/Keq(j));
end
% Rate of conversion of each species
rConv=par.rhoCat * (nu*rate);
end
```

Note the third line of the function. The function accepts the inputs in SI units (Kelvin and Pascal). Since the reaction rate requires partial pressure in bar, this conversion is done within the surface reaction code itself. Different sources in the literature compute the reaction rate expressions in different ways (e.g., based on concentration, partial pressures in Pa, mole fraction, etc.). Hence, a good programming practice is that the main code must consistently use SI units, whereas the unit conversion must be done in the surface reaction code. In this case, R_k is in mol/m$^3 \cdot$s.

9.4.1.2 Input Parameters and Initial Processing

Modular and flexible codes are often scalable, more efficient, easier to understand, and easier to extend to different systems. In this example, I want to reuse as many parts of the code as possible for steady state and transient solutions. With this in mind, a separate function is written for defining and processing input parameters.

```
function param=methanolParam
% Computes parameters for methanol synthesis
```

```
%% Geometric Parameters
d=0.016;
Acs=pi/4*d^2;    % Cross-sectional area
L=0.15;          % Length
dp=0.0005;       % Particle diameter
epsi=0.5;        % Void fraction

%% Operating conditions
Tin=523;
Pin=5e6;
mFlow=2.8e-5;

%    CO,  H2,  CO2,  CH3OH, H2O, N2
Xin=[4;   82;  3;    0;     0;   11]/100;
nu =[0    -3   -1    1      1    0;
     1    -1   -1    0      1    0]';
MW=[28;   2;   44;   32;    18;  28]/1000;
% Calculating Inlet Parameters
Mavg=Xin'*MW;
Ftot=mFlow/Mavg;
Fin=Ftot*Xin;
rhou0=mFlow/Acs;
%% Ergun's Equation Parameters
ergun1=150*visc/dp^2 * (1-epsi)^2/epsi^3;
ergun2=1.75*rhou0/dp * (1-epsi)/epsi^3;
param.ergun1=ergun1;
param.ergun2=ergun2;

%% Store parameters in a structure
param.L=L;
param.Acs=Acs;
% Species / Reactions
param.MW=MW;
param.stoich=nu;
param.visc==2e-5;         % Viscosity (approx.)
param.R=8.314;
param.rhoCat=1775;
% Inlet conditions
param.rhouIn=rhou0;
param.Pin=Pin;
param.Tin=Tin;
param.Fin=Fin;
param.Xin=Xin;
```

Most of the parameters computed in this function are not dependent on the model assumptions, which will get incorporated in the model function file. The discussion on Ergun's equation for pressure drop will follow in the next section.

9.4.1.3 Steady State PFR Model

The steady state model for a PFR is given by

$$\frac{1}{A_{cs}}\frac{dF_k}{dz} = \rho_{cat}\sum_j \nu_{kj}r_j \tag{9.46}$$

The mole fractions and partial pressures are calculated as

$$X_k = \frac{F_k}{\sum_i F_i}, \quad p_k = P \tag{9.47}$$

In the examples so far, the total pressure was assumed to be constant. If the pressure drop along the reactor is significant, it needs to be accounted for using the Ergun's equation for pressure drop:

$$-\frac{dP}{dz} = 150\underbrace{\frac{\mu}{d_p^2}\frac{(1-\varepsilon)^2}{\varepsilon^3}v_s}_{\text{Erg}_1} + 1.75\underbrace{\frac{\rho v_s}{d_p}\frac{(1-\varepsilon)}{\varepsilon^3}v_s}_{\text{Erg}_2} \tag{9.48}$$

Ergun's equation may be used for a packed bed. For the second constant above, we use the fact that mass flowrate remains constant:

$$\rho v = \left(\rho v\right)_{in} = G_{in} \tag{9.49}$$

which can therefore be precomputed.

If instead the reaction is carried out in a tubular reactor, the pressure drop will instead be

$$\frac{dP}{dz} = -\frac{2}{d}\rho v^2 f_f \tag{9.50}$$

where the Blasius equation

$$f = \frac{0.079}{\text{Re}^{0.25}} \tag{9.51}$$

can be used to compute the friction factor for turbulent flow.

The superficial velocity, v_s, is computed from the molar flowrates as

$$v_s = \frac{1}{A_{cs}}\frac{RT}{P}\sum_{i=1}^{n}F_i \tag{9.52}$$

Thus, the steady state model for a pseudohomogeneous model of tubular reactor consists of Equation 9.46, the reaction kinetics are given by Equations 9.41 and 9.42, the pressure drop by Equation 9.48, and the velocity by Equation 9.52.

The following example demonstrates the steady state simulation of gas-phase catalytic methanol synthesis in a packed bed reactor.

Example 9.4 Steady State Model for Methanol Synthesis

Problem: Solve the steady state model for methanol synthesis.

Solution: As described above, we adopt a modular approach in developing this code. The reaction rate is computed using the function surfRxn.m and the initial parameter computation is also given above. The main driver script for steady state simulation is given below:

```
% Steady-State PFR for methanol synthesis
%% Parameters and Setup
param=methanolParam;
opt=odeset('abstol',1e-10,'reltol',1e-10);
Y0=[param.Fin; param.Pin];

%% Solve the ODE
[zSol,YSol]=ode15s(@(z,Y) methanolFun(z,Y,param),...
    [0 param.L], Y0, opt);

%% Output and Plot
figure(1)
XSol=YSol(:,1:6);
for i=1:length(XSol)
    XSol(i,:)=XSol(i,:)/sum(XSol(i,:));
end
plot(zSol,100*XSol(:,[1:5]));
```

Note that the constants used in computing the pressure drop using Ergun's equation were already computed in the inlet parameter function in the previous section. Since the flowrates are in the order of 10^{-4}, the default tolerances are not sufficient. Hence, the tolerance values are specified as $\sim 10^{-10}$.

Function file for solving the PFR problem: The next task is to write the function file for computing dYdz. Here, Y represents the flowrates for all species and the pressure:

```
function dYdz=methanolRxtrFun(t,Y,par)
% Function to calculate dY/dx for methanol reactor
% Isothermal reactor with pressure drop
%% Parameters and variables
T=par.Tin;    R=par.R;
F=Y(1:6);
```

```
P=Y(7);
X=F/sum(F);
pPress=P*X;
Mavg=par.MW'*X;
rho=P*Mavg/(R*T);

%% Model
RConvRxn=surfRxn(T,pPress,par);
dF=par.Acs*RConvRxn;
% Pressure Drop
vs=par.rhouIn/rho;
dP=-(par.ergun1+par.ergun2)*vs;
% Final Model
dYdz=[dF;dP];
end
```

Figure 9.15 shows the steady state profile of various species in the reactor. The reactor attains equilibrium mole fractions for all the reacting species. This can be verified by comparing the final steady state values of mole fractions with the ones computed using the equilibrium conditions. This forms one check on the overall model performance. On the same lines, another sanity check is given by the flowrate of nitrogen, F_{N2}. Since it is inert, the molar flowrate of nitrogen at the reactor exit should equal the molar flowrate at the reactor entrance. The final check is by comparing the results of Figure 9.15 with that of the original paper. The qualitative match indicates that the steady state results are valid.

Figure 9.16 shows the effect of temperature and pressure on methanol mole fraction along the length of the reactor. The inlet consists of 25% of CO, 50% H_2, 10% CO_2, and the rest nitrogen, which represents one of the input conditions from Graaf and coworkers (1989). Under both cases, the pressure drop in the reactor is about 100 Pa, that is, 0.001 bar. The low pressure drop is due to the low velocity in the bed. Hence, pressure drop may be neglected in the rest of the simulations.

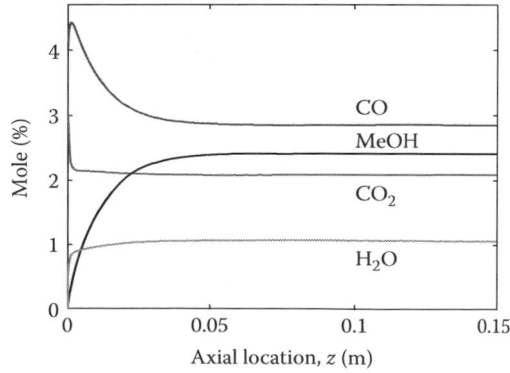

FIGURE 9.15 Steady state profiles of mole fractions of all species in a methanol synthesis reactor.

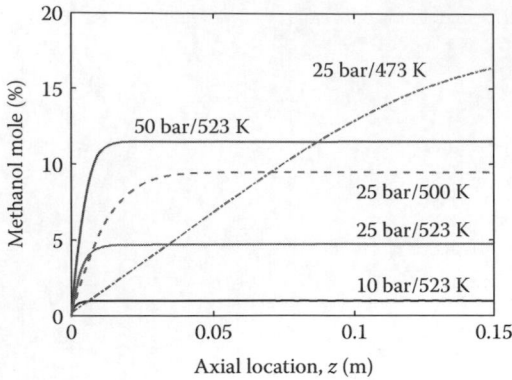

FIGURE 9.16 The effect of temperature and pressure on methanol mole fraction.

The methanol synthesis reaction is a slightly exothermic reaction with a decrease in the number of moles. Thus, lower temperature and higher pressures increase the equilibrium conversion. The thin solid line in Figure 9.16 represents the base case with 523 K and 50 bar pressure. When the pressure is decreased to 25 bar and 10 bar (solid lines) at the same temperature, the equilibrium methanol concentration decreases. Due to high enough temperature, complete conversion is obtained within 2–3 cm of the reactor length. On the other hand, when the temperature is decreased (dashed lines), equilibrium conversion increases. At 500 K temperature, conversion is higher than the corresponding case with 523 K. When the temperature is further decreased to 473 K, equilibrium conversion further increases. However, due to low temperature, the reactor length is not sufficient to ensure that the reaction approaches equilibrium conversion.

With this, the steady state analysis is complete. Next, we focus on the transient reactor model.

9.4.2 Transient Model

The transient model for the pseudohomogeneous reactor is given by

$$\frac{\partial}{\partial t}C_k + \frac{1}{A_{cs}}\frac{\partial}{\partial z}F_k = \rho_{cat}\sum_j \nu_{kj}r_j \tag{9.53}$$

When we solved the *steady state model*, we related concentration C_k to the flowrate F_k as

$$C_k = \frac{F_k}{\sum_i F_i}\left(\frac{P}{RT}\right) \tag{9.54}$$

Thus, the steady state model was converted to a molar flowrate basis and solved in the previous section. However, this strategy gets complicated for a transient model because flowrates F_i and pressure P are all changing with time. An alternative way is to solve the equations in concentration terms. To do so, we write

$$F_k = (A_{cs}\nu)C_k \tag{9.55}$$

where $(A_{cs}v)$ is the volumetric flowrate. This leads us to the familiar equation we encountered in Chapter 4:

$$\frac{\partial}{\partial t}C_k + \frac{\partial}{\partial z}vC_k = \rho_{cat}\sum_j \nu_{kj}r_j \tag{9.56}$$

When a liquid-phase system or a gas-phase system with no volume change was considered, the velocity was constant and could be taken out of the spatial differential. However, since this is not the case here, we need to compute v_s. From the overall mass conservation

$$\rho v = (\rho v)_{in} = G_{in} \tag{9.49}$$

we can compute

$$v = \frac{G_{in}}{\rho} = G_{in}\left(\frac{RT}{P\bar{M}}\right) \tag{9.57}$$

In the above expression

$$\bar{M} = \sum_i X_i M_i$$

is the average molecular weight.

We will solve the model using method of lines. The species balance equation (9.56) can be discretized using the backward difference (first-order upwind) approximation.

The various species include CO_2, H_2, CO, H_2O, CH_3OH, and N_2 (diluent). These are specified at each location. Discretizing the spatial domain into n intervals will result in $6n$ solution vectors and $6n$ ODEs. To do so, we define

$$Y_p = \begin{bmatrix} C_{CO_2,p} \\ C_{H_2,p} \\ C_{CO,p} \\ C_{H_2O,p} \\ C_{CH_3OH,p} \\ C_{N_2,p} \end{bmatrix}$$

at each node.

With this knowledge, we are equipped to formulate and solve the methanol synthesis problem.

Example 9.5 Transient Model for Methanol Synthesis

Problem: Solve the steady state model for methanol synthesis.

Solution: The function to compute the reaction rate, `surfRxn.m`, and model parameters, `methanolParam.m`, remains the same as before. The function file is modified for the transient model as given below.

Function file for solving the PFR problem:

```
function dY=methanolTFun(t,Y,par)
% Transient model for methanol synthesis reactor
%% Variables and Parameters
n=par.n;
h=par.h;
R=par.R;
C=reshape(Y,6,n);
dC=zeros(6,n);

%% Model
P=par.Pin;        % Neglect pressure drop
T=par.Tin;        % Isothermal
for i=1:n
    % Reaction term
    X=C(:,i)/sum(C(:,i));
    pPress=P*X;
    rxnTerm=surfRxn(T,pPress,par);

    % Convection term
    Mavg=par.MW'*X;
    rho=P*Mavg/(R*T);
    if (i==1)
        uC_In=par.Fin/par.Acs;
        u=par.rhouIn/rho;
        convTerm=(u*C(:,i)-uC_In)/h;
    else
        uPrev=u;
        u=par.rhouIn/rho;
        convTerm=(u*C(:,i)-uPrev*C(:,i-1))/h;
    end
    dC(:,i)=-convTerm+rxnTerm;
end
dY=reshape(dC,6*n,1);
```

Notice the first two lines in the model section are `T=par.Tin` and `P=par.Pin`, for the isothermal PFR with negligible pressure drop. In case of pressure drop, the appropriate balance equation would be used to compute the pressure at each location, whereas energy balance will be included for a nonisothermal reactor.

Since velocity is not constant, the convection term is given by

```
(u*C(:,i)-uPrev*C(:,i-1))/h
```

where uPrev is the velocity computed at the previous axial location, i-1.

Driver script for methanol synthesis: The driver script required for methanol synthesis is given below:

```
% Simulation of catalytic synthesis of methanol
% in a transient isothermal reactor
param=methanolParam;
n=100;
h=param.L/n;
param.n=n;
param.h=h;
%% Initial Conditions and setup
opt=odeset('abstol',1e-10,'reltol',1e-10);
Ctot=param.Pin/(param.R*param.Tin);
Cin=Ctot*param.Xin;
param.Cin=Cin;
C0=Cin;                 % Initial condition
Y0=repmat(C0,n,1);

%% Solving the ODE
[tSol,YSol]=ode15s(@(t,Y) methanolTFun(t,Y,param), ...
    [0 100], Y0, opt);
figure(1)
CSS=reshape(YSol(end,:),6,n);
XSS=zeros(size(CSS));
for i=1:n
    XSS(:,i)=CSS(:,i)/sum(CSS(:,i)) * 100;
end
XSS=[param.Xin,XSS];
plot(0:h:param.L, XSS, '--');
```

Figure 9.17 shows a comparison between steady state and transient simulations for the same operating conditions. The thin solid lines represent the profiles from the steady state model. The transient model is simulated for a long enough time (100 s, in this example) for the system to attain steady state. The results from the transient model at this end time are plotted as thick dashed lines. The two overlap, indicating that the transient model is likely to provide a reasonable prediction of methanol synthesis trends.

Figure 9.18 plots the axial profiles of methanol at various times from the start. Initially, there is no methanol in the reactor (thick dashed line). As time progresses, the amount of methanol along the reactor length increases and eventually reaches the steady state value. The steady state profile is represented with thick lines.

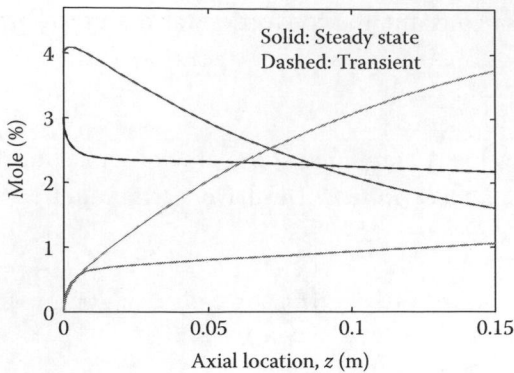

FIGURE 9.17 Comparison between the steady state solution computed using Example 9.4 and the same solution from the transient model.

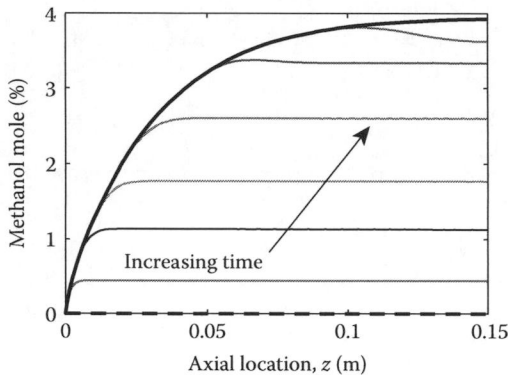

FIGURE 9.18 Axial profiles of methanol mole fraction at various times from the start obtained using the transient solver.

9.5 TRAJECTORY OF A CRICKET BALL

A ball is hit with an initial velocity and at an angle with the ground. The aim is to find the horizontal location from the starting point where the ball lands on the ground. This is a standard numerical problem in several texts. The numerical methods book by Chapra and Canale uses this example for solving a nonlinear equation for the trajectory of a projectile, with gravity as the only force acting on the projectile. In some problems, the air drag is explicitly included in calculations. For example, in the book on numerical methods using MATLAB, Fausett includes air drag into the formulation to determine where a golf ball lands. In their numerical problem solving book, Cutlip and Shacham formulate a problem of computing the trajectory of a baseball in Denver, Colorado, and New York. The problem described in this section is closest to the latter example.

This problem statement originated as a class project in an undergraduate class on simulation that I taught at IIT-Madras several years back. The Indian Premier League is a popular cricket league. After winning a must-win cricket match, the Indian cricket team captain Mahendra Singh Dhoni remarked that it was easier to "hit sixers" in Dharamsala cricket ground than at Mohali. Dharamsala is at an elevation of approximately 2000 m above mean sea level. The students modified a homework problem to a project problem to find out how gravitational acceleration and air drag influence where the ball lands. We will consider the first part of the problem: to obtain the trajectory of a cricket ball hit with a certain velocity and at an angle and, hence, to find the location where the ball hits the ground.

9.5.1 Solving the ODE for Trajectory

The model equations for computing the ball trajectory comprise force balance in x- and y-directions. At any point, the velocity of the ball has two components:

$$\mathbf{v} = u\hat{\mathbf{i}} + v\hat{\mathbf{j}}$$

Since the initial speed and angle are given, the two components u_0 and v_0 are known at $t = 0$. Air drag acts against the direction of motion of the ball, and gravity acts downward. Thus

$$\begin{aligned} mu' &= -\kappa|\mathbf{v}|u \\ mv' &= -g - \kappa|\mathbf{v}|v \end{aligned} \tag{9.58}$$

form the overall balance equations for the ball. The net air drag is given by $\kappa|\mathbf{v}|\mathbf{v}$, which acts against the direction of motion of the ball. Here, $|\mathbf{v}|$ is the magnitude of the velocity. The two components of the air drag are, therefore, $-\kappa|\mathbf{v}|u$ and $-\kappa|\mathbf{v}|v$ (the negative sign indicating the direction is opposite to the corresponding velocity component) in x- and y-directions, respectively. Thus, the overall ODE-IVP is given by

$$\frac{d}{dt}\begin{bmatrix} x \\ y \\ u \\ v \end{bmatrix} = \begin{bmatrix} u \\ v \\ -\kappa|\mathbf{v}|u \\ -g - \kappa|\mathbf{v}|v \end{bmatrix} \tag{9.59}$$

with the initial conditions

$$\mathbf{Y}(0) = \begin{bmatrix} 0 \\ 0 \\ V_0\cos(\theta) \\ V_0\sin(\theta) \end{bmatrix} \tag{9.60}$$

The following example shows the basic code for solving the ODE-IVP.

Example 9.6 Trajectory of a Cricket Ball

Problem: Given an initial velocity and initial angle, compute the trajectory of a cricket ball by solving the ODE (9.59).

Solution: This is a rather straightforward problem as posed. The function `ballFun.m` is given below:

```
function dY = ballFun(t,Y,param)
% Function used with ODE solver to calculate trajectory
% of a cricket ball hit with a particular velocity
% Y(1)     x (Horizontal displacement)
% Y(2)     y (Vertical displacement)
% Y(3)     u (Horizontal velocity component)
% Y(4)     v (Vertical velocity component)
u=Y(3);                % X-velocity
v=Y(4);                % Y-velocity
vel=sqrt(u^2+v^2);     % Velocity magnitude
% Model Parameters
g=param.g;
c=param.kappa;
%% Model equations
dY    =zeros(4,1);
dY(1)=u;
dY(2)=v;
dY(3)=-c*vel*u;
dY(4)=-c*vel*v - g;
```

The driver code for this problem is straightforward and will be skipped in this section. The next section provides a modified driver script.

The problem is solved with the initial velocity of 35 m/s and angle of $\pi/4$. Figure 9.19 shows the variation in x and y locations with time.

The next aim will be to find the location and time where the ball hits the ground.

9.5.2 Location Where the Ball Hits the Ground

The approximate time and location where the ball hits the ground can be obtained from Figure 9.19. The ball hits the ground when $y=0$. Under these conditions, the time is approximately 4.5 s and the horizontal location where the ball lands is approximately 80 m. Let us now modify the code above for obtaining these results numerically, by stopping the ODE solver when the algebraic equation

$$y(t)=0 \tag{9.61}$$

is satisfied.

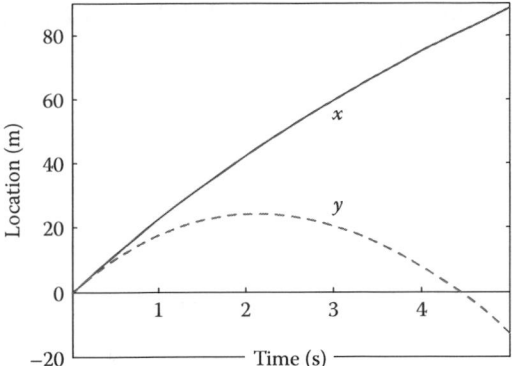

FIGURE 9.19 x and y locations of the ball with time.

An analytical case: To understand how the above equation works, consider the case where the air drag is neglected. In such a case, the ODE (9.59) can be solved analytically to get

$$u(t)=u_0, \quad v(t)=v_0 - gt$$
$$x(t)=u_0 t, \quad y(t)=v_0 t - \frac{1}{2}gt^2 \qquad (9.62)$$

Solving Equation 9.61 implies finding the value of t that forms the root of the equation. In the case of no air drag

$$t_{\text{landing}} = \frac{2v_0}{g} = \frac{2V_0 \cos(\theta_0)}{g} = 5.05\,s$$

In the case of the nonlinear problem, Equation 9.59 along with (9.61) forms a higher-order differential algebraic equation (DAE). This cannot be converted into an index-1 DAE. Thankfully, MATLAB provides *Events Options* in ODE solvers, which allows us to specify the events. This can be activated using the option

```
opt=odeset('events',@eventFun);
```

where `eventFun(t,y)` is a function that returns three output arguments: (i) event function of the type (9.61), (ii) a flag to indicate whether the ODE solver should be halted, and (iii) a flag to indicate the direction of change of the event function. The use of the events option to obtain the location of ball landing is shown in the example below.

Example 9.6 (Continued) Location of Ball Landing

Problem: M.S. Dhoni hits a cricket ball with an initial velocity of 35 m/s and angle of 45° with the horizontal. The boundary line is at 75 m from the batsman. Find the location from the batsman where the ball lands on the ground, and hence determine whether Dhoni scores a six. (A sixer is scored if the x-location exceeds 75 m.)

Solution: We will use the event function in MATLAB to obtain the location where the ball lands. The ODE function, `ballFun.m`, remains the same as in the previous example. We will first write an event function, as given below:

```
function [pos,isTerminal,dir]=ballLanding(t,Y)
pos=Y(2);        % Check if Y-location is zero
isTerminal=1;    % Stop simulation on true
dir=0;           % Ignore direction
```

The first value determines the condition to be checked. This is nothing but the left-hand side of Equation 9.61. The second line indicates that the simulation should be stopped when the event condition $y(t) = 0$ is met.

The overall driver function for solving this problem is given below:

```
%% Define Parameters and Initial Conditions
param.g=9.81;         % gravitational acceleration
param.kappa=0;        % air drag coefficient
initialVel=35;        % initial speed
initialAngle=pi/4;    % initial angle
u0 = initialVel*cos(initialAngle);
v0 = initialVel*sin(initialAngle);

%% Setting up and Solving the problem
X0 = [0; 0;         % starting position: origin
      u0; v0];      % starting velocity
tSpan = [0 20];
opt=odeset('events',@ballLanding);
[tOut,XOut]=ode45(@(t,X) ballFun(t,X,param),...
    tSpan,X0,opt);

%% Displaying the results
figure(1);
plot(tOut,XOut(:,1),'-b',tOut,XOut(:,2),'--r');
xlabel('Time (s)'); ylabel('Location (m)');
figure(2);
plot(XOut(:,1),XOut(:,2),'bo');
xlabel('x (m)'); ylabel('y (m)');
%% Animating results
exitCode = ballAnimation(tOut,XOut);
```

The result of running the code is similar to that in the earlier part of this problem. The main difference is that the code stops when the event is reached. The time and location at which the ball lands can be obtained as

```
>> disp(tOut(end))
4.4430
>> disp(XOut(end,1))
81.3707
```

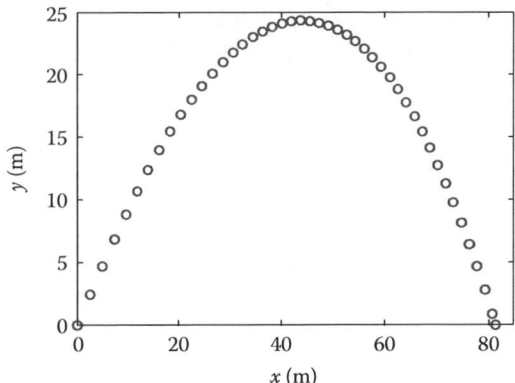

FIGURE 9.20   Trajectory of the ball hit in the *X–Y* plane of the cricket field.

Clearly, since the ball lands beyond 75 m, Dhoni has scored a six. The trajectory of the ball is plotted in Figure 9.20. The origin is where Dhoni hit the ball. Locations in the *X–Y* plane that the ball traverses are shown as circles.

Let us consider a slightly different problem: Find the height of the ball from the ground when the ball reaches the boundary. In other words, we need to obtain the height *y* when the horizontal location is 75 m.

The event function can be written as

$$x(t) - 75 = 0 \tag{9.63}$$

Thus, we change the function `ballLanding.m` as below:

```
function [pos,isTerminal,dir]=ballLanding(t,Y)
pos=Y(1)-75;    % Check if X-location is 75
isTerminal=1;   % Stop simulation on true
dir=0;          % Ignore direction
```

With this change, the ODE code is executed again. The time required for the ball to reach the boundary is found to be approximately 4 s. The height of the ball at that time is 8.16 m. Thus, if there was a 6 ft tall fielder at the boundary, he would not be able to catch the ball and Dhoni would still score a six!

9.5.3 Animation

Finally, let's get to the fun stuff: animation. The way we will animate motion of the ball is to obtain the location of the ball at every 0.1 s. This is already done in Figure 9.20. However, all the symbols are plotted at the same time. In animation, what we need instead is to plot the first circle at time 0, then replace that circle with the second circle at time 0.1, replace that with a third circle at time 0.2, and so on. If we use the `plot` command at each time, the previous plot will get overwritten and the animation will show a flicker. Hence, we are not going to plot a separate circle; instead, we will only change its *X*- and *Y*-coordinates. We do so using

```
set(gca, 'xData',XOut(i,1),'yData',XOut(i,2));
pause(0.1);
```

The last line ensures MATLAB will pause for 0.1 s before changing the location of the ball for the animation.

There is, however, one more issue. As the location changes, MATLAB changes the limits of the two axes automatically. To avoid this, we need to fix the *X*- and *Y*-limits using

```
set(gca,'xLim',[0 85], 'yLim',[0 25]);
```

This will ensure that the two axes will always remain fixed to the range of 0–85 and 0–25 m, respectively.

Finally, you may notice the animation to still be a bit jerky. To make it smooth, we will animate it every 0.02 s. The MATLAB code `ballAnimation.m` that does this is given below:

```
function exitCode = ballAnimation(tOut,XOut)
% To animate trajectory of a ball
% Interpolate data every 0.02 s
tAnim = 0:0.02:tOut(end);
Xanim = spline(tOut,XOut(:,1),tAnim);
Yanim = spline(tOut,XOut(:,2),tAnim);

% Plot a blue ball at the origin and
% fix the limits of X- and Y-axes
figure(4);
oHnd=plot(Xanim(1),Yanim(1),'bo');
set(gca,'xLim',[0 85], 'yLim',[0 25]);
xlabel('x (m)','fontsize',18);
ylabel('y (m)','fontsize',18);
set(oHnd,'markerFaceColor','blue');
pause(1);   % Pause 1 second before start

for i = 2:length(tAnim)
    set(oHnd,'xData',Xanim(i),'yData',Yanim(i));
    pause(0.02);
end
exitCode = 1;
```

The steps explained the following:

> We first set up the data for plotting using interpolation, if required.
> Next, we plot the ball at the origin and fix the *X*- and *Y*-axes limits.
> Finally, we animate the ball by changing its *X*- and *Y*-locations in a loop.

9.6 WRAP-UP

This chapter wraps up Part II of this book by bringing in various tools to analyze and simulate complex processes. The first three sections of this chapter presented nonlinear analysis using several examples. The first tool in nonlinear analysis is used to obtain various steady state solutions (Sections 9.1.1 and 9.2.1), also known as *fixed points*. The model can be linearized, and the linear stability analysis (Section 5.2) around the fixed point is used to study the local stability behavior of the system. Phase-plane analysis is a useful tool to study the overall dynamic behavior. Such an analysis is also useful in studying oscillating systems (Section 9.3), both harmonic oscillations and limit cycle behavior. Indeed, nonlinear dynamics (which also includes study of chaos, logistical maps, etc.) is a rich field itself; this chapter barely scratches the surface.

This was followed by the simulation of tubular reactor in Section 9.4. A steady state PFR is an ODE system, whereas a transient PFR is a hyperbolic PDE. As the velocity changes due to reaction, careful attention needs to be paid to the overall problem formulation. Finally, Section 9.5 simulates the trajectory of a projectile (cricket ball, in this case), where the event-tracking feature of MATLAB ODE solvers was also introduced. This section concluded with a fun example of animation in MATLAB to visualize the trajectory of the cricket ball.

EXERCISES

Problem 9.1 Perform the bifurcation analysis for a chemostat, as shown in Section 9.1 for a larger value of saturation constant, $K = 1.0$. All other parameters are kept the same as in Section 9.1.

Problem 9.2 Transcritical bifurcation: Perform bifurcation analysis for the following systems:

$$x' = (1-x)\left(a - \frac{x}{1-x}\right)$$

$$x' = x - ax(1-x)$$

Problem 9.3 Derive Equation 9.19:

$$\bar{f}_2(T) = (-\Delta H)\frac{k\tau}{1+k\tau}(FC_{A,\text{in}}) + F\rho c_p(T_{\text{in}} - T) - \frac{UA}{1+\beta}(T - T_{j,\text{in}}) = 0$$

Problem 9.4 (*c.f.*, Strogatz's book) Perform bifurcation analysis for the following autocatalytic series reactions:

$$A + X \rightleftharpoons 2X, \quad X + B \rightarrow \text{Products}$$

Assume concentrations of A and B are kept constant at a and b, respectively. Let k_1, k_{-1}, and k_2 be the rate constants of the three reactions. Show that the concentration of X can be given by $X' = c_1 X - c_2 X^2$. What are the conditions for stability of the various steady states of the system?

Problem 9.5 Repeat the ball animation problem to find the time required to reach the highest height, after which the ball starts falling.

III

Modeling of Data

CHAPTER 10

Regression and Parameter Estimation

10.1 GENERAL SETUP

Several models of various types were discussed in the preceding two parts of this book. These included differential equations, algebraic equations, expressions for certain properties, and differential algebraic equations. The aim of these models is to represent a relationship between independent and dependent quantities and/or to predict the behavior of a system. These models had several *parameters*, for example, rate constants to express reaction rate, thermal conductivity in heat transfer problem, drag coefficient, etc. The parameter values were assumed to be known, which allowed us to set up and solve the problems of interest.

The aim of this chapter is to focus on the process of obtaining the model parameters that "best represent" the observed data. This is known by various terms: regression or parameter estimation. We will use these terms to mean the same process: Given a finite number of observations, how do we fit the parameters of a chosen model that best represent the observed data?

Such a model may be phenomenological. For example, the Arrhenius rate law for temperature dependence on reaction rate

$$k(T) = k_0 \exp\left(-\frac{E}{RT}\right) \tag{10.1}$$

has a basis in statistical mechanics. Some others may represent a convenient form of representing correlation observed in the data. For example, dependence of specific heat on temperature

$$c_p = a + bT + cT^2 + dT^3 \tag{10.2}$$

is a convenient polynomial form that represents the observed trends for a wide range of temperatures. Then there are other examples, such as Antoine's relationship between saturation pressure of a pure liquid and its temperature. Antoine's equation

$$\ln\left(p^{\text{sat}}\right) = A - \frac{B}{T+C} \tag{10.3}$$

is a semiempirical relationship, since it is derived from Clausius-Clapeyron relation and modified to represent a majority of pure species. Irrespective of the theoretical, semiempirical, or empirical nature of the relationship, regression aims to compute parameters that best represent the observed data.

The earliest example of regression that we encounter, perhaps in high school or earlier, is fitting of a straight line. In our undergraduate laboratory, we have often plotted experimental observations on a graph in an appropriate way and tried to fit a straight line. In all these cases, we attempted to fit a straight line of the form

$$y = mx + c$$

The abscissa x may be considered an "independent variable," whereas the ordinate y may be considered a "dependent variable." The two may either be directly measured variables or may be derived quantities. For example, you may recall taking a logarithm of reaction rate

$$\ln\left(r\right) = \ln\left(k_0\right) + \left(-\frac{E}{R}\right)\left(\frac{1}{T}\right) \tag{10.4}$$

and plotting $\ln(r)$ vs. $1/T$.

We may also recall that the data did not lie exactly on the straight line. There was some amount of error in the experimental data. If the error is only in the *measurements* (called measurement noise), the data will lie neatly spread around the straight line. However, there are other sources of errors, including systematic errors, and the fact that the model does not *truly* represent the data. The topic of regression is intended to not only determine the model parameters that best fit the data but also to determine whether the model is adequate or appropriate to represent the observed data. The latter is often termed as *hypothesis testing*.

10.1.1 Orientation

A set of observations—either from an experiment, any other measurement, or from a more complex simulation—are available at certain discrete points. Let us consider that the data is available at N different points, x represents the independent variable, and y represents the dependent variable. In the three examples, (10.1) through (10.3), the choice of x and y were clear. Let the data be represented as

$$\left(x_1, y_1\right), \left(x_2, y_2\right), \ldots, \left(x_N, y_N\right) \tag{10.5}$$

pairs at these N data points.

Let us represent the model that we try to fit as

$$y = f\left(x; \Phi\right) \tag{10.6}$$

where $\Phi^T = [a_1 \ a_2 \ \cdots \ a_m]$ is the m-dimensional vector of model parameters.

Let us represent model *predictions* as $\hat{y}$. Given a value of parameters Φ,

$$\hat{y}_i = f\left(x_i; \Phi\right) \tag{10.7}$$

represents the model prediction of the ith data point when the corresponding observed value of the independent variable x_i is substituted in the model (10.6). The difference

$$e_i = y_i - \hat{y}_i \tag{10.8}$$

is the model error or *residual* at the ith data point.

Model fit, regression, or parameter estimation are the terms used to imply the procedure of finding the values of Φ that will minimize, in some way, the errors e_i across all the observed data.

We now review some of the basic statistics that will be useful in this chapter.

10.1.2 Some Statistics

Let us recap some definitions. The arithmetic mean of the data is

$$\bar{x} = \frac{\sum_{i=1}^{N} x_i}{N} \tag{10.9}$$

Let us represent the numerator as Σ_x. Then, the mean is written as

$$\bar{x} = \frac{\Sigma_x}{N} \tag{10.9}$$

The sum of square deviation from the mean is represented as S_x, variance is represented as $\mathrm{var}(x)$, and the standard deviation in the data as s_x:

$$S_x = \sum_{i=1}^{N}\left(x_i - \bar{x}\right)^2 \tag{10.10}$$

$$\mathrm{var}\left(x\right) = \frac{S_x}{N-1}$$

$$s_x = \sqrt{\frac{S_x}{N-1}} = \sqrt{\frac{\sum_i\left(x_i - \bar{x}\right)^2}{N-1}} \tag{10.11}$$

The denominator is var(x) and s_x is the degrees of freedom, ($N - 1$). For example, if there is a single data point, the variance of that data around its mean does not have any meaning!

Next, we come to an important definition: the sum of squares of the residuals:

$$S_\varepsilon = \sum_i \left(y_i - \hat{y}_i \right)^2 \tag{10.12}$$

Since the modeling error $e_i = \left(y_i - \hat{y}_i \right)$, the above expression forms a convenient merit function or objective function for parameter fitting process.

Finally, let us use the example of specific heat data for methane (given in Table 10.3) to further understand some of these concepts. Since this data is obtained at various temperatures, there is an underlying trend in the data, as seen in Panel (a) of Figure 10.1. The aim of regression is to find that trend. Let us next consider the case where the specific heat is calculated five times at the temperature of 400 K. The two cases are shown in Figure 10.1, with the data plotted against the observation index in the abscissa. While there is a clear trend in Panel (a), the data in Panel (b) is well spread around the value of 40.6 J/mol·K.

The data in Panel (a) was obtained from NIST WebBook of thermochemical data, whereas that in Panel (b) is "synthetic" data that I generated randomly. The latter was generated from an underlying normal or Gaussian distribution (also popularly known as "the bell-shaped" curve of probability distribution) with a mean of $\mu = 40.6$ and a standard deviation of $\sigma = 1$.

The data is given as: $c = [40.00 \quad 41.09 \quad 41.14 \quad 42.31 \quad 40.31]^T$. The underlying mean and standard deviation (which are often unknown) are μ and σ. We distinguish them from the *sample mean* $\bar{y} = 40.97$ and the *sample standard deviation* $s_y = 0.897$. Notice that since we have a small number of data points, the sample mean and standard deviation is different from the inherent statistics behind the data. The data shown here is just *one realization* of the stochastic process. After the measurements are taken, this *realization is fixed*. If we were to take the measurements again, we will get a different set of values, even though the underlying process is the same.

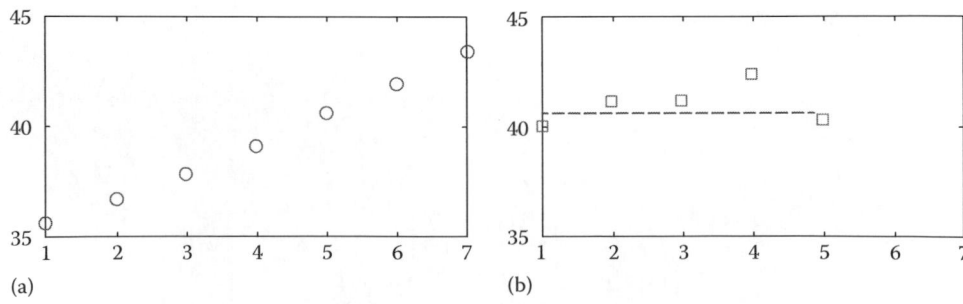

FIGURE 10.1 Specific heat of methane (a) at various temperatures and (b) at 400 K obtained with five different experiments.

10.1.3 Some Other Considerations in Regression

Unlike the previous chapters in this book, the data for x and y are available; the parameters Φ are unknown. The model may either be linear or nonlinear in parameter space. In this chapter, we are not concerned whether the model is linear or nonlinear in x and y; we are concerned about linearity in parameters, $a_1, \ldots, a_m$.

Models may have more than one input, where the model may be written as

$$y = a_1 + a_2 x + a_3 w + a_4 u \tag{10.13}$$

The aim is to find parameters for the above *linear* model, for which we use multilinear regression. As we shall show soon, multilinear regression is a straightforward extension of linear regression in multiple variables.

The model for specific heat

$$c_p = a + bT + cT^2 + dT^3 \tag{10.2}$$

although nonlinear in T, is linear in parameters a, b, c, d. Since T, T^2, and T^3 are linearly independent, this has the same form as Equation 10.13, with $x \equiv T$, $w \equiv T^2$, $u \equiv T^3$. The regression for polynomial data as in Equation 10.2 is called polynomial regression.

Likewise, if we define $a_1 = \ln(k_0)$ and $a_2 = -E/R$, model (10.4) is linear in parameters a_1 and a_2. Recall that Equations 10.1 and 10.4 represent the same equation. However, Equation 10.1 is nonlinear in parameter space. Obtaining the values of k_0 and E, with the model expressed in the original form of Equation 10.1, requires nonlinear regression. It may sometimes not be easy to convert a nonlinear regression into a linear regression. Equation 10.3 provides an example where a nonlinear regression is recommended.

All these examples and analyses will be covered in the remainder of this chapter. The problem of fitting a straight line to observed data will be discussed in Sections 10.2.1 and 10.2.2, and statistical properties related to goodness of fit will be discussed in Section 10.2.3. The linear regression problem will be extended to multiple variables in Section 10.3. Section 10.4 will discuss nonlinear regression, including conversion of a nonlinear regression problem to linear regression (as we indicated in reaction rate Equation 10.2 being converted to linearize form (10.4)). Finally, Section 10.5 will present several case studies.

10.2 LINEAR LEAST SQUARES REGRESSION

This section will be dedicated to problems involving regression in single variable. We will start with fitting a straight line, reinterpret the results in a matrix least squares form, and discuss some statistics to determine *goodness of fit*. It is not sufficient to just compute the best-fit parameters for a model; we would also like to estimate some error bounds around these parameters, as well as get some information about how well the model fits the observed data.

10.2.1 Fitting a Straight Line

The problem of fitting a straight line of the form

$$y = a_1 + a_2 x \tag{10.14}$$

is manageable to understand, while still having all the key features of regression and obtaining goodness of fit estimates. It also covers a rather broad spectrum of regression problems. Given the data at N discrete points, the model predictions are

$$\hat{y}_i = a_1 + a_2 x_i \tag{10.15}$$

Since the error is defined as

$$\begin{aligned} e_i &= y_i - \hat{y}_i \\ &= y_i - (a_1 + a_2 x_i) \end{aligned} \tag{10.16}$$

the sum of squared errors is written as

$$\begin{aligned} S_\varepsilon &= \sum_{i=1}^{N} e_i^2 \\ &= \sum_{i=1}^{N} (y_i - (a_1 + a_2 x_i))^2 \end{aligned} \tag{10.17}$$

Least squares solution: The aim of least squares solution is to find the values of a_1 and a_2 for which S_ε is minimum, that is

$$(a_1, a_2) = \min_{a_1, a_2} S_\varepsilon \tag{10.18}$$

This is said to be a least squares problem because we minimize the sum of squared errors. For the linear mode, the least squares parameter values can be obtained by differentiating Equation 10.17 with respect to a_1 and a_2 and equating them to zero. Differentiating with respect to a_1

$$\frac{\partial S_\varepsilon}{\partial a_1} = \sum_{i=1}^{N} \frac{\partial}{\partial a_1} (y_i - (a_1 + a_2 x_i))^2 = 0 \tag{10.19}$$

yields

$$\sum_{i=1}^{N} 2(y_i - (a_1 + a_2 x_i)) = 0 \tag{10.20}$$

Thus, we get the first equation in parameters a_1 and a_2:

$$a_1 N + a_2 \sum_{i=1}^{N} x_i = \sum_{i=1}^{N} y_i \tag{10.21}$$

The other equation may be derived in a similar manner, with

$$\frac{\partial S_\varepsilon}{\partial a_2} = \sum_{i=1}^{N} \frac{\partial}{\partial a_2} \left(y_i - \left(a_1 + a_2 x_i \right) \right)^2 = 0 \tag{10.22}$$

yielding

$$\sum_i -2x_i \left(y_i - \left(a_1 + a_2 x_i \right) \right) = 0 \tag{10.23}$$

Thus, we get the second equation in the two unknowns:

$$a_1 \sum_{i=1}^{N} x_i + a_2 \sum_{i=1}^{N} x_i^2 = \sum_{i=1}^{N} x_i y_i \tag{10.24}$$

Using the shorthand notation, Equations 10.21 and 10.24 are written as

$$\begin{bmatrix} N & \Sigma_x \\ \Sigma_x & \Sigma_{xx} \end{bmatrix} \Phi = \begin{bmatrix} \Sigma_y \\ \Sigma_{xy} \end{bmatrix} \tag{10.25}$$

to form two linear equations. The above equation can be solved simultaneously to obtain least squares estimates of the parameters, a_1 and a_2.

10.2.2 General Matrix Approach

Before moving to statistical analysis of the estimation error, we will rewrite the above problem in matrix operations form, in line with the discussion in Chapter 2. It is easy to see that the error at the ith point is

$$e_i = y_i - \begin{bmatrix} 1 & x_i \end{bmatrix} \underbrace{\begin{bmatrix} a_1 \\ a_2 \end{bmatrix}}_{\Phi} \tag{10.26}$$

Let us enumerate the above for all N data points:

$$\begin{aligned} e_1 &= y_1 - \begin{bmatrix} 1 & x_1 \end{bmatrix} \Phi \\ e_2 &= y_2 - \begin{bmatrix} 1 & x_2 \end{bmatrix} \Phi \\ &\vdots \\ e_N &= y_N - \begin{bmatrix} 1 & x_N \end{bmatrix} \Phi \end{aligned} \tag{10.27}$$

Writing the above in a matrix form

$$
\underbrace{\begin{bmatrix} e_1 \\ e_2 \\ \vdots \\ e_N \end{bmatrix}}_{E} = \underbrace{\begin{bmatrix} y_1 \\ y_2 \\ \vdots \\ y_N \end{bmatrix}}_{Y} - \underbrace{\begin{bmatrix} 1 & x_1 \\ 1 & x_2 \\ \vdots & \vdots \\ 1 & x_N \end{bmatrix}}_{X} \begin{bmatrix} a_1 \\ a_2 \end{bmatrix}_{\Phi} \tag{10.28}
$$

allows us to define errors in a compact form, $E = Y - X\Phi$. Recall that the model equation

$$
\hat{Y} = X\Phi \tag{10.29}
$$

is in the familiar form encountered in Chapter 2. In this example, the matrices X and Y consist of N rows, corresponding to the N data points. Singular value decomposition (SVD) on X matrix

$$
X = U\Sigma V^T \tag{10.30}
$$

gives two nonzero singular values. When there are two equations in two unknowns, a unique solution is found by inverting the matrix X. However, in this case with N rows, since the matrix X has two nonzero singular values, its left inverse, given by $(X^TX)^{-1}X^T$, exists and will provide a least squares solution.

Let us derive this result using a procedure similar to the one used earlier. The sum of squared errors is given by

$$
\begin{aligned}
S_\varepsilon &= E^T E \\
&= (Y - X\Phi)^T (Y - X\Phi)
\end{aligned} \tag{10.31}
$$

Recall that if we have a quadratic function, $f(x) = hx^2 - 2gx$, with $h > 0$, then the value $x = g/h$ is minima. Analogously, the vector $\mathbf{x}$ that minimizes

$$
\underset{\mathbf{x}}{\text{argmin}} \, \mathbf{x}^T \mathcal{H} \, \mathbf{x} - 2\mathbf{g}^T \mathbf{x}
$$

is $\mathbf{x} = \mathcal{H}^{-1}\mathbf{g}$. This can be easily derived using rules of matrix differentiation.

The least squares solution Φ can be obtained by expanding Equation 10.31

$$
\Phi = \underset{\Phi}{\text{argmin}} \left\{ \Phi^T \left(X^T X \right) \Phi - 2Y^T X\Phi \right\} \tag{10.32}
$$

where the constant term Y^TY does not affect the value of Φ. By comparison, $\mathcal{H} \equiv X^TX$ is a positive definite matrix. Thus, the least squares solution that minimizes the sum of squared errors, S_ε, is

$$
\Phi = \left(X^T X \right)^{-1} X^T Y \tag{10.33}
$$

The following example demonstrates linear regression using Equations 10.25 and 10.33.

Example 10.1 Fitting Specific Heat Data to a Straight Line

Problem: The specific heat data for methane is given in Table 10.3 as a function of temperature. Fit a straight line to the data at the lower range of temperatures, from 298 to 500 K.

Solution: The following code computes the linear least squares solution for specific heat.

```
% Linear regression for specific heat of methane
T=[298; 325; 350; 375; 400; 425; 450; 475; 500];
cp=[35.64; 36.69; 37.85; 39.13; ...
     40.63; 41.93; 43.39; 44.87; 46.63];
```

```
%% Method-1
N=length(T);
Sx=sum(T);          Sy=sum(cp);
Sxx=sum(T.*T);      Sxy=sum(T.*cp);
A=[N Sx; Sx Sxx];
b=[Sy; Sxy];
Phi1=inv(A)*b;
```

```
%% Method-2: Matrix Method
Y=cp;
X=[ones(N,1), T];
Phi2=inv(X'*X)*X'*Y;
Phi3=X\Y;
```

Using the above code, the specific heat of methane (in $J/mol \cdot K$) is obtained as the following linear function of temperature (in Kelvin):

$$c_p = 18.92 + 0.0546T \tag{10.34}$$

Figure 10.2 shows a comparison between the data and the straight-line model fit obtained above. One can visually see from this figure that the model does a reasonable job of predicting the experimental data.

One can also verify that `Phi1`, `Phi2`, and `Phi3` values are exactly the same. Furthermore, both A and `X'*X` give the same value:

```
      9           3598
   3598        1476304
```

The example above was a recap of fitting a straight line to data. This also showed numerically the equivalence between Equations 10.25 and 10.33. In Problem 10.1, you are asked to prove this equivalence for general N-dimensional X and Y matrices.

Next, we will calculate some statistics and analyze how well the model fits the data.

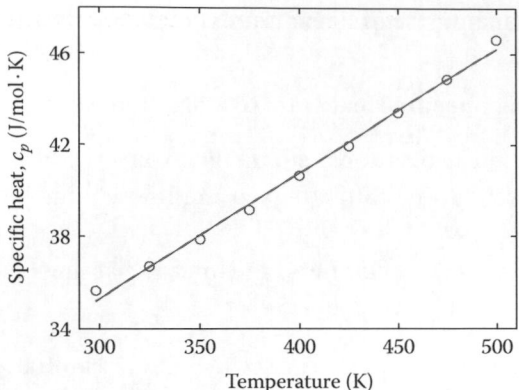

FIGURE 10.2   Comparison of specific heat data (symbols) and a straight-line model fit (lines).

10.2.3 Goodness of Fit

10.2.3.1 Maximum Likelihood Solution*

Linear least squares described above implicitly assumes that the measurement errors draw from a normal distribution. The discussion here simply follows the one in the *Numerical Recipes* book by Press et al. The reader is referred to that book for more details. Let us assume that there exist some true models:

$$y_{\text{true}}\left(x; \Phi\right), \quad \Phi = \begin{bmatrix} a_1 & \cdots & a_m \end{bmatrix}^T \tag{10.35}$$

from which the experimental data y_i are derived. The data has a certain measurement error associated with it. These measurement errors are independent, that is, the measurement error in the ith data point does not influence the measurement error in the jth data point. Furthermore, let us assume the measurement errors to draw from a normal (Gaussian) distribution.

Consider the problem of finding the parameters, Φ, that are most likely to produce the experimental data $((x_1, y_1), (x_2, y_2), \ldots, (x_N, y_N))$. If σ is the actual standard deviation of the measurement errors, then

$$P_i \propto \exp\left(-\frac{1}{2}\left(\frac{y_i - y_{\text{true}}\left(x_i; \Phi\right)}{\sigma}\right)^2\right) \tag{10.36}$$

is the probability at each data point. Since the measurement errors are independent, the probability of all the data is a product of the individual probabilities. The most likely set of parameters Φ are the ones that maximize this probability:

$$\Phi^{\text{mle}} = \max_{\Phi} \prod_{i=1}^{N} \alpha \exp\left(-\frac{1}{2}\left(\frac{y_i - y_{\text{true}}\left(x_i; \Phi\right)}{\sigma}\right)^2\right) \tag{10.37}$$

* This portion may be skipped without loss of continuity.

where "mle" represents that Φ^{mle} is the maximum likelihood estimate. Taking a logarithm

$$\Phi^{\text{mle}} = \max_{\Phi}\left\{N\ln(\alpha) - \frac{1}{2\sigma^2}\sum_{i=1}^{N}\left(y_i - y_{\text{true}}\left(x_i;\Phi\right)\right)^2\right\} \tag{10.38}$$

Since N, σ, α are constants, the maximum likelihood estimator is given by

$$\Phi^{\text{mle}} = \max_{\Phi}\left\{-S_\varepsilon\right\} \tag{10.39}$$

which is exactly the least squares estimation (10.18).

In summary, if the residuals, $e_i = y_i - \hat{y}_i$, are independent and draw from a normal (Gaussian) distribution, the least squares solution is also the maximum likelihood estimate.

It should be mentioned that the assumption of residuals being normally distributed is rather stringent. Measurements may have gross errors or outliers, systematic errors, or process drifts, or they may be correlated. Furthermore, measurements may not draw from normal distribution, for example, due to physical constraints (e.g., mole fractions lie between 0 and 1). Finally, the model may not adequately represent the data. Note that the model $y_{\text{true}}(x)$ is often unknown. However, *if* the residuals are independent and Gaussian, the least squares solution is indeed the maximum likelihood estimate, even for a nonlinear model.

10.2.3.2 Error and Coefficient of Determination

It is not sufficient to obtain the parameters. We also need to know how well the model predicts the data. The sum of squared errors, S_ε is an important parameter to determine the modeling errors. However, a specific value of S_ε does not itself give an indication of how good a fit is.

The *sample standard deviation* in the residual is computed as

$$s_\varepsilon = \sqrt{\frac{S_\varepsilon}{N-m}} \tag{10.40}$$

Since the data has been used to compute m fitting parameters, the degrees of freedom remaining are $(N-m)$. For example, a unique line will pass exactly through two data points. So, if $N=2$ and $m=2$, the variance is not defined.

The *coefficient of determination* is a popular way of quantifying how well the model fits data. It is more popularly known as r^2 value. It is commonly calculated as

$$r^2 = \frac{S_y - S_\varepsilon}{S_y} \tag{10.41}$$

where
  S_y represents the variance in the data
  S_ε represents the variance *not captured* by the model

If the latter is low, $r^2 \to 1$ and the model is said to have a good fit. The coefficient of determination is a convenient way to quantify the performance of a model.

Another approach to *qualitatively* determine the goodness of fit is to plot the original data and model fit on the same figure. Usually, the model is shown as a curve, whereas the data are shown as points. Alternatively, a parity plot between the data y_i as the abscissa and the model prediction $\hat{y}_i$ as the ordinate is obtained. If the data lies along the 45° line, we have a good fit. Finally, the errors e_i may be plotted against the data y_i. Any trends in the errors may become clear with such a plot. However, problems with "overfitting" the data (by using many more parameters than necessary) are difficult to ascertain using this approach. Likewise, it fails to compare two putative models to select the better one.

Quantitative information about the goodness of fit is provided by determining the variance in the estimates of the parameters a_i. The diagonal elements of $M = (X^T X)^{-1}$ give the variance of the corresponding parameters. Thus

$$\sigma_{a_i} = \sqrt{M_{i,i}}\, \sigma_\varepsilon \tag{10.42}$$

However, since the actual value of σ_ε is not available, it needs to be computed from the data. Thus, the standard deviation can only be estimated from the data.

If the actual standard deviation σ_ε is known, the standard results for normal distribution may be used. For example, the range of parameter values with 95% confidence interval is given by $(a_i \pm 2\sigma_{ai})$. However, since we have a limited amount of data, Student's t-distribution

$$\delta_{a_i}^{ci} = \sqrt{M_{i,i}}\, s_\varepsilon t_{N-m} \tag{10.43}$$

may be used to calculate the confidence interval for the parameters. Thus, the 95% confidence interval for the parameters is

$$a_i \pm \delta_{a_i}^{ci} \tag{10.44}$$

The following example demonstrates these calculations.

Example 10.2 Calculating Statistics for Specific Heat Problem

Problem: Compute the various statistical information for Example 10.1 to verify the fitted linear model.

Solution: The MATLAB® code of Example 10.1 can be expanded to calculate some statistics.

Error Variance: Sum of squared errors and the estimate of error variance are calculated as

```
err=cp-cpHat;
sse=err'*err;
var_e=sse/(N-2);
stdErr=sqrt(var_e);
```

From the above

$$S_\varepsilon = 0.5382, \quad \mathrm{var}(e) = 0.0769, \quad s_\varepsilon = 0.2773$$

Coefficient of determination ("*R-squared*"): From the variance in the c_p data, the term S_y is calculated as

$$S_y = (N-1)\,\mathrm{var}(c_p) = 113.555$$

The two can be combined to give the coefficient of determination (R-squared):

$$r^2 = 1 - \frac{S_\varepsilon}{S_y} = 0.9953$$

Visualizing the model fit: The term r^2 indicates that 99.5% of the variance in the original data can be explained by the linear model. This can be visually seen from Figure 10.2. Parity plot and error plots are two other ways of visualizing how well the model represents the data (Figure 10.3). The top panel is the parity plot of c_p (data) vs. $\hat{c}_p$ (model). The closeness of the data points to the 45° line indicates a good model fit. Another way of visualizing the same data is to plot the residuals (errors), $e_i = c_{p,i} - \hat{c}_{p,i}$ as the ordinate vs. $c_{p,i}$ as the abscissa. The errors are spread around the X-axis.

Although the errors are small, they form a convex curve, which may indicate the existence of a higher-order polynomial term.

Confidence interval on least squares parameter values, Φ: The final task is to obtain confidence intervals on the least squares parameters Φ. This will contain two parts: Estimate the

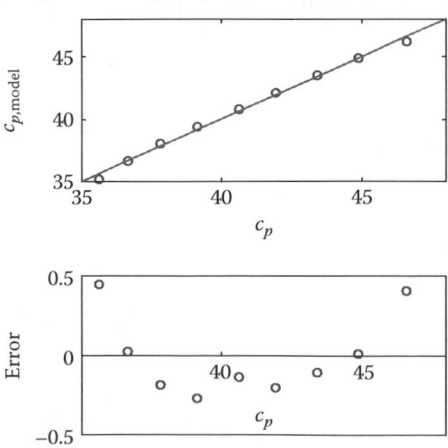

FIGURE 10.3 Parity plot and error e_i vs. $c_{p,i}$ for analyzing model fit.

standard deviation (σ_{ai}) of the parameters and use it to compute the confidence interval. We will use Equation 10.42 to compute σ_{ai}, whereas Student's two-sided t-distribution will use σ_{ai} to provide the 95% confidence intervals on Φ.

The diagonal elements of $(X^T X)^{-1}$ are

```
>> sqrt(diag(inv(X'*X)))
ans =
    2.0803
    0.0051
```

The two elements are indicators of the standard deviations, σ_{ai}, of the two parameters, as given by Equation 10.42. However, in the absence of information about the measurement errors, we replace the standard deviation of measurement error, σ_{yi}, with the value s_ε computed from the data. Thus, the estimated value

$$\sigma_{a_i}^{est} \sim \sqrt{M_{ii}}\, s_\varepsilon$$

If we had a very large number of data points, $s_\varepsilon \to \sigma_\varepsilon$ (i.e., s_ε approaches the true standard deviation of measurement noise sequence). Under this assumption that s_ε is the true value, the 95% confidence interval is given by "mean $\pm 2 \times$ standard deviation."

However, since we have a limited amount of data, the Student's t-distribution is appropriate. The multiplier for 95% confidence interval with a degree of freedom $(N-m) = 7$ is 2.365. The t-distribution tables are easily available. For example, please see Wikipedia link https://en.wikipedia.org/wiki/Student's_t-distribution for this information. If you have the Statistics toolbox in MATLAB, the value t_{N-m} for 95% confidence interval can be calculated as*

```
>> tVal=tinv(0.975,7);
```

Once the `tVal` is obtained, the confidence intervals can be computed using (10.43):

```
>> stdPhi=sqrt(diag(inv(X'*X)))*stdErr;
>> confValue=stdPhi*tVal;
>> confInterval=[Phi-confValue,  Phi+confValue]
confInterval =
    17.5573    20.2852
     0.0512     0.0580
```

Thus, based on the above information, the 95% confidence interval for the two parameters is $17.56 \le \delta_{a_1}^{ci} \le 20.29$ and $0.0512 \le \delta_{a_2}^{ci} \le 0.058$.

---

* The above form is because MATLAB uses one-sided t-distribution. Please see the table in the above link for more information. Thus, `tinv(p,N-m)` will compute the probability.

Since the variance in both the parameters is rather small, the confidence interval computed above is narrow. This indicates a good model fit to the data.

10.3 REGRESSION IN MULTIPLE VARIABLES

Let us now focus on regression in multiple variables.

10.3.1 General Multilinear Regression

Recall that we used the following matrix form:

$$\underbrace{\begin{bmatrix} e_1 \\ e_2 \\ \vdots \\ e_N \end{bmatrix}}_{E} = \underbrace{\begin{bmatrix} y_1 \\ y_2 \\ \vdots \\ y_N \end{bmatrix}}_{Y} - \underbrace{\begin{bmatrix} 1 & x_1 \\ 1 & x_2 \\ \vdots & \vdots \\ 1 & x_N \end{bmatrix}}_{X} \underbrace{\begin{bmatrix} a_1 \\ a_2 \end{bmatrix}}_{\Phi} \tag{10.28}$$

for linear regression in two variables. The sum of squared errors was obtained as

$$S_\varepsilon = E^T E \tag{10.31}$$

In the derivation of least squares solution, we made no assumption about the number of parameters, m. The derivation is more general and is applicable for a larger dimensional Φ as well. In general, if Φ is an $m \times 1$ vector, then X will be a $N \times m$ matrix. The objective function can be expanded to give the following expression for the least squares solution for Φ:

$$\Phi = \underset{\Phi}{\text{argmin}} \left\{ \Phi^T \left(X^T X \right) \Phi - 2 Y^T X \Phi \right\} \tag{10.32}$$

Thus, the expression for least squares solution

$$\Phi = \left(X^T X \right)^{-1} X^T Y \tag{10.33}$$

which minimizes the sum of squared errors, is the same as before.

This result can be expanded to linear regression in multiple *linearly independent* input variables. Specifically, we are interested in parameter estimation for a linear model:

$$y = a_1 + a_2 x + a_3 w + \cdots + a_m u \tag{10.45}$$

with m parameters. The N data points are in the form $(x_i, w_i, \ldots, u_i; y_i)$. The prediction using the above linear model is

$$\hat{y}_i = \begin{bmatrix} 1 & x_i & w_i & \cdots & u_i \end{bmatrix} \begin{bmatrix} a_1 \\ a_2 \\ a_3 \\ \vdots \\ a_m \end{bmatrix} \tag{10.46}$$

Following the same procedure as above, the model predictions for all the data points can be obtained, and the error equation written as

$$
\underbrace{\begin{bmatrix} e_1 \\ e_2 \\ \vdots \\ e_N \end{bmatrix}}_{E} = \underbrace{\begin{bmatrix} y_1 \\ y_2 \\ \vdots \\ y_N \end{bmatrix}}_{Y} - \underbrace{\begin{bmatrix} 1 & x_1 & \cdots & u_1 \\ 1 & x_2 & \cdots & u_2 \\ \vdots & \vdots & \ddots & \vdots \\ 1 & x_N & \cdots & u_N \end{bmatrix}}_{X} \underbrace{\begin{bmatrix} a_1 \\ a_2 \\ \vdots \\ a_m \end{bmatrix}}_{\Phi}
\tag{10.47}
$$

With this definition, the same solution

$$
\Phi = \left(X^T X \right)^{-1} X^T Y
\tag{10.33}
$$

gives the linear least squares solution for an m-dimensional parameter vector Φ.

10.3.2 Polynomial Regression

Consider the example of fitting the specific heat to a wider range of temperatures.

A straight-line fit does not work.

Polynomial of the form

$$
c_p = a_1 + a_2 T + a_3 T^2 + \cdots + a_m T^{m-1}
\tag{10.48}
$$

is often fitted to the c_p vs. T data.

This is still *linear regression*, because the expression (10.48) is linear in the unknown parameters that are to be estimated.

Above equation could be written in the familiar form:

$$
c_p = \begin{bmatrix} 1 & T & \cdots & T^{m-1} \end{bmatrix} \underbrace{\begin{bmatrix} a_1 \\ a_2 \\ \vdots \\ a_m \end{bmatrix}}_{\Phi}
\tag{10.49}
$$

With the above definition

$$
\Phi = \left(X^T X \right)^{-1} X^T Y
$$

$$
\text{where } X = \begin{bmatrix} 1 & T_1 & T_1^2 & \cdots & T_1^{m-1} \\ 1 & T_2 & T_2^2 & \cdots & T_2^{m-1} \\ \vdots & \vdots & \ddots & \ddots & \vdots \\ 1 & T_N & T_N^2 & \cdots & T_N^{m-1} \end{bmatrix}, \quad Y = \begin{bmatrix} c_{p1} \\ c_{p2} \\ \vdots \\ c_{pN} \end{bmatrix}
\tag{10.50}
$$

The next example aims to fit a higher-order polynomial to the data ranging from low to high temperature in steps of ~ 100 K.

Example 10.3 Specific Heat of Methane for Full Range of Temperature

Problem: Fit a polynomial of the form (10.48) to the specific heat data in Table 10.3 for the full range of temperatures from 298 to 1300 K in steps of ~100 K.

Solution: We will use Equation 10.50 to compute the parameters, as shown in the code below:

```
% Linear regression for specific heat of methane
%% Data
T=[298; 400; 500; 600; 700; 800; ...
    900; 1000; 1100; 1200; 1273; 1300];
cp=[35.64; 40.63; 46.63; ...
    52.74; 58.60; 64.08; 69.14; ...
    73.75; 77.92; 81.68; 84.1; 85.07];
N=length(T);

%% Matrix Method
Y=cp;
X=[ones(N,1), T, T.^2];
m=size(X,2);
M=inv(X'*X);
Phi=M*X'*Y;

%% Plotting and comparing
cpHat=Phi (1)+Phi(2)*T+Phi(3)*T.^2;
err=cp-cpHat;
plot(T,cp,'o'); hold on;
plot(T,cpHat);
xlabel('Temperature (K)'); ylabel('c_p (J/mol.K)');

plot(cp,err,'o');
xlabel('c_p (J/mol.K)'); ylabel('error');

plot(cp,cpHat,'o');       % Parity Plot
plot([0 100],[0 100]);    % 45-deg line

%% Error analysis
sse=err'*err;
Yvar=Y-mean(Y);
sst=Yvar'*Yvar;
Rsquare=1-sse/sst;

% Parameter std.dev. and confidence interval
stdPhi=sqrt(diag(M))*stdErr;
tVal=tinv(0.975,N-m);
ciRange= stdPhi*tVal;
CI=[Phi+ciRange, Phi-ciRange];
```

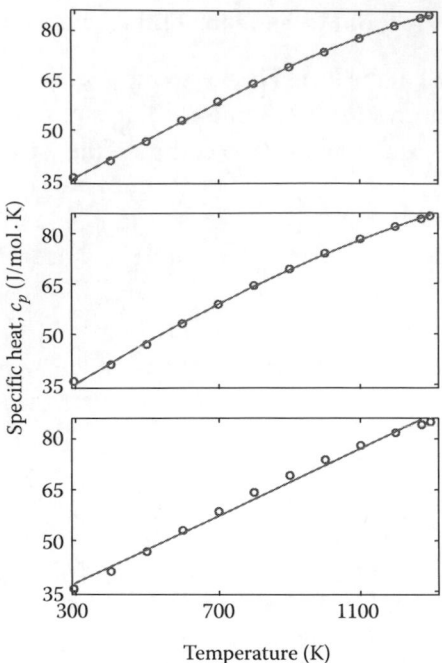

FIGURE 10.4   Polynomial fit to c_p vs. T data for first-, second-, and third-order polynomials.

Figure 10.4 shows the fit for polynomials of first-, second-, and third-order to the c_p data. The corresponding R-squared values are 0.9923, 0.9992, 0.9998, respectively. All the three R-squared values are excellent.

Figure 10.5 shows the parity and error plots for the three polynomial fits. Although the straight line fit gives an excellent R-squared value, a higher-order polynomial gives a better qualitative fit with regard to the spread of the errors across the X-axis.

Based on "eyeballing the errors" from the graphs, a second-order polynomial seems a reasonable fit for the data.

The above procedure of eyeballing the errors is not very reliable. We need some quantitative measure to determine the goodness of fit. If the variance of measurement error is known, the Chi-squared value

$$\chi^2 = \sum_{i=1}^{N} \left(\frac{y_i - \hat{y}_i}{\sigma_i} \right)^2 = \frac{\sum_{i=1}^{N} \left(y_i - \hat{y}_i \right)^2}{\sigma^2} \tag{10.51}$$

provides a good estimate of goodness of fit. However, in the example above, we have no idea of the estimate of measurement error. In such a case, the standard deviation of the parameters becomes a useful criterion for analyzing the model fit. For the straight line

$$\Phi = \begin{bmatrix} 22.27 \\ 0.0499 \end{bmatrix}, \quad \sigma_{\hat{\Phi}} = \begin{bmatrix} 1.254 \\ 0.0014 \end{bmatrix}$$

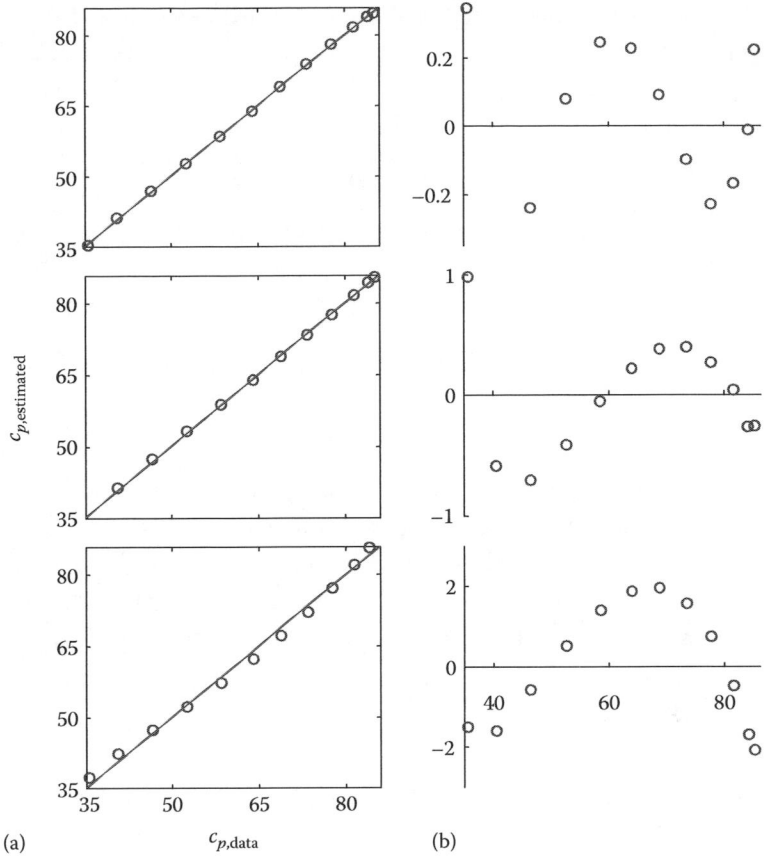

FIGURE 10.5 Parity plots (a) and error plots (b) for first-, second-, and third-order polynomials for fitting c_p vs. T data.

The parameters for the second-order curve are

$$\Phi = \begin{bmatrix} 13.70 \\ 0.0749 \\ -1.52\times10^{-5} \end{bmatrix}, \quad \sigma_{\hat{\Phi}} = \begin{bmatrix} 1.036 \\ 0.0028 \\ 1.68\times10^{-6} \end{bmatrix}$$

In both these cases, the standard deviation of parameters is an order of magnitude lower than the parameter values. However, when a third-order polynomial is considered

$$\Phi = \begin{bmatrix} 20.004 \\ 0.0452 \\ 2.542\times10^{-5} \\ -1.680\times10^{-8} \end{bmatrix}, \quad \sigma_{\hat{\Phi}} = \begin{bmatrix} 1.436 \\ 0.0064 \\ 8.57\times10^{-6} \\ 3.52\times10^{-9} \end{bmatrix}$$

The variance in third-order polynomial is about one-third to one-fourth of the parameter value for the parameter a_3. Consequently, we could either choose second-order (preferred) or third-order polynomial fit. Any higher-order would probably lead to overfitting the data.

When the code in Example 10.3 is modified for a third-order polynomial (with four parameters), the matrix X becomes ill-conditioned. MATLAB gives the following warning:

```
Warning: Matrix is close to singular or badly scaled. Results may
be inaccurate. RCOND = 2.492404e-21.
```

Methods to handle this are discussed next.

10.3.3 Singularity and SVD

The equations of the form Equation 10.50 are known as normal equations.

A practical difficulty in solving the linear least squares problem is that as the number of parameters increase, a larger amount of data is needed, and the matrix X becomes ill-conditioned. Inverting an ill-conditioned matrix must be avoided, because such an operation is prone to large round-off errors and small measurement errors in the data manifest as large errors in computation of the parameter Φ. SVD, discussed in Chapter 2, provides a way around this. The SVD on matrix X can be written as

$$X = U\mathcal{S}V^T \tag{10.52}$$

where $\mathcal{S} = \mathrm{diag}\begin{pmatrix} \mathcal{S}_1 & \cdots & \mathcal{S}_m & 0 & \cdots & 0 \end{pmatrix}$. We have used $\mathcal{S}_i$ to represent singular values because σ has already been used to represent standard deviation in this chapter. The matrix

$$X^T X = V\mathcal{S}^T \mathcal{S} V^T$$
$$\left(X^T X\right)^{-1} = V\left(\mathcal{S}^T \mathcal{S}\right)^{-1} V^T \tag{10.53}$$

Since $\mathcal{S}$ is an $N \times m$ matrix, $\left(\mathcal{S}^T \mathcal{S}\right)$ is an $m \times m$ matrix and

$$\left(\mathcal{S}^T \mathcal{S}\right)^{-1} = \begin{bmatrix} 1/\mathcal{S}_1^2 & & \\ & \ddots & \\ & & 1/\mathcal{S}_m^2 \end{bmatrix}$$

Thus, the least squares parameter estimates may be written as

$$\Phi = V\left(\mathcal{S}^T \mathcal{S}\right)^{-1} \left(\mathcal{S}^T\right) U^T Y \tag{10.54}$$

The term $\left(\mathcal{S}^T \mathcal{S}\right)^{-1} \left(\mathcal{S}^T\right)$ is given by

$$\left(\mathcal{S}^T \mathcal{S}\right)^{-1} \mathcal{S}^T = \begin{bmatrix} 1/\mathcal{S}_1 & & & 0 & \cdots & 0 \\ & \ddots & & \vdots & \ddots & \vdots \\ & & 1/\mathcal{S}_m & 0 & \cdots & 0 \end{bmatrix}$$

Thus, according to Press et al., the least squares parameter estimate is given by

$$\Phi = \sum_{i=1}^{m} \mathbf{v}_i \left(\frac{\mathbf{u}_i^T Y}{\mathcal{S}_i} \right) \tag{10.55}$$

where $\mathbf{v}_i$ and $\mathbf{u}_i$ are the ith column vectors of V and U, respectively. Furthermore, Press et al. suggest that if certain singular values are too small, they may be dropped. Let us say that only first r singular values of X are considered to be important; then

$$\Phi = \sum_{i=1}^{r} \mathbf{v}_i \left(\frac{\mathbf{u}_i^T Y}{\mathcal{S}_i} \right) \tag{10.56}$$

gives the desired estimate.

10.4 NONLINEAR ESTIMATION

10.4.1 Functional Regression by Linearization

A classic example is that of temperature-dependent Arrhenius rate law:

$$r = k_0 \exp\left(-\frac{E}{RT} \right) C^n \tag{10.57}$$

Taking logarithm on both sides

$$\ln(r) = \ln(k_0) + \left(-\frac{E}{R} \right)\left(\frac{1}{T} \right) + n\ln(C) \tag{10.58}$$

Defining $y = \ln(r)$, $x = 1/T$ and $w = \ln(C)$, we get

$$y = a_1 + a_2 x + a_3 w \tag{10.59}$$

with $k_0 = \exp(a_1)$, $E = -Ra_2$, $n = a_3$.

The above strategy is useful for both exponential models

$$y = ae^{bx} \tag{10.60}$$

as well as power-law models

$$y = ax^b \tag{10.61}$$

Another classic example is that of Michaelis-Menten kinetics for enzyme-catalyzed reactions. The reaction rate is given by

$$r = \frac{VC_S}{K_m + C_S} \tag{10.62}$$

This expression is nonlinear in parameters, as written. The so-called Lineweaver-Burke plot is a "double reciprocal" plot, where the plot of $y = 1/r$ vs. $x = 1/C_S$ gives a straight line. Inverting the above equation yields

$$\frac{1}{r} = \underbrace{\frac{K_m}{V}}_{a_1}\frac{1}{C_S} + \underbrace{\frac{1}{V}}_{a_2} \tag{10.63}$$

Since this equation is written in a slightly different form

$$y = a_1 x + a_2$$

the matrix X also must reflect this appropriately. The parameters can be obtained by linear least squares regression:

$$\Phi = \left(X^T X\right)^{-1} X^T Y$$

$$\text{where } X = \begin{bmatrix} 1/C_{S1} & 1 \\ 1/C_{S2} & 1 \\ \vdots & \vdots \\ 1/C_{SN} & 1 \end{bmatrix}, \quad Y = \begin{bmatrix} 1/r_1 \\ 1/r_2 \\ \vdots \\ 1/r_N \end{bmatrix} \tag{10.64}$$

Monod kinetics in the presence of substrate inhibition is given by

$$r = \frac{\mu^{\max} C_S}{K + C_S + aC_S^2} \tag{10.65}$$

Taking the reciprocal yields us

$$\frac{1}{r} = \frac{K}{\mu^{\max}}\left(\frac{1}{C_S}\right) + \frac{1}{\mu^{\max}} + \frac{a}{\mu^{\max}}C_S \tag{10.66}$$

which is of the form required for linear regression.

Example 10.4 Linear Regression for Monod Kinetics

Problem: The data for the rate of microbial reaction (r) was obtained at different values of substrate concentration and at a constant biomass concentration. For this data shown in Table 10.1, use linear regression and hence compute the parameters K and $\mu^{\max}$.

Solution: We can invert the above equation to obtain

$$\frac{1}{r} = \frac{K}{\mu^{\max}}\left(\frac{1}{C_S}\right) + \frac{1}{\mu^{\max}}$$

TABLE 10.1   Reaction Rate vs. Substrate Concentration Data for Monod Kinetics

$[S]$	0.24	0.32	0.70	1.37	1.58	2.04	2.26	2.28	2.39	2.41
r	0.3839	0.3935	0.911	1.4975	1.5735	1.849	1.9355	1.893	1.970	1.924

which is of the same form as Equation 10.63. Thereafter, the procedure is as described in Equation 10.64. Specifically, we define $y = 1/r$ and $x = 1/C_s$ and perform the regression. Once the least squares values are obtained, the original parameters are calculated as

$$\mu^{max} = \frac{1}{a_2}, \quad K = \mu^{max} a_1$$

The values of parameters and their expected variance are obtained in a manner similar to Example 10.1 as

$$\Phi = \begin{bmatrix} 0.6210 \\ 0.2530 \end{bmatrix}, \quad \sigma_\Phi = \begin{bmatrix} 0.0376 \\ 0.0662 \end{bmatrix}$$

The coefficient of determination, R-squared, equals 0.9716.

The original parameter values computed are

$$\mu^{max} = 3.95, \quad K = 2.45$$

Caution!

One must, however, exercise caution in using a nonlinear model, either directly or linearized. For example, consider the model

$$y = a \cdot \exp(bx + c) \tag{10.67}$$

The parameters a and c cannot be obtained independently. This is because the above equation may be written as

$$y = \underbrace{ae^c}_{a_1} \cdot e^{bx} \tag{10.68}$$

The parameters a_1 and b can be obtained using linear or nonlinear regression. However, the parameters a and c are not independently obtainable from x vs. y data. A practical example of this will be discussed in Section 10.5.3.

Another example: Antoine's equation

$$\log(P^{sat}) = A + \frac{B}{T + C} \tag{10.69}$$

Multiplying by $(T + C)$ and rearranging, we get

$$T \log\left(P^{sat}\right) = AT + \left(AC + B\right) - C \log\left(P^{sat}\right) \tag{10.70}$$

There is nothing wrong with the above formulation, and the resulting X matrix is also nonsingular. However, an important assumption made in our analysis of parameter estimation was that of uncorrelated and independent Gaussian noise sequence. In the above equation, $y = T\log(P^{sat})$ does not satisfy that condition of Gaussian noise; more importantly, the term $w = \log(P^{sat})$ and y are now correlated. Thus, one should avoid a formulation where any experimentally measured quantity appears in both definition of dependent variable y and independent variable(s) x/w.

10.4.2 MATLAB® Solver: Linear Regression

MATLAB has several solvers for linear and nonlinear regression. However, most solvers require a toolbox to be installed. The *Statistics and Machine Learning Toolbox* in MATLAB gives a broad spectrum of tools for data analysis, regression, and hypothesis testing. The command `tinv` used in Example 10.2 is a part of the Statistics toolbox.

The linear regression and related statistics provided in Example 10.2 can be handily calculated using the Statistics toolbox command `regress`. The specific syntax for using `regress` is

```
[Phi,confInterval,err,~,STATS]=regress(Y,X);
Rsquare=STATS(1);
stdErr=sqrt(STATS(4));
```

The above command can replace the calculations in Example 10.2. The various outputs are explained below:

Phi	Estimated parameter vector, $\Phi = [a_1 \ \cdots \ a_m]^T$
confInterval	Confidence interval for each parameter, Equation 10.44
err	Model error (residual) at each data point, $e_i = y_i - \hat{y}_i$
Rsquare	Coefficient of determination (or R-squared value), Equation 10.41
stdErr	Standard deviation of residuals, Equation 10.40

The advantage of using `regress` is that it can handle ill-conditioned X matrix better than the direct matrix method of Example 10.3. For example, the matrix X in Example 10.3 for a third-order polynomial fit was ill-conditioned. The round-off error problem is avoided by `regress` since it uses QR factorization for solving the problem. The other SVD-based approach was presented in Section 10.3.3.

PERSONAL EXPERIENCE: DON'T TAKE SHORTCUTS!

It is often the tendency of beginners to directly use a tool such as `regress` and skip the harder steps of stepwise solving a problem. This is especially true when there is any mention of statistics in an engineering course. I have tried to keep statistics in this chapter to a minimum. However, model fitting is not an easy task. Getting the parameters Φ is easy; figuring out whether the model obtained is a good model is not easy.

Example 10.2 is not intended to be a complete word in linear regression and analysis. We barely scratched the surface! However, this chapter is intended to equip the reader to get started with reading advanced material on regression and model fitting.

The *Optimization Toolbox* provides another set of tools for linear and nonlinear regression. These are optimization-based solvers. The solver useful for linear regression is `lsqlin`. If the parameter estimation problem is unconstrained, then the method described in Example 10.3 or Example 10.4 gives good results. However, if there are physical constraints on the parameter values, `lsqlin` provides a good option, since it solves constrained optimization of the form*

$$\min_{\Phi} \frac{1}{2}\|X\Phi - Y\|_2$$
$$\Phi^{\min} \leq \Phi \leq \Phi^{\max} \tag{10.71}$$

The syntax is as follows:

```
[Phi,sse,err]=lsqlin(C,d,[],[],[],[],PhiMin,PhiMax);
```

Phi	Estimated parameter vector, $\Phi = [a_1 \ \cdots \ a_m]^T$
sse	Sum of squared errors, S_ε
err	Model error (residual) at each data point, $e_i = y_i - \hat{y}_i$

Example 10.5 MATLAB® Solvers

The reader may verify the results of using `regress` and `lsqlin` and compare them with those from Examples 10.1 and 10.2.

Using <u>regress</u>:

```
[Phi1,confInterval1,err1,~,STATS]=regress(Y,X);
Rsquare1=STATS(1);
stdErr1=sqrt(STATS(4));
```

---

* In addition to bounds on Φ, `lsqlin` also handles linear inequality and equality constraints. I favor this approach to keep things simple and pertinent. Advanced readers may refer to MATLAB `help` or documentation of Optimization toolbox for more information.

The reader can verify that the values of `Phi1`, `confInterval1`, `err1`, and `Rsquare1` obtained here are the same as those obtained in Examples 10.1 and 10.2.

Using `lsqlin`:

```
[Phi2,sse2,err2]=lsqlin(C,d,[],[],[],[],PhiMin,PhiMax);
```

The readers can verify that `Phi2` and `sse2` are same as those in previous examples, and the residual error, `err2`, is just negative of that in previous examples (since `lsqlin` computes $(X\Phi - Y)$).

10.4.3 Nonlinear Regression Using Optimization Toolbox

The task of obtaining a model that fits the given data is an involved one. First, we choose candidate models based on domain information that could represent the underlying physics of the system or process that generated the data. The next task is to estimate the parameters, including some statistical information regarding the same. The final task is to use goodness of fit measured to determine how well the model represents the data. This task gets quite complex if we do not have a clear idea of what underlying physics dictated the observations, as it often happens in real-world examples. We use hypothesis testing to determine whether there is statistical evidence that the observed data is represented by the model and its parameters. This is a cyclic process.

Throughout this chapter, we took examples where the first part was known. We know with a *reasonable* confidence that a polynomial can capture temperature effects on specific heat data, that the Michaelis-Menten or Monod kinetics are reasonable for simple enzymatic or microbial reactions, that Antoine's equation represents pure component VLE (vapor-liquid equilibrium), and so on. Later, in Sections 10.5.3 and 10.5.4, we will visit an example where the data generated from the same system is used to fit the Langmuir-Hinshelwood model and power-law kinetics, respectively. Nonetheless, the first part of choosing the model structure that represents data reasonably was assumed to be known in this chapter. We gave a brief glimpse of the third aspect: to use goodness of fit measures to determine how well the model represents the data. This chapter, by no means, is a definitive introductory text on this rather complex issue. The primary focus of this chapter, however, was on tools to obtain the model parameters.

While we focused a fair deal on error bounds and confidence intervals for linear models, this section on nonlinear parameter estimation will focus only on the process of obtaining least squares fit for nonlinear models.

The optimization-based solvers (from Optimization toolbox in MATLAB) include `lsqnonlin` and `lsqcurvefit`. These solve a nonlinear least squares problem. A generic optimization solver, `fmincon`, is also available to use from the toolbox. The three methods use similar sets of optimization algorithms but go about the task of finding the least squares parameter fit in a slightly different manner.

Recall that we are interested in finding the parameters, Φ, for a general nonlinear model

$$y = f\left(x; \Phi\right) \tag{10.6}$$

where x and y are measured at N different data points. The aim of the least squares fitting is to find the parameters that minimize the least squares objective function:

$$\operatorname*{argmin}_{\Phi}\left\{\underbrace{\sum_{i=1}^{N}\left(y_i - f\left(x_i;\Phi\right)\right)^2}_{S_\varepsilon}\right\} \tag{10.72}$$

Recall that we define our model prediction for a nonlinear model as

$$\hat{y}_i = f\left(x_i;\Phi\right) \tag{10.73}$$

The three MATLAB solvers, `lsqnonlin`, `lsqcurvefit`, and `fmincon` require different functions to be defined. The first solver, `lsqnonlin`, requires us to supply the residual error *as a vector*:

$$E = \begin{bmatrix} e_1 \\ e_2 \\ \vdots \\ e_N \end{bmatrix} = \begin{bmatrix} y_1 - f\left(x_1;\Phi\right) \\ y_2 - f\left(x_2;\Phi\right) \\ \vdots \\ y_N - f\left(x_N;\Phi\right) \end{bmatrix} \tag{10.74}$$

with Φ being the adjustable parameter that `lsqnonlin` needs to determine. The syntax for this code is

```
Phi = lsqnonlin(@(X) fittingFun(X), Phi0);
```

where `Phi0` is the initial guess. The function, `fittingFun.m`, takes in the parameter vector as input argument and computes the residuals $\left(e_i = y_i - \hat{y}_i\right)$. The $N \times 1$ vector of residual errors must be returned by the function. Since `fittinFun` will also require the data to calculate the residuals, the more appropriate syntax aligned with our way of using MATLAB is

```
Phi=lsqnonlin(@(X) fittingFun(X,xData,yData), Phi0);
```

The second solver, `lsqcurvefit`, requires us to provide y_i and $\hat{y}_i$, with the latter being calculated in a function that is supplied to the solver. In other words, we will write a function `modelFun.m` that will calculate $\hat{y}_i = f\left(x_i;\Phi\right)$ and return a vector

$$\hat{Y} = \begin{bmatrix} f\left(x_1;\Phi\right) \\ f\left(x_2;\Phi\right) \\ \vdots \\ f\left(x_N;\Phi\right) \end{bmatrix}$$

The syntax for using `lsqcurvefit` is

```
Phi=lsqcurvefit(@(X,xData) modelFun(X,xData),Phi0,xData,yData);
```

Personally, I find not much to choose between the two solvers. Since `lsqnonlin` has a more general structure, I prefer it over `lsqcurvefit`.

The final option is to use any generic optimization solver to solve the least squares problem. If a solver, such as `fmincon` or `fminunc` is used, the objective function should provide the least squares objective, that is, S_ε from Equation 10.72. The syntax for using them is

```
Phi=fmincon(@(X) estimObj(X,xData,yData), Phi0);
```

and the function `estimObj.m` should return the sum of squared errors.

The next example demonstrates the three methods for solving nonlinear regression problem.

Example 10.6 Solve Example 10.4 Using Nonlinear Regression

Problem: To solve Example 10.4 using nonlinear regression.

Solution: The nonlinear model for Monod kinetics is

$$r = \frac{\mu^{max} C_S}{K_s + C_S}$$

`lsqcurvefit`: An external function `modelFun.m` is written to compute $r(C_{si})$ and return the result as a vector $\hat{Y}$. The function is given as

```
function Rate=modelFun(Phi,CsData)
mu=Phi(1); K=Phi(2);
Rate=mu*CsData./(K+CsData);
```

`lsqnonlin`: An external function `fittingFun.m` is written to compute $e_i = (r_i - \hat{r}_i)$ and return the result as a vector E. The function is given as

```
function err=fittingFun(Phi,CsData,rData)
mu=Phi(1); K=Phi(2);
Rate=mu*CsData./(K+CsData);
err=rData-Rate;
```

fmincon / fminunc: An external function `estimObj.m` is written to compute the scalar objective function, which is the sum of squared errors:

```
function sse=estimObj(Phi,CsData,rData)
mu=Phi(1); K=Phi(2);
Rate=mu*CsData./(K+CsData);
err=rData-Rate;
sse=err'*err;
```

Now that these functions are written, the following driver script is used to compute the estimates Φ using these three methods:

```
% Nonlinear regression for Monod Kinetics
%% Data
CsData=[0.24; 0.32; 0.70; 1.37; 1.58; ...
    2.04; 2.26; 2.28; 2.39; 2.41];
rData=[0.3839; 0.3935; 0.911; 1.4975; 1.5735; ...
    1.849; 1.9355; 1.893; 1.970; 1.924];
N=length(rData);

%% Nonlinear curve fit
Phi0=[1;1];
Phi1=lsqcurvefit(@(X,xData) modelFun(X,xData),...
    Phi0,CsData,rData);
% Verify the fit
mu=Phi1(1); K=Phi1(2);
rHat=mu*CsData./(K+CsData);
plot(CsData,rData,'o', CsData,rHat,'-');
xlabel('Concentration, C_s'); ylabel('Rate');

%% Nonlinear least squares
Phi2=lsqnonlin(@(X) fittingFun(X,CsData,rData), Phi0);

%% Generic Minimization
Phi3=fminunc(@(X) estimObj(X,CsData,rData), Phi0);
```

The parameters estimated using all these three methods are $\Phi = \begin{bmatrix} 3.80 & 2.23 \end{bmatrix}^T$.

Compare these with the parameter values found in Example 10.4 using linear regression:

$$\mu^{max} = 3.95, \quad K = 2.45$$

Figure 10.6 compares the model predictions from nonlinear regression (top) and linear regression (bottom) for this example. The coefficient of determination, R-square, was 0.996 for nonlinear regression. Thus, both the models provide reasonable performance.

10.5 CASE STUDIES AND EXAMPLES

A few examples are discussed in this section to highlight the procedure of parameter estimation followed in this chapter.

10.5.1 Specific Heat: Revisited

The example of fitting the specific heat of methane as a function of temperature was used to demonstrate various concepts in this chapter. First, a straight line was fitted to low-temperature specific heat data (Example 10.1) in the temperature range of 298–500 K, and

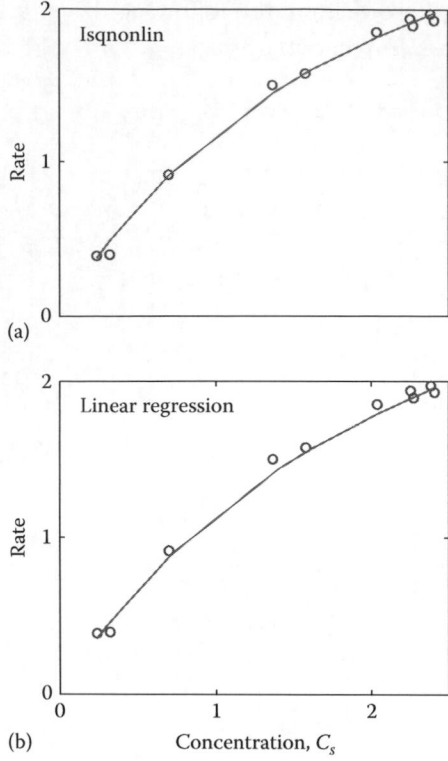

FIGURE 10.6   Model fit (lines) for Monod kinetics data (symbols). (a) Shows results for lsqnonlin and (b) for linear regression from Example 10.4.

goodness of fit based on R-squared values and 95% confidence intervals was calculated (Example 10.2). In Example 10.3, polynomials of various orders were fitted to the specific heat data for the full range of temperature from 298 to 1300 K. A quadratic or cubic polynomial was found to give a good fit to the data. Interested readers will find it useful to review these three examples to understand linear regression and goodness of fit estimates. Finally, Example 10.5 showed that the same results are obtained from in-built MATLAB functions from statistics and optimization toolboxes.

10.5.2 Antoine's Equation for Vapor Pressure

Antoine's equation

$$\ln\left(p^{\text{sat}}\right) = A - \frac{B}{T+C} \tag{10.3}$$

is a semiempirical correlation between pure component vapor pressure and temperature. In the above equation, p^{sat} is in bars and temperature is in °C. The pure component vapor pressure data for benzene and ethylbenzene is given in Section 10.6.2 (see Table 10.4).

10.5.2.1 Linear Regression for Benzene

In case of Benzene, we will assume that the constant C is known and equal to $C = 250$. With this, the problem becomes a linear regression problem between $y = \ln(p^{sat})$ and $x = 1/(T + C)$. This can be easily solved with the tools from Section 10.2.

Example 10.7 Fitting Antoine's Equation Coefficients: Benzene

Problem: Given $C = 250$, use the vapor pressure data for benzene to estimate the values of parameters A and B for Antoine's equation.

Solution: By now, the linear regression problem should be familiar to the readers. The following is the code for solving this problem:

```
%% Observed data
T=[50:10:140];
pSatB=[0.3751, 0.5027, 0.7112, 0.9883, 1.2529,...
     1.8758, 2.3648, 2.9474, 3.9049, 4.5455];
%% Obtain Antoine's coefficients
C=250;     % Coefficient C is given
Y=log(pSatB');
X=[ones(10,1), -1./(T'+C)];
PHI=inv(X'*X)*X'*Y;
A=PHI(1);
B=PHI(2);
pSatModel=exp(A-B./(T+C));
%% Plot results
plot(T,pSatB,'bo',T,pSatModel,'-r');
xlabel('Temperature (^oC)'); ylabel('p^{sat} (bar)');
figure(2);
plot(T,(pSatB-pSatModel),'bo');
xlabel('Temperature (^oC)'); ylabel('Residual, e_i');
```

The values of the parameters are $A = 10.12$ and $B = 3343.0$. The coefficient of determination is $r^2 = 0.998$. Figure 10.7 shows the comparison between the model fit and observed data (top panel) as well as how the model errors are spread.

Let us also look at 95% confidence intervals for the two parameters, which can be computed as discussed in Section 10.2.3 to yield $A = 10.12 \pm 0.358$ and $B = 3343 \pm 122$.

Note how narrow the 95% confidence intervals are for the parameters A and B. Very narrow confidence intervals means that the model is very likely to capture the actual variance underlying the observed data. This could either be due to overfitting, or if the measurement errors are low and all the measurements nicely line up as we expect from the model. If these values sound too good to be true, they really are! The vapor pressure data in Table 10.4 is not actual observed data; it is synthetic data that I generated using the same model.

Thus, the statistical analysis of model parameters can reveal additional information about the model fit and the data, which is not revealed by merely looking at Figure 10.7 or

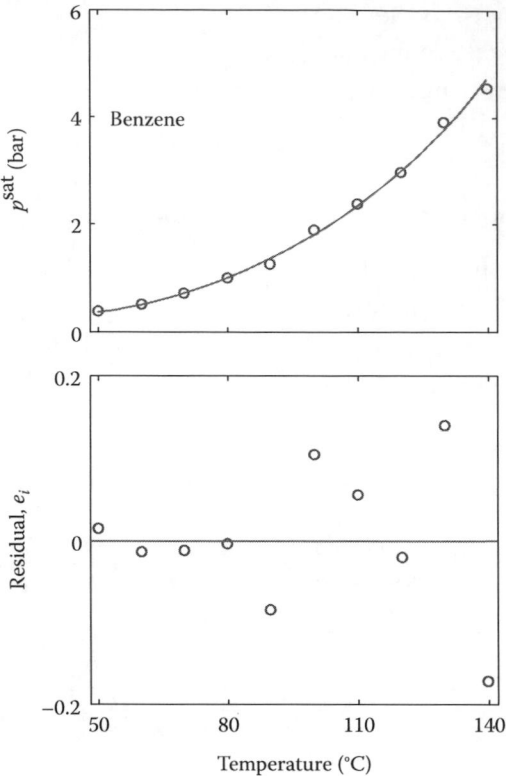

FIGURE 10.7   Observed data (symbols) and model fit (lines) for the saturation pressure of benzene and the residual errors at various temperatures.

the R-squared value. This is merely a brief introduction to this field. Interested readers are referred to advanced texts on regression and data analysis for more information.

10.5.2.2 Nonlinear Regression for Ethylbenzene

We now use nonlinear regression to fit the model parameters for ethylbenzene.

Example 10.7 (Continued) Antoine's Coefficients: Ethylbenzene

Problem: Use the vapor pressure data for ethylbenzene to estimate the values of parameters *A*, *B*, and *C* for Antoine's equation.

   Solution: The following code is used to estimate coefficients using nonlinear regression. We will use `lsqnonlin` in this example. The following function computes the residual error vector required by the solver:

```
function errVal=antoineFun(Phi,xData,yData)
T=xData';
lnPsat=log(yData');
% Model calculation
A=Phi(1); B=Phi(2); C=Phi(3);
```

```
modelVal=A-B./(T+C);
errVal=lnPsat-modelVal;
end
```

The following script uses `lsqnonlin` to obtain model parameters:

```
%% Observed data
T=[50:10:140];
pSatEB=[0.0489, 0.0703, 0.1082, 0.1626, 0.2182, ...
    0.3595, 0.4786, 0.6281, 0.8885, 1.0714];

%% Regression
PhiGuess=[10; 2500; 200];
PHI=lsqnonlin(@(X) antoineFun(X,T,pSatEB),PhiGuess);
A=PHI(1); B=PHI(2); C=PHI(3);

%% Verify model fit
pSatModel=exp(A-B./(T+C));
figure(3)
plot(T,pSatEB,'bo',T,pSatModel,'-r');
xlabel('Temperature (^oC)'); ylabel('p^{sat} (bar)');
figure(4)
plot(T,pSatEB-pSatModel);
xlabel('Temperature (^oC)'); ylabel('Residual, e_i');
```

The values of the three parameters are $A = 11.3604$, $B = 4583.9$, $C = 267.9$, whereas Figure 10.8 shows the model fit for ethylbenzene.

10.5.3 Complex Langmuir-Hinshelwood Kinetic Model

A Langmuir-Hinshelwood-Hougen-Watson (LHHW) model for catalytic reaction

$$A \rightarrow B$$

is given by the following rate expression:

$$r = \frac{kC_A}{\left(1 + K_A C_A + K_B C_B\right)^2} \tag{10.75}$$

A series of experiments were run and the reaction rate vs. concentration data was obtained.[*] We will do this problem in two parts. In the first case, the reactor was operated with only A in the feed at an initial concentration of $C_{A0} = 1$. With only this data available, we would be unable to fit all the three parameters to the data. Next, we will use experimental data at various initial concentrations of A and B to obtain all the rate constants.

---

[*] This is also synthetic data used for demonstration purpose. It was generated with the same underlying model to highlight the issues discussed in this example.

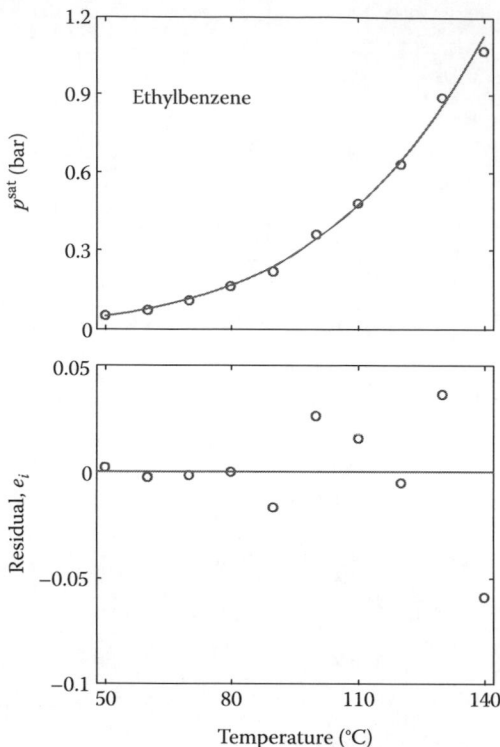

FIGURE 10.8   Observed data (symbols) and model fit (lines) for saturation pressure of ethylbenzene and the residual errors at various temperatures.

10.5.3.1 Case 1: Experiments Performed at Single Concentration of B

The following table shows experimental data obtained from kinetic experiments:

C_A	1	0.9	0.8	0.7	0.6	0.5	0.4
C_B	0	0.1	0.2	0.3	0.4	0.5	0.6
r	5.55e−3	5.58e−3	5.45e−3	5.31e−3	5.14e−3	4.96e−3	4.52e−3

Since the only reaction taking place is $A \rightarrow B$, the concentration of B is obtained from overall mass balance, noting that $C_A + C_B = C_{A0}$ holds for the duration of the experiment. As a first try, let us naïvely use the above data for performing linear regression of the LHHW model. The nonlinear expression (10.75) can be converted to a linear form by inverting and taking a square root:

$$\frac{1}{\sqrt{r}} = \underbrace{\frac{1}{\sqrt{k}}}_{\phi_1}\left(\frac{1}{\sqrt{C_A}}\right) + \underbrace{\frac{K_A}{\sqrt{k}}}_{\phi_2}\left(\sqrt{C_A}\right) + \underbrace{\frac{K_B}{\sqrt{k}}}_{\phi_3}\left(\frac{C_B}{\sqrt{C_A}}\right) \tag{10.76}$$

The next example demonstrates this attempt.

Example 10.8 Fitting an LHHW Model—Part 1

Problem: For the experimental data given above, use linear regression to fit an LHHW kinetic model (10.75).

Solution: Given the data, we construct the required matrices and solve as below:

```
rateData=[5.55e-3, 5.58e-3, 5.45e-3, 5.31e-3, ...
    5.14e-3, 4.96e-3, 4.52e-3];
CaData=1:-0.1:0.4;
CbData=1-CaData;
%% Linear least squares
Y=sqrt(1./rateData');
rootCa=sqrt(CaData');
X=[1./rootCa, rootCa, CbData'./rootCa];
phi=X\Y;
k=1/phi(1)^2;
Ka=phi(2)*sqrt(k);
Kb=phi(3)*sqrt(k);
```

When the above code is executed, MATLAB shows the following warning message:

```
Warning: Rank deficient, rank = 2, tol =  5.144832e-15.
```

This message **should not** be ignored. Since we have used the left division operator "\" to compute Φ, MATLAB will give a solution. The software does not know whether the rank deficient condition was intended. Some students would look at the result vector:

```
>> phi
phi =
    6.7348
    6.6996
         0
```

and accept it as answer without further scrutiny.

If the importance of the above example is not yet clear, let me write it again in bold font: Do not ignore warnings in MATLAB.

Let's see why this happens—due to the material balance, $C_B = 1 - C_A$. Substituting this in Equation 10.75, we get

$$r = \frac{kC_A}{\left(1 + K_A C_A + K_B \left(1 - C_A\right)\right)^2}$$

$$= \frac{kC_A}{\left(\underbrace{\left(1 + K_B\right)}_{a_1} + \underbrace{\left(K_A - K_B\right)}_{a_2} C_A\right)^2} \tag{10.77}$$

As seen in the above expression, only a_1 and a_2 can be identified independently with the data used in this example. When we write the rate expression in this form, the problem perhaps becomes obvious. But what about the modified expression, Equation 10.76:

$$\frac{1}{\sqrt{r}} = \underbrace{\frac{1}{\sqrt{k}}\left(\frac{1}{\sqrt{C_A}}\right)}_{\phi_1} + \underbrace{\frac{K_A}{\sqrt{k}}\left(\sqrt{C_A}\right)}_{\phi_2} + \underbrace{\frac{K_B}{\sqrt{k}}\left(\frac{1-C_A}{\sqrt{C_A}}\right)}_{\phi_3} \tag{10.78}$$

Consider the three modified independent variables:

$$x = \frac{1}{\sqrt{C_A}}, \quad u = \sqrt{C_A}, \quad w = \left(\frac{1}{\sqrt{C_A}} - \sqrt{C_A}\right)$$

When written this way, we can immediately observe that x, u, and w are linearly dependent. While this problem is fairly clear when we use linear regression, naïve use of nonlinear regression will mask these problems. Recall our statement in Section 10.4.1: One must exercise caution when using regression.

10.5.3.2 Case 2: Experiments Performed at Different Initial Concentrations of B

The three kinetic parameters can be independently determined if the data in C_A and C_B is not linearly dependent. In fact, this is something that is done in our CRE (chemical reaction engineering) books in the "Determining kinetic constants" chapter without us realizing it. The experiments are repeated with different initial conditions, introducing different amounts of B each time. Such a data set is shown in Section 10.6.2 in Table 10.5. Three different sets of experiments were run, with concentration of A kept constant at $C_{A0} = 1$. In the first set, only A was introduced; in the second set, $C_{B0} = 0.1$ was also introduced; and in the third set, $C_{B0} = 0.25$ was introduced. The data collected during these experiments are tabulated in Table 10.5.

We randomly split this data into two parts: training set and testing set. Four of the eighteen data points (highlighted cells in the table) were selected as test set; the model parameters were fit to the remaining fourteen data points. Thereafter, linear regression was used as before.

Example 10.8 (Continued) Fitting an LHHW Model—Part 2

```
%% Linear estimation of LHHW kinetics
load LHdata
Y=sqrt(1./rateData');
rootCa=sqrt(CaData');
X=[1./rootCa, rootCa, CbData'./rootCa];
phi=X\Y;
k=1/phi(1)^2;
Ka=phi(2)*sqrt(k);
Kb=phi(3)*sqrt(k);
```

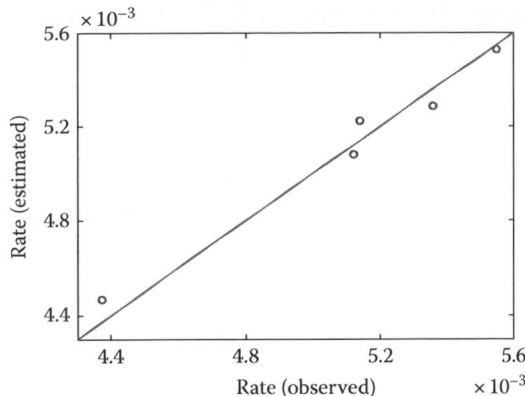

FIGURE 10.9   Parity plot for the test data, with estimated values plotted vs. observed values.

```
%% Testing results
NTest=length(rateTest);
rHatTest=(k*CaTest)./(1+Ka*CaTest+Kb*CbTest).^2;
err=rateTest-rHatTest;
sse=err*err';
ssy=var(rateTest)*(NTest-1);
Rsquare=1-sse/ssy;
plot(rateTest,rHatTest,'bo','markersize',3);
```

The rate parameters obtained are $k = 0.051$, $K_A = 2.03$, and $K_B = 0.49$. Figure 10.9 shows a comparison between observed data and model predictions. The symbols lie close to the $45°$ line, indicating a reasonable fit. The R-squared value of $r^2 = 0.97$ with testing data indicates a good fit. Statistical information about confidence intervals also indicates a good fit for the kinetic model. Thus, including additional data at different initial C_{B0} concentrations improves the model performance.

10.5.4 Reaction Rate: Differential Approach

Contrary to the examples of estimating reaction rate constants in this chapter, the reaction rate r is not directly measured. Instead, it is the concentration and temperature that are directly measured and the reaction rate is computed from these measurements. It is in this context that "differential approach" and "integral approach" are introduced in reaction engineering texts. In this example, we will apply a differential approach for catalytic reactions taking place in a batch reactor. The reactor contains the reactant at a concentration of $C_{A0} = 1$ mol/L at initial time. The change in concentration with time is tracked. The observations are shown in Table 10.2. The objective is to find a power-law model of the type

$$r = kC_A^n \qquad (10.79)$$

to fit the observed data.

TABLE 10.2 Concentration vs. Time Data for Determining Kinetic Rate Constants

t	5	30	60	90	120	150	180	210	240
C_A	0.988	0.828	0.651	0.525	0.379	0.253	0.160	0.086	0.063

There are two different approaches that chemical engineers are familiar with. The first one is the differential approach. Here, we use the definition of the reaction rate *in constant volume batch reactor* as

$$r = \frac{1}{V}\frac{dN_A}{dt} = \frac{dC_A}{dt} \tag{10.80}$$

Given the above data, we may be tempted to use numerical differentiation to compute the reaction rate defined above. However, this approach is discouraged for two reasons. First, the experimental data is available at sparse intervals, making the rate computations highly approximate. However, much more importantly, there are measurement errors in the data. High frequency errors (such as measurement errors) get highly accentuated in numerical differentiation. For example, at time $t = 120$, numerical derivative gives

$$r_{120} = \frac{0.253 - 0.525}{60} = -4.48 \times 10^{-3} \text{ mol/L·s}$$

However, if there was up to 2.5% error in the experimental data (which is actually on the lower side of what one can expect in experiments)

$$r_{120} = \frac{0.246 - 0.535}{60} = -4.82 \times 10^{-3} \text{ mol/L·s}$$

may have 7% or higher error.

Hence, an alternative is to fit a polynomial to the reaction rate data and hence compute the rate of reaction. As will be shown in the next example, we find that second-order polynomial is adequate to represent the data.

Example 10.9 Differential Method of Kinetic Analysis

Problem: Use the differential method of kinetic analysis to obtain rate parameters for the power-law model (10.79) and fit it to the observed data in Table 10.2.

Solution: We will use the differential approach, where we will first calculate the rate of reaction. We will do so by fitting polynomials of order 1 through 4. After doing this, we choose polynomial of order 2:

$$r_A^{\text{poly}}(t) = a_1 + a_2 t + a_3 t^2$$

to fit the observed data. The following part of the code does that:

```
%% 1. Fit a polynomial to original data
load batchData
Y=CaData;
```

```
n=length(CaData);
X=[ones(n,1), tData, tData.^2];
p=X\Y;
% Plot and verify the fit
Y=p(1)+p(2)*tData+p(3)*tData.^2;
plot(tData,CaData,'bo',tData,Y,'-r');
```

Figure 10.10 shows that the quadratic polynomial represents the data quite well. We are not done yet. This is just a correlation that we fitted for convenience. We want to use it for obtaining reaction rate parameters. It is easy to see that for the fitted polynomial, the reaction rate is

$$r = -\left(a_2 + 2a_3\right)$$

where the negative sign is because A is a reactant. The following part computes the corresponding reaction rate:

```
%% 2. Calculate reaction rate
rData=-(p(2)+2*p(3)*tData);
disp([tData, CaData, rData*1000]');
```

Let us now update Table 10.2 as below:

0	30.0	60.00	90.00	120.0	150.0	180.0	210.0	240.0
0.988	0.828	0.651	0.525	0.379	0.253	0.160	0.086	0.063
6.459	5.841	5.224	4.607	3.990	3.372	2.755	2.138	1.521

The last row is multiplied by 1000 to display. Now that we have reaction rate and concentration data for each, we can proceed with parameter estimation.

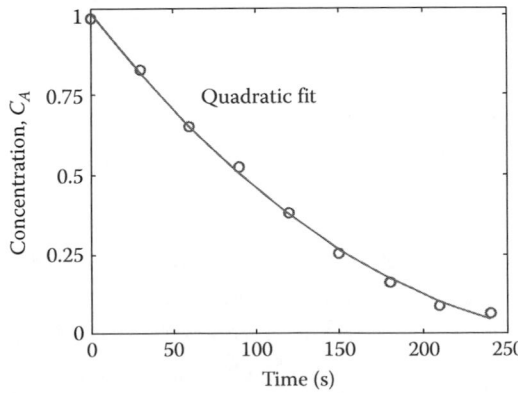

FIGURE 10.10 Observed data (symbols) and quadratic polynomial fit (line) to the data.

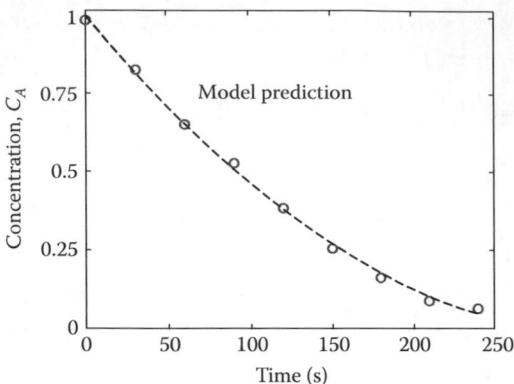

FIGURE 10.11 Observed data (symbols) and the fit for kinetic rate expression (dashed line).

```
%% 3. Compute parameters using linear regression
Y=log(rData);
X=[ones(n,1), log(CaData)];
par=X\Y;
k0=exp(par(1));
mu=par(2);
% Model comparison
tSol=tData;
ySol=(1-k0*(1-mu)*tSol).^(1/(1-mu));
plot(tSol,ySol);
```

The overall code for using differential method for rate analysis consists of the above three parts, solved together in the same script file. Figure 10.11 shows that the model fits the overall data very well. The kinetic rate expression obtained through this approach is

$$r = 6.46 \times 10^{-3}\, C_A^{0.487}$$

10.6 EPILOGUE

10.6.1 Summary

In this chapter, we covered methods for obtaining model parameters that fit a set of observations. We also provided a basic introduction to analyzing statistical properties to verify goodness of fit. I would like to impress upon the reader that obtaining least squares parameters is the *easy part* of the regression process. These parameters are not useful unless accompanied by some goodness of fit measures. Three different ways were introduced: (i) qualitative method by plotting experiments and model fits together, (ii) simple statistics such as coefficient of determination ("R-squared") and sum of squared errors, and (iii) statistical bounds on parameters and confidence intervals.

I close this chapter by mentioning matrix-based and singular value decomposition–based approaches as my recommended approaches for linear least squares. If the parameter estimation problem is not linearizable, then the nonlinear estimator may be used. I would recommend lsqnonlin as a preferred algorithm (Optimization toolbox).

If you have access to the Statistics toolbox, using the full power of the toolbox is highly recommended. The treatment in this chapter will help you build the foundation for understanding more advanced material in statistical analysis.

10.6.2 Data Tables

Table 10.3 shows the data for specific heat of methane at two different temperature ranges (298–500 K and 600–1300 K). The data for pure component vapor pressures of benzene and ethylbenzene are shown in Table 10.4, whereas reaction rate data is shown in Table 10.5.

TABLE 10.3   Specific Heat of Methane at 1 bar Pressure

Normal Temperature Range		High Temperature Range	
T (K)	c_p (J/mol·K)	T (K)	c_p (J/mol·K)
298	35.64	600	52.74
325	36.69	700	58.6
350	37.85	800	64.08
375	39.13	900	69.14
400	40.63	1000	73.75
425	41.93	1100	77.92
450	43.39	1200	81.68
475	44.87	1273	84.1
500	46.63	1300	85.07

Sources: Data obtained from NIST WebBook, http://webbook. nist.gov/cgi/cbook.cgi?ID=C74828&Mask=1; Gurvich, L.V. and Alcock, C.B., *Thermodynamic Properties of Individual Substances*, Hemisphere, New York, 1989.

TABLE 10.4   Pure Component Vapor Pressure Data for Benzene and Ethylbenzene

Temperature (°C)	Vapor Pressure (Bar)	
	Benzene	EB
50	0.3751	0.0489
60	0.5027	0.0703
70	0.7112	0.1082
80	0.9883	0.1626
90	1.2529	0.2182
100	1.8758	0.3595
110	2.3648	0.4786
120	2.9474	0.6281
130	3.9049	0.8885
140	4.5455	1.0714

Sources: Benzene: NIST WebBook, http://webbook.nist.gov/cgi/ cbook.cgi?ID=C71432&Mask=4#Thermo-Phase; Ethyl benzene: NIST WebBook, http://webbook.nist.gov/cgi/ cbook.cgi?ID=C100414&Mask=4#Thermo-Phase.

TABLE 10.5   Reaction Rate Data for Langmuir-Hinshelwood
Kinetics for Section 10.5.3

C_A	C_B	r	C_A	C_B	r
0.9	0.1	5.58e−3	1	0.25	5.08e−3
0.8	0.2	5.45e−3	0.8	0.45	5.05e−3
0.7	0.3	5.31e−3	0.7	0.55	4.88e−3
0.5	0.5	4.96e−3	0.6	0.65	4.72e−3
1.0	0.1	5.37e−3	1.0	0	5.55e−3
0.9	0.2	5.34e−3	0.6	0.4	5.14e−3
0.7	0.4	5.19e−3	0.8	0.3	5.36e−3
0.6	0.5	5.01e−3	0.9	0.35	5.12e−3
0.5	0.6	4.81e−3	0.5	0.75	4.37e−3

The model parameters are fitted to the testing data (unshaded rows
in the table), whereas the shaded rows indicate the test data to test
the model performance.

EXERCISES

Problem 10.1 Show that Equations 10.25 and 10.33 are equivalent. In other words, show
that

$$X^T X = \begin{bmatrix} N & \Sigma_x \\ \Sigma_x & \Sigma_{xx} \end{bmatrix}, \quad X^T Y = \begin{bmatrix} \Sigma_y \\ \Sigma_{xy} \end{bmatrix}$$

Problem 10.2 Write your code for Example 10.4 and derive the parameter values and the
standard deviations of these parameters.

For the data in Example 10.4, fit substrate inhibited rate law:

$$r = \frac{\mu^{max} C_S}{K + C_S + a C_S^2}$$

Appendix A: MATLAB® Primer

MATLAB® is a portmanteau of *matrix lab*oratory. It is a high-level computing language, widely used in computational analysis, control, signal processing, communication, and related fields. This appendix provides a nonexhaustive primer for the reader to get started with MATLAB. The treatment here will be necessarily succinct. This appendix may also be used by people who have used MATLAB in the past but would like to orient their way of using MATLAB to the one followed in this book.

There are several resources available to get started with MATLAB. The MathWorks website has video tutorial on Getting Started with MATLAB.* I have a video course on NPTEL[†] that focuses on MATLAB.

Each individual set of concepts are discussed in sections below. Each section starts with a set of problems that a user can solve, followed by the solution. Users of MATLAB are encouraged to try solving these problems on their own before checking the solutions provided in the same section.

A.1 SOME SPECIAL VARIABLES

The following constants and special variables are used in MATLAB:

ans	Captures the last evaluated value that was not assigned to a variable
pi	The value of π
i or j	The imaginary unit $\left(\sqrt{-1}\right)$
eps	Machine precision (see Chapter 1)
inf	Infinity ∞
nan	Not a number (e.g., $\dfrac{0}{0}$ evaluates as nan)

If your code replaces any of the above variables, then that new assigned value stays with that variable and it loses its original (default) value. Thus,

```
>> i = 2
i =
    2
```

---

* http://in.mathworks.com/videos/getting-started-with-matlab-68985.html (August 1, 2016).

[†] National Programme for Technology Enhanced Learning (NPTEL) is a Government of India initiative, which provides online web and video courses free. Video lectures on MATLAB and related material can be accessed at: http://nptel.ac.in/courses/103106118/ (August 1, 2016).

From this point onward, i will not represent the imaginary unit but instead shall represent the value assigned to it.

Note that the value of i above is echoed (i.e., displayed) on the screen. The echo can be suppressed by ending the statement with a semicolon:

```
>> i=2;     % This will suppress echo on the screen
```

The variable ans is a special variable. When an expression is executed but the value is not assigned to any variable, the value is captured in the variable ans. For example

```
>> 3*i
ans =
     6
```

Since the expression 3*i is not assigned to any variable, the latest such value is captured in ans and the value is echoed on the screen (since there is no semicolon to end the statement).

A.2 MATRIX OPERATIONS AND VECTOR FUNCTIONS

Users who are new to MATLAB may wish to review video lectures on NPTEL and MathWorks.*

A.2.1 Creating and Accessing Vectors and Matrices

A set of problems that I will use to introduce students to matrix and vector operations is listed below. The readers can try to solve these problems before proceeding to the discussions.

1. Create the following 3×4 matrix A:

$$A = \begin{bmatrix} -4 & 2 & 3 & 4 \\ 3 & 1 & 4 & 7 \\ 4 & 0 & 5 & 8 \end{bmatrix}$$

2. Create a vector $p = [1 \ 2 \ \cdots \ n]$. Choose n = 3.
   In other words, I want you to use " : " command instead of typing p= [1,2,3];

3. Create a 3×3 identity matrix and assign it to B.

4. Type help diag to understand the diag command.
   If you type diag(A), what do you get? Try and check.

5. What will you get if you type diag(diag(A)). Check. Does it make sense?

6. If you type A(2,p), what will you get?

---

* http://in.mathworks.com/videos/working-with-arrays-in-matlab-69022.html (August 1, 2016).

7. Obtain the second row of A and assign it to vector d.

8. Obtain E as a 2×2 subarray of A starting from row 1–column 2. In other words

$$E = \begin{bmatrix} 2 & 3 \\ 1 & 4 \end{bmatrix}$$

However, E should be obtained from array A, *and not* by entering a new matrix.

A.2.1.1 Building Up Matrices
Let us now look at MATLAB commands that achieve the above. Matrix A can be assigned as

```
>> A = [-4, 2, 3, 4; 3, 1, 4, 7; 4, 0, 5, 8];
```

The semicolons separate multiple rows. Individual cells on each row can be separated by comma (or spaces). It is important to know a couple of things: First, like math operations follow brackets–exponent–multiplication–addition, MATLAB parses row–columns. For example

```
>> A = [4; 3; 4, 2; 1; 0, 3; 4; 5, 4; 7; 8];
Dimensions of matrices being concatenated are not consistent.
```

Note that MATLAB gives an error because it parses the command as

```
A = [4;
3;
4, 2  etc.
```

The error signifies that I am trying to concatenate the first two rows with one element each and the third row with two elements, hence the error. However, one may use brackets to create the matrix:

```
>> A = [[4; 3; 4], [2; 1; 0], [3; 4; 5], [4; 7; 8]]
A =
     4     2     3     4
     3     1     4     7
     4     0     5     8
```

Here, each bracket creates a 3×1 vector; four such vectors are concatenated to give a 3×4 matrix. Note that square brackets enclose a vector/matrix/array. On the other hand, parentheses are used to query a location in the array. Thus, the command

```
>> A(3,2)
ans =
     0
```

returns the element in the third row and second column of A.

The next problem brings us to colon notation. In MATLAB, a : b is read as "from a to b in steps of 1," whereas a : h : b is read as "from a to b in steps of h." Thus, 1 : 3 (or 1 : 1 : 3) will result in a *row vector* [1 2 3]. You may try the following commands in MATLAB, thinking about what the result would be before you execute the command in MATLAB (and hence check your understanding):

```
>> p=1:3
>> q1=1:3:5
>> q2=1:1.5:5
>> q3=5:-1:1
>> q4=5:1
```

Perhaps the last command needs explanation. It says "5 to 1 in steps of 1." However, the first value, 5, is greater than the final value. Hence, no value can be assigned and q4 becomes an empty matrix. Note that this *does not* give an error; instead an empty matrix is returned.

A.2.1.2 Some Special Matrices

The commands for creating identity matrix, matrix of zeros, and matrix of ones are eye(m,n), zeros(m,n), and ones(m,n), respectively. The command eye(m) is equivalent to eye(m,m). Clearly then, B = eye(3); returns a 3×3 identity matrix.

There are some further commands to create matrices:

magic(3);	Creates a 3×3 magic matrix
rand(m,n);	Creates an $m \times n$ matrix of random numbers from uniform distribution
randn(m,n);	Creates an $m \times n$ matrix of random numbers from normal distribution
diag(A);	Extracts diagonal elements from A and stores them as a vector
diag(vec);	Creates a diagonal matrix with elements of vec forming the diagonal
tril(A);	Extracts lower triangular matrix from A
triu(A);	Extracts upper triangular matrix from A

The next command I will use is diag, as follows:

```
>> diag(A)
ans =
     4
     1
     5
```

The above result is fairly straightforward—the vector returned contains diagonal elements of A:

```
>> diag(diag(A))
ans =
     4     0     0
     0     1     0
     0     0     5
```

From the above description of the diag command, does your understanding of diag match MATLAB results?

A.2.1.3 Querying Elements of a Matrix

Elements of a matrix, as described earlier, can be queried as A (k, l). If k and l are scalars, this returns the element in the kth row and lth column. If k and/or l are vectors, all elements corresponding to all the row-column pairs are returned:

```
>> A(2,p)
ans =
     3     1     4
```

The above command returns from the second row of A, the elements corresponding to column numbers indicated by p vector, [1 2 3]. In other words, it returns the first three elements of the second row of A.

The command to extract the entire second row of A is

```
>> d=A(2,:)
d =
     3     1     4     7
```

Note that simply using colon implies "all." Thus, the above command implies elements in A at row 2—and all elements in that row.

It should be fairly clear how to get matrix E that contains entries from first two rows and columns 2 and 3 of matrix A:

```
>> E  =  A(1:2,2:3)
E  =
     2     3
     1     4
```

Finally, here is a command for the reader to try. It is perhaps a good idea for you to think of what result you would expect and then execute the command in MATLAB:

```
>> F  =  A([3,2],[1,3])
```

Does the result match your reasoning? If still confused, think what the commands below will give:

```
>> A(3,[1,3])
```

Next, consider

```
>> A(2,[1,3])
```

The result F above is nothing but the result of the above two commands placed in the first and second rows of F, respectively.

Finally, there is a special keyword, end, to represent the last element of a vector/matrix:

```
>> d(end)
ans =
     7
```

Since d represents the second-last row of matrix A, it can also be obtained as

```
>> A(end-1,:)
ans =
     3     1     4     7
```

A.2.2 Basic Mathematical Operations on Arrays

In this section, we focus on basic math operations on arrays, using the following questions to expound the concepts:

1. Perform scalar addition, multiplication, and division of A and verify.

2. Get transpose of matrix A. Assign it to Atrans.

3. Compute $Cr = AA^T$ and $Cl = A^T A$.

4. We computed d in Item #7 above. What will the following commands yield:
   sum(d)   cumsum(d)   prod(d)   cumprod(d)   diff(d)?

5. Calculate inverse of Matrix E (see Item #8 above) and assign it to F.

6. Compute 1/E. What do you get? Why? Now compute G=1./ E

7. Compute H=E^(-1) and I=E.^(-1). Compare with F and G.
   Do you understand the meaning of .^, ./, and .*?

```
>> 2*A;
>> A/2;
>> A+2;
```

In the commands above, each element of A is doubled and halved, and 2 is added to each element of A, respectively. Note that in MATLAB, $(A + 2)$ is not equivalent to $(A + 2I)$.*

Transpose in MATLAB is obtained using single quotation (apostrophe) key. Thus, the reader can verify the following commands:

```
>> Atrans=A';
>> Cr=A*A';
>> Cl=A'*A;
```

---

* The origin of this misunderstanding among my students is not clear to me. In matrix algebra, one does not add a scalar to a matrix. In this book, however, we will understand $(A+2)$ in the context of the MATLAB definition: 2 added to each element of A.

Since A is a 3×4 matrix, the above three matrices have size 4×3, 3×3, and 4×4, respectively. The variable Cr is

```
Cr =
     45      54      63
     54      75      88
     63      88     105
```

The commands sum(d) and prod(d) compute the sum and product of all elements of the vector, d, respectively.* Since the vector d= [3 1 4 7], the two values are 15 and 84, respectively. The cumulative sum is computed using cumsum(d), the result being a vector of the same dimension as d itself. The first element of the resulting vector is the first element of d, the second element is the sum of the first two elements, the third element is the sum of the first three elements, and so on:

```
>> cumsum(d)
ans =
     3      4      8     15
```

The command cumprod works in a similar way and gives the cumulative product.

The command to compute the inverse is inv. The following two commands are equivalent:

```
>> F=inv(E)
F =
     0.8000    -0.6000
    -0.2000     0.4000
>> H=E^(-1)
H =
     0.8000    -0.6000
    -0.2000     0.4000
```

The latter is just the matrix E to the power -1.

At this stage, I end this section by describing element-wise operations in MATLAB. The element-wise operator† has a dot preceding the regular operator, for example, .*, ./ , and .^, and here is what they give:

```
>> E.*E
ans =
     4      9
     1     16
```

---

* The sum (or prod) command operates on arrays as well, where the sum (or product) of all elements in each of the columns of the array is computed.

† MATLAB calls element-wise operator as array operator to distinguish it from matrix operator. Thus, E*E is "*matrix multiplication*," whereas E.*E is "*array multiplication*" of individual elements of the first array with the corresponding elements of the second array.

Note that the product operates on each element of E individually. Since E = [2 3; 1 4], the result of the above operation is a matrix of the same dimension, with each individual element squared. Contrast that with the regular matrix product:

```
>> E*E
ans =
       7      18
       6      19
```

In the same context:

```
>> G=1./E
G =
    0.5000     0.3333
    1.0000     0.2500
>> I=E.^(-1)
I =
    0.5000     0.3333
    1.0000     0.2500
```

The element-by-element power operation (array operation) not only works to get scalar power of a vector, it can also be employed as follows:

```
>> 2.^d
ans =
       8       2      16     128
```

Note that each element of the resulting ans is the scalar 2 raised to the power of the corresponding element of the vector d.

A.2.3 Matrix Divisions and the Slash Operators

I left Item #14 for last. It deserves its own subsection since matrix division is a more involved concept. Let us try the command below:

```
>> 1/E
Error using  /
Matrix dimensions must agree.
```

The forward slash operator is the so-called right division. Roughly speaking, A/B evaluates as A*inv(B) if B is a square matrix.* Likewise, the backward slash operator is the left division, with A\B roughly evaluating as inv(A)*B.

---

* A right division is a least-squares solution of a linear equation $xB = A$, whereas a left division is a least-squares solution of a linear equation $Ax = B$. Please refer to Chapter 10 for more information on least-squares solutions.

A.2.4 Test Yourself

Example A.1 Test Yourself with Array Operations

Test yourself with the following questions:

1. Generate a vector $v = [1, \; 3, \; 5, \; 7, \ldots, \; n]$ (where n is an odd number).
2. Given a scalar x, calculate $f = [x^1, \; x^3, \; x^5, \ldots, x^n]$ using vector v.
3. Calculate g such that each element is the inverse of the corresponding element of f.
4. Compute the value of $\dfrac{1}{x} + \dfrac{1}{x^3} + \cdots + \dfrac{1}{x^n}$. Save all the steps as a single script file.

Here is the MATLAB code to compute the above:

```
% MATLAB Array Operations
n=11; x=2;      % Assign values to variables
v=1:2:n;
f=x.^v;
g=1./f;
res=sum(g);
```

The percentage sign (%) signifies that all text following it is a comment, which is ignored by MATLAB. The vector v is defined using the colon notation described in Section A.2.1. The array operations described in Section A.2.2 give the vectors f and g, and the final result is calculated using the command sum. The result is

```
>> res
res =
     0.6665
```

The above code can be saved in a MATLAB file. As discussed in Section A.5, such a file is known as "*MATLAB Script.*"

A.3 MATHEMATICAL FUNCTIONS

I have already covered a few vector functions (sum, cumsum, etc.) and a couple of matrix functions (inv and slash). Let us now cover a few more functions.

A.3.1 Common Matrix and Array Functions

One of the important things to query in a vector or array is its size:

```
>> l=length(d);
>> [m,n]=size(A);
```

The first command returns the length of the vector d, whereas the second command returns the number of rows (in variable m) and columns (in variable n). Thus, l=4, m=3, and n=4. Applying the `length` command to an array returns the larger between m and n:

```
>> length(A)
ans =
    4
```

Power, logarithm, and exponent are other important functions in MATLAB, which have both matrix and array versions. In the previous section, I explained matrix multiplication, division, and power and contrasted them with array (i.e., element-wise) versions of the same. An alternate command for power include mpower (matrix power, E^n) and power (array power, E.^n):

```
>> power(E,2)
ans =
    4     9
    1    16
```

This is same as E.^2. The next command is equivalent to E^2:

```
>> mpower(E,2)
ans =
    7    18
    6    19
```

In the same manner, the `sqrtm` and `sqrt` functions are used to find the matrix square root and element-wise square root of an array. The commands E^0.5, mpower(E,0.5), and sqrtm(E) all give the same results.

The functions for finding element-wise exponent or natural logarithm of an array are:

exp(A)	Array exponent of any array A
log(A)	Array logarithm of any array A
expm(E)	Matrix exponent of a square matrix E
logm(E)	Matrix logarithm of a square matrix E

Self-Test Question: What do you expect if you use sqrtm(A)?

```
>> sqrtm(A)
Error using sqrtm
Expected input to be a square matrix.
```

The matrix functions mpower, expm, logm, and sqrtm can be applied to square matrices only.

A.3.2 Trigonometric and Other Array Functions

The trigonometric functions `sin(theta)`, `cos(theta)`, `tan(theta)` and their inverses `asin(x)` (stands for $\sin^{-1}(x)$), `acos(x)`, and `atan(x)` are all array functions, in that they operate on individual elements of an array. It needs to be emphasized that the argument, θ in trigonometric functions, and the result of inverse trigonometric functions are in radians. Hyperbolic functions are similarly defined as `sinh`, `cosh`, and `tanh` and inverse hyperbolic functions as `asinh`, `acosh`, and `atanh`.

It bears emphasizing that `sin(theta)^-1` implies $\dfrac{1}{\sin(\theta)}$ and `sin^-1(theta)` will return an error.

Some other important array functions include

`min(d)` or `max(d)`	Minimum or maximum value in vector d.
`mean(d);`	Average of all values in vector d.
`std(d);`	Standard deviation of vector d.
`round(d,2);`	Round off each element of d to two decimal places.
`ceil(d);`	Round up each element of d to the next higher integer.
`floor(d);`	Round down each element of d to the previous integer.

A.4 OUTPUT DISPLAY AND PLOTTING

A.4.1 Text Display on Command Window

After a MATLAB script completes execution, we would want to display some of the key results on the command window. This was shown in Example A.1: Typing the variable of interest ("`res`" in this example) without the trailing semicolon made MATLAB echo the variable value on the screen. Echoing MATLAB variables without the trailing semicolon is one way to display the results, which can be used to display any variable or expression from a MATLAB file. Another equivalent way* of doing this is to use the command `display`:

```
>> display(res)
res =
    0.6665
```

Either echoing a variable or using `display` causes MATLAB to display the name of the variable as well. Moreover, the entire display takes three lines.† Alternatively, the `disp` command is a more succinct method since the variable name is suppressed:

```
>> disp(res)
    0.6665
```

---

* In reality, when one does not use a trailing semicolon, MATLAB internally calls the `display` command itself. This is hidden to the user but is the reason why one sees the same result.
† The example shown above is exactly how MATLAB echoes a variable or uses `display`. A blank line is left before the displayed information and another blank line between the variable name and values. In all of this text, these superfluous blank lines are not shown for brevity.

Since the above method does not print any other information than the variable value, MATLAB string operations are used instead. For example

```
>> str = ['first string', 'second string'];
```

will yield a string `str = 'first stringsecond string'`. Note the lack of space between the two strings that are concatenated. The above string can be displayed using `disp(str)`. Let us now add a space at the end of the first string and replace the second string with a numeric string:

```
>> disp(['first string ', '0.6665']);
first string 0.6665
```

Here are two important points to note: (i) the use of square brackets to define a concatenated string and (ii) the use of single quotes enclosing `'0.6665'`. The command is `disp(s1)`, where `s1` and `s2` are string variables. If the square brackets are eliminated, MATLAB reads this as `disp(s1,s2)`. Since `disp` takes only one input argument, this will result in an error: `Too many input arguments`. Since the two strings need to be printed side by side, horizontal catenation is used (exactly in a similar way as that with numeric variables). Thus, the command used is `disp([s1,s2])`. The second point is regarding the difference between `0.6665` and `'0.6665'`: The former is a number, whereas the latter is a string. Numbers cannot be concatenated with strings.

So, what do we do if we have the result in numerical variable, `res`? This number is converted to a string using the `num2str` command. Thus, a better way of displaying the result is

```
>> disp(['The result is ',num2str(res)]);
The result is 0.6665
```

A recap of the various things just discussed: (i) using `num2str` to convert numeric variable `res` to a string; (ii) concatenation of this to a string `'The result is '`; (iii) enclosing the strings to be concatenated within square brackets; and (iv) inclusion of a space *within* the prefix-string to ensure there is a space between the string and displayed value.

Next, consider the command:

```
>> disp(['Matrix E is: ',num2str(E)])
Error using horzcat
Dimensions of matrices being concatenated are not consistent.
```

The reason for this error is that `num2str(E)` results in a string array with two rows; this cannot be concatenated with a string (`'Matrix E is: '`) having a single row.

An alternative way of displaying results is to use the command `sprintf`. Since this command is not used much in this book, it will be skipped for brevity.

A.4.2 Plotting Results

Try the following problems yourself:

1. We want to plot $y = 2x^2 + 0.5$ for the values of $0 \leq x \leq 1$ in steps of 0.01. Generate x using colon " : " notations. This line should be *dashed-red line.*

2. Please use `help plot` to learn how to plot in MATLAB, and to find out how to plot with a dashed-red line instead of solid blue.

3. Label the two axes as "*x*" and "quadratic," respectively, using `xlabel` and `ylabel` commands in MATLAB.

4. On the same figure, plot $z = x^2 + 2x$ as a default line

The following video on the MathWorks site gives an introduction to basic plotting functions: http://in.mathworks.com/videos/using-basic-plotting-functions-69018.html

The following example solves the above problem.

Example A.2 Basic Plotting Functions

The MATLAB code below plots the above requirements:

```
% Define variables and plot
x=0:0.01:1;
y=2*x.^2+0.5;
plot(x,y,'--r');
xlabel('x'); ylabel('quadratic');
% Define and plot the second line
z=x.^2+2*x;
hold on;    % To keep the previous line
plot(x,z)
```

The first plot command tells MATLAB the format for the line to be plotted: `'--r'` implies dashed line (– –) with red color (r). The resulting plot is shown in Figure A.1.

Note the use of the command `hold on`. Without the use of this command, the current plot will be overwritten. Since I wanted to plot the second line on the same plot as the first one, I used this command to *hold* the previous plot. The `hold on` command needs to be used only once for each figure. To toggle back* to the default mode of overwriting a previous plot, use `hold off`.

Note that symbols may be displayed by using the appropriate plot formatting string. The last item in the format string signifies the symbol. Thus, `'-b.'` indicates solid line, blue color, with dot symbols, whereas `'ko'` implies no line, black color with circle symbols.

---

* Simply using hold will toggle between on and off states. I personally discourage such use in favor of using hold on and hold off explicitly.

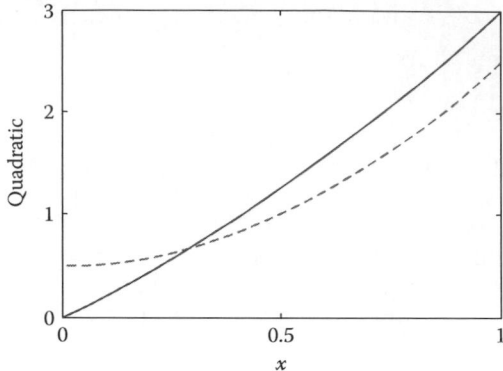

FIGURE A.1   Demo on using plot command to plot two lines.

In addition to `plot`, `xlabel`, and `ylabel`, other useful commands include

`loglog`	Make a log-log plot.
`semilogx`	Semilog plot, logarithmic scale for X-axis.
`semilogy`	Semilog plot, logarithmic scale for Y-axis.
`mesh, surf`	3D mesh or surface plots.
`figure`	Opens a new figure (or selects existing figure).
`subplot`	Make multiple axes on same figure in tiled position.
`title`	Insert a title for the plot.
`legend`	Insert a plot legend.

A.5 WORKING WITH MATLAB® FILES

MATLAB files are of two types: scripts and functions. A good video tutorial on using files is provided on MATLAB website.* Lecture 1–4 of my NPTEL course† discusses MATLAB scripts and functions, and when to use them. All MATLAB files are in plain text and have extension ".m." MATLAB comes with a powerful editor. Although MATLAB m-files can be read by any text editor, the MATLAB editor has very powerful and useful features, and is used almost exclusively to program using MATLAB files.

A MATLAB script is a sequence of commands written sequentially. MATLAB treats these commands in the script file as if it were executed sequentially at the command prompt (or within a calling function that calls this script file). The variables in a script file share the workspace from where they were called. When the script files are called from the command prompt, as will be the case in most of this book, they will use the MATLAB's command workspace. Consider the program in Example A.1. As mentioned in the example, the entire

---

* Please see: http://in.mathworks.com/videos/writing-a-matlab-program-69023.html.

† Please visit: http://nptel.ac.in/courses/103106118/ and click on Lecture 1–4. In this section, I summarize the details discussed in this online video lecture.

code can be written in a MATLAB file (say "`arrayOperations1.m`"). The code can be executed by typing

```
>> arrayOperations1
```

at the command prompt. All the variables declared and used in the script file are available in the MATLAB workspace after the script completes its execution.

MATLAB function, on the other hand, is a file that takes certain sets of inputs, executes a sequence of steps defined in that file, and returns the outputs to the calling entity. The set of inputs and outputs are known as input / output *arguments*. A MATLAB function file (usually) has the same name as the name of the function itself. The first line of a function (`myFirstFun.m`) will be

```
function [out1,out2,...] = myFirstFun(in1,in2,...)
```

There can be zero or more input arguments. These input arguments (`in1`, `in2`, etc.) are written after the function name, and are comma-separated and enclosed within parentheses. Likewise, there can be zero or more output arguments (`out1`, `out2`, etc.), which are comma-separated and enclosed within square-brackets and written before the function name. Each MATLAB function has its own workspace and exchanges information through these input and output arguments. Any other variables defined or changed within the function have a local scope, meaning that they are not accessible outside the function and are "lost" once the function finished execution.*,†

A THOUGHT EXERCISE

Consider a thought exercise: You and your friend have a set of instructions on how to solve a problem. Your friend will assist you in solving the problem. You have a notepad where you have written down several values that are required for solving the problem. You have access to your own set of instructions, but not to those of the other person. After executing some of your instructions, you are told to seek your friend's help.

Scenario 1: Consider the first scenario, where you and your friend use the same shared notepad. At this stage, you pass on this notepad to your friend, who then executes their instructions and does all the computations on this notepad. On completion, your friend returns the notepad to you. Since you shared the notepad, you have access to not only the result but also to all the number-crunching work done by your friend. In this case, *your friend is a MATLAB Script*.

Scenario 2: Consider another scenario, where your friend has their own notepad that is not accessible to you. For helping you, they only need a couple of pieces of information from you. You write them on a yellow post-it note and give the post-it to your friend. Your friend then implements instructions and does all the computation on their own notepad. On completion, the information you seek is returned to you written on another green post-it note. In this case, *your friend is a MATLAB function*, their notepad is the function's own workspace, the yellow post-it note is a set of input arguments and the green post-it note is a set of output arguments.

---

* Exception to this rule is `global` variables, which have to be declared as `global` in the function. They are accessible to any function or script using `global` declaration.

† Additionally, MATLAB also allows defining `persistent` variables, which are not lost once the function finishes execution. They are mentioned here for the sake of completeness, but will not be discussed in this book.

Let us now consider an example where we will use MATLAB files. In Chapter 10, the specific heat of methane gas at various temperatures is obtained in the form of the following quadratic function:

$$c_p = 13.7 + 0.075T - 1.52 \times 10^{-5} T^2 \tag{A.1}$$

where
   c_p is the specific heat in J/mol·K
   T is the temperature in Kelvin

Let us consider two cases: (i) where we want to compute the specific heat for a given temperature and (ii) where this computation of $c_p(T)$ needs to be used as a part of a larger code. The next example distinguishes the two cases.

Example A.3 Specific Heat of Methane: Use of MATLAB® Files

Problem: Write a MATLAB file to compute the specific heat of methane using Equation A.1 at a given temperature in Kelvin.

   Solution: The following MATLAB script computes the specific heat of methane:

```
% Compute specific heat of methane
T=input('Please enter temperature (K): ');
% Specific heat parameters and computation
a1=13.7; a2=0.075; a3=-1.52e-5;
cp=a1+a2*T+a3*T^2;
disp(['At T = ', num2str(T),...
    ' K, c_p = ', num2str(cp), ' J/mol.K']);
```

The above lines are typed in MATLAB editor, and the file is saved as spHeat.m. This file is MATLAB script. It can be executed by typing the file name on the command prompt (or in any other file). This gives

```
>> spHeat
Please enter temperature (K): 450
At T = 450 K, c_p = 44.372 J/mol.K
```

Note that the value of 450 (underlined for highlighting) was entered by the user. Specific heat was computed at that temperature and assigned to the variable cp.

A script is a perfect way to write a sequence of MATLAB statements that one needs to execute for a particular job. In examples in Chapters 4 and 7, the specific heat will be used for other computations. In such a case, a MATLAB function will be more appropriate.

Example A.3 (Continued) Specific Heat of Methane: Use of MATLAB® Files

Solution: Here, we will write a MATLAB function that takes temperature as an input and returns specific heat as the output. The MATLAB code written above is modified as below and used as a MATLAB function:

```
function cp=spHeat(T)
%% Compute specific heat at a given temperature
a1=13.7;
a2=0.075;
a3=-1.52e-5;
cp=a1+a2*T+a3*T.^2;
end
```

The computation of cp is *vectorized* by using array operand ". ^" so that the specific heat can be computed at multiple temperature values if T is a vector. The above commands were saved as a function. The function needs to be invoked from the command line by providing input arguments and capturing the results as output arguments:

```
>> clear all
>> T_gas=450;
>> cp=spHeat(T_gas)
cp =
    44.3720
```

The first command clears all the workspace variables, whereas the final command calls the function to compute specific heat. The input argument may either be a variable or just a value (scalar, vector, or array, as required in the function).

The first difference between the script and function is based on *scope* of the variables. Variables of a function are local to that function itself. So, after executing the function, if we were to check variables in our workspace, we will find that there are only two variables in the workspace, T_gas and cp, because the parameters a1, a2, and a3 are local to the function spHeat and hence unavailable in the MATLAB workspace.

Another difference is the way the two are invoked. If the function is simply called by the file name (without providing the adequate number and type of input arguments), MATLAB will return an error:

```
>> spHeat
Not enough input arguments.
Error in spHeat (line 7)
cp=a1+a2*T+a3*T^2;
```

The final thing to note is that the file name should be the same as the name of the function.

This completes our discussion on MATLAB scripts and functions. Before I close this section, I would like to draw the attention of the user to the variables used. Note that the temperature was named in the MATLAB workspace as `T_gas`. This value was passed on to the function. As described in the pullout text box, this is equivalent to the "`workspace`" writing the *value* of T_gas on a post-it note and handing it to "`spHeat`." The function "`spHeat`" refers to this value as variable `T`, does its computations using its own copy of the variable (equivalent to doing this "on its own notepad"), and has no access to any variable from the MATLAB workspace.

A.6 LOOPS AND EXECUTION CONTROL

A.6.1 The `for` and `while` Loops

MATLAB provides two constructs for executing a set of tasks repeatedly in a loop. The first, and more commonly used, is the `for` loop. The command `for` indicates start of the loop, whereas the `end` statement indicates termination of the loop. The structure of the `for` loop is

```
for index=<list of values>
    <commands that are repeated>
end
```

The most common use is when a block of commands need to be repeated a certain number of times. For example, if the block is to be repeated five times, the first line written as

```
for i=1:5
```

so that the loop is repeated five times, and the index `i` is incremented every time the terminal `end` statement is encountered. The increment of the index need not be in steps of 1; any other increment is also possible, such as

```
for j=1:2:10
```

In this case, the index j takes the values 1, 3, 5, 7, and 9 in each iteration; clearly, the loop is executed five times. Note that the value of the index is available within the loop, but it may not be modified when the loop is being executed. When the loop completes execution, the index variable takes the last value used in the loop. This value is accessible outside the loop. Thus, after the two loops mentioned above finish execution, the variables `i` and `j` have the values of 5 and 9, respectively.

With this background and the discussion in previous sections, the reader may try answering the following questions. For the several `for` statements below, what are the values that the index will take each time the loop executes?

```
for p=0:0.2:1
for q=5:1
for r=5:-1:1
for s=[1 -2 0 2]
```

In each of the above example, the following lines of code may be executed:

```
for p=0:0.2:1
    disp(p);
end
```

with the first and second line replaced appropriately.

1. The first case should be fairly clear based on the preceding discussions. Here, the loop will execute six times; the variable p takes the values of 0, 0.2, 0.4, 0.6, 0.8, and 1 each time the statements in the loop get executed.

2. Since the command [5:1] returns a blank matrix, the loop will not get executed. In other words, the commands in the for block, before the corresponding end statement, will be skipped, and variable q will become an empty variable (i.e., q= [ ]).

3. The loop will get executed five times, with the value of r starting at 5 and decrementing with each iteration. In other words, in each of the five iterations of the for loop, r will consecutively take the values of 5, 4, 3, 2, and 1.

4. The final example is a uniquely MATLAB-specific implementation of the for loop. In each iteration, the index variable s consecutively takes the values from each column of the argument. In other words, the loop will get executed four times. During the first iteration, s takes the value of the first element, i.e., s=1; in the second iteration, it takes the value of s=-2; in the third iteration, s=0; and in the fourth iteration, s=2.

While the last example shows the versatility of the for loop in MATLAB, it can make the code somewhat confusing to debug. Instead, I suggest using the following, rather traditional, approach:

```
A=[1 -2 0 2];
for i=1:length(A)
    s=A(i);
    disp(s);
end
```

Although the variable s takes the same values in each iteration as in the previous case, the latter code is more readable. The readers can compare the two codes and verify that they give the same results.

What do you expect when the following code is used?

```
for s=A
    disp(s);
end
```

The reader may verify that the above code also gives the same results.

Next, we will write a small code that uses the `for` loop for computing the terms of a Fibonacci series. A Fibonacci series is a series comprising [1 1 2 3 5 8 13 21 ...], where each element is the sum of the preceding two elements.

Example A.4 Fibonacci Series Using `for` Loop

Problem: Write a code that computes the first *n* terms of a Fibonacci series.

Solution: The following script computes the Fibonacci series:

```
% Compute n terms of a Fibonacci series
n=10;
fSeries=ones(n,1);   % Initialize the series
for i=3:n
    fSeries(i)=fSeries(i-1)+fSeries(i-2);
end
disp(fSeries')
```

The result of executing the above code is the first ten terms of the series:

```
>> fibo
1    1    2    3    5    8    13    21    34    55
```

Astute readers will notice that the `for` loop starts with an initial value of 3. This is done because the first two terms of the series are given to be 1, and the calculation of the series starts from the third term.

The other construct that can be used for iteratively executing a block of code is the `while` loop:

```
while <condition>
    <block of commands that are repeated>
end
```

The `while` loop keeps executing as long as the abovementioned <condition> is true. If the condition becomes false at any time, MATLAB finishes executing the block of the code, reaches the `end` line, and exits the loop. If the <condition> evaluates as false during the first instance, the block of commands within the `while` construct are not executed. If the condition never becomes `false`, we get an infinite loop (which needs to be interrupted using `Ctrl-C`).

The reader is encouraged to use the next example as a test example to check their understanding of the `while` loop.

Example A.5 The while Loop

Problem: For each of the three cases below, identify the response from MATLAB to test your understanding of the while loop.

Case 1: How many times is the while loop expected to run?

```
i=1;
while i<5
    disp(i);
    i=i+1;
end
```

Case 2: How many times is the while loop expected to run?

```
i=10;
while i<5
    disp(i);
    i=i+1;
end
```

Case 3: How many times is the while loop expected to run?

```
i=1;
while i<5
    disp(i);
    i=i-1;
end
```

Solutions:

1. The loop will run four times. In the fourth iteration, the value of i is incremented to 5; the condition i<5 computes as false, and the loop does not execute the fifth time.
2. The commands within the loop will not execute at all because, at the first instance, the condition i<5 is false.

Here, it will run in an infinite loop because the condition i<5 always evaluates as true since i is decremented.

This completes our discussion of iterative loops.

A.6.2 Conditional if Block

The syntax for the if-then-else block is

```
if <condition>
    <statements-1 evaluated for true>
else
    <statements-2 evaluated for false>
end
```

Here, <condition> refers to any comparison (including Boolean expression) that evaluates as either true or false. If the condition is true, the first block, statements-1, is executed; otherwise, the second block, statements-2, is executed. Nesting of multiple if-blocks is permitted. Let us write a code for printing all prime numbers between 1 and *n*.

Example A.6 Using if Statement

Problem: Write a code to display prime numbers between 1 and *n*.
  Solution:

```
n=20;
for i=1:n
    if isprime(i)
        disp([num2str(i), ' is a prime number']);
    end

end
```

The command to display is only executed when isprime(i) evaluates as true (i.e., when i is a prime number), giving the following result:

```
2 is a prime number
3 is a prime number
5 is a prime number
7 is a prime number
11 is a prime number
13 is a prime number
17 is a prime number
19 is a prime number
```

This concludes our introduction to a conditional if statement. If there are multiple cases to be analyzed, the switch-case statements are more efficient and readable. Interested readers may look up MATLAB help for the switch statement, if required.

A.7 FUNCTION HANDLES

MATLAB functions were discussed in Section A.5. A function to calculate specific heat was written in Example A.3. The function was used to calculate the specific heat at any desired temperature as

```
>> cp=spHeat(T_gas);
```

There are several occasions, as we will see throughout this book, where a function needs to be provided to another function for performing numerical computations. *Function handle* is MATLAB's way to provide or pass on a function to another function. As per MATLAB help documentation, a function handle is a "data type that stores an association to a function." The function handle for the `spHeat` function may be written in one of the following two ways:

1. More convenient form: `@spHeat`

2. More efficient form (which I use in this book): `@(T) spHeat(T)`

Let us deconstruct the second approach, which is called the anonymous function method and was introduced in MATLAB quite recently.[*] The symbol @ indicates function handle and is immediately followed by comma-separated input arguments in round brackets and then the function name (note that in the `spHeat` example, the function name is the same as the file name). This function handle can then be passed to another function for further computations.

For example, the enthalpy at a temperature, say 450 K, is computed as

$$H_{450} = H_{298}^0 + \int_{298}^{450} c_p(T)\, dT \tag{A.2}$$

The integral in the above equation can be computed using the MATLAB command `integral` (see Appendix D). This command requires one to pass the function handle to the function $c_p(T)$. This can be done as follows:

```
>> IntCp = integral(@(T) spHeat(T), 298,450);
```

The anonymous function handle passed on to the MATLAB solver `integral` is underlined above.

This brings us to the end of this primer on MATLAB.

---

[*] For this reason, students and practitioners still prefer the `@spHeat` form of function handle. However, the anonymous function method is more modern and versatile and will be followed in this book. Readers are however free to follow the method of their choice.

Appendix B: Numerical Differentiation

B.1 GENERAL SETUP

Differentiation is one of the essential tools in engineering on which the foundation of calculus rests. For a function $y = f(t)$, it is defined as

$$\frac{dy}{dt} = \lim_{t \to 0} \frac{f(t + \Delta t) - f(t)}{\Delta t} \tag{B.1}$$

The shorthand representation of the first derivative is $f'(t)$. The value of the derivative evaluated at a certain point, $t = t_i$, is represented as $f'(t_i)$. The following notations will be used throughout:

$$f'(t_i) \equiv \frac{d}{dt} f(t) \Big|_{t_i}$$

whereas

$$\frac{d}{dt} f(t_i) = 0$$

The former indicates the value of the derivative of $f(t)$ at $t = t_i$, whereas the latter indicates derivative of the (constant) value that the function achieves at t_1.

The second derivative is defined as

$$f''(t) \equiv \frac{d^2 y}{dt^2} = \frac{d}{dt}\left(\frac{dy}{dt}\right) \tag{B.2}$$

Likewise, higher-order derivatives are also defined.

Geometrically, the first derivative is the slope of tangent of the curve $f(t)$ vs. t at the point $t = t_i$. It represents the *rate* at which the function $f(t)$ is changing at t_i. The second derivative is the rate at which the slope of $f(t)$ changes.

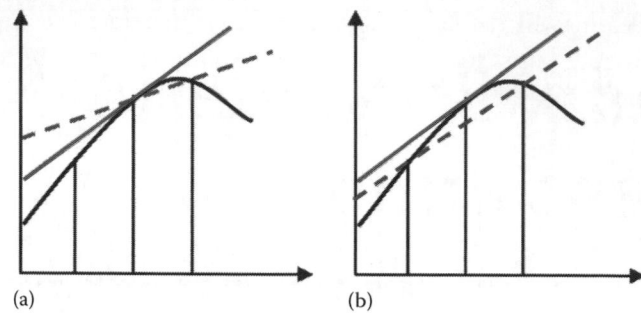

(a)　　　　　　　　　　(b)

FIGURE B.1　Schematic of the first derivative, $f'(t)$ and numerical approximation using (a) forward and (b) central difference formulae.

Numerical differentiation involves approximating the derivative with an appropriate finite difference formula. The simplest one seems evident from Equation B.1—choose a step-size, Δt or h, small enough and apply the difference formula:

$$f'(t_i) \approx \frac{f(t_i + h) - f(t_i)}{h}$$

This is the *forward difference formula* for finding the first derivative of a function. A formal derivation, error analysis, and trade-off between round-off and truncation errors (also see Chapter 1) will be presented in the rest of this chapter.

Figure B.1 shows a schematic representation of derivative as a (slope of) tangent to the curve in the Y–T plane. The slope of the line connecting the points a and $(a+h)$ is the approximation using forward difference formula. The approximation can be improved by reducing h. Alternatively, a different method (called *central difference formula*) may be used, where the derivative is approximated as the slope of the line connecting $(a-h)$ and $(a+h)$. As indicated in the schematic in Figure B.1, the central difference formula better approximates the true derivative.

Numerical differentiation finds several applications. A large number of numerical techniques for solving nonlinear algebraic equations (Chapter 6), implicit ODEs and differential algebraic equations (Chapter 8), and regression (Chapter 10) requires derivatives (or Jacobian) to be calculated. For example, in the Newton-Raphson method of Chapter 6

$$x^{(i+1)} = x^{(i)} - \frac{f(x^{(i)})}{f'(x^{(i)})} \tag{B.3}$$

if the derivative $f'(x)$ is not available, numerical differentiation may be used instead.

Solutions to partial differential equations (Chapters 4 and 10) using finite difference approximation discussed in this chapter is popular for generic engineering problems.

Another application is in the calculation of heat or mass flux. If the walls of a heat exchanger are at a higher temperature than the fluid in the channel, a temperature gradient exists. For even slightly complex geometries, the net heat flux into the fluid

$$J = -\lambda_f \frac{dT}{dz} \tag{B.4}$$

often needs to be calculated numerically. Other applications include finding the rate-determining step in a complex network of reactions, analyzing sensitivity of system performance to various parameters, as well as in data analytics where gradient (first derivative) and curvature (second derivative) can reveal critical information about the qualitative nature of systems.

B.2 NUMERICAL DIFFERENCE FORMULAE

Numerical differentiation rules are computed from Taylor's series approximation, which provides both differentiation formulae as well as error estimates. As discussed in Chapter 1, these refer to *truncation errors* since they result from truncating an infinite series to a finite number of terms. Usually, retaining a larger number of terms in the series expansion results in a more accurate approximation of numerical derivatives. Derivation and error analysis of first and higher derivatives will be presented in this section.

B.2.1 First Derivatives

Consider the Taylor's series expansion of $f(a+h)$:

$$f(a+h) = f(a) + hf'(a) + \frac{h^2}{2} f''(a) + \frac{h^3}{3!} f'''(a) + \cdots \tag{B.5}$$

Retaining the term in $f'(a)$ and moving the other terms to the left-hand side, we get

$$f'(a) = \frac{f(a+h) - f(a)}{h} - \left[\frac{h}{2} f''(a) + \frac{h^2}{6} f'''(a) + \cdots \right] \tag{B.6}$$

The first term on the right-hand side is the *forward difference approximation* of $f'(a)$, whereas the terms in square brackets represent the error. Here, $\frac{h}{2} f''(a)$ is the leading error term. Hence, the forward difference formula is $\mathcal{O}(h^1)$ accurate. Using the mean value theorem, the above equation may be written as

$$f'(a) = \underbrace{\frac{f(a+h) - f(a)}{h}}_{\text{Fwd. diff}} - \underbrace{\frac{h}{2} f''(\xi)}_{\text{Error}} \tag{B.7}$$

There exists a point $\xi \in [a, (a+h)]$ such that the terms in the brackets of Equation B.6 may be equivalently written in the above form.

Additional nodal points are required to derive a more accurate formula for the first derivative. The central difference formula uses $f(a-h)$, in addition to $f(a)$ and $f(a+h)$, to approximate the derivative $f'(a)$ (hence the name). Again, Taylor's series expansion is

$$f(a-h) = f(a) - hf'(a) + \frac{h^2}{2}f''(a) - \frac{h^3}{3!}f'''(a) + \cdots \tag{B.8}$$

Subtracting from Equation B.5

$$f(a+h) - f(a-h) = 2hf'(a) + \frac{h^3}{3}f'''(a) + \cdots \tag{B.9}$$

This leads to the following central difference formula:

$$f'(a) = \underbrace{\frac{f(a+h) - f(a-h)}{2h}}_{\text{Central diff}} - \underbrace{\frac{h^2}{6}f'''(\xi)}_{\text{Error}} \tag{B.10}$$

Comparing the forward and central difference formulae in Equations B.7 and B.10 demonstrates what I argued in the schematic of Figure B.1. Since the forward difference formula scales as $\mathcal{O}(h)$, whereas the central difference formula scales as $\mathcal{O}(h^2)$, the latter is more accurate for the same choice of step-size h. This has already been observed in Chapter 1 and will be demonstrated later in this section.

Similar to the forward difference formula, it is easy to observe that the *backward difference formula* for the first derivative is

$$f'(a) = \underbrace{\frac{f(a) - f(a-h)}{h}}_{\text{Bkd difference}} + \underbrace{\frac{h}{2}f''(\xi)}_{\text{Error}} \tag{B.11}$$

The backward difference formula has the same accuracy as the forward difference formula.

In certain situations, a forward (or backward) difference formula with a higher accuracy is needed, for example, to calculate the heat flux at the boundary with higher accuracy or in solving PDEs (see Chapter 4). A three-point forward difference formula would be used in such a case. Here, the values $f(a), f(a+h)$ and $f(a+2h)$ are used to calculate the derivative $f'(a)$. This is in contrast to $f(a-h), f(a)$ and $f(a+h)$ used in the central difference formula. Taylor's series expansion of $f(a+2h)$ is

$$f(a+2h) = f(a) + 2hf'(a) + \frac{4h^2}{2!}f''(a) + \frac{8h^3}{3!}f'''(a) + \cdots \tag{B.12}$$

Note that subtracting $4f(a+h)$ from the above equation eliminates the term in $f''(a)$, and the third-order term becomes the leading term. This can indeed be done by observation.

However, as the functions become more complex, or if higher-order formulae are required, a more versatile method called *method of undetermined coefficients* may be used to compute the approximation. For example, the three-point forward difference formula is written as a weighted sum:

$$f'(a) = w_1 f(a) + w_2 f(a+h) + w_3 f(a+2h) \tag{B.13}$$

Using the Taylor's series expansion, the above may be written as

$$f'(a) = w_1 f(a) + w_2 \left[f(a) + hf'(a) + \frac{h^2}{2} f''(a) + \frac{h^3}{3!} f'''(a) + \cdots \right]$$

$$+ w_3 \left[f(a) + 2hf'(a) + \frac{4h^2}{2} f''(a) + \frac{8h^3}{3!} f'''(a) + \cdots \right] \tag{B.14}$$

Collecting terms in $f(a), f'(a)$, etc., together

$$f'(a) = (w_1 + w_2 + w_3) f(a) + (w_2 + 2w_3) hf'(a) + (w_2 + 4w_3) \frac{h^2}{2} f''(a)$$

$$+ (w_2 + 8w_3) \frac{h^3}{3!} f'''(a) + \cdots \tag{B.15}$$

Comparing with the left-hand side, the coefficient of the second term on the right-hand side should equal 1, whereas the others should be zero. Thus, we obtain three equations in three unknowns:

$$\begin{aligned} w_1 + w_2 + w_3 &= 0 \\ w_2 + 2w_3 &= 1/h \\ w_2 + 4w_3 &= 0 \end{aligned} \tag{B.16}$$

The above equations can be solved to obtain

$$w_1 = \frac{-3}{2h}, \quad w_2 = \frac{2}{h}, \quad w_3 = \frac{-1}{2h} \tag{B.17}$$

The coefficient of the leading error term is

$$w_2 + 8w_3 = -\frac{2}{h}$$

Thus, one can derive that the error term is given by

$$E = -\frac{h^2}{3} f'''(\xi) \qquad \text{(B.18)}$$

The three-point forward difference formula is thus given by

$$f'(a) = \frac{-3f(a) + 4f(a+h) - f(a+2h)}{2h} - \frac{h^2}{3} f'''(\xi) \qquad \text{(B.19)}$$

This is more accurate than the standard forward difference formula.

A three-point backward difference formula may be derived in a similar manner. One can verify that substituting $h \leftarrow (-h)$ in Equation B.19 will yield a three-point backward difference formula.

B.2.2 Second Derivative

Having derived the formula for $f'(a)$, let us now determine higher-order differentiation formulae. Unlike first derivatives, at least three nodes are required for computing the second derivative. The central difference formula for $f''(a)$ can be derived by inspection from Equations B.5 and B.8. Note that adding the two eliminates the term in $f'(a)$:

$$f(a+h) + f(a-h) = 2\left[f(a) + \frac{h^2}{2} f''(a) + \frac{h^4}{4!} f''''(a) + \cdots \right] \qquad \text{(B.20)}$$

The odd-powered terms of the Taylor's series vanish. The fourth-order term becomes the leading error term. Thus, the formula and error estimate for the central difference is given by

$$f''(a) = \underbrace{\frac{f(a+h) - 2f(a) + f(a-h)}{h^2}}_{\text{Formula}} - \underbrace{\frac{h^2}{12} f''''(\xi)}_{\text{Error}} \qquad \text{(B.21)}$$

Thus, the central difference formula for $f''(a)$ is $\mathcal{O}(h^2)$ accurate. This formula can also be derived using method of undetermined coefficients, discussed earlier, by writing

$$f''(a) = w_1 f(a-h) + w_2 f(a) + w_3 f(a+h)$$

An interested reader can verify that this method yields the same formula and error information as derived in Equation B.21.

In a similar manner, higher-order derivatives can also be derived. Since these formulae are not required in this book, I will not discuss them further. Still, the method of undetermined coefficients forms a useful template to derive these formulae, including cases with unequal grid size.

Example B.1 Numerical Differentiation

Problem: Find $f'(x)$ for the function $f(x) = \tan^{-1}(x)$ at $x = 1$. Also compute the error.
Solution: The following code is used for numerical differentiation:

```
% Problem Setup
a=1; h=0.01;
trueVal=1/(1+a^2);
%% Compute numerical derivatives and errors
% Forward difference for f'(x)
d1_fwd=(atan(a+h)-atan(a))/h;
err1_fwd=abs(d1_fwd-trueVal);
% Central difference for f'(x)
d1_ctr=(atan(a+h)-atan(a-h))/(2*h);
err1_ctr=abs(d1_ctr-trueVal);
% 3-point forward difference for f'(x)
d1_3pt=(-3*atan(a)+4*atan(a+h)-atan(a+2*h))/(2*h);
err1_3pt=abs(d1_3pt-trueVal);
% Display Results
disp([d1_fwd, d1_ctr, d1_3pt]);
disp([err1_fwd, err1_ctr, err1_3pt]);
```

With the step-size $h = 0.01$, the forward difference formula gave the numerical derivative value as 0.4975. For all other cases, the displayed value was 0.5000. Hence, it is more instructive to see the errors. The results below show the errors for forward, central, and 3-point forward difference formulae for various values of h:

h	err1_fwd	err1_ctr	err1_3pt
0.01	0.0025	8.333e-6	1.666e-5
0.001	2.499e-4	8.333e-8	1.667e-7
1e-4	2.500e-5	8.33e-10	1.666e-9
1e-5	2.500e-6	8.83e-12	2.276e-12
1e-6	2.501e-7	4.11e-11	1.522e-10

Clearly, the central and three-point forward difference formulae are more accurate than the standard forward difference formula.

Moreover, when step-size is reduced by one order of magnitude, the error in forward difference formula decreases by one order of magnitude, whereas the errors in central and three-point forward difference formulae decrease by two orders of magnitude. This is a consequence of the fact that the latter two are $\mathcal{O}(h^2)$ accurate, whereas the standard forward difference formula is $\mathcal{O}(h)$ accurate.

For the initial few step-sizes, the error in the central difference formula is approximately half of that in three-point forward difference formula. This approximate relationship is expected, comparing the error terms for the two formulae from Equations B.10 and B.19.

Finally, notice that the error in central and three-point forward difference formulae is greater at $h = 10^{-6}$ than it is at $h = 10^{-5}$. This will be the topic of discussion in the next section.

B.2.3 Effect of Round-Off Error

The trade-off between truncation and round-off errors is apparent from Example B.1. It is also demonstrated in Chapter 1. In the previous section, we observed that the truncation error decreased as the step-size was reduced. The round-off error, on the other hand, increases as the step-size is reduced. Calculation of the numerical derivatives requires subtraction between two (or more) numbers that are very close to each other. The smaller the step-size, h, for accuracy of the numerical scheme, the closer are the two numbers and, hence, the greater is the effect of round-off errors. There is an optimal value of h for which the total error is minimized. It is apparent from Example B.1 that this optimum lies somewhere in the vicinity of $h \sim 10^{-5}$.

Consider the forward difference formula from Equation B.7:

$$f'(a) = \frac{f(a+h) - f(a)}{h} - \frac{h}{2} f''(\xi) \tag{B.7}$$

The computer representation of a value differs from its true value due to machine precision. So, the true value of $f(a)$ differs from the value represented in the computer, $\overline{f(a)}$.

Consider the example of $\tan^{-1}(1)$. The true value is 0.785398… (additional digits are ignored for this example). Presuming that there is a computer using the decimal number system (as discussed in Chapter 1), which uses a four-digit representation of the mantissa, this number is represented in the "decimal computer" as $\overline{\tan^{-1}(-1)} = 0.7853$. The relative error between the two is bounded by the machine precision. Thus, one can get the following relationship between true value and the computer representation:

$$\overline{f(a)} = f(a)(1 + \varepsilon) \tag{B.22}$$

The numerical derivative is given by

$$\overline{f'(a)} = \frac{\overline{f(a+h)} - \overline{f(a)}}{h} \tag{B.23}$$

Subtracting the two, the net error is given by

$$E_{net} = \left| f'(a) - \overline{f'(a)} \right| \tag{B.24}$$

$$= \left| \frac{f(a+h) - f(a)}{h} \varepsilon \right| + \left| \frac{h}{2} f''(\xi) \right| \tag{B.25}$$

$$\leq \frac{2\varepsilon_{mach}}{h} \left| f(\xi_1) \right| + \frac{h}{2} \left| f''(\xi) \right| \tag{B.26}$$

where both $\xi, \xi_1 \in [a, a+h]$. The first term is the round-off error, and the second term is the truncation error. The optimum value of h that minimizes the net error is obtained by differentiating the above equation and equating it to zero:

$$\frac{d}{dh} E_{net} = -\frac{2\varepsilon_{mach}}{h^2} \left| f(\xi_1) \right| + \left| f''(\xi) \right| = 0$$

Thus

$$h_{opt} = 2\sqrt{\left| \frac{f(\xi_1)}{f''(\xi)} \right| \varepsilon_{mach}} \tag{B.27}$$

A practical result for forward difference would then be

$$h_{opt} \sim \left[\varepsilon_{mach} \right]^{1/2} \tag{B.28}$$

Arguing on similar lines, the truncation error in the central difference formula is

$$E_{trunc} = c_1 h^2$$

whereas the round-off error is

$$E_{r-off} = \frac{c_2}{h}$$

Since $E_{net} = |E_{trunc}| + |E_{r-off}|$, it is easy to show that the optimum value of step-size for the central difference formula is

$$h_{opt} \sim \left[\varepsilon_{mach} \right]^{1/3} \tag{B.29}$$

Since the machine precision in MATLAB is 2×10^{-16}, the optimum value of step-size is of the order of $h_{opt} \sim 0.6 \times 10^{-5}$. The observation in Example B.1 is thus in agreement with the optimal step-size value derived herein.

Appendix C: Gauss Elimination for Linear Equations

C.1 BASIC GAUSS ELIMINATION

C.1.1 A Simple Example

Gauss Elimination is a popular method for solving linear equations. The basic Gauss Elimination is discussed here using the following example:

$$
\begin{aligned}
x_1 + x_2 + x_3 &= 4 \\
2x_1 + x_2 + 3x_3 &= 7 \\
3x_1 + 4x_2 - 2x_3 &= 9
\end{aligned}
\tag{C.1}
$$

The aim in this example is to solve the system of linear equations given above. This can be expressed in the following standard form:

$$
\underbrace{\begin{bmatrix} 1 & 1 & 1 \\ 2 & 1 & 3 \\ 3 & 4 & -2 \end{bmatrix}}_{A}
\underbrace{\begin{bmatrix} x_1 \\ x_2 \\ x_3 \end{bmatrix}}_{x}
=
\underbrace{\begin{bmatrix} 4 \\ 7 \\ 9 \end{bmatrix}}_{b}
\tag{C.2}
$$

The above system of equations has a unique solution if the determinant of A is nonzero.

A more general way of expressing the same statement is that a unique solution exists if A is a full-rank matrix $\text{rank}(A) = n$. If the rank of A is not n, then the system of equations may have infinitely many solutions or no solution. If $\text{rank}(A) = m < n$, it implies that only m rows of A are linearly independent, whereas the other rows can be written as a linear combination of these m rows. If the vector b also satisfies the same linear combination, then there are infinitely many solutions to the equations. Consider the following equations:

$$
x_1 + x_2 = 1
$$

$$
2x_1 + 2x_2 = a
$$

Clearly, the left-hand side of the second equation is twice that of the first. Thus, rank$(A) = 1$ and the system of equations does not have a unique solution. If the right-hand side, a, is also twice that of the right-hand side of the first equation, then the two lines are collinear and there are infinitely many solutions. If $a \neq 2$, then there is no solution. Summarizing,

- The system of equations has a unique solution if rank$(A) = n$.

- Else, there are infinitely many solutions if rank$([A \quad b]) = \text{rank}(A)$.

- There are no solutions otherwise.

Consider how we solve the above set of three equations given in Equation C.1. We will use the first equation to eliminate x_1 from the second and third equations, that is, $R_2 - 2R_1$ and $R_3 - 3R_1$:

$$\begin{aligned} x_1 + x_2 + x_3 &= 4 \\ 0 - x_2 + x_3 &= -1 \\ 0 + x_2 - 5x_3 &= -3 \end{aligned} \tag{C.3}$$

Now, we can use the last two equations to eliminate x_2 and obtain the value of x_3:

$$-4x_3 = -4$$

Thus, we now have the following:

$$\begin{aligned} x_1 + x_2 + x_3 &= 4 \\ 0 - x_2 + x_3 &= -1 \\ 0 + 0 - 4x_3 &= -4 \end{aligned} \tag{C.4}$$

The above equation has reduced the overall system of equations to one where the matrix $\bar{A}$ has an upper triangular form. The process we followed is called *Gauss Elimination*. The final solution is obtained using a procedure called *back-substitution*. The final equation is used to obtain the value of $x_3 = 1$. Thereafter, this value can be substituted in the second equation. This will yield the value of $x_2 = 2$. Substituting these values in the first equation yields $x_1 = 1$.

Let us revisit the steps used in solving the above set of equations:

- In the first sequence of steps, the first equation was used to eliminate x_1 from all the remaining equations.

- Thereafter, the second equation was used to eliminate x_2 from the remaining equations. No change was made to the first equation.

- For a larger system of equations, the procedure continues until the right-hand side involves an upper triangular matrix. In each sequence of steps, the ith equation is used to eliminate x_i from all subsequent equations.

The following code demonstrates the use of Gauss Elimination with back-substitution for this problem. This is not a generalized code. It simply follows the steps laid out in the example above.

Example C.1 Naïve Gauss Elimination* Only

Problem: Write a code for naïve Gauss Elimination with back-substitution for the linear system of Equation C.1.

Solution: The first step is to formalize the steps given above. The first step is to obtain an augmented matrix, $Ab = [A \; b]$.

The first row is used to eliminate x_1 from rows 2 and 3. Here row-1 is called the pivot row. Thereafter, the second row is used to eliminate x_2 from the third row. With this, the Gauss Elimination procedure is complete. The code for the same is given below:

```
% Solve Ax = b using Naive Gauss Elimination
A = [1 1 1; 2 1 3; 3 4 -2];
b = [4;7;9];
n = length(A);
%% Gauss Elimination
% Get augmented matrix
Ab = [A, b];

% With A(1,1) as pivot Element
k=1;
for i=2:3
    alpha=Ab(i,1)/Ab(1,1);
    Ab(i,1:end)=Ab(i,1:end) - alpha*Ab(1,1:end);
end

% With A(2,2) as pivot Element
k=2;
for i=3:3
    alpha=Ab(i,2)/Ab(2,2);
    Ab(i,2:end)=Ab(i,2:end) - alpha*Ab(2,2:end);
end
```

Let us verify the matrix *Ab* obtained after Gauss Elimination:

```
Ab
    1    1    1    4
    0   -1    1   -1
    0    0   -4   -4
```

This is same as the result obtained in Equation C.4.

---

* The reason why we call this naïve Gauss Elimination will be come clear in Section C.3.

Now that the above results are in the reduced upper triangular form of Equation C.4, we can back-substitute the values to obtain **x**. First, x_3 can be calculated from the last row:

$$x_3 = \frac{Ab_{3,end}}{Ab_{3,3}} = 1 \tag{C.5}$$

Thereafter, it can be used to calculate x_2 as

$$x_2 = \frac{Ab_{2,end} - \left(Ab_{2,3}x_3\right)}{Ab_{2,2}} = 2 \tag{C.6}$$

followed by calculation of x_1 as

$$x_1 = \frac{Ab_{1,end} - \left(Ab_{1,2}x_2 + Ab_{1,3}x_3\right)}{Ab_{1,1}} \tag{C.7}$$

This is shown in the continuation of the code in the example below.

Example C.1 (Continued) (Back-Substitution)

The following code implements back-substitution, which is part of the same naïve Gauss Elimination code.

```
%% Back-Substitution
x=zeros(3,1);
for i=n:-1:1
    x(i)=( Ab(i,end)-Ab(i,i+1:n)*x(i+1:n) ) / Ab(i,i);
end
```

The result of running the above code with back-substitution is

```
x =
     1
     2
     1
```

which is the solution obtained analytically earlier.

The strategy I have followed in Example C.1 is the following: The Gauss Elimination part of the code is written for a specific 3×3 system example, whereas the back-substitution part is written as a general-purpose code for any $n \times n$ system. This is intentional. Converting the Gauss Elimination code to a general $n \times n$ system of equations is left as an exercise for the reader. The building blocks are already present in the above code, and the algorithm for a generic solver is discussed presently.

C.1.2 Algorithm for Naïve Gauss Elimination

A few definitions are in order before we formalize the steps in Gauss Elimination. In the kth sequence of steps, the kth equation is used to eliminate the coefficients of x_k in subsequent equations. In this kth step, the diagonal element $A_{k,k}$ is called the *pivot element*, and the kth row is the *pivot row*. The kth column is called the *pivot column*, with the procedure focusing on elements below the diagonal only. Thus, the procedure for naïve Gauss Elimination is as follows:

- Obtain an augmented matrix Ab = [A, b];.

- Set $k = 1$ so that $Ab_{k,k}$ is the pivot element.

- Use the pivot element to eliminate coefficients $Ab_{i,k}$ in the pivot column through row operations for $i = (k+1)$ to n:

  - Set $i = k + 1$

  - Set

$$\alpha_{ik} = Ab_{i,k}/Ab_{k,k} \tag{C.8}$$

  - Perform the row operation $R_i \leftarrow R_i - \alpha_{ik}R_k$ where "$\leftarrow$" represents the assignment operator.* Here, R_i represents the ith row, that is, Ab(i,:). Note that since the pivot row elements $Ab_{k,1}, Ab_{k,2}, \dots, Ab_{k,k-1}$ are already zero due to previous row operations, the above needs to be performed only for elements after the diagonal element. In other words

$$Ab_{i,j} = Ab_{i,j} - \alpha_{ik}Ab_{k,j}, \quad \text{for } j = k \text{ to end} \tag{C.9}$$

    In MATLAB, the above row operation can be done efficiently using the colon notation to indicate entire row:

    ```
    Ab(i,k:end)=Ab(i,k:end)-alpha*Ab(k,k:end);
    ```

    Compare this with the underlined statements in Example C.1 to understand how the row operation is coded to simplify the calculation of Equation C.9.

- Increment i and repeat the previous two steps until all the elements in the pivot column are eliminated, that is, until $i = n$.

- Increment k so that the next row is the pivot row and $A_{k,k}$ is the pivot element. Repeat the elimination steps until Ab is an upper triangular matrix, that is, until k reaches $(n-1)$.

---

* I use the assignment operator $\leftarrow$ in linear equations chapters to avoid ambiguity. This is to avoid readers interpreting $R_2 = R_2 - \alpha R_1$ as a linear equation. When I need to use the notation $R_2^{(i+1)} = R_2^{(i)} - \alpha R_1^{(i)}$, the assignment operator is not required because there is little scope of ambiguity.

After the Gauss Elimination steps, Ab is an upper triangular matrix. Let us say that this reduced matrix represents the following modified equation:

$$U\mathbf{x} = \beta \tag{C.10}$$

Note that the matrix modified after row eliminations is related to the above as $\overline{Ab} = [U \quad \beta]$. Row eliminations do not affect the desired solution, $\mathbf{x}$, as we have seen that in the reduction for the example in Equation C.2. For this example, the reader may verify that

$$\underbrace{\begin{bmatrix} 1 & 1 & 1 \\ 0 & -1 & 1 \\ 0 & 0 & -4 \end{bmatrix}}_{U} \underbrace{\begin{bmatrix} x_1 \\ x_2 \\ x_3 \end{bmatrix}}_{\mathbf{x}} = \underbrace{\begin{bmatrix} 4 \\ -1 \\ -4 \end{bmatrix}}_{\beta} \tag{C.11}$$

Note the equivalence between row operations described in the algorithm above and the familiar equation solving procedure through elimination that I described to reduce a set of equations (C.1 through C.4).

C.1.3 Back-Substitution

Due to the upper triangular form of the matrix in Equation C.11, the solution is obtained by starting from the last row. Specifically

$$x_3 = \frac{\beta_3}{U_{3,3}} = \frac{Ab_{3,\text{end}}}{Ab_{3,3}} = 1$$

Thereafter

$$x_2 = \frac{Ab_{2,\text{end}} - \left(Ab_{2,3}x_3\right)}{Ab_{2,2}} = 2 \tag{C.12}$$

and

$$x_1 = \frac{Ab_{1,\text{end}} - \left(Ab_{1,2}x_2 + Ab_{1,3}x_3\right)}{Ab_{1,1}} \tag{C.13}$$

Generalizing, the steps in back-substitution include

$$x_n = \frac{Ab_{n,\text{end}}}{Ab_{(n,n)}} \tag{C.14}$$

Starting with the end, with $i = (n-1)$ down until $i = 1$

$$x_i = \frac{Ab_{i,\text{end}} - \sum_{j=i+1}^{n} Ab_{i,j} x_j}{Ab_{(i,i)}} \tag{C.15}$$

C.2 LU DECOMPOSITION WITHOUT PIVOTING

The idea of LU decomposition is that any matrix A can be factorized as a product of two matrices, a lower triangular matrix L and an upper triangular matrix U:

$$A = LU \tag{C.16}$$

It should be noted that the decomposition is not unique. Several different factorizations into L and U matrices can be obtained. If we were to multiply Equation C.11 by L on both sides

$$\underbrace{LU}_{A} \mathbf{x} = \underbrace{L\beta}_{\mathbf{b}} \tag{C.17}$$

This is the algorithm behind the *Doolittle method* for LU decomposition. The upper triangular matrix U is obtained using Gauss Elimination. It turns out that the elements of the lower triangular matrix are the values of α_{ik} used during factorization, as per Equation C.8. The lower triangular matrix to complete the LU decomposition is then given by

$$L = \begin{bmatrix} 1 & 0 & 0 & \cdots & 0 \\ \alpha_{21} & 1 & 0 & \cdots & 0 \\ \alpha_{31} & \alpha_{32} & 1 & \cdots & 0 \\ \vdots & \vdots & \ddots & \ddots & \vdots \\ \alpha_{n1} & \alpha_{n2} & \alpha_{n3} & \cdots & 1 \end{bmatrix} \tag{C.18}$$

The following demonstrates LU decomposition for the previous example.

Example C.2 LU Decomposition

Problem: Perform LU decomposition for the problem in Example C.1.

Solution: The main idea of LU decomposition using the Doolittle algorithm is to store the values of α_{ki} and store them in the correct location of the L matrix. This is shown in the following code:

```
% Solve Ax = b using Naive Gauss Elimination
% Modified for LU decomposition
A = [1 1 1; 2 1 3; 3 4 -2];
b = [4;7;9];
n = length(A);
```

```
%% Gauss Elimination
% Get augmented matrix
Ab=[A, b];
L=eye(n);

% With A(1,1) as pivot Element
k=1;
for i=2:3
    alpha=Ab(i,1)/Ab(1,1);
    Ab(i,1:end)=Ab(i,1:end) - alpha*Ab(1,1:end);
    L(i,k)=alpha;
end

% With A(2,2) as pivot Element
k=2;
for i=3:n
    alpha=Ab(i,2)/Ab(2,2);
    Ab(i,2:end)=Ab(i,2:end) - alpha*Ab(2,2:end);
    L(i,k)=alpha;
end

%% Back-Substitution
x = zeros(3,1);
for i=n:-1:1
    x(i)=( Ab(i,end)-Ab(i,i+1:n)*x(i+1:n) ) / Ab(i,i);
end
```

The core of the code has remained the same, with the addition of the underlined statements: First, the matrix L is initialized as identity, followed by inserting values of α_{ik} in the first column and then inserting α_{ik} in the second column. This completes the L matrix:

```
L =
    1      0      0
    2      1      0
    3     -1      1
```

It is easy to verify that the product L*U indeed gives back the matrix A:

```
>> U=Ab(1:n,1:n);
>> Acheck=L*U;
```

Indeed, the matrix Acheck is equal to the original matrix A.

It should be noted that LU decomposition really needs to work on only the matrix A and not the augmented matrix Ab. I retained the original code, which used the Ab matrix for pedagogical reasons. LU decomposition is actually performed on the matrix A itself.

LU decomposition can be used to solve a linear equation $Ax = b$. It consists of three steps. First, LU decomposition is used to obtain matrices L and U. Since $L\beta = b$, as per Equation C.17, the second step is to obtain β from L and b. This is done using *forward substitution*, which works exactly like backward substitution method but starting with the first row. Note that the vector β resulting after forward substitution step is nothing but the last column of the Ab matrix. Finally, the third step is backward substitution to solve $Ux = \beta$, which was discussed earlier.

The utility of LU decomposition is not really in solving a single equation, $Ax = b$. Instead, it is very useful if we need to solve a series of equations $Ax_{[i]} = b_{[i]}$, where the matrix A remains constant while vector b changes. The greatest effort in solving linear equation is in factorization/elimination. Hence, factorizing a matrix once and using forward/backward substitutions multiple times reduces the effort. An example of a system where $Ax_{[i]} = b_{[i]}$ is solved multiple times is covered in Chapter 7, where the right-hand side is a nonlinear term; $b_{[i+1]} = g(x^{(i)})$ is nothing but the function evaluated at the current guess, followed by solving the nonlinear equation to obtain the new guess, $x^{(i+1)} \equiv x_{[i+1]}$.

Another use from a numerical methods perspective is that sparse matrices, of the type discussed in Chapter 7, can sometimes be factorized into L and U using efficient algorithms. Thus, this method becomes a computationally useful means to solve such equations.

The MATLAB code for obtaining LU decomposition is lu(A). The result of this code is different from the algorithm discussed here, since it uses partial pivoting (see next section). Thus, the final result is of the form $LU = PA$, where P is a permutation matrix that accounts for row exchanges done in the partial pivoting method.

C.3 PARTIAL PIVOTING

I have referred to partial pivoting method a couple of times. Let us look into what this means. As a simple, motivating example, consider that the linear system of equations was instead given as

$$
\begin{aligned}
x_1 + x_2 + x_3 &= 4 \\
2x_1 + 2x_2 + 3x_3 &= 9 \\
3x_1 + 4x_2 - 2x_3 &= 9
\end{aligned}
\tag{C.19}
$$

The first set of row operations using first row as the pivot row yields

$$
Ab = \begin{bmatrix} 1 & 1 & 1 & 4 \\ 0 & 0 & 1 & 1 \\ 0 & 1 & -5 & -3 \end{bmatrix}
\tag{C.20}
$$

The naïve Gauss Elimination, as discussed earlier, cannot continue from this stage because the next pivot element $A_{2,2} = 0$. However, if we were to solve the equations manually, we would just switch the last two equations and use the second equation to obtain the value of x_3. The matrix after exchange of the last two rows becomes

$$Ab = \begin{bmatrix} 1 & 1 & 1 & 4 \\ 0 & 1 & -5 & -3 \\ 0 & 0 & 1 & 1 \end{bmatrix} \tag{C.21}$$

This is the crux of Gauss Elimination with *partial pivoting*. Partial pivoting is basically a row exchange operation carried out at each step. Unlike the example mentioned here, the aim of partial pivoting is broader. A premature end to Gauss Elimination is not the only problem. Another issue is round-off errors. In the case of examples where two large numbers are subtracted, pivoting improves the performance of Gauss Elimination. For example, if the second equation was

$$2x_1 + (2+\delta)x_2 + 3x_3 = 9.\times\times\times$$

the diagonal element, $A_{2,2}$, would become $\delta \ll 1$. This can lead to round-off errors because of the subtraction of a close number and/or division by a small number. Recall that in back-substitution, we compute $x_2 = (\text{numerator})/A_{2,2}$.

Theoretical guarantees exist for Gauss Elimination with full pivoting, where both row and column interchanges are allowed. However, partial pivoting, where only row interchanges are executed, is found to be equally useful in practice. Hence, all the numerical solvers use Gauss Elimination with partial pivoting.

In the algorithm given in Section C.1.2, at any kth step, $A_{k,k}$, is the pivot element. In naïve Gauss Elimination, the pivot element is used to eliminate coefficients $A_{i,k}$ in the pivot column below the diagonal. In case of Gauss Elimination with pivoting, an additional step is added before this:

- When k is incremented, $A_{k,k}$ becomes the pivot element.

- Compare the values of all the subdiagonal elements, $A_{i,k}$, in the same column ($i \geq k$). Exchange row k with the row that has highest absolute value of $A_{i,k}$ in that column. This is the pivoting step.

- Thereafter, continue with elimination as before, using the (new) pivot element for the elimination of elements in the pivot column below the pivot element.

Consider the original example from Equation C.1. In the first step, with $k = 1$, we perform the pivoting step. The largest absolute value in the pivot column (column 1) is $A_{3,1}$.

Therefore, rows 3 and 1 are interchanged so that the new pivot element is the dominant element in that column. The matrix *Ab* after the row interchange is

```
Ab  =
       3       4      -2       9
       2       1       3       7
       1       1       4
```

The elimination is then performed with $\alpha_{21} = 2/3$ and $\alpha_{3,1} = 1/3$. After appropriate row operations, the matrix becomes

```
Ab  =
    3.0000     4.0000    -2.0000     9.0000
         0    -1.6667     4.3333     1.0000
         0    -0.3333     1.6667     1.0000
```

Next, $A_{2,2}$ is the pivot element. The elements in column 2 at and below the diagonal as −1.6667 and −0.3333. Since the former has the largest absolute value, no row interchange needs to be performed. The algorithm continues with $\alpha_{3,2} = 0.2$ and yields

```
Ab  =
    3.0000     4.0000    -2.0000     9.0000
         0    -1.6667     4.3333     1.0000
         0          0     0.8000     0.8000
```

The matrix is now in upper triangular form. We also note that the α values used were

$$\alpha = \begin{bmatrix} * & * & * \\ 0.6667 & * & * \\ 0.3333 & 0.2 & * \end{bmatrix}$$

The following code is a modification of Example C.1 with partial pivoting included.

Example C.3 Gauss Elimination with Partial Pivoting

Problem: Solve Example C.1 using Gauss Elimination with partial pivoting.

  Solution: The code in Example C.1 is modified to include verification of dominance of the diagonal pivot element and column exchange. The other parts of the code remain unchanged:

```
% Solve Ax = b using Naive Gauss Elimination
A = [1 1 1; 2 1 3; 3 4 -2];
b = [4;7;9];
n = length(A);
```

```
%% Gauss Elimination
% Get augmented matrix
Ab = [A, b];

% With A(1,1) as pivot Element
k=1;
pivotCol=Ab(k:n,k);
[~,idx]=max(abs(pivotCol));
idx=(k-1)+idx;
if (idx~=k)       % Row interchange
    pivotRow=Ab(k,:);
    Ab(k,:)=Ab(idx,:);
    Ab(idx,:)=pivotRow;
end
for i=2:3
    alpha=Ab(i,1)/Ab(1,1);
    Ab(i,1:end)=Ab(i,1:end) - alpha*Ab(1,1:end);
end

% With A(2,2) as pivot Element
k=2;
pivotCol=Ab(k:n,k);
[~,idx]=max(abs(pivotCol));
idx=(k-1)+idx;
if (idx~=k)       % Row interchange
    pivotRow=Ab(k,:);
    Ab(k,:)=Ab(idx,:);
    Ab(idx,:)=pivotRow;
end
for i=3:3
    alpha=Ab(i,2)/Ab(2,2);
    Ab(i,2:end)=Ab(i,2:end) - alpha*Ab(2,2:end);
end

%% Back-Substitution
x = zeros(3,1);
for i = 3:-1:1
    x(i) = ( Ab(i,end)-Ab(i,i+1:n)*x(i+1:n) ) / Ab(i,i);
end
```

The underlined block has been added. The first line in that block extracts the pivot column from the diagonal to the last row into a vector $pivotCol$. The next line uses the max function to find the index of the largest value in $pivotCol$. Since $pivotCol$ contained elements from row k to n, the original index in Ab matrix is $(k-1)+idx$. The next three lines perform the row interchange: the kth row is stored in a dummy variable $pivotRow$, the idxth row overwrites the kth row, and the stored $pivotRow$ overwrites the idxth row.

The readers can verify that the results from this code match the one obtained from hand calculations and described immediately preceding this example.

Recall that I had mentioned that the MATLAB function `lu` gives LU decomposition with partial pivoting, in the form $PA = LU$. Permutation matrix P takes care of the row interchanges due to pivoting. In this example, only rows 3 and 1 were exchanged. Therefore

$$P = \begin{bmatrix} 0 & 0 & 1 \\ 0 & 1 & 0 \\ 1 & 0 & 0 \end{bmatrix}$$

As described earlier in this appendix, the values of L and U would be

$$U = \begin{bmatrix} 3 & 4 & -2 \\ 0 & -1.6667 & 4.3333 \\ 0 & 0 & 0.8 \end{bmatrix}, \quad \text{and} \quad L = \begin{bmatrix} 1 & 0 & 0 \\ 0.6667 & 1 & 0 \\ 0.3333 & 0.2 & 1 \end{bmatrix}$$

One can now verify the `lu` command in MATLAB:

```
>> [L,U,P]=lu(A)
L =
    1.0000          0          0
    0.6667     1.0000          0
    0.3333     0.2000     1.0000
U =
    3.0000     4.0000    -2.0000
         0    -1.6667     4.3333
         0          0     0.8000
P =
    0     0     1
    0     1     0
    1     0     0
```

Most commercial linear algebra solvers use the Gauss Elimination method with partial pivoting. It is not recommended to use naïve Gauss Elimination. In any case, linear algebra is a very strong suite of MATLAB. Unless you have reasons otherwise, I recommend only using the inbuilt MATLAB methods for linear algebra since they are well optimized.

C.4 MATRIX INVERSION

Matrix inversion is done by writing the following linear equations:

$$Ac_1 = e_1, \quad Ac_2 = e_2, \dots, Ac_n = e_n$$

where $\mathbf{e}_i$ is the ith unit vector (i.e., column vector with all zeros, except ith element is 1). The inverse of the matrix A contains vector $\mathbf{c}_i$ as its ith column. Thus, $\texttt{iA=[c1, c2, ...,}$ $\texttt{cn]}$. The inverse is obtained by writing

$$A^{Aug} = \begin{bmatrix} A & | & I_{n \times n} \end{bmatrix}$$

Two methods are popular among students and practitioners to solve a linear problem $A\mathbf{x}=\mathbf{b}$ in MATLAB:

```
>> x=inv(A)*b;
```

or

```
>> x=A\b;
```

If solving a linear equation is the aim, one should always prefer the second method.* Calculating inverse, as described above, takes a lot more computational effort than solving a linear equation using the slash operator.

---

* Note that slash operator also solves a linear least squares problem (see Chapter 10) if A is a non-square "tall" matrix with $n_{\text{rows}} > n_{\text{cols}}$. So be careful!

Appendix D: Interpolation

D.1 GENERAL SETUP

Consider a car moving on a city road. The speed of the car recorded at every 10 s interval is given in Table D.1.

Interpolation is used if one is interested in knowing the speed of the car at any time between 0 and 90 s. Interpolation involves "joining the dots" with a smooth curve and reading off the data on this fitted curve at any time of interest.

The above example of interpolating vehicle speed might be closer to chemical engineering than you think. It is indeed used in simulating exhaust emissions performance of a vehicle under test conditions. For example, the U.S. FTP-75 cycle is U.S. Federal Test Protocol for emission testing of light-duty vehicles.[*] It involves testing a vehicle under simulated urban driving conditions. Numerical simulation of a process under U.S. FTP-75 cycle would involve simulating the behavior at any arbitrary computational point within the entire cycle. Since discrete measurements are limited to the measurement frequency, interpolation is used to obtain data as a smooth continuous function of time. Figure D.1 shows the U.S. FTP-75 test cycle. As can be seen from the figure, the vehicle speed changes frequently in the 1833 s test cycle. The data may be sampled at a rate of 10 to 0.1 Hz. If simulation requires data between the sampling instances, interpolation is used to fill in that data.

Another example would be in the modeling of solar or wind power sources. Solar radiation or wind speeds may be available at discrete time instances, whereas numerical simulation would require the information as a continuous function of time. Similar to the example above, the intermediate data may be filled in using *interpolation*.

Let the tabulated data above be represented as n pairs: $(t_1, y_1), (t_2, y_2), \ldots, (t_n, y_n)$. Here, t is the independent variable, y is dependent variable, and t_i and y_i represent the data points. Furthermore, we will try to fit some functional form, $y = \bar{p}(t)$ to the above data, such that the fitted function passes through all the data points. In other words, each of the n data points exactly satisfies the equation $y_i = \bar{p}(t_i)$.

---

[*] Retrieved from: https://www.dieselnet.com/standards/cycles/ftp75.php (June 2016).

TABLE D.1   Data for Speed of a Moving Vehicle in Moderate Traffic Urban Conditions

Time (s)	0	10	20	30	40	50	60	70	80	90
Speed (km/h)	45	32	0	0	7	12	20	15	29	55

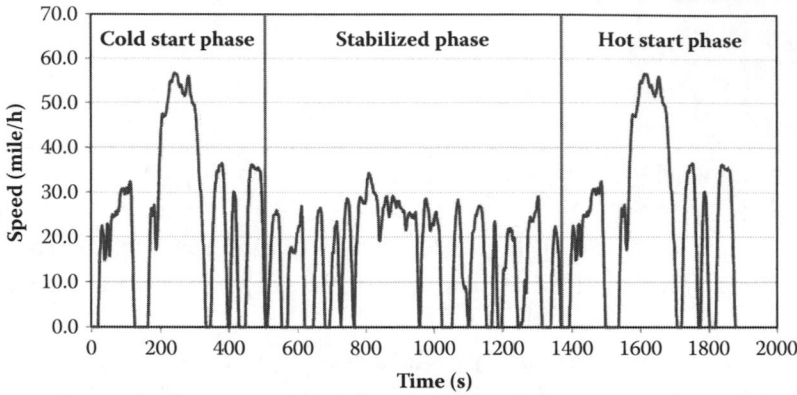

FIGURE D.1   Vehicle speed vs. time in a U.S. FTP-75 test cycle.

D.2 INTERPOLATING POLYNOMIALS

One of the first attempts at interpolation was the fit a polynomial function for $\bar{p}(t)$. When there are n data points, one can fit an $(n-1)$th order polynomial exactly to these data points. Indeed, we can choose to write this polynomial as

$$\bar{p}(t) = a_0 + a_1 t + a_2 t^2 + \cdots + a_{n-1} t^{n-1}$$

However, more efficient forms of polynomial function have been introduced so that the coefficients a_i can be obtained in a more straightforward manner. Newton's interpolating polynomials take the form

$$\bar{p}_N(t) = a_0 + a_1(t-t_1) + a_2(t-t_1)(t-t_2) + \cdots + a_{n-1}(t-t_1)\cdots(t-t_{n-1}) \qquad \text{(D.1)}$$

Thus, at the first point, only the first (constant) term on the right-hand side is retained; at the second point, the first two terms are retained; and so on. The coefficients a_0 to a_{n-1} can be obtained as Newton's divided difference polynomials. While this formula is more generic, it will not be discussed in this section. Instead, conceptually simpler Lagrange interpolating polynomials will be discussed, followed by Newton's forward difference formula. The latter is applicable when the independent variable, t, is equispaced. In spite of this strong requirement, this formula is useful for the derivation of integration and differential equation schemes discussed elsewhere.

D.2.1 Lagrange Interpolating Polynomials

The function in Lagrange interpolation is written as a sum of n polynomials, each of which are of order $(n-1)$:

$$\bar{p}_L(t) = a_1 \frac{(t-t_2)(t-t_3)\cdots}{(t_1-t_2)(t_1-t_3)\cdots} + a_2 \frac{(t-t_1)(t-t_3)(t-t_4)\cdots}{(t_2-t_1)(t_2-t_3)(t_2-t_4)\cdots} + \cdots \quad \text{(D.2)}$$

$$\bar{p}_L(t) = \sum_{i=1}^{n} a_i L_{i,n}(t) \quad \text{(D.3)}$$

Consider the denominator of $L_{i,n}$: It is a product of all terms, $(t_i-t_1), (t_i-t_2)$, and so on until (t_i-t_n), except the term (t_i-t_i). The numerator is the same product, with the variable t replacing t_i. Thus

$$L_{i,n}(t) = \prod_{\substack{j=1 \\ j\neq i}}^{n} \frac{t-t_j}{t_i-t_j} \quad \text{(D.4)}$$

Since the polynomial $\bar{p}_L(t)$ passes through the first point, (t_1, y_1) satisfies Equation D.2. When we substitute $t = t_1$, the numerator of the first term becomes exactly equal to the denominator. Thus, the first term becomes a_1. In all the other terms, the numerator is a product of (t_1-t_1) with other terms, whereas the denominator is nonzero. Thus the other terms, except the first term, vanish. Thus, $y_1 = a_1$. One can easily see the same pattern repeating for other polynomials as well:

$$L_{i,n}(t_i) = \prod_{\substack{j=1 \\ j\neq i}}^{n} \frac{t-t_j}{t_i-t_j} = 1$$

whereas

$$L_{i,n}(t_m) = \frac{(t_m-t_1)\cdots(t_m-t_m)\cdots(t_m-t_n)}{\text{Denominator}} = 0, \quad i \neq m$$

Consequently, the Lagrange interpolation formula is given by

$$\bar{p}_L(t) = \sum_{i=1}^{n} y_i L_{i,n}(t) \quad \text{(D.5)}$$

$$\text{where } L_{i,n}(t) = \prod_{\substack{j=1 \\ j \neq i}}^{n} \frac{t - t_j}{t_i - t_j} \tag{D.6}$$

The application is demonstrated in the following example. This example (for divided difference formula) is also shown in video lectures on NPTEL.

Example D.1 Lagrange Interpolating Polynomial

Problem: Use Lagrange interpolating polynomials for the data given in Table D.1 and obtain the vehicle speed at 45 s.

Solution: There are 10 data points. The first step is to calculate and store the values of polynomials $L_1(45)$ to $L_{10}(45)$ as per Equation D.6. This is easy to calculate using MATLAB's `prod` command within a for loop. The result is stored in a 10D vector `lagPoly`. The interpolated value can then be computed using `dot(yData,lagPoly)`.

The value of $\bar{p}_L(45)$ obtained using Lagrange interpolation in MATLAB is 9.0107.

D.2.2 Newton's Forward Difference Formula

Newton's forward difference formula is applicable to equally spaced data, that is, the independent variables, $t_1, t_2, \ldots$ are equally spaced. Let $h = t_{i+1} - t_i$ be the interval length. Furthermore, let the normalized distance on the t-axis be defined as

$$\alpha = \frac{t - t_1}{h} \tag{D.7}$$

With this definition, at $t = t_1, \alpha = 0$; at $t = t_2, \alpha = 1$; and so on at $t = t_n, \alpha = (n - 1)$. Considering an individual term in the polynomial (D.1)

$$t - t_i = (t - t_1) - (t_i - t_1) = h(\alpha - (i - 1))$$

Similar to Equation D.1, the interpolating polynomial is now written as follows:

$$\bar{p}_F(\alpha) = a_0 + a_1 h\alpha + a_2 h^2 \alpha(\alpha - 1) + \cdots + a_{n-1} h^{n-1}(\alpha(\alpha - 1)\cdots(\alpha - n + 2)) \tag{D.8}$$

The above polynomial passes through the point (t_1, y_1), which corresponds to $\alpha = 0$. At this value of α, all the terms on the right-hand side, except the constant term, drop out. Thus

$$y_1 = a_0 \tag{D.9}$$

The polynomial also satisfies the condition $y = y_2$ at $\alpha = 1$. At this condition, the first two terms on the right-hand side are retained, whereas the other terms drop out:

$$y_2 = a_0 + a_1 h \tag{D.10}$$

Subtracting Equation D.9 from this equation yields

$$\Delta y_1 = a_1 h \tag{D.11}$$

where Δ is a forward difference operator such that $\Delta \phi_i = \phi_{i+1} - \phi_i$.

Continuing the process for other data points,

$$y_3 = a_0 + a_1 h(2) + a_2 h^2 (2)(1) \tag{D.12}$$

$$y_4 = a_0 + a_1 h(3) + a_2 h^2 (3)(2) + a_3 3! \tag{D.13}$$

$$\vdots$$

$$y_i = a_0 + a_1 h(i-1) + a_2 h^2 (i-1)(i-2) + \cdots + a_{i-1} h^{i-1} (i-1)! \tag{D.14}$$

$$y_{i+1} = a_0 + a_1 h(i) + a_2 h^2 (i)(i-1) + \cdots + a_{i-1} h^{i-1} \frac{i!}{1!} + a_i h^i i! \tag{D.15}$$

As before, the first-order forward differences can be computed by subtracting the consecutive equations. From Equations D.12 and D.10, we obtain

$$\Delta y_2 = a_1 h + 2 a_2 h^2 \tag{D.16}$$

Repeating this procedure yields us the remaining first-order forward differences:

$$\Delta y_3 = a_1 h + a_2 h^2 (2)(2) + 3! a_3 h^3 \tag{D.17}$$

$$\vdots$$

$$\Delta y_i = a_1 h + a_2 h^2 2(i-1) + a_3 h^3 3(i-1)(i-2) + \cdots + i! a_i h^i \tag{D.18}$$

So far, the values of the first two parameters are known from the original data (see Equation D.9) and first-order forward difference Δy_1 (see Equation D.11). The next Newton's forward difference parameters can be computed from the higher-order forward difference formula. Specifically,

$$\Delta^2 y_1 = (\Delta y_2 - \Delta y_1)$$
$$= 2 a_2 h^2 \tag{D.19}$$

Continuing further,

$$\Delta^2 y_2 = 2 a_2 h^2 + 3! a_3 h^3 \tag{D.20}$$

and so on. Without going into further details, one can observe that a clear pattern emerges:

$$y_1 = a_0 \tag{D.9}$$

$$\Delta y_1 = a_1 h \tag{D.11}$$

$$\Delta^2 y_1 = 2a_2 h^2 \tag{D.19}$$

$$\Delta^3 y_1 = 3! a_3 h^3 \tag{D.21}$$
$$\vdots$$

$$\Delta^i y_i = i! a_i h^i \tag{D.22}$$

Newton's forward difference formula of Equation D.8 may now be written as

$$\bar{p}_F(\alpha) = y_1 + \Delta y_1 \alpha + \frac{\Delta^2 y_1}{2!} \alpha(\alpha-1) + \cdots + \frac{\Delta^{n-1} y_1}{(n-1)!} \prod_{k=0}^{n-2}(\alpha-k) \tag{D.23}$$

Now consider that there exists some function $f(t)$ from which the data points for the interpolation were generated. The difference between a value generated from the "true" function $f(t)$ and the interpolated polynomial $\bar{p}_F(\alpha)$ is the error in Newton's forward difference formula. Consider that a new point, (t_{n+1}, y_{n+1}) also becomes available. The error in ignoring this additional data while computing $\bar{p}_F(\alpha)$ is given by the additional term

$$R_F = \frac{\Delta^n y_1}{n!} \alpha(\alpha-1)\cdots(\alpha-n+1) \tag{D.24}$$

Recall that from mean value theorem

$$\frac{d^n}{dt^n} f(\xi) = \frac{\Delta^n y_1}{h^n}, \quad t_1 \le \xi \le t_n \tag{D.25}$$

Thus, the residual term in computing Newton's forward difference formula from Equation D.23 is

$$R_F = \frac{h^n}{n!} f^{(n)}(\xi)\big(\alpha(\alpha-1)\cdots(\alpha-n+1)\big) \tag{D.26}$$

Consider the interpolation problem again as below.

Example D.2 Newton's Forward Difference Interpolation

Problem: Solve Example D.1 using Newton's forward difference interpolation formula

Solution: As before, the data is in the form of two vectors, `tData` and `yData`, of size 10. The value of α is computed from the definition (D.7). The forward differences can be computed using the MATLAB command `diff`. The first use of `diff(yData)` yields $\Delta y_1 \cdots \Delta y_9$, whereas each subsequent use gives the higher-order forward differences.

Once the forward differences have been calculated, the interpolated value at $\alpha = 4.5$ can be calculated using Equation D.23.

The value of $\overline{p}_F$ obtained using Newton's interpolation in MATLAB is `9.0107`. Note that this value is the same as the one obtained using Lagrange interpolating polynomials.

D.3 SPLINE AND INTERPOLATION FUNCTIONS IN MATLAB®

MATLAB provides various options for interpolation, which can be invoked using the function `interp1`. The syntax for this command can be obtained by typing

```
>> help interp1
```

The syntax for `interp1` is as follows:

```
yq = interp1(tData,yData,tq,'<method>');
```

where, `tData` and `yData` form the original data, and the interpolated value `yq` is returned for the query point, `tq`. From the help text for `interp1`, the following information is displayed about the available methods for interpolation:

```
'linear'    - (default) linear interpolation
'nearest'   - nearest neighbor interpolation
'next'      - next neighbor interpolation
'previous'  - previous neighbor interpolation
'spline'    - piecewise cubic spline interpolation (SPLINE)
'pchip'     - shape-preserving piecewise cubic interpolation
```

The first option fits a piecewise linear curve to the datapoints. For the data in Table D.1, the value at $t_q = 45$ is the average between the values at 40 and 50, as shown below:

```
>> yq = interp1(tData,yData,45,'linear')
yq =
    9.5000
```

The method `'previous'` provides the so-called *zero-order hold* (also known as piecewise constant) approximation, wherein the value of y_q is kept constant at the value y_i for the entire interval $t_i \leq t < t_{i+1}$. Thus, the value of y_q at $t_q = 45$ is the same as the value at $t = 40$, that is, `yq=7`. The most popular options, though, are cubic spline and piecewise cubic Hermite interpolating polynomial (PCHIP). The preferred MATLAB functions for cubic spline and PCHIP algorithms are `spline` and `pchip`, respectively. The following example illustrates their use.

Example D.3 Spline and PCHIP Interpolation

Problem: Solve Example D.1 using Newton's forward difference interpolation formula.

Solution: First, define vectors tData and yData to contain the data in Table D.1. Thereafter, the following commands may be used:

```
>> y_sp = spline(tData,yData,45)
y_sp =
      9.0185
>> y_pchip = pchip(tData,yData,45)
y_pchip =
      9.4599
```

It is interesting to note that for this data, cubic spline happens to give interpolated value close to that obtained using Lagrange and Newton's interpolating polynomials, whereas PCHIP provides a significantly different value.

Spline interpolation fits piecewise polynomial functions to successive points. The simplest one is "linear spline." Here, two successive points are connected by straight line. The interpolating function is just a combination of these lines, given by the following piecewise linear function:

$$
\bar{f}(t) = \begin{cases} y_1 + (t - t_1)\dfrac{y_2 - y_1}{t_2 - t_1}, & t_1 \leq t < t_2 \\ y_2 + (t - t_2)\dfrac{y_3 - y_2}{t_3 - t_2}, & t_2 \leq t < t_3 \\ \quad \vdots \end{cases}
\tag{D.27}
$$

The first function above is clearly a straight line connecting (t_1, y_1) and (t_2, y_2). Any point between t_1 and t_2 is then interpolated based on this equation. The same applies to the rest of the data intervals.

The term spline is typically not used for the function above. Instead, this is referred to as linear interpolation, or more appropriately, *piecewise linear interpolation*. The linear functions result in a curve that is nonsmooth. The term "spline" evokes an idea similar to using the Bezier curves in an art class to smoothly connect dots on a canvas. The connecting curves are smooth. While different splines exist, the practically most relevant is cubic spline interpolation. When the term "spline" is used without any qualifier, it usually means cubic spline, where piecewise cubic polynomials are used for interpolation. Thus, a (cubic) spline would be represented as

$$
\bar{f}(t) = \begin{cases} q_1(t), & t_1 \leq t \leq t_2 \\ q_2(t), & t_2 \leq t \leq t_3 \\ \quad \vdots \\ q_{n-1}(t), & t_{n-1} \leq t \leq t_n \end{cases}
\tag{D.28}
$$

$q_i(t)$ is a cubic function. The choice of $q_i(t)$ can be any cubic function; however, as this function connects points (t_i, y_i) and (t_{i+1}, y_{i+1}), they satisfy the equation. Borrowing idea from Lagrange interpolation, let us define

$$\ell_i = \frac{t - t_i}{t_{i+1} - t_i} \tag{D.29}$$

so that at x_i, $\ell_i = 0$ and at x_{i+1}, $(1 - \ell_i) = 0$. The cubic polynomial may then be written as

$$q_i = y_i\left(1 - \ell_i\right) + y_{i+1}\ell_i + \ell_i\left(1 - \ell_i\right)\left[a_i\ell_i + b_i\left(1 - \ell_i\right)\right] \tag{D.30}$$

The above function has an advantage that two of the four coefficients of a cubic polynomial are conveniently defined. Note that when $\ell_i = 0$, only the first term remains and the other three are zero; whereas when $\ell_i = 1$, only the second term is nonzero.

A total of $(n-1)$ polynomials of the form (D.30) define the cubic spline. Hence, $2(n-1)$ parameters have to be obtained. There are two boundary nodes, t_1 and t_n and $(n-2)$ internal nodes. In order to ensure that the curve is smooth, the slope and curvature of the two curves $q_{i-1}(t)$ and $q_i(t)$ should be equal at each internal node t_i. This leads to the following $2(n-2)$ equations:

$$q_i'\left(t_i\right) = q_{i+1}'\left(t_i\right), \quad i = 1 \text{ to } (n-1) \tag{D.31}$$

$$q_i''\left(t_i\right) = q_{i+1}''\left(t_i\right), \quad i = 1 \text{ to } (n-1) \tag{D.32}$$

whereas the final two equations are obtained by imposing no curvature condition at the two boundary nodes:

$$q_1''\left(t_1\right) = 0 \tag{D.33}$$

$$q_{n-1}''\left(t_n\right) = 0 \tag{D.34}$$

The derivation of the above equations is straightforward, though tedious. I refer an interested reader to any numerical methods text for the derivation. The intent of this discussion was to give a flavor of cubic spline interpolation.

The cubic spline option is available in the `interp1` function discussed earlier in this section. Rather than the generic `interp1` function, the preferred means of using cubic spline is the function `spline`, as demonstrated in Example D.3. It should be noted that all these functions allow the independent query variable `tq` to be a vector. In such a case, the code returns a vector `yq` of the same length such that each element of `yq` is the interpolated value at the corresponding element of the vector `tq`. Thus, if interpolated values are required at multiple points, the function `spline` or `pchip` may be used with a vector-valued `tq`.

The primary advantage of `spline` function is in the cases where interpolants are required to be computed multiple times in any MATLAB application. Typically, if the values of independent variables are known *a priori*, a vector-valued `tq` can be used. However, when the query points for interpolation may be generated within the code, the function `spline` (or `pchip`) needs to be used multiple times. Since calling these functions involves computing the parameters of Equation D.30 each time for the same data sets `tData` and `yData`, this would be computationally inefficient. A better way to do this is to use the `spline` function to define the piecewise cubic polynomials and store them:

```
>> splineParam = spline(tData,yData)
splineParam =
      form: 'pp'
    breaks: [0 10 20 30 40 50 60 70 80 90]
     coefs: [9x4 double]
    pieces: 9
     order: 4
       dim: 1
```

Invoking `spline` without the query vector `tq` returns a pp structure, which defines the node points (original values of *t* in Table D.1) as well as coefficients of the nine cubic polynomials; it also indicates that this pp structure is cubic spline interpolation (`order: 4`) for 1D interpolation (`dim: 1`). Once this structure is obtained, it may be used multiple times through the `ppval` command:

```
>> yq = ppval(splineParam,45)
yq =
    9.0185
```

Note that this result is the same as that in Example D.3. The next example demonstrates the use of spline interpolation through multiple applications.

Example D.4 Cubic Spline Interpolation

Problem: Obtain the interpolated values of vehicle speed for the data in Table D.1 using cubic spline at every randomly generated interval between 0 and 1 s. Use the command `rand` to generate the next interval point.

Solution: The code below demonstrates the use of `spline` and `ppval` functions:

```
% Given data
tData = [0:10:90]';
yData = [45 32 0 0 7 12 20 15 29 55]';
plot(tData,yData,'bo'); hold on;   % Plot original data

% Obtain interpolating polynomials
splineParam = spline(tData,yData);
```

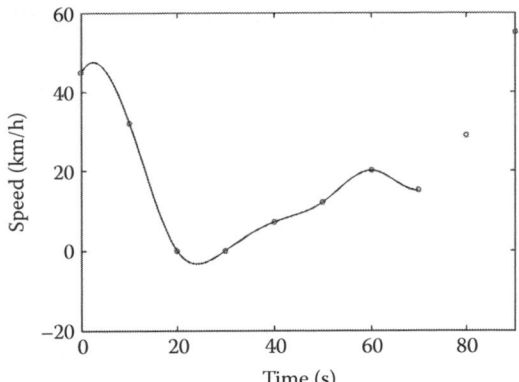

FIGURE D.2   Cubic spline interpolation for vehicle speed data.

```
% Interpolate and plot
tq(1)=0;
yq(1)= ppval(splineParam,tq(1));
for i=2:100
    tq(i)=tq(i-1)+rand;
    yq(i)=ppval(splineParam,tq(i));
end
plot(tq,yq,'-b');
```

Figure D.2 shows the result of the above code. Spline interpolation results in a smooth curve fitting to the initial data. The smoothness is due to the conditions of (D.31) and (D.32) imposed at all interior nodes.

Spline interpolation is not guaranteed to be monotonic. In other words, the interpolated values within an interval can lie outside the range of the y values at the bounding nodes. An interesting thing to note is that the interpolated values of velocity fall below 0 between 20 and 30 s. However, the vehicle speed has to be nonnegative. This is where a "shape-preserving" or *monotonic* interpolation is useful. The piecewise cubic Hermite interpolation polynomials can be used if the dependent variable has to meet the requirement of monotonicity. MATLAB provides the function pchip for this purpose. The next example demonstrates this, and a comparison between spline and pchip is shown in Figure D.3.

Example D.5   Comparison between spline and pchip

Problem: Compare spline and pchip interpolation for the vehicle speed data
   Solution: The two functions will be used to obtain interpolated values of vehicle speed at a frequency of 1 s. Assuming that tData and yData are already available, the functions spline and pchip can be called using a vector-valued tq = 0:90.

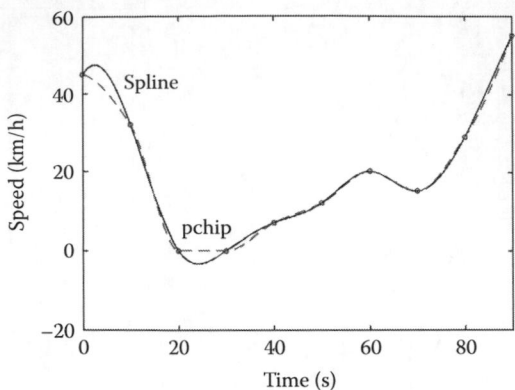

FIGURE D.3   Comparison of cubic spline (solid line) and monotonicity shape preserving PCHIP interpolation (dashed line) for vehicle speed data.

The resulting interpolated values are plotted in Figure D.3. The dashed line shows that the `pchip` interpolation preserves monotonicity. Between $t = 20$ and $t = 30$, the interpolated values remain at 0 km/h because the interpolant does not allow y_q to get values outside the range of the bounding points.

This completes our discussion on interpolation. While Newton's difference formulae were derived in this chapter, the practical implementation of interpolation in MATLAB is through `spline` and `pchip` functions.

Appendix E: Numerical Integration

E.1 GENERAL SETUP

Numerical integration aims to find an approximate solution to the problem:

$$I = \int_a^b f(x)\,dx \qquad (E.1)$$

where $f(x)$ (known as an *integrand*) is an arbitrary function of the independent variable x. The integral, I, is to be found between the limits of integration, that is, for $x \in [a, b]$. The discussion in this section will be limited to integration in a single independent variable. First, the topic of (numerical) integration will be introduced, followed by a discussion on some numerical techniques, and finally ending with MATLAB® implementation of numerical integration.

The geometric interpretation of integration is that the integral, I, is the area under the curve $f(x)$ between the points $x = a$ and $x = b$. This is represented as the shaded area in Figure E.1. This region is split into multiple intervals, as delineated by vertical dotted lines, to obtain integral I to the desired accuracy. Integration is the counterpart of differentiation (covered in Appendix B). If we write

$$f(x) = \frac{d}{dx} y(x) \qquad (E.2)$$

then integral I is equal to $I = y(b) - y(a)$. Numerical methods for computing the integral can be used if $f(x)$ either is known as an explicit function of x, or can be calculated numerically or indirectly (for any given value of x), or is available as a tabulated pair $(x, f(x))$. Additionally, note that Equation E.2 is an ordinary differential equation. The aim of an ODE solver is to obtain $y(x)$ given an initial value $y(a)$ at a point $x = a$. Thus, evaluating the integral I is equivalent to solving the ODE (E.2) for $y(a) = 0$. An equivalence between numerical integration and solving ODE is provided in a plug flow reactor (PFR) case study in Chapter 3.

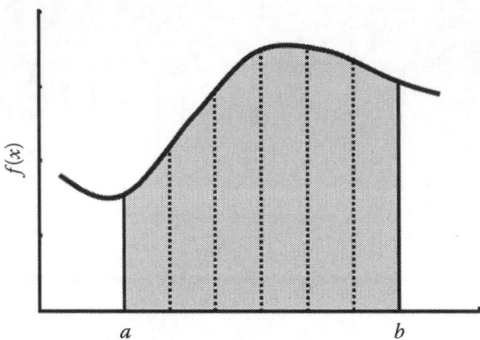

FIGURE E.1 Geometric interpretation of integration.

E.1.1 Some Exemplary Applications

Numerical integration finds several applications in science and engineering, in general, and in chemical engineering in particular. The first course in thermodynamics introduces students to enthalpy, where enthalpy at any temperature, T, is calculated as

$$H_T^0 = H_{T_0}^0 + \int_{T_0}^{T} c_p \, dT \tag{E.3}$$

where c_p is commonly expressed as a polynomial function of temperature.

Another application of integration is the inverse of the application covered in Appendix B: to calculate net quantity when flux is known. For example, if the heat flux through a surface is known, the net heat exchange is calculated by integrating the flux over the entire surface. Likewise, in case of membrane separation, the flux of permeate through the membrane can be computed; surface integral of the flux over the entire membrane surface gives the net quantity of permeate flowing across the membrane.

Finally, an application unique to chemical engineering is the design equation of a PFR. The volume of a PFR required to achieve a target conversion X_{set} is given by

$$V_{pfr} = \int_{0}^{X_{set}} \frac{F_{A0}}{\left(-r_A\left(X\right)\right)} \, dX \tag{E.4}$$

In the above equation, V_{pfr} is the desired volume of the PFR, whereas the rate of chemical reaction taking place in the system depends on the concentration. It can be converted into a function of conversion, X, and is given by the function $-r_A(X)$. In the rest of this chapter, the example of PFR design equation will be used to demonstrate the concepts of numerical integration.

E.2 NEWTON-COTES INTEGRATION FORMULAE

Newton-Cotes integration formulae are so-called *closed* integration formulae, where we attempt to find the area under the curve by dividing the region $x \in [a, b]$ into several intervals, as seen in Figure E.1. The integration formulae are then applied to individual intervals. Higher-order formulae use multiple function evaluations and yield more accurate results. Since a single application of the formula (even higher-order ones) has numerical errors, usually multiple applications of the integral formulae are sought. The most popular Newton-Cotes integration formulae are discussed below. First, single application of the integration formulae is discussed, followed by numerical derivation of the formulae. Thereafter, multiple applications of the integration formulae to improve the numerical accuracy will be discussed.

E.2.1 Single Application of Integration Formulae

E.2.1.1 *Trapezoidal Rule*

Consider a single application of the trapezoidal rule, where the integral (E.1) is approximated as the area under a straight line joining two endpoints. The interval size for the trapezoidal rule is the entire domain length, that is, $h = (b - a)$. The two endpoints, (a, f_a) and (b, f_b), can be connected by a unique straight line. Figure E.2 shows the schematic of a single application of the trapezoidal rule applied to the $f(x)$ curve of Figure E.1. The straight line, denoted as T in Figure E.2, is this line joining the two points. The value of integral, I, is approximated as the area under the line T.

This is nothing but the area of the trapezoid bounded by $x = a$, $x = b$ and line T, which is given by

$$I \approx \frac{h}{2}(f_a + f_b) \tag{E.5}$$

Clearly, the area of trapezoid given in Equation E.5 is an approximation of integral, I.

All Newton-Cotes integration formulae can be derived by first finding the equation of the polynomial curve connecting the points and then integrating this equation to approximate

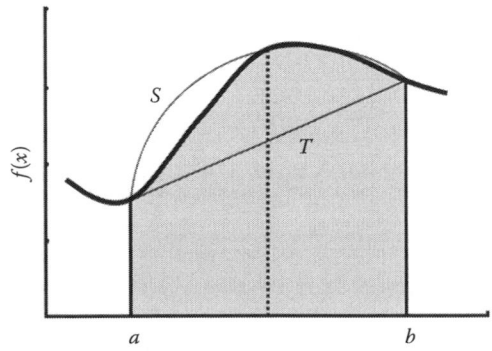

FIGURE E.2   Schematic showing single application of the trapezoidal and Simpson's 1/3rd rules.

the value of I. In case of trapezoidal rule, the polynomial connecting the two endpoints is a straight line, $T(x)$. The equation of this line is given by

$$\frac{T(x) - f_a}{x - a} = \frac{f_b - f_a}{b - a}$$

Rearranging the above leads to the equation of the line:

$$T(x) = f_a + \frac{f_b - f_a}{h}(x - a) \tag{E.6}$$

Integrating the above equation yields the approximate value of I using the trapezoidal rule:

$$I \approx \int_a^b T(x) \, dx \tag{E.7}$$

$$I = f_a \left[x \right]_a^b + \frac{f_b - f_a}{h} \left[\frac{x^2}{2} - ax \right]_a^b$$

$$I = f_a h + \frac{f_b - f_a}{h} \left[\frac{b^2 - a^2}{2} - a(b - a) \right]$$

Noting that $(b^2 - a^2) = (b - a)(b + a) = h(a + b)$, the final equation above can be rearranged to give

$$I \approx f_a h + (f_b - f_a) \underbrace{\left(\frac{a + b}{2} - a \right)}_{\frac{h}{2}} \tag{E.8}$$

It is easy to verify that this is the same as the trapezoidal rule written in Equation E.5. ■

Thus, the trapezoidal rule has been derived in two different ways: geometrically using the analogy with the area of a trapezoid and algebraically integrating the equation of a straight line connecting the two points. While these derivations are elegant, they still do not reveal the numerical error in computing the approximate value of integral using Equation E.5. A formal derivation of the trapezoidal rule including the error estimates will be discussed presently in Section E.2.2. The overall trapezoidal rule may be written as follows (also see Equation E.17):

$$I = \frac{h}{2}(f_1 + f_2) - \frac{h^3}{12} f''(\xi) \tag{E.9}$$

This means that the error in the trapezoidal rule scales as a cube of the interval size. In other words, the error reduces by a factor of 8 if the interval size, $(b - a)$, is halved. Moreover, the integral is exact if $f''(\xi) = 0$, $\xi \in [a, b]$. Obviously, when $f(x)$ is a straight line, then the trapezoidal rule is exact. Practical implementation of the trapezoidal rule, including demonstration of the error analysis is presented in Example E.1.

Example E.1 Trapezoidal Rule

Problem: For various values of b, compute the following integral, as well as the numerical error, using a single application of the trapezoidal rule:

$$\int_1^b \left(2 - x + \ln(x)\right) dx$$

Solution (direct): The solution to the problem will involve computing f_a, f_b and hence use Equation E.5 to compute the integral. The true value of integral is

$$y(x) = \int \left(2 - x + \ln(x)\right) dx = x \ln(x) + x - \frac{x^2}{2}$$

Thereafter, the true value is computed as $y(b) - y(1)$ and the absolute error can be found. The MATLAB code to do this is given below:

```
% Single application of Trapezoidal Rule for Integration
% f(x) = 2 - x + ln(x)
a=1;
b=2;
trueVal = (b-b^2/2+b*log(b)) - (a-a^2/2+a*log(a));
%% Trapezoidal Rule (single application)
h = (b-a);
fa= 2-a+log(a);
fb= 2-b+log(b);
I_trap=h/2*(fa+fb);
err_trap=abs(trueVal-I_trap);
```

The value of integral for $b = 2$ is 0.8466, whereas the numerical error is $3.9721E-2$. Thus, using the trapezoidal rule with a large step-size of 1 results in nearly 4% error.

Next, the error behavior is analyzed for various values of b. The results are given below:

```
b     h      error
2     1      3.972e-2
1.5   0.5    6.831e-3
1.1   0.1    7.569e-5
```

Thus, when the interval width is decreased by one order of magnitude, from 1 to 0.1, the error decreases by almost three orders of magnitude. This is because, as seen in Equation E.9, the trapezoidal rule has an accuracy of $\mathcal{O}\left(h^3\right)$.

While the above solution works, it can be made more elegant using MATLAB function. A separate MATLAB function `fun4Int` may be defined to calculate $f(x) = 2 - x + \ln(x)$ as a function of independent variable x. There are two ways to do this:

Using Anonymous Function: `fun4Int = @(x) 2-x+log(x);`
Using Function File (`fun4Int.m`) :
```
    function fval = fun4Int(x)
        fval = 2-x+log(x);
    end
```

Once the function has been defined, the code in Example E.1 may be modified as follows :

Solution to Example E.1 using functions:

```
% Problem Setup
a=1; b=2;
h = (b-a);
% Trapezoidal Rule (single application)
I_trap=h/2*(fun4Int(a)+fun4Int(b));
err_trap=abs(trueVal-I_trap);
```

The solution above is more elegant than the "direct" solution. It is modular, in that it allows the core integration code to be used for integrating other functions with little change. Since the function $f(x)$ is written once and called multiple times as needed, it is prone to lesser errors.

E.2.1.2 Simpson's 1/3rd Rule

The trapezoidal rule was derived using only the two endpoints and joining them with a straight line. The accuracy of the numerical integration method can be improved by including additional intermediate points. The next method in Newton-Cotes sequence of methods is known as Simpson's 1/3rd rule. This is shown schematically in Figure E.2, where in addition to the two endpoints, a and b, a third point is chosen as the midpoint. A quadratic curve is fitted to the three points. The step-size for a single application of Simpson's 1/3rd rule is now half of the original domain length. Thus, the following three points are used:

$$
\begin{aligned}
t_1 &= a, \quad f_1 = f(t_1) \\
t_2 &= \frac{a+b}{2}, \quad f_2 = f(t_2) \\
t_3 &= b, \quad f_3 = f(t_3)
\end{aligned}
\tag{E.10}
$$

Simpson's 1/3rd rule along with the numerical error is given by the following expression:

$$
I = \frac{h}{3}(f_1 + 4f_2 + f_3) - \frac{h^5}{90} f''''(\xi)
\tag{E.11}
$$

Simpson's 1/3rd rule requires, for each interval, one additional function evaluation than the trapezoidal rule. However, the accuracy of the numerical integration formula increases by two orders of step-size, from $\mathcal{O}(h^3)$ in case of the trapezoidal rule to $\mathcal{O}(h^5)$ in case of Simpson's 1/3rd rule. The derivation of Simpson's 1/3rd rule and error estimate will be discussed in Section E.2.2. Use of this method for numerical integration will be demonstrated in the next example.

Example E.2 Simpson's 1/3rd Rule

Problem: Solve Example E.1 using Simpson's 1/3rd rule.

  Solution: The code for Simpson's 1/3rd rule is similar to that of the trapezoidal rule. It involves calculating step-size, h=(b-a)/2, and the midpoint, t2=a+h. The three function values are fun4Int(a), fun4Int(t2), and fun4Int(b), which are plugged in Equation E.11 to compute the integral.

  The result of applying Simpson's 1/3rd rule is given below:

```
b     h       error
2     0.5     4.598e-4
1.5   0.25    2.772e-5
1.1   0.05    1.718e-8
```

In this example, when the interval width is decreased by one order of magnitude, from 1 to 0.1, the error decreases between four and five orders of magnitude: The ratio of the two errors is 3.7×10^{-5}, consistent with the observation that Simpson's 1/3rd rule is $\mathcal{O}(h^5)$ accurate.

Direct comparison between the trapezoidal and Simpson's 1/3rd rules reveals that the latter is more accurate. This rather significant improvement in accuracy comes at a cost of only one additional function evaluation. Another example, discussed in Section E.2.3, will demonstrate the significance of the order of accuracy: The additional computational requirement is justified by highly significant reduction in accuracy of the numerical method.

E.2.1.3 Simpson's 3/8th Rule

The trapezoidal rule uses the two endpoints of the region of integration, fits a straight line, and approximates the integral as the area under the line. In Simpson's 1/3rd rule, an additional point, midpoint of the region, is introduced and the integral is approximated as the area under a second-order curve passing through these three points. Continuing this idea further, Simpson's 3/8th rule finds an area under a cubic curve passing through four equi-spaced points. The step-size is $h=(b-a)/3$ and the three points include $t_1=a, t_2=a+h$, $t_3=a+2h$, and $t_4=b$. The function values $f(t_i)$ are represented as f_i. Then, the integral is approximated using Simpson's 3/8th rule as

$$I = \frac{3h}{8}\left(f_1 + 3f_2 + 3f_3 + f_4\right) - \frac{3h^5}{80} f''''(\xi) \tag{E.12}$$

Equation E.12 gives Simpson's 3/8th rule. Comparing with Equation E.11, Simpson's 3/8th rule has the same order of accuracy as the 1/3rd rule, since the error in both these methods scales as $\mathcal{O}(h^5)$. Note that Simpson's 1/3rd rule gives the same order of accuracy as the 3/8th rule, but with one less function evaluation in each application.

Since Simpson's 3/8th rule follows the same general pattern as the trapezoidal and Simpson's 1/3rd rules, a demonstration will be skipped for brevity. The reader can verify that the error using Simpson's 3/8th rule for Example E.1 is of the same order of magnitude as the 1/3rd rule, albeit slightly lower.

E.2.2 Derivation of Integration Formulae

The derivation of Newton-Cotes integration formulae can be done using Newton's interpolating polynomial (see Appendix D for more details). The interpolating polynomial is fitted to an appropriate number of equispaced data, and the resulting polynomial integrated between the limits, $x = a$ to $x = b$. Thus, the integral may be approximated as

$$I = \int_a^b f(x)\,dx \approx \int_a^b \bar{p}(x)\,dx \tag{E.13}$$

In case of trapezoidal rule, a first-order interpolating polynomial is fitted to two data points. Thus, the polynomial $\bar{p}$ takes the following form:

$$\bar{p}(\alpha) = f_1 + \Delta f_1 \alpha + \frac{f''(\xi)}{2}\alpha(\alpha-1)h^2 \tag{E.14}$$

Recall from Appendix D that for Newton's forward difference interpolation, we define

$$\alpha = \frac{x-a}{h}, \quad \text{with } h = (b-a) \tag{E.15}$$

From the above equation, it is clear that

$$dx = h\,d\alpha$$

Change the variable of integration to α, the limits of integration are from $\alpha = 0$ to $\alpha = 1$:

$$I = \int_0^1 \bar{p}(\alpha)\cdot h\,d\alpha$$

Substituting

$$I = h\int_0^1 \left(f_1 + \Delta f_1 \alpha + \frac{f''(\xi)}{2!}\alpha(\alpha-1)h^2 \right)d\alpha$$

$$= h\left[f_1\alpha + \Delta f_1 \frac{\alpha^2}{2} + \frac{f''(\xi)}{2!}h^2\left(\frac{\alpha^3}{3} - \frac{\alpha^2}{2} \right) \right]_0^1 \tag{E.16}$$

$$I = h \left[\frac{f_1 + f_2}{2} \right] + h^3 \frac{f''(\xi)}{2} \left[-\frac{1}{6} \right] \tag{E.17}$$

$$\underbrace{\qquad\qquad}_{\text{Formula}} \quad \underbrace{\qquad\qquad}_{\text{Error}}$$

This demonstrates the $\mathcal{O}(h^3)$ accuracy of the trapezoidal rule mentioned in Section E.2.1 and demonstrated in Example E.1.

In a similar manner, Simpson's 1/3rd rule may also be derived. Here, the step-size is given by

$$h = \frac{b - a}{2}$$

With the same definition of α as before, the limits of integration, $x = a$ and $x = b$, correspond to $\alpha = 0$ and $\alpha = 2$, respectively. Thus, integral I in Simpson's 1/3rd rule is written as

$$I = \int_0^2 \bar{p}(\alpha) \cdot h \, d\alpha$$

Substituting second-order polynomial expansion yields

$$I = h \int_0^2 \left(f_1 + \Delta f_1 \alpha + \frac{1}{2} \Delta^2 f_1 \alpha (\alpha - 1) + \frac{f'''(\xi)}{3!} \alpha (\alpha - 1)(\alpha - 2) h^3 \right) d\alpha$$

$$= h \left[f_1 \alpha + \Delta f_1 \frac{\alpha^2}{2} + \Delta^2 f_1 \left(\frac{\alpha^3}{6} - \frac{\alpha^2}{4} \right) + \frac{f'''(\xi)}{6} h^3 \left(\frac{\alpha^4}{4} - \alpha^3 + \alpha^2 \right) \right]_0^2 \tag{E.18}$$

$$I = h \left[2 f_1 + 2 \Delta f_1 + \Delta^2 f_1 \left(\frac{4}{3} - 1 \right) \right] + h^4 \frac{f'''(\xi)}{6} \left[4 - 8 + 4 \right] \tag{E.19}$$

$$\underbrace{\qquad\qquad\qquad\qquad}_{\text{Formula}} \quad \underbrace{\qquad\qquad}_{\text{Is Error = 0?}}$$

The above equation gives a surprising result. The first part gives Simpson's 1/3rd rule. However, notice that in the final term in the above equation, the expression in brackets equates to zero. Does that mean that Simpson's 1/3rd rule is accurate and does not have any numerical error? As demonstrated in Example E.2, there is a numerical error in implementing Simpson's 1/3rd rule. So, what the above derivation implies is that the leading error term is not the third-order term; instead, the analysis needs to be done including an additional term in Newton's forward difference polynomial used in (E.18). Doing so yields

$$I = h \int_0^2 \left(f_1 + \Delta f_1 \alpha + \frac{1}{2!} \Delta^2 f_1 \alpha (\alpha - 1) + \frac{1}{3!} \Delta^3 f_1 \alpha (\alpha - 1)(\alpha - 2) \right.$$

$$\left. + \frac{f''''(\xi)}{4!} \alpha (\alpha - 1)(\alpha - 2)(\alpha - 3) h^4 \right) d\alpha \tag{E.20}$$

Note that the first four terms are the same as the one obtained in Equation E.19. Thus

$$I = h \left[2f_1 + 2\Delta f_1 + \Delta^2 f_1 \frac{1}{3} + \Delta^3 f_1 \cdot 0 + \frac{f'''(\xi)}{24} h^4 \left(\frac{\alpha^5}{5} - \frac{3}{2}\alpha^4 + \frac{11}{3}\alpha^3 - 3\alpha^2 \right) \right]_0^2$$

Substituting the expressions for Δf_1 and $\Delta^2 f_1$ yields

$$I \approx h \left[2f_1 + 2(f_2 - f_1) + \frac{1}{3}(f_3 - 2f_2 + f_1) \right]$$

and

$$E = h^5 \frac{f'''(\xi)}{24} \left(\frac{32}{5} - 24 + \frac{88}{3} - 12 \right) = -h^5 \frac{f'''(\xi)}{90}$$

Thus, Simpson's 1/3rd rule may be written as

$$I = \underbrace{\frac{h}{3} \left[f_1 + 4f_2 + f_3 \right]}_{\text{Formula}} - \underbrace{h^5 \frac{f'''(\xi)}{90}}_{\text{Error}} \tag{E.21}$$

The above equation shows Simpson's 1/3rd rule along with the error estimate. This justifies the observation in Example E.2 regarding the effect of interval size on the accuracy of Simpson's 1/3rd rule.

E.2.3 Multiple Applications of Newton-Cotes Formulae

A single application of Newton-Cotes integration formulae can have significant errors. In Example E.1, the error in computing integral for $b = 2$ using the trapezoidal rule was nearly 4%. While the error drops significantly when Simpson's 1/3rd rule is used, error of the order of $\sim 10^{-4}$ may still be high for a number of applications. One way to improve the accuracy of numerical integration is to divide the region of integration into multiple intervals and apply the method of choice multiple times to obtain an appropriate numerical integral. As shown schematically in Figure E.1, the region of integration may be divided into n intervals. Thus

$$h = \frac{b-a}{n}, \quad x_1 = a, \quad x_{n+1} = b$$

and $(x_1, f_1), (x_2, f_2), \dots, (x_{n+1}, f_{n+1})$ form the nodes used in obtaining the integral.

Multiple implementation of the trapezoidal rule can be performed for any choice of number n of intervals. The value of total integral may then be written as

$$I = I_1^{\text{trap}} + I_2^{\text{trap}} + \cdots + I_n^{\text{trap}} \tag{E.22}$$

where I_i^{trap} is the integral obtained at the ith interval using data at ith and $(i + 1)$th nodes:

$$I_i^{\text{trap}} = \frac{h}{2}\left(f_i + f_{i+1}\right) + \frac{-h^3}{12} f''\left(\xi_i\right) \tag{E.23}$$

where $\xi_i \in [x_i, x_{i+1}]$. Substituting in Equation E.22

$$I = \sum_{i=1}^{n} \frac{h}{2}\left(f_i + f_{i+1}\right) + \frac{-h^3}{12} f''\left(\xi_i\right) \tag{E.24}$$

$$I = \frac{h}{2}\Big[\left(f_1 + f_2\right) + \left(f_2 + f_3\right) + \cdots + \left(f_n + f_{n+1}\right)\Big] - \frac{h^3}{12} \sum_{i=1}^{n} f''\left(\xi_i\right) \tag{E.25}$$

$$I = \frac{h}{2}\Big[f_1 + 2\left(f_2 + \cdots + f_n\right) + f_{n+1}\Big] - \frac{h^3}{12}\left(nf''\left(\bar{\xi}\right)\right) \tag{E.26}$$

The last term in the above equation is derived using mean value theorem, with $\bar{\xi} \in [a, b]$. The first sum gives the integral using multiple applications of the trapezoidal rule. Using the relationship of h, the error term may be written as

$$E_{\text{net}} = -\frac{\left(b-a\right)^3}{12n^3} nf''\left(\bar{\xi}\right) \tag{E.27}$$

This may then be written as

$$E_{\text{net}} = -\frac{\left(b-a\right)}{12} h^2 f''\left(\bar{\xi}\right) \tag{E.28}$$

There are a few things to note from the error analysis of multiple applications of the trapezoidal rule. The first thing is that the net error is proportional to h^2 and not h^3. This means that the error reduces by two orders of magnitude when the step-size is reduced by one order of magnitude. This issue of local vs. global truncation errors (GTEs) is discussed at length in Chapter 1.

Another important point to note is regarding this drop in order of accuracy for GTE. Throughout this book, I have discussed the importance of the order of accuracy, that $\mathcal{O}\left(h^n\right)$ method would often be a preferred numerical method over $\mathcal{O}\left(h^m\right)$, when $n > m$. So, what does it mean when the local truncation error (LTE) scales as $\mathcal{O}\left(h^3\right)$ while the GTE scales as $\mathcal{O}\left(h^2\right)$? Does that mean that multiple applications of the trapezoidal rule are not useful? In reality, this is not the case because for a single application, $h = (b - a)$, whereas for multiple applications of the trapezoidal rule, $h = (b - a)/n$. So, the comparison between two options is valid *only if* the step-size h refers to the same quantity. In LTE, h refers to the entire interval size, whereas in GTE it refers to individual step-size. On the other hand, comparison between the trapezoidal and Simpson's rules is still valid because in both these cases h refers to the same thing.

So, how does one compare single and multiple applications of the trapezoidal rule? In case of the former, $n = 1$ in Equation E.27. Thus the error in a single application of the trapezoidal rule is

$$E_{net} = -\frac{(b-a)^3}{12} f''(\bar{\xi}) \tag{E.29}$$

whereas that in multiple applications is

$$E_{net} = -\frac{(b-a)^3}{12 n^2} f''(\bar{\xi}) \tag{E.27}$$

Thus, the error in multiple applications of the trapezoidal rule is expected to improve by a factor of 4 when two intervals are used, and by two orders of magnitude when $n \sim 10$ intervals are used. The next example demonstrates this.

Example E.3 Multiple Applications of the Trapezoidal Rule

Problem: Compute the following integral, as well as the numerical error:

$$\int_1^2 (2 - x + \ln(x)) dx$$

using multiple applications of the trapezoidal rule.

Solution: The following code may be used for multiple applications of the trapezoidal rule. First, the array of nodal points, $[x_1, \ldots, x_{n+1}]$, is used to obtain the corresponding function values, $[f_1, \ldots, f_{n+1}]$, followed by the implementation of Equation E.26.

```
% Multiple applications of Trapezoidal Rule
% f(x) = 2 - x + ln(x)
%% Problem Setup
a=1; b=2;
trueVal = (b-b^2/2+b*log(b)) - (a-a^2/2+a*log(a));
n=20;
%% Trapezoidal Rule (multiple applications)
h=(b-a)/n;
xVec=a:h:b;
fVec=fun4Int(xVec);
I_trap=(h/2)*(fVec(1)+2*sum(fVec(2:n))+fVec(n+1));
%% Results
err=abs(trueVal-I_trap);
disp(['For h=', num2str(h), ', Error=', num2str(err)]);
```

When the code is executed for $n = 10$, the error in the trapezoidal rule is 4.164×10^{-4}. Comparing with Example E.1, the error has fallen by nearly two orders of magnitude, as predicted in Equation E.27.

Next, the step-size was halved (with $n = 20$). The resulting error was

```
For h=0.05, Error=0.00010415
```

Finally, decreasing the step-size to 0.01 (i.e., $n = 100$) yields

```
For h=0.01, Error=4.1666e-06
```

In summary, the error decreased by a factor of 4 when the step-size was reduced by half, whereas it decreased by two orders of magnitude when the step-size was reduced by one order of magnitude. This observation is a consequence of the fact that the GTE of the trapezoidal rule scales as $\mathcal{O}\left(h^2\right)$.

Another interesting observation on comparing the above results with Example E.2 is that the error in a single application of Simpson's 1/3rd rule happened to be similar to the error using 10 applications of the trapezoidal rule. This demonstrates the power of using higher-order formulae: Three function evaluations in Simpson's 1/3rd rule were sufficient to provide similar accuracy as eleven function evaluations in the trapezoidal rule.

Consequently, it is advantageous to use multiple applications of Simpson's 1/3rd rule as well. Unlike the trapezoidal rule, there is a restriction that n has to be an even number. This is because single application of Simpson's 1/3rd rule involves two intervals. Thus, first application of the rule will involve nodes (x_1, x_2, x_3), and the next application would involve (x_3, x_4, x_5), followed by nodes (x_5, x_6, x_7) and so on. Figure E.3 shows a schematic to further explain this.

As in the trapezoidal rule, multiple applications of Simpson's rule can be derived as

$$I = \frac{h}{3}\left[\left(f_1 + 4f_2 + f_3\right) + \left(f_3 + 4f_4 + f_5\right) + \cdots + \left(f_{n-1} + 4f_n + f_{n+1}\right)\right] + E_{net} \qquad (E.30)$$

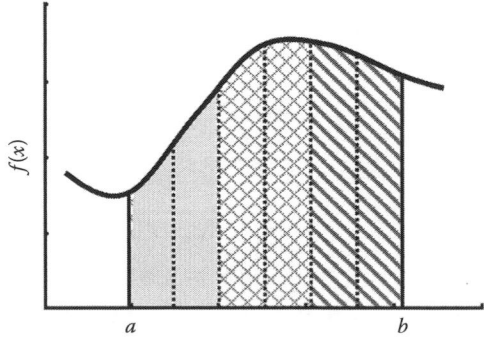

FIGURE E.3   Schematic of multiple applications of Simpson's 1/3rd rule.

The reader can verify that the formula for and the error in Simpson's 1/3rd rule is given by

$$I = \frac{h}{3}\Big[f_1 + 4\big(f_2 + f_4 + \cdots + f_n \big) + 2\big(f_3 + f_5 + \cdots + f_{n-1} \big) + f_{n+1} \Big] + E_{net} \quad\quad \text{(E.31)}$$

$$E_{net} = -h^4 \frac{(b-a)}{90} f'''\big(\bar{\xi}\big) \quad\quad \text{(E.32)}$$

where $h = (b-a)/n$. Thus, the GTE in Simpson's 1/3rd rule scales as $\mathcal{O}\big(h^4 \big)$.

In the same manner, a formula for multiple applications of Simpson's 3/8th rule can be derived. An interested reader may read advanced text on numerical techniques to learn about higher-order Newton-Cotes integration formulae that use a larger number of nodes to provide higher accuracy.

Example E.4 Comparison of the Trapezoidal and Simpson's Rules

Problem: Compare the three methods described above for the same step-size.

Solution: For this comparison, choose $n = 6$, so that $h = 1/6$. This corresponds to six applications of the trapezoidal rule, three applications of Simpson's 1/3rd rule, and two applications of Simpson's 3/8th rule. As before, choose

```
h=(b-a)/n;
xVec=[a:h:b];
fVec=fun4Int(xVec);
```

With `fVec` available, Equation E.26 is used for the trapezoidal rule, Equation E.31 for Simpson's 1/3rd rule, and a similar equation (derivation is left as an exercise to the reader) for Simpson's 3/8th rule. With the true value of the integral known, the errors can also be computed. The results are given below.

Value of integral using the three methods

```
    0.8851      0.8863      0.8863
```

Error using the three methods

```
    1.1555e-3      7.20e-6      1.56e-5
```

This completes the comparison of the three methods.

E.3 QUADRATURE AND MATLAB® FUNCTIONS

MATLAB provides several functions to compute an integral. The trapezoidal rule can be invoked using the function `trapz`:

```
>> I_trap=trapz(xVec,fVec);
```

will give the exact same result as the trapezoidal rule implemented in Example E.3 earlier. Simply replacing the third line from the end from Example E.3 will provide the same results as multiple implementation of the trapezoidal rule.

A more powerful computation of integral is provided by the command `quad`. The syntax for `quad` is

```
>> I=quad(@(x) fun4Int(x),a,b);
```

where the first argument passes the function `fun4Int` that calculates $f(x)$, while the last two arguments are limits of integration. This uses Simpson's quadrature method.

In later versions of MATLAB, `quad` will be superseded by an improved algorithm `integral`. This uses global adaptive quadrature method. The regular implementation is similar:

```
>> I=integral(@(x) fun4Int(x),a,b);
```

There are two main differences between `trapz` and the two quadrature methods mentioned here. The first one is that `trapz` takes actual values $[x_1 \ x_2 \ \cdots \ x_{n+1}]$ and corresponding function values $\begin{bmatrix} f_1 \ f_2 \ \cdots \ f_{n+1} \end{bmatrix}$ as input arguments. On the other hand, `quad` and `integral` require the actual function $f(x)$, in the form of a MATLAB function file (or MATLAB anonymous function) as an input argument, and compute the integral for the specified integration limits (also input arguments). The second difference is that `trapz` simply calculates the numerical integral value as such; `quad` and `integral` calculate the numerical integral such that the error $E < \text{tol}$. In other words, if I_{num} is the numerical integration value and I_{true} is the true value of the integral, then the step-size h is chosen such that the error $|I_{\text{true}} - I_{\text{num}}| < \varepsilon_{\text{tol}}$.

The error tolerance value ε_{tol} can be specified by the user. If it is not specified (as in the example above), a default value of $\varepsilon_{\text{tol}} = 10^{-6}$ is used.

Recall that according to Equation E.32, it is possible to know the error E_{net}. However, in most practical examples, it is either not possible or not convenient to know $f'''' (\overline{\xi})$. Hence, there needs to be an alternate method to *estimate* the error.

Simpson's quadrature method, implemented in `quad`, utilizes the fact that although the error cannot be known exactly, there are ways to estimate it even when $f'''' (\overline{\xi})$ is unknown. I will use Simpson's 1/3rd rule to explain the concept behind this.

The true value of integral and its relationship with Simpson's 1/3rd rule is

$$I_{\text{true}} = I_{\text{simp}}(h) + c_{1,h} h^4 \tag{E.33}$$

where $c_1 = -\dfrac{1}{90}(b-a) f''''(\overline{\xi})$ is a constant.

If the same calculation is repeated using a step-size of $h/2$

$$I_{\text{true}} = I_{\text{simp}}(h/2) + c_{1,h/2} \frac{h^4}{2^4} \tag{E.34}$$

where again $c_1 = -\dfrac{1}{90}(b-a) f''''(\overline{\xi})$.

Usually, $f''''(\bar{\xi})$ from Equation E.33 and $f''''(\bar{\xi})$ from Equation E.34 are reasonably constant for most practical problems. Thus, the difference between the two will provide an estimate of the error:

$$0 = I_{\text{simp}}(h) - I_{\text{simp}}(h/2) + c_1 \frac{15h^4}{16} \tag{E.35}$$

Since

$$E_{\text{net}} = c_1 h^4$$

hence, the term

$$E_{\text{net}} \sim \frac{16}{15} \left| I_{\text{simp}}(h) - I_{\text{simp}}(h/2) \right| \tag{E.36}$$

gives an estimate of the error. Simpson's quadrature method involves calculating the integral at a chosen value of h, then repeating the calculations with half step-size $h/2$. Thereafter, the error is estimated as per Equation E.36. If this value is above the required tolerance, the algorithm sets $h \leftarrow h/2$ and the procedure is repeated until the error tolerance is met.

When this condition is met, the more accurate of the two values is returned as the integral.

The command integral uses a different algorithm. Since it is not aligned with the objective of this textbook, it will not be discussed further. This chapter ends with the following example of implementation of quad and integral.

Example E.5 Demonstrate the Use of quad and integral Functions

Problem: Demonstrate the use of quad and integral functions for the same problem.

Solution: Command-line use of the two functions is very straightforward. First, integration using quad is shown as follows:

```
>> I1=quad(@(x) fun4Int(x),a,b)
I1 =
    0.8863
>> err = abs(trueVal-I1)
err =
   2.6783e-08
```

Similarly, integral may be used as follows:

```
>> I2=integral(@(x) fun4Int(x),a,b)
I2 =
    0.8863
>> err = abs(trueVal-I2)
err =
   0
```

Bibliography

Bequette, B.W., *Process Dynamics: Modeling, Analysis, and Simulation* (1998), Prentice Hall PTR, Upper Saddle River, NJ.

Butcher, J.C., *Numerical Methods for Ordinary Differential Equations*, 2nd edn. (2008), Wiley, West Sussex, England.

Carslaw, H.S. and Jaeger, J.C., *Conduction of Heat in Solids*, 2nd edn. (1986), Oxford University Press, Oxford, U.K.

Chapra, S.C. and Canale, R., *Numerical Methods for Engineers*, 5th edn. (2006), McGraw-Hill, Inc., New York.

Cutlip, M.B. and Shacham, M., *Problem Solving in Chemical and Biochemical Engineering with POLYMATH, Excel, and MATLAB*, 2nd edn. (2008), Prentice Hall, Upper Saddle River, NJ.

Fausett, L.V., *Applied Numerical Analysis Using MATLAB*, 2nd edn. (2007), Pearson Inc., New York.

Fogler, H.S., *Elements of Chemical Reaction Engineering*, 5th edn. (2008), Prentice Hall, Pearson Education, Boston, MA.

Golub, G.H. and Van Loan, C.F., *Matrix Computations*, 3rd edn. (1996), Johns Hopkins University Press, Baltimore, MD.

Gupta, S.K., *Numerical Methods for Engineers*, 3rd edn. (2013), New Academic Press, New Delhi, India.

Petzold, L.R. and Ascher, U.M., *Computer Methods for Ordinary Differential Equations and Differential-Algebraic Equations* (1998), Society for Industrial and Applied Mathematics, Philadelphia, PA.

Press, W.H., Teukolsky, S.A., Vetterling, W.T., and Flannery, B.P., *Numerical Recipes: The Art of Scientific Computing*, 3rd edn. (2007), Cambridge University Press, New York.

Pushpavanam, S., *Mathematical Methods in Chemical Engineering* (2004), Prentice-Hall India Pvt. Ltd (Verlag), New Delhi, India.

Shampine, L.F. and Reichelt, M.W., The MATLAB ODE suite, *SIAM Journal on Scientific Computing*, **18** (1997), 1–22.

Shampine, L.F., Reichelt, M.W., and Kierzenka, J.A., Solving index-1 DAEs in MATLAB and Simulink, *SIAM Review*, **41** (1999), 538–552.

Skogestad, S., Dynamics and control of distillation columns: A tutorial introduction, *Transactions of IChemE (UK)*, **75**, Part A (1997), 539–562.

Strang, G., *Linear Algebra and Its Applications*, 4th edn. (2006), Cengage Learning—India Edition, New Delhi, India.

Index

A

Adams-Bashforth (AB) methods
  coefficients, 321–322
  definition, 320–321
  second-order, 92
  stability, 325
Adams-Moulton (AM) methods
  first-order method, 318
  higher-order method, 319–320
  second-order method, 318–319
  stability, 325
Adaptive step-size methods, 329–330
Antoine's equation, 410, 431
  linear regression, 439–440
  nonlinear regression, 440–442
  vapor-liquid equilibrium, 253–254
Array functions, 461

B

Backward difference formula (BDF) methods
  first-order method, 322
  formula for, 323
  NDF, 324
  second-order method, 322–323
  stability, 325–326
Banded system, 277–278, 288–289
Batch distillation model, 356–357
  Antoine's equation, 354
  benzene and ethylbenzene, 353–354
  definition, 355
  flowrates, 358–359
  process, 354–355
  Raoult's law, 354
  temporal variations, 358
  vapor fraction, 354–355
BDF, *see* Backward difference formula methods
Belusov-Zhabotinsky (B-Z) reaction, 384
Bifurcation point, 367

Bisection method
  convergence order, 229
  failure, 231
  function values, 228
  iteration value, 228
  MATLAB® code, 232–233
  Redlich-Kwong equation, 229–231
Blasius equation, 391
Boundary value problems (BVPs), 71–72
Bracketing method, 228–233, 235
Brent's method, 235–236
Broyden's method, 246
Brusselator system, 384–387
Butcher's tableau, 87, 94
BVPs, *see* Boundary value problems

C

Cartesian coordinate system, 341
Catalyst pellet
  Langmuir-Hinshelwood kinetics, 308–311
  Thiele modulus, 305–307
Chemostat analysis
  eigenvalues and eigenvectors, 188–189
  Jacobian value, 188
  multivariate Taylor's series expansion, 187
  phase portrait, 190
  stability condition, 189–190
  trivial steady state, 190–191
Chemostat model, 261
  bifurcation analysis, 366–368
  multivariate Newton-Raphson technique,
    243–244
  phase-plane analysis, 365–366
  steady state multiplicity and stability,
    364–365
  stiff system, 120–124
  transcritical bifurcation
    bifurcation plot, 371
    dilution rate, 369

graphical method, 368
linear stability analysis, 370–371
steady state, 368–370
Complex kinetics, 266–268
Constants and special variables, 451–452
Continuous stirred tank reactor (CSTR)
DAEs, 338
nonisothermal
bifurcation diagram, 378
extinction point, 377
graphical method, 372–374
inlet temperature, 115–116
jacket temperature, 112–113
parameters, 111
parametric continuation, 378–379
phase-plane analysis, 376–377
stability analysis, 374–376
steady state multiplicity, 257–261
T-axis, 377
transient simulations, 113–115
Convective systems
finite differences, space and time, 136
FTCS differencing, 138–139
higher-order upwind method, 140
Lax-Friedrichs scheme, 139
Lax-Wendroff method, 140
leapfrogging method, 139–140
MOL
central difference in space, 142–145
driver script, 144–145
upwind difference in space, 145–148
numerical diffusion, 148–150
second-order implicit method, 140–141
stability methods, 157
upwind difference in space, 136–138
Crank-Nicolson method, 316
analytical solution, 336
central difference formula, 331–332
exploiting sparse structure, 337
function argument, 333
i_{th} equation, 331
MATLAB function, 333–334
ODEs, 331
outlet concentration, 336
overall driver function, 335–336
PER, 332–333
pfrCN_Stepper function, 334
pfrCrankNicolFun function, 334
CSTR, *see* Continuous stirred tank reactor

D

Differential algebraic equations (DAEs)
CSTR, 338
direct substitution, 338
formulating and solving, 339
heterogeneous reactor models
algebraic equation, 342
concentration driving force, 341
differential equation, 342
mass transfer coefficient, 341
single complex reaction, 351–353
index of, 340–341
ode15s, 348–350
pendulum model, 341
solution methodology
algebraic variable, 342
combined approach, 345–348
first-order kinetics, 342
nested approach, 343–345
Differential equations
Adams-Bashforth methods, 320–322
Adams-Moulton methods, 318–320
BDF, 322–324
non-self-starting property, 318
RK methods, 317
Diffusive systems
Crank-Nicolson method, 153–154
MOL using MATLAB® ODE solvers, 154–157
packed bed reactor, 151
Dirichlet boundary condition, 274, 280, 285, 304–305
Discrete-time model, 212–215
Doolittle method, 491
Dynamical systems, 3

E

Eigenvalues and eigenvectors
applications, 58–62
characteristic equation, 55–56
chemostat analysis, 188–189
complex eigenvalues, 56
decomposition, 56–58
definitions, 54–56
linear difference equations, 64–65
linear differential equations, 63–64
similarity transform, 62–63
Elliptic PDEs, 276–277, 289–290
Equation of state (EOS), 253
Ergun's equation, 391
Error analysis
convergence and stability, 20–21
global truncation error, 21–23

Euler's explicit method, 74–76, 315
Euler's implicit method, 76–77, 325

F

Finite element methods (FEM), 135
Finite volume methods (FVM), 134–135
First-order kinetics, 263–265
First same as last (FSAL) method, 105
Fixed point iteration, 236–238
Flash separation, *see* Batch distillation model
Forward in time central in space (FTCS), 138–139,
          152–153
Fredholm integral equations, 262
fsolve method, 250–252
Function handles, 472–473
fzero method, 249–250

G

Gauss Elimination
     back-substitution, 486, 488, 490–491
     LU decomposition, 491–493
     naive Gauss Elimination, 487–490
     partial pivoting method, 493–497
Gauss-Siedel method
     diagonal dominance, 292–293
     fixed point iteration, 292
     *k*th equation, 291
     tridiagonal system, 294–295
Ghost-point method, 154
Global truncation errors (GTEs), 521–522

H

Harmonic oscillations, 382–383
Heat conduction, 304–305
Héron's method, 10–11
Heterogeneous reactor models
     algebraic equation, 342
     concentration driving force, 341
     differential equation, 342
     mass transfer coefficient, 341
     single complex reaction, 351–353
Heun's method, 82, 84–86
Hybrid two-tank heater system, 116–120
Hyperbolic functions, 461
Hyperbolic PDEs
     finite differences in space and time, 136
     FTCS differencing, 138–139
     higher-order upwind method, 140
     Improve stability methods, 157
     Lax-Friedrichs scheme, 139

Lax-Wendroff method, 140
leapfrogging method, 139–140
MOL
     central difference in space, 142–145
     driver script, 144–145
     upwind difference in space, 145–148
numerical diffusion, 148–150
second-order implicit method, 140–141
upwind difference in space, 136–138

I

Implicit methods
Adams-Bashforth methods
     coefficients, 321–322
     definition, 320–321
     stability, 325
Adams-Moulton methods
     first-order method, 318
     higher-order method, 319–320
     second-order method, 318–319
     stability, 325
batch distillation model, 356–357
     Antoine's equation, 354
     benzene and ethylbenzene, 353–354
     definition, 355
     flowrates, 358–359
     process, 354–355
     Raoult's law, 354
     temporal variations, 358
     vapor fraction, 354–355
BDF
     first-order method, 322
     formula for, 323
     NDF, 324
     second-order method, 322–323
     stability, 325–326
Crank-Nicolson method
     analytical solution, 336
     central difference formula, 331–332
     exploiting sparse structure, 337
     function argument, 333
     i_{th} equation, 331
     MATLAB function, 333–334
     ODEs, 331
     outlet concentration, 336
     overall driver function, 335–336
     PER, 332–333
     pfrCN_Stepper function, 334
     pfrCrankNicolFun function, 334
DAEs
     algebraic variable, 342
     combined approach, 345–348

CSTR, 338
definition, 316–317
direct substitution, 338
first-order kinetics, 342
formulating and solving, 339
heterogeneous reactor models, 341–342,
351–353
index of, 340–341
nested approach, 343–345
ode15s, 348–350
pendulum model, 341
distributed parameter
systems, 316
Euler's implicit method, 325
MATLAB® nonstiff solvers, 326
MATLAB® stiff solvers, 326
non-self-starting property, 318
RK methods, 317
stiff system
linearized model, 314
ODE-IVP method, 315
phase-plane plot, 314
Runge-Kutta (RK) methods, 313
single variable, 315
trapezoidal methods, 325
adaptive step-sizing, 329–330
AM-2, 327–329
multivariable, 330
Integral equations
complex kinetics, 266–268
first-order kinetics, 263–265
Fredholm integral equations, 262
Interpolation
polynomial function
Lagrange interpolation, 501–502
Newton's forward difference formula,
502–505
speed data, 499–500
spline
cubic spline interpolation,
508–509
methods, 505
vs. pchip interpolation, 506–510
U.S. FTP-75 test cycle, 499–500
Inverse quadratic interpolation, 236
Iterative methods
Gauss-Siedel method, 293–297
with under-relaxation, 297

J

Jordan canonical form, 63

L

Langmuir-Hinshelwood-Hougen-Watson (LHHW)
model
different initial concentrations, 444–445
rate expression, 441
single concentration, 442–444
Langmuir-Hinshelwood (LH) kinetics, 308–311
Lax-Friedrichs scheme, 139
Lax-Wendroff method, 140
Linear algebra
eigenvalues and eigenvectors
applications, 58–62
characteristic equation, 55–56
complex eigenvalues, 56
decomposition, 56–58
definitions, 54–56
linear difference equations, 64–65
linear differential equations, 63–64
similarity transform, 62–63
overview, 28–30
SVD
condition number, 47
directionality, 51–54
MATLAB command, 45–46
orthonormal vectors, 41–42
rank of a matrix, 47
sensitivity of solutions, 47–51
system of equations, 27–28
vector space
blending operations, 40–41
change of basis, 34–35
definition and properties, 30–32
image space and column space, 39
linear independence, 32–33
natural basis, 34
null and image spaces in MATLAB, 39–40
null and image spaces of matrix, 35–38
span, 32
subspace, 33
Linear difference equations, 64–65
Linear differential equations, 63–64
Linear least squares regression
error and coefficient
quantitative information, 420
sample standard deviation, 419
statistics, 420–423
general matrix approach, 415–416
maximum likelihood solution, 418–419
straight line
error, 414–415
N discrete points, 414
specific heat data, 417–418

Linear multi-step methods
    Adams-Bashforth methods, 320–322
    Adams-Moulton methods, 318–320
    BDF, 322–324
    non-self-starting property, 318
    RK methods, 317
Linear systems
    catalyst pellet
        Langmuir-Hinshelwood kinetics, 308–311
        thiele modulus, 305–307
    chemostat analysis
        eigenvalues and eigenvectors, 188–189
        Jacobian value, 188
        multivariate Taylor's series expansion, 187
        phase portrait, 190
        stability condition, 189–190
        trivial steady state, 190–191
    convective/radiative heat losses, 304–305
    elliptic PDEs, 274–275
    mass-spring system, 197
    ODE-BVPs, 274
    ODEs
        mass-spring system, 197
        qualitative responses, 197–198
        real and distinct eigenvalues, 192–197
    transient growth
        normal and nonnormal matrices, 198–199
        normal and nonnormal systems, 199–201
Line search method, 247–249
Lineweaver-Burke plot, 430
Local truncation error (LTE), 521–522
Loops and execution control
    conditional if block, 471–472
    for and while loops, 468–471
LU decomposition, 491–493

M

Mass-spring-damper system, 379–380
Mass-spring system, 197
Mathematical functions
    array functions, 461
    common matrix, 459–460
    trigonometric functions, 461
MATLAB® code
    array operations, 9–10
    functions, 7–9
    loops and execution control, 10–11
    script, 5–6
MATLAB file, 466–468
MATLAB® functions, 465–466, 524–526
MATLAB script, 464–465
MATLAB® Solver ode23, 100–101

Matrix inversion, 497–498
Matrix operations and vector functions
    array operations, 459
    basic mathematical operations, 456–458
    building, 453–454
    creation, 452–453
    matrix divisions and slash operators, 458
    querying elements, 455–456
    special matrices, 454
Methanol synthesis
    copper/zinc catalyst, 387
    equilibrium constants, 388
    input parameters and initial processing,
        389–390
    reaction kinetics, 388–389
    steady state PFR model, 391–394
    transient reactor model, 396–397
        axial profiles, 397–398
        driver script, 397
        gas-phase system, 395
        pseudohomogeneous reactor, 394
        steady state and transient simulations,
            397–398
        volumetric flowrate, 395
    water-gas shift reaction, 387
Method of lines (MoL), 134
    central difference in space, 142–145
    driver script, 144–145
    upwind difference in space, 145–148
    using MATLAB® ODE solvers, 154–157
Mixed boundary condition, 274

N

Naive Gauss Elimination, 487–490
Neumann boundary condition, 274, 285,
    290, 304
Newton-Cotes integration formulae
    derivation, 518–520
    multiple applications, 520–524
    Simpson's 1/3rd rule, 516–517
    Simpson's 3/8th rule, 517–518
    trapezoidal rule, 513–516
Newton-Raphson method, 476
    line search method, 247–249
    multivariate, 241–245
    numerical methods, 240–241
    secant method, 245–246
    in single variable, 238–240
Nonlinear algebraic equations
    chemostat, 261
    EOS, 253

integral equations
    complex kinetics, 266–268
    first-order kinetics, 263–265
    Fredholm integral equations, 262
multiple variable solver, 250–252
single variable
    bisection method, 228–233
    Brent's method, 235–236
    fixed point iteration, 236–238
    Newton-Raphson method, 238–240
    numerical methods, 240–241
    Redlich-Kwong EOS method, 234–235
    Regula-Falsi method, 235
single variable solver, 249–250
steady state, CSTR, 257–261
VLE
    bubble temperature calculation, 254
    dew temperature calculation, 254
    T–x–y diagram, 255–257
Nonlinear analysis
    chemostat model
        bifurcation analysis, 366–368
        phase-plane analysis, 365–366
        steady state multiplicity and stability,
            364–365
        transcritical bifurcation, 368–371
    cricket ball trajectory
        animation, 403–404
        ODE, 399–400
        time and location, 400–403
    limit cycle oscillations
        Brusselator system, 384–387
        B-Z reaction, 384
        linear systems, 379–381
        van der Pol oscillator, 381–383
    methanol synthesis
        copper/zinc catalyst, 387
        equilibrium constants, 388
        input parameters and initial processing,
            389–390
        reaction kinetics, 388–389
        steady state PFR model, 391–394
        transient reactor model, 394–398
    nonisothermal CSTR
        bifurcation diagram, 378
        extinction point, 377
        graphical method, 372–374
        parametric continuation, 378–379
        phase-plane analysis, 376–377
        stability analysis, 374–376
        T-axis, 377
        temperature, 377–378

Nonlinear banded systems; *see also* Linear systems
    `fsolve`, sparse systems, 304
    Gauss-Siedel method, 304–305
    linearization–based approach, 298–302
    ODE-BVP model, 296–297
Nonlinear estimation
    Lineweaver-Burke plot, 430
    Michaelis-Menten kinetics, 429
    monod kinetics, 430–432
    optimization toolbox
        error bounds and confidence intervals, 434
        least squares fitting, 435
        `lsqnonlin`, `lsqcurvefit` and
            `fmincon`, 435–437
        prediction model, 435
    power-law models, 429
Nonnormal matrices, 198–199
Normal matrix, 198–199
Numerical difference formula (NDF), 324
Numerical differentiation
    first derivative
        backward difference formula, 478
        central difference formula, 476, 478
        error term, 479–480
        forward difference formula, 476–477, 480
        representation, 475
        Taylor's series expansion, 477–479
    heat flux, 477
    Newton-Raphson method, 476
    round-off errors, 482–483
    second derivative, 481
        central difference formula, 480
        definition, 475
        forward difference formula, 481–482
Numerical integration
    geometric interpretation, 511–512
    Newton-Cotes integration formulae
        derivation, 518–520
        multiple applications, 520–524
        Simpson's 1/3rd rule, 516–517
        Simpson's 3/8th rule, 517–518
        trapezoidal rule, 513–516
    quadrature and MATLAB® functions,
        524–526
    volume of PFR, 512
Numerical methods
    errors, 12
    machine precision, 12–14
    round-off errors, 14–15
    Taylor's series, 15–16
    trade-off, 18–20
    truncation error, 16–18

O

Optimization toolbox, 434–437
Ordinary differential equation (ODE)
  BVPs, 69–70
  chemostat, 120–124
  classical RK-4 method, 102–103
  Euler's explicit method, 74–76
  Euler's implicit method, 76–77
  fourth-order RK method, 101–102
  geometric interpretation, 72–73
  Higher-order RK methods
    Butcher's tableau, 94
    error estimation, 97–100
    explicit RK methods, 95–96
    MATLAB® Solver ode23, 100–101
    step-size adaptation, 100
  Kutta's 3/8th rule RK-4 method, 103
  MATLAB® Solver ode23, 100–101
  multivariable case, 80–81
  nonlinear case, 81
  PFR
    comparison, 110–111
    idealized reactor, 106
    as ODE-IVP, 106–108
  second-order methods
    Adams-Bashforth methods, 92
    BDF formula, 93
    overview, 82
    predictor-corrector method, 92
    Richardson's extrapolation method,
      89–91
    RK-2 method, 83–89
    trapezoidal rule, 91–92
  stability and step-size, 78–80
Ordinary differential equation–
      boundary value problems
      (ODE-BVPs), 276
  boundary conditions, 285–288
  convective/radiative heat losses, 304–305
  nonlinear banded systems, 296–297
Ordinary differential equation–initial value problem
      (ODE-IVP), 179
Ordinary differential equations (ODEs)
      model
  binary distillation column
    description, 181–183
    equations and simulation, 183–185
    parameters effect, 185–186
  mass-spring system, 197
  qualitative responses, 197–198
  real and distinct eigenvalues, 192–197
Orth command, 40

Output display and plotting
  basic plotting functions, 463–464
  results, 463
  text display, 461–462

P

Parabolic PDEs
  Crank-Nicolson method, 153–154
  MOL using MATLAB® ODE solvers, 154–157
  packed bed reactor, 151
  polymer curing
    driver script, 205–206
    1D transient heat equation, 202
    spatial temperature profile, 206
    temporal variations, 207–208
    thermosetting polymer, 203–205
Partial differential equations (PDEs)
  Dirichlet condition, 132–133
  elliptic, 129–130
  hyperbolic, 130–132; see also Convective
      systems
  initial conditions, 132
  linear vs. nonlinear, 128
  mixed condition, 133
  Neumann condition, 133
  nonisothermal plug flow reactor, 157–164
  numerical methods
    FEM, 135
    finite difference techniques, 134–135
    FVM, 134–135
    MOL, 134
  packed bed reactor with multiple reactions,
      164–170
  parabolic, 132; see also Diffusive systems
  Steady Graetz problem
    heat transfer, 170–174
    Nusselt number calculation, 174–176
    parabolic velocity profile, 174
Partial pivoting method, 493–497
PCHIP, see Piecewise cubic Hermite interpolating
      polynomial
PDEs, see Partial differential equations
Pendulum model, 341
Peng-Robinson EOS, 253
PFR, see Plug flow reactor
pfrCrankNicolFun function, 334
Phase-plane analysis
  chemostat model, 365–366
  nonlinear CSTR, 376–377
Piecewise cubic Hermite interpolating polynomial
      (PCHIP), 505–508

Plug flow reactor (PFR), 2, 332–333
  boundary constraints, 216–219
  comparison, 110–111
  idealized reactor, 106
  integral equations
    complex kinetics, 266–268
    first-order kinetics, 263–265
    Fredholm integral equations, 262
  as ODE-IVP, 106–108
Polymer curing process
  driver script, 205–206
  1D transient heat equation, 202
  spatial temperature profile, 206
  temporal variations, 207–208
  thermosetting polymer, 203–205
Polynomial function
  Lagrange interpolation, 501–502
  Newton's forward difference formula, 502–505
Polynomial regression, 413
  Chi-squared value, 426
  definition, 424
  parity and error plots, 426–427
  polynomial fit, 425–426
  second-order curve, 427
  third-order polynomial, 427–428

Q

Quadrature methods, 524–526

R

Radiative heat losse, 304–305
Redlich-Kwong (RK) EOS method, 226, 234–235, 249–250
Redlich-Kwong (RK) equation, 229–231
Reflux ratio (RR), 185–186
Regression and parameter estimation
  Antoine's equation, 410, 431
    linear regression, 439–440
    nonlinear regression, 440–442
  differential approach
    concentration *vs.* time data, 446
    constant volume batch reactor, 446
    kinetic analysis, 446–448
    power-law model, 445–446
  general multilinear regression, 423–424
  LHHW model
    different initial concentrations, 444–445
    rate expression, 441
    single concentration, 442–444
  linear least squares regression
    error and coefficient, 419–423

  general matrix approach, 415–416
    maximum likelihood solution, 418–419
    straight line, 414–415, 417–418
  linear regression, 432–434
  methane, specific heat, 437–438
  nonlinear estimation
    Lineweaver-Burke plot, 430
    Michaelis-Menten kinetics, 429
    monod kinetics, 430–432
    optimization toolbox, 434–438
    power-law models, 429
  orientation, 410–411
  polynomial regression, 413
    Chi-squared value, 426
    definition, 424
    parity and error plots, 426–427
    polynomial fit, 425–426
    second-order curve, 427
    third-order polynomial, 427–428
  singularity and SVD, 428–429
  statistics, 411–412
  temperature dependence, 409
Regula-Falsi method, 235
Relative volatility, 185–186
Relaxation oscillations, 382–383
Richardson's extrapolation method, 89–91
Runge-Kutta (RK) methods, 313, 317
  Butcher's tableau, 87, 94
  classical RK-4 method, 102–103
  error estimation, 97–100
  explicit RK methods, 95–96
  fourth-order method, 101–102
  Heun's method, 84–86
  Kutta's 3/8th rule RK-4 method, 103
  MATLAB® Solver ode23, 100–101
  Ralston's method, 86
  step-size adaptation, 100
  Step-Size Halving, 87–89

S

Secant method
  Brent's method, 235–236
  Newton-Raphson method, 245–246
  Redlich-Kwong EOS, 234–235
  Regula-Falsi method, 235
Simpson's 1/3rd rule, 516–517
Simpson's 3/8th rule, 517–518
Singular value decomposition (SVD)
  condition number, 47
  directionality, 51–54
  MATLAB command, 45–46
  orthonormal vectors, 41–42

rank of a matrix, 47
sensitivity of solutions, 47–51
spHeat function, 473
Stability analysis, *see* Linear systems
Steady Graetz problem
    heat transfer, 170–174
    Nusselt number calculation, 174–176
    parabolic velocity profile, 174
Stiff system
    chemostat model, 120–124
    linearized model, 314
    ODE-IVP method, 315
    phase-plane plot, 314
    Runge-Kutta (RK) methods, 313
    single variable, 315
Successive over-relaxation, 295

T

TDMA, *see* Tridiagonal matrix algorithm
Thomas algorithm, 277–284
Time-varying inlet conditions
    chemostat, 208–212
    zero-order hold reconstruction, 212–215
Transcritical bifurcation
    bifurcation plot, 371
    dilution rate, 369
    graphical method, 368
    linear stability analysis, 370–371
    steady state, 368–370
Transient reactor model, 396–397
    axial profiles, 397–398
    driver script, 397
    gas-phase system, 395
    pseudohomogeneous reactor, 394
    steady state and transient simulations, 397–398
    volumetric flowrate, 395
Trapezoidal methods, 325
    adaptive step-sizing, 329–330
    AM-2, 327–329
    multivariable, 330

Trapezoidal rule, 513–516
trapz function, 524
Tridiagonal matrix algorithm (TDMA), 277–284
Tridiagonal system
    elliptic PDEs, 289–290
    Gauss-Siedel method, 294–295
    ODE-BVP, 287–290
    TDMA, 277–281
Trigonometric functions, 461
Turning-point bifurcation
    bifurcation diagram, 378
    extinction point, 377
    graphical method, 372–374
    parametric continuation, 378–379
    phase-plane analysis, 376–377
    stability analysis, 374–376
    T-axis, 377
    temperature, 377–378
T-x-y diagram, 255–257

V

van der Pol oscillator, 381–383
Vapor-liquid equilibrium (VLE)
    bubble temperature calculation, 254
    dew temperature calculation, 254
    T–x–y diagram, 255–257
Vector spaces
    blending operations, 40–41
    change of basis, 34–35
    definition and properties, 30–32
    image space and column space, 39
    linear independence, 32–33
    natural basis, 34
    null and image spaces in MATLAB, 39–40
    null and image spaces of matrix, 35–38
    span, 32
    subspace, 33

Z

Zero-order hold reconstruction, 212–215